SCOTLAND'S FRESHWATER FISH

ECOLOGY, CONSERVATION & FOLKLORE

Peter S Maitland

Order this book online at www.trafford.com
or email orders@trafford.com

Most Trafford titles are also available at major online book retailers.

Print information available on the last page.

ISBN: 978-1-4251-1064-2 (sc)

Trafford rev. 01/16/2019

www.trafford.com

North America & international
toll-free: 1 888 232 4444 (USA & Canada)
fax: 812 355 4082

SCOTLAND'S FRESHWATER FISH

ECOLOGY, CONSERVATION & FOLKLORE

Peter S Maitland

*'But the streams of Scotland are incomparable in themselves -
or I am only the more Scottish to suppose so -
and their sound and colour dwell for ever in the memory.'*

Robert Louis Stevenson (1913) *Pentland Essays*

For Kathleen, Eileen, Fergus & Amanda

The author, then and now (photos: J.S. Maitland & K.R. Maitland)

'The spirit of interested fun and play in small boys is not to be disregarded.'

Frank Fraser Darling (1948) *The Highlands and Islands*

SCOTLAND'S FRESHWATER FISH
ECOLOGY, CONSERVATION & FOLKLORE

CONTENTS PAGE

PART 3 - THE ISSUES

INTRODUCTION

*'The Allander above Millguy is a bonnie burn,
loitering here and there in quiet backwaters
and anon swiftly eddying through a narrow channel
into deep and dark pools, where trout were wont to hide ...'*

T.C.F. Brotchie (1914) *Glasgow Rivers and Streams*

AS a boy, growing up during the 1940s in the countryside around the village of Milngavie to the north of Glasgow, the local rivers and lochs seemed inspiring places. Even more exciting were their inhabitants, some of which I was able to take home and keep in a variety of receptacles. The River Allander was on my doorstep - but dead below what T.C.F. Brotchie described as 'various bleach and dye works and laundries, which in their turn have been responsible for destroying the virgin purity of the moorland burn.' Thus there were no migratory fish in the river, which was, however, clean and fresh above Milngavie, with a good variety of fish in the various cascades and pools - in one of which I learned to swim (an important attribute for any budding freshwater biologist). Brotchie, whose sketch is shown above, describes walking up the Allander from Milngavie to its source. 'Our way lies through a broad valley, where tree-clad slopes melt into sunny meadowlands and green knowes: quiet nooks, where doubtless "fairies linger still"; and from the wooded aisles and bird-haunted hollows

we wander into the heart of great hills, into sombre solitudes of gorse and bracken, and over moorlands rich with their mantle of gorgeous purple.'

The fish in the river included Brown Trout, Stone Loach, Brook Lampreys, Three-spined Sticklebacks and Minnows - the latter being the 'baggies' and their young the 'pinheids' of my boyhood. Just a few hundred yards from my house was Tannoch Loch, scene of my first skating episodes, and full of Perch and Pike, which I and my friends pursued with very primitive angling equipment. The outflow from Tannoch Loch was on my route to and from primary school, and so was a regular stopping place to watch young Trout and get my feet wet.

It was impossible to avoid developing an interest in fresh waters and their denizens under such circumstances, and my initial interest grew to a passion. My elderly neighbour, John Boyd, was one of the few people who had kept Goldfish from before the War - I still remember the thrill of my first look over the fence into his shed to see these incredible shining creatures - some gold, some yellow and some white - swimming around in one of his aquaria. He was kind enough later to give me my first tank and my first Goldfish. Just outside the foot of our garden was a small seepage pool known locally as 'the Mucky Corner', a convenient source of sticklebacks and - so exciting at the time - my first newts! I remember too the pleasure, many years later, of seeing this pool featured in a botanical study of the area, published in the *Journal of Ecology* in the 1930s - a world of science of which I had no conception as young boy.

The pursuit of information about anything aquatic became a driving force soon after I was able to read, starting with 'Ditchfields Little Wonder Books' and progressing to other volumes, some more erudite than others. One of the strange things about some of the fish described or implied in these books as occurring all over Britain - Bullhead, Spined Loach, Ruffe, Bream and others which were unknown but sounded exciting to me - was that no matter how hard I tried to find them in the lochs and rivers of the West of Scotland, I was unsuccessful. It was a decade or more before I was to discover that the fault lay, not with me but with my books, which were all written from a southern perspective, and largely ignorant of the true nature of the fish fauna of Scotland. As a magpie where literature is concerned, I still have some of these volumes - now of historical rather than scientific value.

The realisation, after I became a professional ecologist that, in the 1960s, we were still very ignorant about the real occurrence of freshwater fish in the British Isles, and perhaps especially in Scotland, led me to initiate the first scheme to map their distribution, and by the early 1970s the first distribution maps were published. Since the 1960s too there have been several books dealing with the fish fauna of the British Isles and of Europe; I have

been pleased to have had a part in some of these. However, with perhaps only one exception that I am aware of, there has never been a comprehensive book about the freshwater fish of Scotland, and so it occurred to me recently that this would be an appropriate time to produce one. The new Millennium, the new Scottish Parliament, and, at last - after many years of campaigning - the possibility of a national scheme for managing Scotland's diverse and valuable fish and fisheries, seemed to indicate that this was a good time to produce a book about Scotland's freshwater fish: one that would include their origins, natural history and intriguing relationships - many now in the past, unfortunately - with the people of Scotland.

This is that book. It has many gaps - I welcome comments from readers, as well as new distribution records and true fishy experiences.

In producing it, I have a large number of people to thank. John Boyd, for first introducing me to fish and aquaria. My late parents for their toleration, over many years, of containers of water and their denizens around the house. Many colleagues have helped with comments or material on various aspects and I thank Robin Ade, Carlos Assis, David Burnett, Niall Campbell, Ronald Campbell, Ruby Ceron-Carrasco, Alain Crivelli, Ross Doughty, Ann Duncombe, David Grant, Iain Gunn, Marja-Liisa Koljonen, Dan Livingstone, Hazel Macleod, Ann MacSween, Willie Miller, John Mitchell, Donald Scott, Alastair Stephen and Archie Young. Much of my research in recent years was with my colleagues Alex Lyle, Ken East and Ken Morris. I thank the authors of the past, whose words I have quoted and the artists whose classical illustrations I have used. Many friends and colleagues have contributed their fishy experiences for this book and I am grateful for the chance to pass these on to readers. Finally, I am indebted to my wife, Kathleen, for all her help in reading and commenting on the manuscript.

'When I was a boy in Scotland,
I was fond of everything that was wild,
and all my life I've been growing fonder and fonder
of wild places and wild creatures.'

John Muir (1913) *The Story of my Boyhood and Youth*

*****

PART 1

THE BACKGROUND

'... I have ... seen extracts from a treatise, by a Scotchman, on this subject, and in this treatise it is gravely asserted, first, that salmon breed with a fish called par; next that salmon do not become what we call fry ... but have a kind of tadpole existence ... Now, to my mind, these are monstrous doctrines and, I think, incapable of being proved ... '

James O'Gorman (1845) *The practice of angling particularly as regards Ireland*

FISHY LITERATURE

*'Anglers should know something of angling literature,
of the natural history of the fishes;
of gastronomic merits and demerits and nomenclature,
so that they shall not pursue their quarry as mere savages.'*

J.J. Manley (1877) *Fish and Fishing*

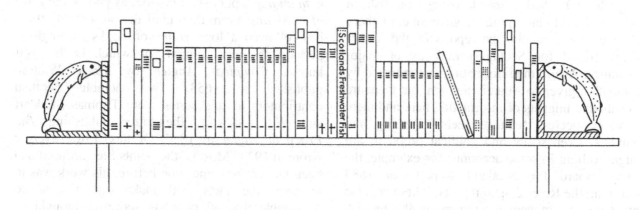

PERHAPS the only general book concerning all our freshwater fish which has been written from a Scottish perspective is Peter Malloch's *Life-history and Habits of the Salmon, Sea-trout, Trout and other Freshwater Fish*, published in 1910 and reprinted in 1912. However, even this excellent volume is not aimed specifically at the entire Scottish fish fauna, and, as with so many other books concerning Scottish fish, deals very largely with Salmon and Trout (276 pages) as opposed to the many other freshwater species (18 pages) which occur here. Peter Malloch was essentially a fisheries manager and his book was naturally written from a salmocentric perspective.

Though there have been some poor books dealing with fish in the British Isles as a whole, there have been many excellent classical volumes, notably those by Pennant (1776), Donovan (1808), Yarrell (1836), Couch (1865), Buckland (1873), Houghton (1879), Maxwell (1904), Regan (1911) and Jenkins (1925). All of these dealt with the Scottish fish fauna to a greater or lesser degree but, with the exception of that by Herbert Maxwell, all have had a southern perspective.

Though there are many gaps, however, the collective literature embracing Scottish fish is actually both extensive and diverse, and a variety of strata have been mined by the author to produce the present volume. As far as the natural history of our freshwater fish in Scotland is concerned there is now an extensive but lop-sided scientific literature available - hundreds of papers on some species (e.g. Atlantic Salmon) but virtually none on several others (e.g. Sea Lamprey). The distillation of recent research has led to several recent books dealing with the fish of the British Isles as a whole, most recently that by the author and Niall Campbell in 1992 and by Nick Giles in 1994. In parallel with these have been several excellent books covering the freshwater fishes of Europe as a whole, and naturally a few aspects of the Scottish fish fauna are included in these.

Fish, especially sea fish, do feature here and there in Gaelic literature but there is, unfortunately, no book in Gaelic which deals with the freshwater fish of Scotland, nor even of Gaeldom.

Researchers

The results of scientific research by fish biologists and others must be the foundation of our information on the biology of freshwater fish in Scotland - and indeed elsewhere. Once again, the Atlantic Salmon has been a major focus of attention here, and some studies in Scotland, such as John

Thorpe's discovery that salmon parr of the same year class divide into two different length classes and 'opt' to smolt one year apart, have been of major importance and attracted attention world-wide. Though there is, however, considerable published research on a few species, there has been relatively little on many others, and much of our knowledge of these must at present be based on work done outwith Scotland.

Alongside published research papers on fish in Scotland, the official reports of government fishery biologists (for example, the reports of the former Fishery Board for Scotland) rank as of major importance. Again, the information is dominated by the attention given to Atlantic Salmon, but there are also valuable nuggets of information here and there on other species, such as Eels, Sparling and Flounders. Even social and political problems are brought to light in some accounts; for example, the Fishery Board for Scotland Report for 1883 concerning the River Annan reports: 'There are also a great number of haaf net fishers on the English side, and many of them come through the channel at low water and take possession of the Scotch side of the channel. These Englishmen defy the men who have the right to fish on the Scotch side, and have even taken fish from before their nets, and were the Scotchmen to attempt resistance a serious breach of the peace would be the result.'

Several of the species the author looked for in vain as a boy in the West of Scotland are now with us, for there has been substantial movement of many southern species northwards during the last half century, mainly due to the transfer of fish by coarse fishermen. Fish like Dace, Chub, Ruffe and others are now well established in waters north of the Border. However, as far as any literature or Scottish traditions relating to such species, there is obviously none, so in the accounts of such species below, reference is made to the north of England or even further afield. Similarly, with alien species from Europe or North America, reference may be made to their biology or traditions surrounding them in their native haunts.

Anglers

Several books covering the whole fish fauna of Britain have been written by anglers, a curious aspect of many of them being the complete lack of any references to other literature - as though all the facts in them had been determined by their authors. In fact, most of these publications have relied heavily, not only on the original foundation research work carried out by fish biologists, but also on the other texts mentioned above, where attribution of source data is the norm.

One of the earliest books on angling in Britain was credited to Juliana Berners: *Treatyse of ffyshynge with an angle* published in 1496 as part of the *Boke of St Albans*. From then until the present day there has followed a long series of books on angling, perhaps the most famous of which is the well known *Compleat Angler* by Izaak Walton, published in 1653. Two notable Scottish contributors to this series were Thomas Stoddart and W.C. Stewart, the latter publishing *The Practical Angler* in 1857. Of this, Eric Taverner wrote in 1929: 'Most of the points Stewart used had been known for some time before; his work was to marshall the facts and make out a case so reasonable that no one has seriously thought of going against the logic of it.'

Original frontispiece in Walton (1653).

Dozens, perhaps even hundreds, of books have been written by anglers who have fished in Scotland and the present author has read many of these as background to the present account. In fact, though there are extensive descriptions of scenery, accounts of the weather and numerous battles with fish large and small, some eventually captured, others not, very few of these books actually have any factual information regarding fish biology. Again, the great majority of such volumes are concerned with Atlantic Salmon and, to a lesser extent Brown (and Sea) Trout. Many anglers consider that Atlantic Salmon is the only species worth bothering about and this, together with the legislation which supports it, has been the curse of holistic fish and fishery management in Scotland for several centuries, as discussed in a later chapter. Some years ago the author enquired of an elderly, but experienced, Loch Lomond angler how many fish he had caught that season. 'Twelve' was the reply. Knowing that he fished very regularly I cautiously asked if that was all. 'Yes,' he said 'of course I've had about a hundred Sea Trout and Brown Trout as well.'

As well as numerous books about angling in Scotland there are also many poems about freshwater fish and fishermen, some in angling texts, others scattered elsewhere. A particularly powerful piece is *Fisher Jamie* by John Buchan from his *Poems Scots and English*, dedicated to his beloved brother who had been killed in the First World War.

> 'Puir Jamie's killed. A better lad
> Ye wadna find to busk a flee
> Or burn a pule or wield a gad
> Frae Berwick to the Clints o'Dee ...
>
> I picter him at gloamin' tide
> Steekin' the backdoor o' his hame
> and hastin' to the waterside
> To play again the auld auld game;
>
> And syne wi' saumon on his back,
> Catch't clean against the Heavenly law,
> And Heavenly byliffs on his track,
> Gang linkin' doun some Heavenly shaw.'

As well as books about angling by anglers there are also a number of books which deal with single fish species - some, again by anglers about angling for them, but others by biologists. Notable among the latter are several excellent volumes dealing with the Atlantic Salmon (for example by William Calderwood, Jock Menzies and Derek Mills) or Trout (for example the classical study of Sea Trout by Herbert Nall), but again, most other species are neglected.

Travellers

Most of the earliest accounts of fish in Scotland have come from the journals of travellers in various parts of the country. Few of these contain much detail, but many are invaluable in confirming the presence or value of particular species at these times. Among the earliest accounts is that of Don Pedro de Ayala in 1498: 'It is impossible to describe the immense quantity of fish. The old proverb says already "piscinata Scotia". Great quantities of salmon, herring, and a kind of dried fish ... are exported. The quantity is so great that it suffices for Italy, France, Flanders and England.'

Other accounts followed, notable among which was the *Description of the Western Isles of Scotland called Hybrides* by Donald Monro, High Dean of the Isles, who travelled through the most of them in the year 1549. Virtually every Hebridean island is mentioned, and in many of them fish were important, for example: 'Duray ...The watter of Lasay ther, the watter of Udergan, the watter of Glongargister, the waters of Knockbraik, Lindill, Caray, Ananbilley; all thir waters salmond slaine upon them. This iyle is full nobell coelts, with certaine fresche water Loches, with meikell of profit.' or another island: 'Mull ... In this ile there is twa guid fresche waters, ane of them are callit Ananva, and the water of Glenforsay, full of salmond, with uther waters that has salmond in them, but not in sic aboundance as the twa foresaid waters.' and again: 'Sky ... Within this ile ther is gud take of salmant upon five watters principally, to wit, the water of Sneisport, Sligachan, Straitswardill, Ranlagallan, and Kilmtyne, with seven or aught uther smallar watters, quherupon salmont are also slayne. In this ile there is ane freshe water loche, callit the loche of Glenmoire, quheron ther is abundance of salmont and kipper slayne.'

A century later, Richard Franck, a Cromwellian soldier and keen angler, travelled through northern England and the Highlands of Scotland and, being an angler, his journal contains much of interest, for example: 'The Firth [of Forth] runs here [Stirling] that washeth and melts the foundations of the city, but relieves the country with her plenty of salmon; where the burgomasters (as in many other parts of Scotland) are compell'd to reinforce an ancient statute, that commands all masters and others, not to force or compel any servant, or an apprentice, to feed upon salmon more than thrice a week.' In spite of Norman Simmond's deplorable comments about Richard Franck's writing skills, he did provide useful information about fish in Scotland during this period.

Perch (from Wood 1863).

Thomas Thornton provided a contemporary view of angling in Scotland in 1804 in his book *Sporting Tour*. This successful angler made several visits to Scotland and caught many fish, some of them very large. Notable among these was a successful catch in Loch Lomond. '... passed through the straits of Loch Lomond, a most likely place for pike: still we had no rise, when, in turning gently round a dark bay, I felt a fish strike ... After much trouble he was secured in the landing-net, and proved to be a perch of ... seven pounds three ounces, or thereabouts. He was very thick about the shoulders, and I regret I did not measure him, as I never saw a fish so well fed.'

Naturalists

Although most accounts of the natural history of parts of Scotland tended to concentrate on the three most popular fffs (fur, feathers and flowers), some authors did mention a fourth - fish. Notable among these was Herbert Maxwell, who had an especial interest in fish (and indeed published a notable text on British freshwater fish), and was a copious writer. Within a series of fascinating volumes (each containing many essays on different aspects of the countryside), published at the turn of the 20th century under the general title *Memories of the Months*, are several useful articles on fish in Scotland. For example, an extract from 1922: 'In the Cree, another of the Solway rivers frequented by sparlings, the members of an angling association which rented the salmon-fishing became aware of

the destruction wrought by the small-meshed nets of the sparling-fishers among the smolts of salmon and salmon-trout descending to the sea in April and May, this being the very period when the sparlings seek the top of the tide to spawn. Consequently, gravid smelts and migrating smolts were hauled out indiscriminately, and however much care conscientious fishermen (and it is rumoured that all net fishermen are not conscientious) might exert in returning the smolts to the water, it was unavoidable that thousands of them should be destroyed for no purpose whatsoever.' Contemporary accounts such as these are invaluable in giving us a picture of the usage and social customs surrounding various fish species at the time.

Among the many accounts of the flora and fauna of different parts of Scotland can be found valuable information on the status of freshwater fish in these areas at the time they were written. Several volumes on the vertebrate fauna of northern Scotland by J.A. Harvie-Brown & T.E. Buckley and faunas of other parts of Scotland by R. Service, W. Evans, and others are of value, some of them particularly so in that they review previous, often unavailable, literature.

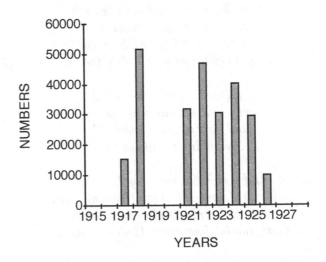

The numbers of Powan caught in Loch Lomond at the end of the First World War (data from Lamond 1931).

There are a few classical accounts of individual Scottish waters which deserve special mention and one such is Henry Lamond's *Loch Lomond* published in 1931. This is an exceptionally fine account of the fish and fisheries of Scotland's largest loch, now the focus of our first National Park, and his study is referred to several times in the present book. He tells us that: 'During the closing months of the War, when there was a shortage in the food supplies of the country, powan fishing was intensively prosecuted by request of the Government.' The minimum annual catch from 1917-26 was 10,000 fish (1916), the maximum catch 51,500 (1918). 'In season 1926 prices fell to fourpence and even threepence per pound, and fishing was abandoned in the second week of July.' These are the only data ever produced on the potential numbers of Powan in Loch Lomond.

Geographers

Geographical accounts of Scotland, or parts of it, are also a very important source of historical information on fish. Notable among such accounts is the wonderful 'Statistical Account of Scotland' edited by Robert Sinclair during the 1770s. The many volumes include accounts of virtually every parish in Scotland and it is common to have entries referring to the presence of various fish species in named local lochs and rivers. So useful is this broad coverage of Scotland that, in 1977, the author was able to base an account, and produce maps, of the fish fauna of Scotland at that time entirely on this one series.

Even before the 1st Statistical Account, MacFarlane's Geographical Collections, produced in the mid-18th Century, included valuable information about the freshwater fish of different waters and gave fascinating insights to the primitive fisheries of those days and the value of fish to local folk. For example, MacFarlane describes St Mary's Loch and the neighbouring, and connected, Loch of the Lowes thus: 'these lochs called St Mary Lochs ... They abound with diverse kinds of fishes as Trouts, Eels, Pearches and Pikes of a greater bigness then are to be found elsewhere, some of them being five or six quarters in length. There is also taken in thir lochs a little fish called by the Countrey people Red Waimbs. It is about the bigness of a Herring, and the belly of it wholly red. It is only to be found about Michaelmass and that only in the little stream that runneth betwixt the two Lochs, but not seen at any other time or in any other part of the Lochs. Yet at that season the Countrey people with Plaids sewed together like a Net have taken such Store of them, that they carried them home and salted them up in Vessels for the food of their families.'

Newspapers

Newspapers, although regarded by many as ephemeral literature, do have an important role to play in relation to freshwater fish, especially relating to local customs or events. Old newspapers are valuable sources of information about unusual or notable items concerning freshwater fish. For example they are one of the few places where catches of Sturgeon are recorded, as this extract from *The Glasgow Herald* of March 19, 1970 indicates: 'A 51 lb royal sturgeon, caught in the Firth of Clyde by Mr Andrew McCrindle, skipper of the Golden Venture, was offered to the royal household by Ayr Lord provost Mr Alex Handiside. It had been bought by an Ayr businessman for £23. The sturgeon was male and "contained no caviar".'

As one of the few public fora for debating topical issues, the letters pages of newspapers have an important role to play, something obviously appreciated by Allan Berry in the opening of his contribution to one of the many debates on fish farming in Scotland in recent years: 'The Herald has done a great service to the cause of arresting the decline of Scottish wild salmon and sea-trout stocks. Articles and letters have outlined the problem and highlighted many of the scientific and political factors. ...'. Sometimes, of course, the correspondence gets very heated, and biased views are blatantly offered. William Crowe of the Scottish Salmon Growers pronounced that 'The Herald ought to do a rain-check on its correspondents before publishing their rantings. Yet again Bruce Sandison has his facts and figures wrong. ...' The letter then goes on to itemise these apparent failings and finishes with the partisan statement that 'The absence of any logical, reasoned or scientific analysis of the many factors affecting wild salmonoids throughout the UK is likely to be rectified by the expertise and knowledge of salmon-farming companies participating in the work, together with riparian owners ...' Many readers must have smiled at this - apparently independent analysis of facts is not something favoured by the salmon farming industry.

Acting as medium for the passage of information concerning current science and conservation is another major role of newspapers. *The Express*, of January 18, 1998, published an excellent article by Colin Calder entitled *Queen Mary's fish Returns*: 'A rare fish is on the brink of making a comeback north of the Border after an absence of 20 years. And a historic breeding programme means the six-inch vendace, regarded as a delicacy by Mary Queen of Scots, could be here to stay. ...' An article by Elizabeth Buie in *The Herald* kept readers up to date with the current moves to fulfil obligations to the EC Habitats Directive: 'The primitive eel-like lamprey - once eaten in such quantities that Henry I and King John supposedly died of a surfeit - is to be protected under a proposed European designation, the Scottish Office has announced. Endrick Water, which ... contains a significant number of the now-rare lamprey, is one of nine new proposed Special Areas of Conservation ...'.

From THE SCOTSMAN 1855

Wednesday, July 25, 1855

VENDACE AT LOCHMABEN.—On Thursday last, at Lochmaben was held the annual gathering of the St Magdalene Club, whose objects are at set seasons to fish the lochs for vendace and other finny tribes, and socially to masticate the same when cooked. When the fishings commenced in the forenoon, a considerable crowd of members and others had collected on the shore of the Mill Loch. The first haul of the net brought up about 28 dozen of vendace —some of them small, but mostly seven or eight inches in length, which is about their full size. Another cast or two of the net added only a few dozen to the previous number, and in the Castle Loch, which was next tried, not a solitary vendace came within range. Meanwhile on the Castlehill were being carried on such pastimes as foot-racing, leaping, wrestling, and putting the stone, and races between grown-up girls and even stout dames for prizes of bonnets, caps, &c. About five o'clock the outdoor recreations terminated, and the members of St Magdalene's adjourned to a dinner at the Commercial Inn, in which vendace formed, as a matter of course, a principal dish. About half-past eight the party broke up, and thus ended the anniversary gathering of the St Magdalene's Fishing Club of Lochmaben.

Thus, in a modern context, and in spite of strong competition from the Internet, newspapers still have a valuable function in recording and keeping the public informed of topical items of interest concerning freshwater fish. For example, most recently, after the Barbel was introduced to Scotland, the first indication in print that this had occurred, and that this species was being caught by anglers in the River Clyde, appeared in local newspapers. The author has a large collection of such clippings, but he just wishes that some journalists would get their facts right before committing an item to print!

Advertisers

Freshwater fish have been used many times in advertising campaigns which are often directed at anglers. One of the best known is the series of 50 cigarette cards issued by John Player and Sons which covered most of the British species and were issued with a small booklet, with text by R.L. Marston, into which to stick the cards. Complete booklets are still available today quite cheaply in the second-hand market.

Some years ago, the author was sent an advert which had appeared in a Dumfriesshire newspaper. It was to publicise a local fishmonger and the illustration accompanying it was of a Vendace - actually a drawing which the author had done some years before. Enquiries revealed that the newspaper had been asked to place the advert and use any available fish illustration to accompany it! The drawing had actually been supplied to the paper by the author some years previously as part of an article on the plight of the Lochmaben Vendace.

Prose writers

Many of Scotland's finest writers have been inspired to include evocative descriptions of freshwater fish and their environment in their works. Some are in Scots, for example, Christopher North in his *Noctes Ambrosianae*: 'His tackle, for bright airless days, is o' gossamere; and at a wee distance aff, you think he's fishin' without ony line ava, till whirr gangs the pirn, and up springs the sea trout, silver bricht, twa yards out o' the water, by a delicate jerk o' the wrist.' Others, like Charles St John, wrote descriptively in English: 'The morning was bright ... every little pool was dimpled by the rising trout ...

Beautiful in its grand and wild solitude is the glen where the Findhorn takes its rise. It is too remote even for the sportsman ... ' Ratcliffe Barnett had a similar style: 'Each day a new adventure in itself ... The plop of trout in a stretch of still water at the gloaming, and the splash of leaping salmon below the fall.'

Neil Gunn was obviously fascinated by fish and mentions them in his writings, some of which are quoted elsewhere. However, he was also an excellent observer of fish habitat, as the following excerpt from *Morning Tide* shows: 'It was a sunny afternoon in late July, and the soft water came over the stony shallows and swirling into the pool's narrow neck in a warm gurgle. Its brown tinge was scarcely perceptible, and already the sleepy stones were gathering a hint of green at the pool-edge. The trees on both side of the glen were heavily leaved and hung in a drowsy silence. ...'

Halliday Sutherland too produced wonderful descriptive pieces in some of his autobiographical accounts: 'On the loneliest loch there is movement by day and by night. In dead calm a rising fish here and there splashes, leaving on the surface circles that expand to infinity. One breath of air, and the calm is broken by ripples of yellow-gold in sunshine, or of shimmering silver under white clouds. With a light wind the water is lapping on pebbles by the shore, and the rushes are moving. ... By day there is the cry of the peewit and the lone curlew, at dusk the haak, haak of wild-duck flying low, and at night the wail of the redshank.'

Poets

Even poetry can tell us something about a local fish fauna and perhaps its relevance to society at the time, for example some of the lines of Tobias Smollett, dealing with the River Leven, which flows out of Loch Lomond:

> *'While lightly poised the scaly brood*
> *In myriads cleave the crystal flood;*
> *The springing trout in speckled pride,*
> *The salmon, monarch of the tide,*
> *The ruthless pike intent on war,*
> *The silver eel, the mottled par,*
> *Devolving from thy parent lake.'*

Not much later, Robert Burns was able to show that he too understood about fish and their habits in this verse from *Tam Samson's Elegy*:

> *'Now safe the stately saumont sail,*
> *And trouts bedropp'd wi' crimson hail,*
> *And eels, weel-ken'd for souple tail,*
> *And geds for greed,*
> *Since, dark in Death's "fish-creel, we wail"*
> *Tam Samson's dead!'*

Even William McGonagall wrote about fish!

> *'Beautiful Loch Leven, near by Kinross*
> *For a good day's fishing the angler is seldom at a*
> *loss*
> *For the loch it abounds with pike and trout*
> *Which can be had for the catching, without any*
> *doubt;*
> *And the scenery around it is most beautiful to be*
> *seen,*
> *Especially the castle wherein was imprisoned*
> *Scotland's ill-starred Queen.'*

Conclusion

This book has been written on the foundation provided by the sorts of literature mentioned above and there are clearly many debts to past authors. In general, detailed references are not given within the text, but those to factual information are indicated at the end of each chapter, and the full bibliography at the end of the book includes all the key writings from which information has been drawn.

Most of the line illustrations are by the author but, here and there, as in this chapter, exceptional and relevant illustrations from some of the old classic titles have been used. The source is acknowledged in each case.

References

The references mentioned above and noted below are just a fraction of the wealth of literature which pertains to freshwater fish and fishing in Scotland. Further details and additional references are listed in the Bibliography at the end of this volume.

Armistead (1895), Barnett (1927), Berners (1496), Buchan (1917), Buckland (1873), Calderwood (1907), Couch (1865), Crandall (1914), de Ayala (1498), Donovan (1808), Evans (1906), Franck

(1694), Giles (1994), Gunn (1931), Harvie-Brown & Buckley (1889), Houghton (1879), Jenkins (1925), Lamond (1931), MacFarlane (1748), Maitland (1972, 1977, 1984, 1992, 2000, 2004, 2006), Maitland & Campbell (1992), Malloch (1910), Manly (1877), Marston (undated), Maxwell (1900, 1904, 1922), Menzies (1931), Mills (1971, 1989), Monro (1549), Moss (1889), Nall (1930), Norman (1943), North (1876), Pennant (1776), Regan (1911), Rintoul & Baxter (1935), Scrope (1843), Service (1902), Simmonds (1997), Sinclair (1779), Stewart (1857), St John (1891), Stoddart (1831), Sutherland (1934), Thornton (1804), Thorpe (1977), Walton (1653), Wheeler (1969), Wood (1863), Yarrell (1836), Young (1843).

'Here and there in the interminable catalogue
are books which it is a privilege to know;
books that it is refreshment to drop into;
books which speak of a world that seems far fresher than our own,
more leisurely, less methodical.'

Herbert Maxwell (1900) *Memories of the Months: Second Series*

*****

WHAT'S IT LIKE TO BE A FISH?

'Have you ever wondered what a strange, eerie world the water-world must be?
... no sound breaks the tuneless gloom of the reedy lairs in which the great fish lurk;
no footfall echoes through the rush-strewn halls of those dim shades.
... all is strangely still and silent,
for are not all the living creatures of this soundless underworld mute?'

Arthur Young (1926) *The story of the stream*

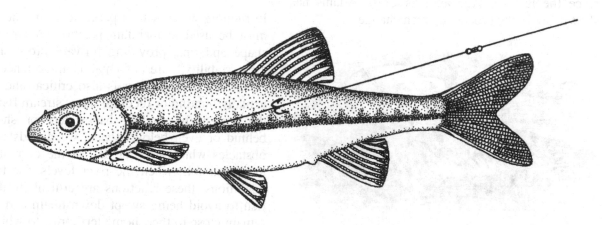

ARTHUR Young's view of life under water is very wide of the mark, for some sounds travel very well there; most fish are receptive to them and many fish species are capable of making a variety of sounds.

One of the reasons that there is less interest in fish compared to, say, birds and mammals is that the latter are warm and often cuddly whereas the former are thought of as cold and slimy. Combined with this perception is the fact that fish live in a completely different medium from humans. This creates an environmental barrier which not only makes fish difficult to observe but also to appreciate how they have evolved to be the efficient creatures that they are in coping with life in the medium of water. By understanding just what it is really like to be a fish we can develop an awareness and appreciation of just how they cope with the good and bad features of their environment.

Many people believe that fish live in a very strange and alien world of the kind depicted by Rupert Brooke in his poem *The Fish*:

'Fantastic down the eternal stream;
An obscure world, a shifting world,
Bulbous, or pulled to thin, or curled,
Or serpentine, or driving arrows,

Or serene slidings, or March narrows.
There slipping wave & shore are one,
And weed and mud. No ray of sun,
But glow to glow fades down the deep
(As dream to unknown dream in sleep);
Shaken translucency illumes
The hyaline of drifting glooms;'

In fact, in some ways, the world of aquatic animals is much more stable and a better medium for life (which started in water) than the terrestrial environment. Fish are well adapted to it, in all kinds of ways, which is why they are the most successful vertebrates on earth - with more species in total than all amphibians, reptiles, birds and mammals put together!

Shape

Though fish can move about in three dimensions within their environment, in most cases their habitat is actually quite restricted. Rivers are linear systems, where downstream movement is usually easy (but for most freshwater fish stops at salt water), but upstream travel may be difficult because of waterfalls and other obstacles above which fish cannot pass. Most lochs are relatively small and there are certainly many more tiny lochans than large lochs in Scotland, often resulting in the isolation of their fish populations from one another. In some ways, fresh water systems can be regarded as ecological islands, where, depending on the length of isolation and the pressures on them, the stocks of fish have started to evolve towards a separate genetic identity from ones nearby but with which they may have had common ancestry. Arctic Charr have been isolated in most Scottish lochs in which they now occur for much of the 10,000 years since the last Ice Age and, as Colin Adams has shown, are in the process of such change.

Pike (from Wood 1863).

Horace Hutchinson noted that 'The fish, of course, is throughout a very perfect swimming machine. He is shaped so as to offer as little resistance as possible to the water, and his scales are so set and made of such material that the friction is reduced to the very minimum. His fins are, no doubt, aids to swimming, but all the hinder part of his slim shapely body is active in the work of his propulsion. It is with his body chiefly that he swims.'

Apart from their streamlined form to aid swimming and resist currents, one of the important adaptive features of fish is the swim bladder. This simple organ means that fish can adjust their buoyancy so that it is neutral to water and thus they need to expend no energy in keeping position in still conditions. It took humans until the 20th Century to work out how to produce such weightlessness in preparation for the first ventures into space! Fish in water therefore are just as free of gravity as spacemen in their capsules - a considerable advantage, meaning that they can eat and sleep away from the bottom if they wish to, without expending any energy. A few fish (e.g. lampreys, sharks and rays) have no swim bladder and must swim continuously or drop to the bottom.

Only some fish are adapted for a life of speed. Many living in open water or in shallow water along shores have different problems and adaptations, but for them too, the possession of a swim-bladder is enormously helpful, and has freed their body fins for a variety of purposes. The Pike has short fins for sculling, moving gently back and forth, so that the fish can compensate for small variations in current and apparently hang motionless in the water waiting for its victims to swim by.

In running water with a permanent current, energy must be used to maintain position. A streamlined shape and fins, providing forward propulsion but also stability, are of major importance here. Nevertheless, behaviour is also critical and, unless actually hunting or feeding, many stream fish avoid the current by finding backwaters or sheltering behind or even below stones, weed beds or other obstacles which slow or deflect the current. After heavy rainfall, when the river levels rise to spate conditions, these reactions are critical in allowing fish to avoid being swept downstream and thus to remain close to their home territories to which they can return when water levels return to normal.

Movement

Most fish require more than one habitat during their life cycle; some require many and if these essential habitats are far apart, then long, arduous and dangerous migrations may be required. The Atlantic Salmon is, of course, the prime example of such a traveller; not only has it to travel long distances, but it also has, physiologically, to tolerate two quite different aquatic media in so doing. Starting life in the poor, fast-flowing waters of a highland nursery stream, after smolting, the young salmon has to travel downstream to the estuary of its natal river. Although there are many hazards on the way, there is relatively little effort involved, thanks to the current, depicted convincingly by William Shakespeare in *Two Gentlemen of Verona*:

'The current that with gentle murmur glides,
Thou know'st, being stopp'd, impatiently doth rage;
But when his fair course is not hindered,
He makes sweet music with the enamelled stones,
Giving a gentle kiss to every sedge
He overtaketh in his pilgrimage;
And so by many winding nooks he strays
With willing sport, to the wild ocean.'

Once safely in the sea, the juvenile Salmon may have to travel many hundreds of miles to its feeding grounds and then back again as it matures and feels the urge to spawn in its natal river. Even as a strong adult, the journey upstream is much more difficult than downstream, thanks to droughts, spates and waterfalls, as neatly encompassed by William Jeffrey in his poem *Salmon*:

> *'Out of the sea with spring he came*
> *Joy-clothed in blue and silver flame,*
> *A vision'd beauty, a strength untame.*
>
> *Now prisoned in this pool scooped out*
> *Of rock by the ageless water-spout*
> *He waits the ending of the drought.*
>
> *Come spate of water, swift he'll leap*
> *On curving flash the foam-dazed steep*
> *And onwards urge till he groove deep.*
>
> *Far natal shallows. There vast hills*
> *Contain the Majesty who wills*
> *The cycle that all life fulfils.'*

Neil Gunn, in his novel *Morning Tide*, covers the same story: 'The river was fairly low. The fish had presumably come up on the tail-end of the last spate and not risked going all the way up to the high pools.'

The main natural factor affecting the distribution of Atlantic Salmon and Sea Trout in Scotland is the disposition of waterfalls which act as natural barriers to upstream migration. Both species will spawn in quite small burns, providing they can reach them, but if they encounter a fall which necessitates a vertical leap of more than four metres, the waters above, no matter how large, will be devoid of these two migratory fish. There has been much speculation and argument about the ability of salmon to surmount waterfalls: how do they do it, and how high can they jump? At one point it was believed that fish could leap by holding their tails in their mouths as this poem *Salmon Leap* by Michael Drayton depicts:

> *'His tail takes in his teeth, and bending like a bow,*
> *That's to the compass drawn, aloft himself doth*
> *throw:*
> *Then springing at his height, as doth a little wand,*
> *That, bended end to end, and flerted from the hand,*

> *Far off itself doth cast, so does the salmon vaut.*
>
> *Here, when the labouring fish does at the foot (of*
> *the fall) arrive,*
> *And finds that by his strength he does but vainly*
> *strive;*
> *His tail takes in his mouth, and bending like a bow*
> *That's to full compass drawn, aloft himself doth*
> *throw.'*

In fact, Tom Stuart was able to show in 1962, with the help of some ping-pong balls, that the leaping behaviour of Salmon and Trout 'can be correlated directly with the hydraulic conditions obtaining in the river channel. ... The stimulus to leap was found to be closely related to the presence of a standing wave (or hydraulic jump) and the location of the standing wave distant to the obstacle influenced the success of the leap.' Hutchinson, many years earlier (1907) had actually been very near the mark with his suggestion: 'It had been argued that possibly the down-rushing force of the water from which the salmon have their "take-off" for these leaps is less formidable than it appears; it has been suggested that although the surface-stream is then hurrying downwards, there may be a back-current below ...'

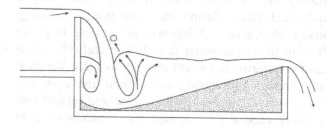

Tom Stuart's ping-pong ball experiment.

Many other, indeed most, fish migrate at some stage in their life history. Some species have short migrations such as the few hundred metres upstream that Brook Lampreys may migrate from their nursery to their spawning beds, while others, like the Salmon, have long distances to swim - as does the European Eel which is a challenger for the long-distance title. Many fish, however, make very short migrations, usually on a daily basis, like Powan and Vendace which stay near the bottom of their lochs during the day but migrate up at dusk to feed near the surface and then swim down again at dawn. One curious daily migration has been described by Harvie-Brown & Buckley: 'Eels are very common in the sluggish rivers and lochs of

Caithness. Large numbers migrate daily at Wick, from the fresh to the salt water and back. They leave the river just before dark and make their way to the harbour to feed on the garbage from the herring refuse thrown overboard there. They keep a regular course, which they invariably follow day after day. A long string or file of three or four deep continues for hours to pour out of the river into the bay, and thence into the harbour.... Such eels as are able to run the gauntlet [of fishermen] ... return to the river before morning.'

Activity

Fish, like amphibians and reptiles, are cold-blooded and so their activity is regulated by the temperature of the water they live in. However, unlike amphibians and reptiles which go into complete hibernation over the winter, many freshwater fish in Scotland are relatively active during the coldest months of the year and, though they may not feed much or grow, they display their arctic ancestry by spawning then - Brown Trout and Arctic Charr in October, Salmon in November, Vendace in December, Powan in January, Grayling in February and Pike in March. However, members of the carp family are of more southern origin and have a quiescent phase during the cold months - hiding under stones (e.g. Minnows) or staying near the bottom in deeper water (e.g. Roach and others). In contrast, during the warm summer months, Vendace and Arctic Charr spend much of their time in deeper water where it is cool, whilst Carp and related fish actively choose the warmest water available. It is very common in many Scottish lochs and rivers in summer to observe shoals of small Minnows crowding into the shallowest (and therefore warmest) water available at the edge, where their chances of growing faster through increased metabolism are improved.

Seton Gordon commented on the situation in the River Dee at Cambus O' May in winter. 'In the shallows where stones are near, or above the surface the drifting ice is pressed and consolidated, and rises, white and glistening, into the air in the form of miniature icebergs. As I stood on a bridge over quiet, clear water, where the river was not entirely frozen across, I saw two salmon, each perhaps twenty pounds in weight, lying beneath six feet of icy water on the river bed; a few yards away from them was a fish of about six pounds. When I passed

that way again later in the day the fish were still in the same place. Although salmon are cold-blooded creatures they undoubtedly feel the cold, although they are not inconvenienced by it the way warm-blooded animals are. It has, rather, the effect of making them torpid and lethargic.'

Some fish virtually hibernate during cold weather. Norman Morrison noted that '... in the course of my researches I have come to the conclusion that eels pass through a form of hibernation ... when fishing with the bait in winter and spring I could never get an eel earlier than about the end of April. ... I questioned several hill drainers ... when following their vocation in winter and spring they often found eels in a torpid state, buried in mud at the bottom of old drains on the hillside, hundreds of feet above sea level. ... Some years ago a gamekeeper near Campbeltown found half a dozen eels at the bottom of a drain which he had opened.' In a similar vein, St John recalled Mr Young's experiments where 'In some of the ponds he had put a number of small eels, which soon grew in size, and became as tame and familiar as the young salmon. As the cold weather came on, the eels all disappeared ... One fine spring day ... he was delighted to see them all issue out from under the stones asking for food, as if a day only, instead of many weeks, had passed since he last had fed them. Does not this most clearly prove that eels lie dormant during cold weather?'

Hearing

Although fish do not have 'ears' as we understand them, they do have a very efficient nervous system near the surface of the skin which is extremely responsive to vibrations in the water. Richard Jefferies, in *The Life of the Fields* , tries to summon up the images: 'Does he hear the stream running past him? Do the particles of water, as they brush his sides and fins, cause a sound, as the wind by us? While he lurks beneath a weed in the still pool, suddenly a shoal of roach rush by with a sound like a flock of birds whose wings beat the air. The smooth surface of the still water appears to cover an utter silence, but probably to the fish there are ceaseless sounds. Water-fowl feeding in the weedy corners, whose legs depend down into the water and disturb it; water rats diving and running along the bottom; water beetles moving about; eels in the mud; the lower parts of flags and aquatic grasses swinging as the breeze ruffles their tips; the thud, thud of horse's hoofs, and now and then the more distant roll of a hay-laden wagon. And thunder - how does thunder sound under the water?'

A.M. Young in *The Story of the Stream* wrongly imagines that all is quiet under water: 'Gaze down into the water and you will see curious, ghost-like forms, gliding like shadow-shapes in and out of their watery Valhalla - as silent as now are the Halls of Aesir. No sound comes from them; so silently do they move that the water is scarcely stirred. In and out they glide, only their prehensile lips opening and shutting as they pass. They do not speak; they do not murmur; they cannot even whisper.... Dumb they are, and speechless we think them.'

In fact, many species of fish are capable of making a noise and of communicating with one another in this way. Equally, sound does travel through water and so many noises originating on land are heard there as well as those produced under water, by waterfalls, etc.

Vision

The eyes of fish are essentially similar to our own and predatory fish such as Salmon and Trout have excellent vision. However, since most fish have an eye on each side of the head, few have binocular vision comparable to ours. Bottom living fish like Stone Loach, whose eyes are positioned high on the head, tend to have poorer sight, but have a highly developed tactile awareness through their barbels.

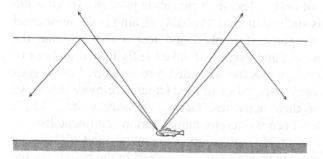

The fish's 'window' on the world above.

Fish can certainly see through the water surface but the extent of their vision there is limited to what is known as the 'fish's window'. Light entering or leaving the surface of still water is bent or deflected unless its path is at exactly 90° to the surface. As the angle relative to the surface is decreased, so is light refracted at a greater angle until at 48.5° there is total reflection and light is reflected from the water surface. This means that a fish looking up to the surface of the water above it can see through a 'window' whose size is determined by the depth of the fish in the water and an angle of 48.5° to the surface all round. Outside this window, the fish sees only the reflection of objects (such as the bottom) below the surface. This 'cone of vision' means that the area through which it can see through the surface decreases as the fish moves upwards.

Water, though it is actually a clear and colourless fluid, is also a medium for chemicals in solution and solids in suspension, thus giving rise to the ambiguous question asked by P.J. Schroevers in one of his papers: 'Is water H_2O?' Many fish live in extremely turbid water - dark peaty waters or rich lochs where there are algal blooms or highland streams brown during spates or lowland rivers with a heavy load of suspended silt washed in from farmland. Under such conditions fish visual awareness requires support from other senses such as their extremely sensitive acoustic lateral line system (which also has many branches on the head), their hearing (which may involve the swim bladder as a sound box) and an acute sense of smell, helped by the fact that they are constantly passing the medium in which they live through the mouth and past their sensitive gills. In human terms, living through a turbid spate might be equated to trying to survive in a hurricane with nil visibility and objects constantly hurtled at you! Fish do it regularly, surviving through conditions similar to those described by Arthur Johnson: 'The stream was rushing and tumbling in a brown flood over its rocky bed, carrying with it sheaves of corn, clusters of rushes, brown and green, branches, and dead driftwood, as it hurried on it way to the sea. The leaves of sycamore and hazel fell upon its murky tide, and were quickly hustled out of sight, or they collected upon the slower backwaters, covering the foamy surface with a carpet of yellow and russet brown.'

With such sensitive systems fish are able to cope with darkness better than many other animals. Indeed, as noted above, many loch fish (for example Arctic Charr and Vendace) spend most of their lives in darkness or semi-darkness, living during the day in the dark depths, but coming up to the surface in the gloaming to feed. At dawn, they gradually return to the haven of dark deep water, where they are safer - and cooler. Most fish are especially active at dawn and dusk, a behaviour pattern that many anglers have come to recognise. However, some fish sleep at night, as Edward Boulenger recognised: 'Curious was the behaviour at night of fish that during the daylight hours swam about in mass formation. As soon as it was dark the shoals broke up and the fish took to the bottom, every individual facing a different point of the compass. As soon as disturbed, however, they awoke and speedily formed shoal again on the surface.'

Ward believed that 'The silvery sides of these fish act as mirrors reflecting their environment and so tending to render them invisible or, at least, inconspicuous. But this protection only avails the fish so long as it maintains its normal posture in the water, with its sides perpendicular. Directly it attempts any antics, the light from above flashes on the burnished surface and the creature becomes conspicuous. ... When one realise how very slightly a fish has to turn on its side to reveal itself, it is remarkable that we so seldom see a silvery glint as he dashes through the water.'

Pain

There has been much past discussion over the question of just how much pain fish can feel and indeed some question as to whether they can feel pain at all. Most anglers, of course, generally argue that fish are largely insensitive to pain, and this justifies their sport as a whole and in particular the use of fish as livebait. This line of argument is often supported by instances of captures of fish which have just recently been detached from a hook and released, or which have actively taken a bait and hook, in spite of being seriously wounded in some way. The belief is that if such fish were in pain they would not be active and caught in this way. In fact, since they must eat to survive, it would be very unadaptive for fish not to eat even though wounded, and parallels can be seen in many birds and mammals in the wild which continue to lead apparently normal lives even though seriously injured. Human beings do not stop eating just because they are in pain.

Fish have well-developed tactile and acoustic sensory systems with associated complex nerve networks and it is virtually certain that they feel pain in much the same way as we do. The present legislation in the United Kingdom seems to agree with this, for it is actually illegal to carry out any experimental work on fish (and other vertebrates) without a licence. Any scientist in a laboratory who angled directly for live fish in tanks or hooked them up as the equivalent of livebait would be prosecuted unless he/she had a licence specifically to carry out such experiments. Yet every day thousands of fish must suffer pain when hooked by anglers, whilst the practice of livebaiting (which can have other disastrous consequences for fish populations, as discussed elsewhere) must cause agonising pain for the individual bait fish involved. If fish could scream then attitudes might be different!

The ethics of catching fish at all, or more often how many it is reasonable to kill, are often debated, but few people follow the logical line taken by Christopher North in his *Noctes Ambrosianae*. 'It's a maist innocent, poetical, moral & religious amusement. Gin I saw a fisher gruppin creelfu' after creelfu' o' troots, and then flingin' them awa among the heather and the brackens on his way hame, I micht begin to suspec that the idiot was by nature rather a savage. But, as for me, I send presents to my freens, & devour dizzens on dizzens every week in the family - maistly dune in the pan, wi' plenty o' fresh butter & roun' meal - sae that prevents the possibility o' cruelty in my fishin', & in the fishin' o'a' reasonable creatures.' The author, a supporter of angling, agrees with these sentiments.

Some anglers may admit that fish do feel some pain, but nevetheless their pursuit is justified. Braithwaite is such an angler: 'Probably fishing is the least cruel of sports; if hooked and lost a fish is none the worse, and may again take the fly in half an hour or so, whereas the wounded bird or animal often goes away crippled or to die.' Others are more defensive of their activities. Gathorne-Hardy writes: 'Much has been written by humanitarian sentimentalists on the cruelty of sport in general and fishing in particular; but I comfort myself in the belief that the sense of feeling in fishes cannot be acute. I have caught a sea-trout with a fly, still bleeding from the fresh mark of the heron's bill, which had transfixed it through the middle of its body; and it would be easy to multiply instances of insensibility to pain.'

This is the same Gathorne-Hardy who shoots seals: '... I fired a successful shot ... and although the seal disappeared, the reddened water showed clearly that the shot had struck home. I feared he was lost; but after a few minutes he rose to the surface ... and we pursued him in the boat, his dives growing shorter and shorter each time he rose. At last we approached sufficiently near to drive a long trident-shaped fish spear into his body, and we had almost got him to the side of the boat when he gave a convulsive struggle, twisted the strong triple barbs into the shape of fish hooks, and once more dropped off into the deep water. I feared that he was lost to me after all, but this was his last effort, and when he arose again we were able to secure him with the spear ... The skin was mine at last, but sadly spoiled by the wounds inflicted by the barbs in his struggle to get off.'

Some anglers have tried to turn the debate using humour. A notable example is William Scrope: 'Much has been said by humane persons about the cruelty of fishing.... Let us see how the case stands. I ... make an imitation of a fly; then I throw it across a river, ... Up starts a monster fish with his murderous jaws, and makes a dash at my little Andromeda. Thus he is the aggressor, not I; his intention is evidently to commit murder. He is caught in the act of putting that intention into execution. I drag him, somewhat loth to the shore, ... I find his stomach distended with flies. ... You see, then, what a wretch a fish is; no ogre is more bloodthirsty, for he will devour his own nephews, nieces, and even his own children, when he can catch them; ... What a bitter fright must the smaller fry live in! ... I relieve them of their apprehension, and thus become popular with the small shoals.'

Ancelot turns this sentiment back again in *L'homme du monde*: 'La ligne est un instrument ou il y a une bete a chaque bout.'

The most contentious part of the argument concerning cruelty in angling relates to the use of livebait - a practice which the author abhors and believes is both indefensible in humanitarian terms and unnecessary for successful fishing. Many anglers are quite blasé in adopting this practice. One well known one (who used the pseudonym Jock Scott) seems quite unconcerned: 'Remains the bait. I will cheerfully use any little fish that I can get, for at times they are plaguey hard to get! If your lakes hold perch, then probably some small ones are to be had; if not, anything that is going, roach, rudd, dace.

Slip your lower hook through the gill cover and the top Ryder hook under the back fin, and swing your bait out. ... After this there is very little left to do. If the bait tows the float along ease out a bit of line and let him travel.'

The practice of livebaiting has been going on for hundreds of years, and no doubt seemed small beer in the days of cock fighting and bear baiting. T. Westwood and Thomas Satchell wrote their advice in the 1600s in *For the Pike or Pearch*:

'Now for to take these kinde of Fish with all,
It shal be needful to haue still in store,
Some liuing baites as Bleiks, & Roches small,
Goodgion, or Loach, not taken long before,
Or yealow Frogges that in the waterscraul,
But all aliue they must be euermore.
For as baites that dead & dull doe lye,
They least esteeme & set but little by.'

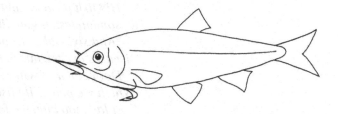

One way to hook up a fish as livebait.

Anyone who doubts the pain inflicted on fish by some anglers need only read Izaak Walton's description of how to mount a fish for live-baiting. 'Of fish, a roach or dace is, I think, best and most tempting (and a perch is longest lived on a hook); and having cut off his fin on his back, which may be done without hurting him, you must take your knife, which cannot be too sharp, and cut or make an incision, or such a scar, as you may put the arming wire of your hook into it, with as little bruising or hurting the fish as art and diligence will enable you to do; and so carrying your arming wire along his back, into or near the tail of your fish, between the skin and the body of it, draw out that wire or arming of your hook at another scar near to his tail: then tie him about it with thread, but of no harder than necessity to prevent hurting the fish; and the better to avoid hurting the fish, some have a kind of probe to open the way, for the more easy entrance of your wire or arming ...' This was published in 1653, but the practice continues to this day.

Recent experimental work by Lynne Sneddon has shown convincingly that Rainbow Trout possess pain receptors (nociceptors) equivalent to those found in higher vertebrates. Other research has shown that Goldfish can learn to avoid parts of their tanks where they receive electric shocks. At present (2006), the Home Office 'is considering the question of whether fish can feel pain or not'.

Henry Lamond, that enthusiastic and erudite Loch Lomond angler, admits in his excellent review of 'statutory cruelty' that one can hardly read Walton's passage 'without a shudder'. He is one of the few anglers who, having considered the question of live-baiting and other angling practices fully, concludes by 'conceding that there is to a certain extent some unavoidable cruelty'; but his hope is that those against angling will not 'harass devotees of the gentle art in spite of the modicum of suffering the pursuit entails upon some of the humbler members of the animal kingdom'.

References

Alexander & Adams (2000), Boulenger (1946), Braithwaite (1923), Gathorne-Hardy (1901), Gordon (1944), Gunn (1931), Harvie-Brown & Buckley (1887), Hutchinson (1907), Jefferies (1879), Johnson (1907), Lamond (1931), Maitland (1990), Morrison (1936), North (1876), Schroevers (1967), Scott (1932), Scrope (1843), Sneddon (2001, 2003), Sneddon *et al.* (2003), St John (1891), Stuart (1962), Walton (1653), Ward (1911), Westwood & Satchell (1883), Wood (1863), Young (1926).

'Death is a fisherman, the world we see
His fish pond is, and we the fishes be.
He, sometimes, angler-like, doth with us play,
And slyly takes us one by one away;
Diseases are the murdering hooks, which he
Doth catch us with, the bait mortality,
Which we poor silly fish devour, til strook,
At last, too late, we feel the bitter hook.
At other times he brings his net, and then
At once sweeps up whole cities full of men,
Drawing up thousands at a draught, and saves
Only some few, to make the others' graves,
His net some raging pestilence; now he
Is not so kind as other fishers be;
For if they take one of the smaller fry,
They throw him in again, he shall not die:
But death is sure to kill all he can get,
And all is fish with him that comes to net.'

Anonymous (1600s) *Death's Trade*

*****

FISH ECOLOGY

'This feckless hairy oubit cam' hirpli' by the linn,
A swirl o' win' cam' doun the glen an' blew that oubit in:
O when he took the water, the saumon fry they rose
An' tigged him a' to pieces sma', by head an' tail an' toes.'

Charles Kingsley (1819-1875) *The Oubit*

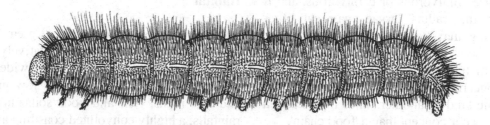

ALTHOUGH fish ecology can be broadly defined by a number of principles, opportunism is often important - so when, as Charles Kingsley describes, a caterpillar accidentally falls into a burn, salmon fry there make the most of this welcome addition to their diet - a common event in highland burns.

Fish ecology may be defined as the relationship between fish and their environment. In this connection it must be remembered that an important part of the environment of fish is other fish! These may function as competitors, predators or prey. Charles Elton, in 1927, was one of the first scientists to use the term ecology and to emphasise the dynamic nature of communities, but showing at the same time that they all adhered to certain broad principles. The basic ideas concerning inter-relationships within ecosystems have evolved from elementary ideas about food chains, through food webs and energy flow pathways, to sophisticated computer simulation models of such systems. The contribution of work from fresh waters and fisheries to this field has been considerable, for the communities involved are often less complex and more self-contained than those of marine or terrestrial ecosystems.

All freshwater ecosystems have certain general features and levels of organisation in common. The ability to support quantities of life of various kinds depends initially on the amount of energy available, linked closely with the degree of complexity of the system. Organisms within the community transfer this energy from one grade to another, through what

are known as trophic levels. There are three main classes of organism concerned with this transfer of energy: (a) Primary producers, which use the energy from solar radiation and available inorganic nutrients to produce more energy-containing (plant) material. The main organisms involved here are those capable of photosynthesis - certain bacteria and most algae and higher plants. (b) Consumers, which cannot synthesise material from inorganic sources only, but have to rely on organic substances already elaborated by primary producers. There are two main types of consumer: herbivores (secondary producers which feed only on plant material), and carnivores (tertiary producers which feed only on animal material). In practice the situation is often more complex, involving additionally omnivores, which feed on both plant and animal material, and various parasites which may utilise organisms at all levels. (c) Decomposers, which attack other organisms (usually after they are dead), breaking them down into simpler compounds and releasing many of the inorganic salts again, making them available to primary producers. The main organisms involved in the decomposer process are bacteria and fungi.

Simple systems, formerly known as food chains, are rarely found in isolation in nature. In a common ecosystem such as a stream, for example, a basic part of the energy flow pattern might be through a benthic alga (utilising available solar radiation and nutrients) which is eaten by a caddis larva, which in turn is eaten by a fish such as a Minnow. This simple chain, however, may be complicated by a variety of cross-links; it is likely that invertebrates other than caddis larvae eat the alga, and that fish other than Minnows eat the caddis larvae. Since too, many animals are opportunists as far as feeding is concerned, and rarely restricted to one type of food, it is likely that the caddis larvae will browse on algae other than the species under consideration, and that the fish concerned will eat various invertebrates additional to the caddis larvae. Many invertebrates are omnivorous or carnivorous, and it is likely that the caddis larvae may eat other invertebrates (e.g. mayfly larvae) which feed on the algae, or themselves be eaten by predacious stonefly larvae, these in turn being eaten by fish. All the organisms concerned may well be attacked by parasites of one kind or another. Thus the idea of a 'food web' is a better concept than a 'food chain'.

A freshwater food web (from Maitland & Crivelli 1996, drawing by Robin Ade)

Even in the most complex situation, however, it is still possible to consider principles common to the structure of the communities and the flow of energy within them. The concept of trophic levels is linked with the idea of a pyramid of numbers; this conception is dependent on the fact that in any energy flow system the quantities involved tend to be less and less between primary producers and top consumers. Charles Elton pointed out that 'animals at the base of a food chain are relatively abundant,

while those at the end are relatively few in number, and there is a progressive decrease between the two extremes'. This simple concept may be modified in different ways, depending on the structure of the community concerned and its stability in space and time. Fish are the top predators in many fresh waters and often the main factor in controlling the dynamics of the system.

Habitat

Scotland is renowned for the great variety of landscapes contained within a relatively small land area. This quality reflects the wide range of contrasting land forms: high and low ground, hard and soft rocks, rich and poor soils, high and low rainfalls, a highly convoluted coastline and a history of various intensive land uses. It follows therefore that this variety has also produced a wide variety of natural freshwater habitats, to which must be added the considerable number of artificial water bodies such as farm ponds, open ditches and canal networks in the lowlands and numerous reservoirs of all shapes and sizes, mostly in the uplands. In recent years the availability of grants to landowners has resulted in the creation of large numbers of small ponds which, to some extent, make up for the ones lost in previous decades due to drainage schemes.

The range of our natural fresh waters encompasses many fish habitats. In the uplands, the high corrie lochs of the Cairngorm plateau in the Central Highlands at around 900m in altitude, with their clear green-tinted waters are ice-free for less than half the year, whereas some of the large, deep and elongated, glaciated loch basins of mountainous areas, with their 'acid', biologically poor (oligotrophic) waters, rarely freeze over. Other waters include the less common shallow limestone lochs of Durness and Lismore and the thousands of acid peat-stained pools in the uplands and the Flowe Country, often the source of moorland burns. In the lowlands, important habitats include the turbid lower reaches of large rivers, whose lowland tributaries are now largely dominated by the run-off from agricultural land and the many lovely estuaries with full ranges of salinity from fresh to salt water. There are also less common habitats around the coast such as the mildly saline water bodies impounded by storm beaches or found in basins at the heads of Hebridean tidal systems and the immense variety of ponds, ditches and marshy pools throughout the lowlands.

The most important qualities of freshwater habitats that, in theory, dictate the density and species composition of the fish present are water velocity, level of dissolved oxygen, summer temperatures and the level of chemical, and therefore biological, richness and degree of pollution. In reality, the situation is often much more complex than this, for other factors of importance (e.g. spawning or nursery substrates).

Many of Scotland's large rivers which rise in upland areas gradually progress from being fast flowing and oligotrophic to slow flowing and eutrophic. There are many classic examples on the Continent (e.g. the Rivers Rhine and Gironde) where the species composition of the fish fauna is closely related to the habitat provided by each particular section of the river, and various schemes of zonation have been based on these fish communities. In Scotland, however, with its relatively short rivers, its impoverished fauna and flora and specialised history of fish colonisation, re-distributions and introductions, any such correlations are only approximate.

There, however, are some features and relationships characteristic of the communities found in fresh waters in Scotland. In the mountainous regions, the mainly hard insoluble rocks and poor soils mean that the acid waters of their streams and lochs are poor in the minerals required to promote growth, and can therefore only support a low level of biological productivity. Their invertebrate life is poor in species and dominated by insects. These habitats, though, favour salmonid fishes, and Atlantic Salmon, Brown Trout and Arctic Charr thrive in the cool, clean, well-oxygenated waters and clean silt-free gravels which are so important for the survival of their eggs and subsequent young stages.

Such habitats are the most vulnerable to the impact of 'acid rain' as they have insufficient buffering capacity to neutralise the acids being deposited from the atmosphere. A few other fish also inhabit such waters but they seldom dominate the salmonids, for most coarse species do not tolerate the environmental poverty or low summer temperatures.

At the other extreme there are typical lowland river systems flowing over soluble mineral-rich strata and influenced by run-off from the rich agricultural soils in the catchment. This results in high pH 'alkaline' waters and a biologically rich, eutrophic, environment. The turbid waters of these systems usually support much aquatic vegetation and a rich and diverse invertebrate fauna with molluscs and crustaceans as important members. Being at a low altitude, the relatively high summer water temperatures provide the conditions essential to so many coarse fish species for successful ova hatching and fry survival. Such habitats suit most members of the carp family and many other purely freshwater fishes. However, where these (especially Pike) are absent, salmonid fishes will thrive as long as there are tributaries with good spawning and nursery areas to provide an adequate recruitment to maintain their populations further downstream.

Carp and Barbel - lowland species introduced to Scotland from England (from Wood 1863).

Through the passage of time, fresh waters naturally tend to become silted up and more eutrophic, but this process has been greatly accelerated by human activities. This applies even to some water bodies in highland areas due to the extensive afforestation that has taken place. This has involved a considerable amount of ploughing and drainage and the application of chemical fertilisers to land newly prepared for tree planting. The rehabilitation and reclamation of hill land for pasture also results in substantial amounts of fertiliser run-off into feeder streams.

An early sign of eutrophication is the appearance of green algae on rocks and stones in or near streams and along the shores of lochs where this has never been seen before. Slight eutrophication probably has little effect on the fishes of these upland areas, but the accumulation of nutrients, along with those from other human sources in lowland lochs or the lower reaches of rivers, can and does produce conditions lethal to fish, especially due to low oxygen levels during hot dry weather and low water flows or in lochs under ice.

Behaviour

The behaviour of fish, like that of other vertebrates, is made up of two components - instinct and learning. Much of the life of a fish is dominated by the former but the learning process should not be underestimated as a visit to a modern fish farm will show. Here, fish such as Rainbow Trout and Atlantic Salmon quickly learn to assemble at particular places to be fed, or even to feed themselves from automatic dispensers which release food when the fish press a lever or some other triggering mechanism.

A simple type of behaviour is shown by shoaling species, such as Roach. Though many species of fish are solitary virtually all of their time (except at spawning time), others spend most of their lives in the company of their own species, forming shoals which may number many thousands of individuals. Most purely solitary species are predatory and often large; the Pike is a good example. Other than on the spawning grounds, members of this species are rarely found together; one of the reasons for this is that large Pike regularly eat small ones and there are several Scottish lochs in which, when suitable food species are rare or absent, they eat little else.

Shoaling species in contrast, tend to be smaller, herbivorous, planktivorous or more commonly omnivorous, fish which keep together in packs - the density of which usually depends on the activity in which they are engaged. Shoaling fish often, but by no means always, tend to be silvery in colour, living in open areas of water. Many of the carp family (e.g. Roach, Dace, Minnows) are typical examples. Almost immediately after hatching, the young start to congregate together and move about as one unit. This is at its most dispersed at night or sometimes when feeding, but when moving about, especially if danger threatens, a very tight pack is formed. Such shoals may break up to form smaller units or join together to form larger ones, but many of the fish may remain together all their lives. Spawning is

likewise a communal shoaling activity, often preceded by a migration in which enormous shoals move to one part of a lake or upstream into running water to reproduce.

Shoaling behaviour is usually thought to be for mutual protection and advantage; it is much more difficult for predators to approach unseen, and when a new source of food discovered by one member of the shoal it very soon engages the attention of most of its other members. Occasionally, fish are found together in numbers and the term shoal is applied, but the congregation here is one of chance and there is no interplay among fish as there is in a true shoal. Thus, during winter, certain fish (e.g. Carp) may congregate in depressions in lakes or rivers and remain together in a torpid state for many weeks.

Loch Eck echo sounding - the short top traces are active Brown Trout, the long bottom traces are inactive Powan.

An important aspect of fish behaviour which has received considerable attention from fish biologists and others is migration. Many more fish undertake migrations than is commonly supposed, but the movements are often of a local nature and do not involve the spectacular distances travelled or homing problems solved by better known species. A good example of small-scale migration is found in Brown Trout, populations of which are found in lochs over most of Scotland. Adult Brown trout in these lochs are territorial and often occupy one relatively small area of the loch for most of the year. In the autumn, however, they move to the mouths of streams entering the loch and migrate upstream to their spawning grounds. Normally, salmonids only migrate into the natal stream from which they themselves originated.

Atlantic Salmon have characteristic patterns of migratory behaviour and the distances travelled and obstacles surmounted are spectacular. It is now known that many Atlantic Salmon reach maturity in the seas off Greenland. In order to reach the European (or North American) coast, these fish have to migrate in a particular direction for many hundreds of kilometres and then locate the mouth of the river from which they originally emigrated. They then have to swim through the estuary with its violent changes of salinity (and in modern times often significant pollution) and finally upstream to the headwaters, sometimes leaping waterfalls 3-4 metres in height. However, when water levels are low, especially in smaller rivers, salmonid fish may have to wait long periods in pools for sufficient rainfall to create a spate and allow further progress upstream. The scene is well depicted in Thomas Stoddart's poem *The Sea Trout Gray*:

> 'The sea-trout gray
> Are now at play,
> The salmon is up - hurra! hurra!
> For the streamlets brown
> Are dancing down;
> So quicken the cup - hurra! hurra!'

In contrast, the migration of the few adult fish which survive spawning and return to the sea, or of the descending smolts (as the young are called) seems a much simpler and more passive affair, though the same problems of pollution and salinity are involved. Although much studied, some of the patterns of behaviour involved in the migrations of these fish are still not clearly understood. It is generally felt that orientation over long distances, say in the sea, is related to physical features (e.g. water currents) but that in the recognition of natal rivers chemistry is more important.

Migratory fish such as Atlantic Salmon, which move into fresh water to spawn but whose progeny then pass down into the sea to grow to maturity, are called anadromous. Lampreys, European Sturgeon, Allis and Twaite Shad, Sparling and several other species come into this category. Catadromous species, on the other hand, show the opposite type of behaviour, growing up in fresh water but migrating downstream to the sea to spawn. Eels and some populations of Flounders are good examples of this type of life history.

The behavioural aspects of fish biology reach their most complicated and bizarre at breeding time. In many species, only then is it possible to distinguish the sexes externally, sometimes simply because the female is plump with eggs, but more often because the male has become brightly coloured or has developed tubercles. These are small white nodules which appear on the head and sometimes the fins and bodies of sexually mature fish during the spawning season. They are most common among the carp family (Cyprinidae), but occur in some other families too, for example the whitefishes (Coregonidae). Many types of behaviour are shown by fish during the breeding season: apart from the spawning act itself, these range from aggression towards males of the same species or predation of the eggs and young, through complicated nest-building activities to fanning the eggs in the nest and actively 'herding' the young fish until they can fend for themselves.

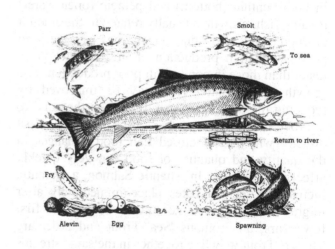

Atlantic Salmon life cycle (from Maitland 1989, drawing by Robin Ade).

The carp family (Cyprinidae) as a whole show relatively little in the way of sophisticated spawning behaviour. In contrast, the Three-spined Stickleback, which becomes very colourful at spawning time has a complex reproductive behaviour which has been studied intensively, and there are many complex interactions between the males and females and their environment.

The feeding behaviour of fish is also of considerable interest to biologists, fish farmers, anglers and others. It is in relation to feeding that fish seem to exhibit their greatest potential for learning, and the opportunistic nature of many species has led to their considerable success in some waters. The feeding patterns of some species, mainly fish predators (e.g. Pike) on the one hand, and filter feeders (e.g. lamprey larvae) on the other are very instinctive and stereotyped. Other fish are more adaptable. Thus Brown Trout in a river may be feeding very actively on benthic invertebrates on the bottom one day, but ignoring these completely the next day to feed on mayflies which have started to emerge at the edge of the river. On the third day, both these sources may not appear in the diet, which then might consist entirely of terrestrial insects blown on to the water surface and thus very easily available. Opportunist fish species in changing circumstances like these prefer the food source which is most easily available to them for the least expenditure of energy.

Food

In both benthic (bottom) and pelagic (open water) feeding fish, growth is usually related to the amount of food available. Thus populations of Roach in ponds with high production of invertebrates grow faster than those in ponds with poor production. The growth of the latter fish may be improved by reducing the number of fish, fertilising the pond to improve invertebrate production or feeding the fish directly with manufactured fish food. Changes in the quality and quantity of food can very quickly affect growth rates. In Atlantic Salmon, a dramatic increase in growth takes place immediately after migration from fresh water to the sea. For their first few years, anadromous 'Sea' Trout and sedentary 'Brown' Trout may live together in the same streams and grow at identical rates. As soon as Sea Trout move into salt water where more food is available and winter temperatures are milder, they start to grow at a very much faster rate than the Brown Trout which have remained permanently in fresh water.

Freshwater fish occupy a major role in aquatic food webs, and are often the top predators except where they are, in turn, eaten by fish-eating birds or mammals. Few fish eat much vegetable matter and none are pure vegetarians. Some of the carp family, such as Rudd and the Minnow, are omnivorous, eating both plant and animal food, but most are completely carnivorous and rely on that great group of small spineless creatures, the invertebrates. Usually aquatic invertebrates are most important but insects falling on to the water surface in summer are of significance in the diet of some fish.

Some water birds occupy the fish role in lochs and rivers by eating invertebrates. These include Dipper, Tufted Duck and Goldeneye, but many others, from Kingfishers and Herons to Cormorants and Goosanders, are fish-eaters, one step further up the food pyramid.

Growth

Egg development time varies greatly among species and within a species it is very dependent on temperature. Usually the eggs of fish which spawn in the spring, and especially of those spawning in the summer, hatch quickly, the incubation period lasting about 14 days in the spring for Three-spined Sticklebacks and only 3-5 days for Carp in the summer. In fish which spawn in the autumn and winter on the other hand, the eggs may take up to 150 days to hatch (e.g. Atlantic Salmon).

The egg itself undergoes profound changes during incubation. Most of its volume is occupied by yolk, the cell resulting from the original union of the sperm and ovum being very small. This cell, however, undergoes a process known as cleavage to give two cells, then 4, then 8, 16, 32 and so on, each group of cells gradually differentiating and growing into a different part of the embryo fish. The yolk is gradually used up during this process and eventually the egg consists of a spherical membrane inside which is curled up a small fish. The pigmented eyes and the rapidly beating heart can usually be seen quite clearly at this stage.

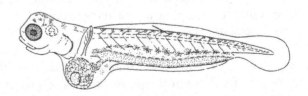

Powan larva with yolk sac (from Slack et al. 1957).

Upon hatching, the young fish may start swimming immediately and continue to do so for virtually the rest of its life (e.g. Whitefish) or, more commonly, it may rest in a protected place for some time until the remains of the yolk sac are fully absorbed. Atlantic Salmon and Brown Trout alevins lie among gravel in their nest (redd) during this period, while Pike have a temporary adhesive organ by means of which they hang, resting, attached to vegetation. At some time during this period, most fry make a rapid visit to the water surface to fill the swim-bladder with air.

Various names, often used rather imprecisely, are given to the young stages of fish. In general, the term larva is used to describe the stage from hatching until the fish is a miniature adult. This may take only a few days, (Carp) or up to year or more (Sea Lamprey and Eel). In some cases, including the two just mentioned, the two stages are so different that they were originally described by taxonomists as different species. The larval stage itself is sometimes divided into prolarval where the yolk sac is still present and postlarval when it has disappeared. In general, larvae are characterised by transparency, absence of scales, presence of large pigment cells and embryonic, undifferentiated fins. Beyond the larval stage, fish tend to look very much more like the adults although many features, particularly colouration and sexual differences, are not evident until full maturity is attained. Most identification keys refer to mature specimens of the species concerned and some difficulty may be met with if only young fish are available. If it is necessary to identify larval specimens, specialist keys must be used.

Age at maturation (i.e. when fish are able to reproduce) varies very much among different species and even among different populations of the same species. In general, sexual maturity is related to size which in turn is dependent on growth, thus fish in fast-growing populations tend to become mature earlier than those in slow-growing ones. Both food and temperature affect growth; fish at higher latitudes and altitudes mature later than those at lower ones. In Finland, Roach do not mature until they are 5-6 years old, whereas in southern Europe the same species may be mature at 2-3 years.

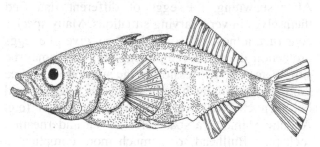

Three-spined Stickleback (drawing by Gordon Reid).

Small species of fish tend to mature and die early. The Three-spined Stickleback, for instance, normally matures and dies within two years, whereas the massive Sturgeon does not mature until it is 15-20 years old and may live for 10 or more years beyond that. Some fish spawn only once: the Atlantic Salmon usually spends 2-6 years in fresh water and a further 1-2 years in the sea before it is mature and comes back to fresh water to spawn; very few fish live beyond this to spawn a second time. In the case of the Allis Shad and Eel there appears to be complete mortality after the first spawning. Other species such as Brown Trout, may spawn several times during their lives.

Atlantic Salmon scale showing growth rings (from Malloch 1910).

The growth of most Scottish fish takes place in annual spurts - usually during the warmer months of the year. These seasonal variations lead to physical and chemical differences in scales and various bones which are subsequently of great value in determining age and growth. Growth itself may be controlled by many factors, among which the most important are food, temperature and genetic constitution. An understanding of these and other factors is essential to the successful management of fish stocks, particularly in fish farms and small closed sport fisheries where success may be determined by the ability of the fish stock to achieve optimum growth relative to available food.

Niall Campbell found that in new impoundments, the sequence of events which takes place is: (a) on impoundment, terrestrial organisms become available to fish as food, (b) within a short time after impoundment increases take place in the numbers of zooplankton (Cladocera) and midge larvae (Chironomidae), (c) a general increase in the numbers of bottom fauna takes place. These events result in an increase in the growth rate of the fish.

Reproduction

With fish, as with other animals, a top priority in the life cycle is to breed. Temperate fresh waters are less productive than the coastal marine environment but they support fewer predators. Some originally marine fish, like the lampreys, salmonids, smelts and sticklebacks, have come to protect their young by leaving the relatively dangerous coastal waters and spawning in fresh water, evolving special adaptations for living in this dilute environment at that time. But this is a compromise, because the relative scarcity of food in fresh water limits growth opportunities. If the young fish grow fast enough to mature there they will stay in fresh water, but if they cannot they will move back to the richer marine world until they are ready to spawn.

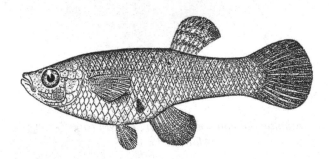

Female Mosquito Fish Gambusia affinis (from Jordan & Evermann 1900).

All British freshwater fish are oviparous, i.e. the sperm and eggs are ejected close together in the water and after fertilisation the egg undergoes development quite independently of its parents, though they may protect it in some way, or keep it clean. This is in contrast to the many livebearing species found in the tropics, e.g. the Mosquito Fish (*Gambusia affinis*) where the anal fin of the male is modified as an elongate penis. This is used to fertilise the eggs inside the female, where they remain protected until hatching. The young are then born free-swimming. However, unlike mammals, they do not receive any food materials from the female subsequent to fertilisation.

Oogenesis in the female parallels spermatogenesis and leads to the development of varying numbers of eggs within her ovary. Like sperm, each egg carries the hereditary characteristics of its parent, but the cell itself is very much larger as it also contains large quantities of yolk and fat. The number of eggs produced by a female fish usually increases with size and varies tremendously among species. Thus small fish like the Three-spined Stickleback and Bullhead, whose eggs and early fry are afforded considerable protection after laying, produce relatively few (50-100) eggs at each spawning. Larger species, however, whose eggs are shed directly into the open water and given no further protection lay very large numbers each year (e.g. up to one million per female in the Flounder). The number of eggs produced by a female is referred to as her fecundity.

In almost all Scottish fish, reproduction is a cyclic process related to the seasons of the year. It is controlled by reproductive hormones whose secretion is in turn dependent on environmental factors - e.g. temperature and daylength. The gonads, particularly the ovaries, may undergo considerable changes in size and appearance over the year, starting at their smallest just after the spawning season when the sexual products are shed. The actual spawning period for any population of a species is short (usually a few weeks); the time of year at which different species spawn can vary tremendously. Thus in Scotland, some fish are spawning somewhere in virtually every month of the year.

After spawning, the eggs of different fish find themselves in very varying situations. Many species construct a nest of some kind to give the eggs protection during development. This 'nest' may be simply an open depression in the substrate (e.g. Sea Lamprey); a similar hole in which the eggs are laid but then covered over with adjacent gravel (e.g. Atlantic Salmon); a space cleared out underneath a rock (e.g. Bullhead) or a much more complicated structure created from pieces of weed (Three-spined Stickleback). Some species protect the eggs and often the young fish in these nests (e.g. sticklebacks); others leave immediately after spawning (e.g. Brown Trout).

Eggs which are spawned without the protection of a nest of some kind can be laid in various ways. A few fish lay long strings of eggs which tangle up among vegetation (e.g. Perch). Many others lay adhesive eggs which may stick to stones (e.g. Sturgeon), to plants (e.g. Carp and Pike) or to sand and plants (e.g. Gudgeon and Stone Loach). Finally, many species simply spawn in the open water and the eggs may then just float (e.g. Flounder in the sea) or sink to the bottom and lie in crevices there (e.g. Vendace and Powan).

Most fish spawn in pairs or small groups, but some are mass spawners. The Sparling is a good example of this, behaviour which makes it very vulnerable at spawning time when the whole adult population of a river system may be simultaneously on the spawning grounds in one short stretch of the river. They are then very vulnerable, as a short-lived species, to commercial fishermen and overfishing is thought to be the main reason for the extinction of this species in many Scottish rivers.

Predators

Not surprisingly, few fish eat birds but an exception that turns the tables is the Pike - which some bird watchers call the 'freshwater wolf'. Pike certainly catch young wildfowl given the chance and, in waters where Pike reach 40 pounds or more, even well grown birds are vulnerable. Young Mallards, Tufted Ducks, Grebes and Moorhens can be a large part of a Pike's diet in summer and at some bird reserves, Pike numbers are controlled. At Loch Leven, a study by K.F. Laughlin showed that a high percentage of young Tufted Duck were eaten by Pike as they tried to make their way across the narrow stretch of loch from the shelter of St Serf's Island, where they nest, to the safety of the mainland shore.

One other fish in British waters reaches enormous proportions and eats birds - the introduced Danube Catfish *Silurus glanis*, a native of central Europe. This can grow to three metres in length and 300 kg in weight - not one to meet fishing from a dark river bank - and birds have few defences against such predators. So far, this damaging species has been introduced to many waters in England. There have been proposals by coarse fishermen to introduce this fish to Scotland and, for obvious reasons, these must be resisted.

Quite a few birds eat fish and some rely on them almost entirely. A few only eat fish when the chance presents itself - Crows, Buzzards and Sparrowhawks and Blackbirds or Robins at a garden or woodland pond. Rarely is there any real threat to wild fish populations, but the impact on the unnatural situations (e.g. at fish farms or put-and-take fisheries) can be significant.

The real freshwater fish-eaters include Grey Herons, Common Terns, Kingfishers, Goosanders, Red-breasted Mergansers, Cormorants, Black-throated Divers, Great-crested and Slavonian Grebes and Ospreys. Some chase fish under water (Cormorant and Goosander), others dive on them from above (Common Tern, Kingfisher, Osprey) and some stand quietly waiting to strike at a fish as it swims by (Grey Heron). Greenshanks, Redshanks and Egrets often dash about, stirring up fish fry with their feet and snatching them as they try to flee.

An Osprey with its prey (from Colquhoun 1938).

Most predators have a bad name with anglers and gamekeepers. This is unfortunate, for predators can be an important driving force in keeping its prey populations healthy. There is much evidence to show that predators select damaged, old and heavily parasitised prey which are easier to catch than healthy individuals. There is also research which shows that fish predators find it easier to catch stocked fish, especially those recently introduced to a water, than wild ones. It is no coincidence that the influx of Cormorants to Loch Leven in recent years coincided with the start of intensive stocking of domestic fish there. Unfortunately, most studies of the diet of predators do not distinguish within the prey items regarding prey quality - age, condition or degree of parasitism. This aspect is important in relation to proper scientific management.

Perhaps surprisingly, many fish species are cannibalistic, and it is very common for adult fish to eat their own eggs and fry. Piscivorous fish rarely discriminate between their own kind and other species and in some situations adult fish rely entirely on their own young for survival. The pure population of Pike in Loch Choin, studied by Bill Munro, is a good example of this - the young Pike feeding on invertebrates and then being eaten in turn by adult Pike. This is the only way in which pure populations of piscivorous fish can maintain themselves in the absence of other fish species as prey.

Per-Arne Amundsen makes the point in his amusing poem *Cannibalism*:

'Cannibalism has developed in an evolutionary sense,
as a gastronomical affection for kin and friends.
The selection is logical and natural as well;
the ultimate adaptation is to consume yourself.'

References

Amundsen *et al.* (1998), Campbell (1963), Colquhoun (1938), Elton (1927), Jordan & Evermann (1900), Laughlin (1974), Lonnrot (1849), Maitland (1972, 1978, 1981, 1989), Maitland & Crivelli (1996), Malloch (1910), Munro (1957), Slack *et al.* (1957), Stuart (1953, 1957), Wood (1863).

'In the gray pike's belly they find the silvery lake trout;
in the belly of the silvery lake trout there was the smooth whitefish.'

Elias Lonnrot (1849) *The Kalevala*

*****

LOCHS AND RIVERS, PONDS AND BURNS

'Numberless springs also well up ...
and give birth to many rivers,
in which Scotia marvellously abounds beyond any other country,'

Henry Lamond (1911) *The Gentle Art*

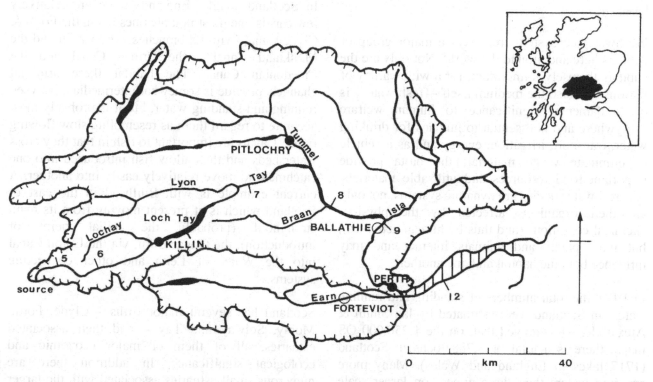

THE catchment of the River Tay, above, the largest river by flow in Great Britain, exemplifies the variety of fresh waters in Scotland. Scottish lochs and rivers, famed all over the world in poetry and song, make up an extensive range of habitats which fish are able to exploit in different ways, depending on their ecological requirements. Thomas Stoddart was enthusiastic about them in his poem: *The Streams of Old Scotland for Me!*

> *'The rivers of Scotland for me!*
> *They water the soil of my birth,*
> *They gush from the hills of the free,*
> *And sing, as they seek the wild sea,*
> *With a hundred sweet voices of mirth.'*

As discussed in an earlier chapter, ecology is the study of animals and plants in relation to their environment. In the case of any particular species of fish this means understanding not only its physical and chemical surroundings, but also its biological surroundings - vegetation, predators and prey. The whole basis of speciation in the biological world is related to the fact, that over generations, species of animals and plants developed differently from one another as a result of one type being particularly successful in at least one environment. This implies that, although living together in the same body of water, different species tend to exploit distinct parts of it.

The fresh waters of Scotland are a major national resource. More than 2% (0.60 % in England, 0.52% in Wales) of the land surface is covered by over 27,000 lochs and more than 50,000 kilometres of rivers and burns. Fresh waters represent a more significant landscape feature in Scotland than in any other part of Britain. In fact, most of the volume of standing fresh water in Great Britain is within Scotland. It should come as no surprise, therefore, that the fresh waters of Scotland comprise a significant and diverse resource.

Scotland's fresh waters have always been of importance to its inhabitants and there is a long tradition in the scientific study of this most valuable resource. The early classical research on lochs in the 1890s by Murray & Pullar and on rivers in the 1930s by McLean were among the first such studies in the world, and provided a firm foundation for the subsequent work on which our present knowledge is based.

Freshwater ecosystems represent a major group of habitats here and around the world. Not only are the habitats themselves important, for a wide variety of reasons, but also the medium itself - fresh water - is of fundamental significance to human welfare everywhere and is essential to humans for drinking water, transport, irrigation, energy and as a vehicle to eliminate waste material; the biota provide important food resources and utilisable materials. All freshwater bodies are dynamic systems, not only are their organisms affected by the physico-chemical conditions (and thus by human activities), but the plants and animals interact and may influence both the habitat and one another.

In 1979, the total numbers of standing and running waters in Scotland were estimated by Ian Smith & Alex Lyle, who showed that, on the 1:250,000 OS maps, there is a total of 3,788 lochs in Scotland (1717 lakes in England and Wales). Many more standing waters than these appear on larger scale maps, so that in Scotland it is estimated that altogether there are over 31,000 lochs, lochans and ponds. The vast majority of these are small waters less than a few hectares in area.

The numbers of rivers and streams in Great Britain were also assessed by Ian and Alex. They showed that, on the 1:625,000 O S maps there is a total of 1445 river systems entering the sea around Great Britain; associated with these are 7,835 stream systems. In Scotland, 950 river systems enter the sea - composed of 563 1st order, 296 2nd order, 69 3rd order, 19 4th order and three 5th order waters. As with lakes, the numbers greatly increase with increasing scale of the map, so that the total number of rivers and burns is probably about 5,000..

Thus the fresh waters of Scotland are a major resource. They include some of the most spectacular waters in Great Britain as well as the largest - Loch Lomond (the greatest in area: 71 km^2), Loch Awe (the longest: 41 km), Loch Morar (the deepest: 310 m) and Loch Ness (the greatest in volume: 7452 million m^3). The River Tay is the largest river by flow (194 m^3/second). Altogether the fresh waters of Scotland, occupying over 2% of the surface area of the country, make up 91% of the volume of fresh water in Great Britain as a whole.

In Scotland (unlike England) there are relatively few canals, the most notable ones being the Forth & Clyde Canal (with its branches - the Union and the Monkland Canals), the Crinan Canal and the Caledonian Canal. The habitat these artificial channels provide is somewhat intermediate between running and standing water, but it is probably most accurate to regard them as resembling slow flowing rivers. Canals are important to fish in that they cross watersheds and thus allow fish introduced into one catchment to move relatively easily into another. A current example is with Bullhead in the east of Scotland which is at present moving from its main stronghold (probably the initial point of introduction), the Gogar Burn, via the Union Canal into the Water of Leith and other connecting systems.

Scotland has several major firths – Clyde, Forth, Moray, Solway and Tay – and their associated estuaries, all of them of major economic and ecological significance. In addition there are numerous small estuaries associated with the larger rivers where they enter the sea.

There are, however, enormous demands on this resource - often conflicting ones. In addition, many activities, especially industry, urbanisation and land use, affect water even before it gets into lochs and rivers. Fertilisers, herbicides and pesticides used in agriculture and forestry may end up in nearby water courses, aided by land drainage which makes water flow off faster. These problems are dealt with in a later chapter.

Water types

In broad terms, fresh waters may be divided into two major categories - running and standing waters. The obvious primary distinction between these is the constant unidirectional flow of water in the former, and its absence in the latter. Each has different size categories. Running waters, for instance, range from small trickles and burns through streams to large rivers. As just described, canals are an important type of man-made running water. Standing waters range from small (often temporary) pools, through ponds and lochans to large lochs. The most important types of man-made systems in this category are reservoirs, which in some areas (e.g. Kilpatrick Hills, Lammermuir Hills, etc.) are the dominant type of water body.

The simple tabular classification of freshwater habitats, shown at the end of this chapter, covers the full range of fresh water types in Scotland. The diagram below is another way of showing this range. The commonest situation is a continuum, for most water bodies interconnect with others. In some large catchments, all these water types are found.

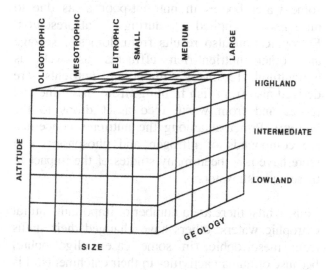

A simplistic classification of fresh waters in Scotland (from Maitland 1979).

Superimposed on the size and other physical attributes of a water (its altitude, geology, etc.) is its chemistry. Apart from the influence of salinity in waters in coastal areas, the chemical nature of a water depends on the geology of its catchment. Broadly speaking, base-rich rocks and soils (limestones, some sandstones, etc.) give rise to chemically rich water while base-poor rocks (granites and other igneous rocks) are associated with poor waters. Often the former are found in lowland areas and the latter in highland areas. The influence of man is important here, and the forms of agriculture carried out in a catchment can substantially affect the chemical quality of the water there. Lowland waters tend to be affected more than highland waters in this way.

Biologically speaking, two major types of water are recognised by freshwater biologists. Rich eutrophic waters are usually rather turbid, with chemically rich water characterised by high pH and alkalinity. The invertebrate fauna is made up of large numbers of worms, leeches, snails, mussels and shrimps, while coarse fish, especially members of the carp family, are the dominant vertebrates. In poor oligotrophic waters on the other hand, the clear water is low in chemicals and is typically slightly acid with a low to neutral pH. The invertebrate fauna is dominated by insects, especially stoneflies, mayflies and midge larvae, while salmonids are the major fish found. A third type of water, known as dystrophic, is found in areas dominated by peatlands. This is characterised by high quantities of humic acids derived from peat which stain the water brownish. Such waters may have no fish.

Lochs

Scotland is exceedingly fortunate in the numbers of beautiful lochs it possesses, especially in the northwest highlands. John Robertson in *Guide to the Highlands* notes that 'The number of lakes here, and in Assynt, especially in the north-west division, is incredible; and, being distinguished either by dark, still water, indicative of great depth, at the foot of rugged rocks, or by green sedgy banks and shallow margins, beautifully ornamented by the stately bulrush and the elegant flowers and handsome leaves of the white water lily, are very pleasing features amid the singular scenery of the district.'

Standing waters are characterised by a lack of unidirectional current, but stratification is a common feature. Conditions in standing waters are normally much more homogeneous than in running waters; they may reach great depths, but most have simple, often broad, basins. Whilst erosion does occur in standing waters, it is only severe on exposed shores and eroded materials usually remain within the same basin. In addition, materials from inflowing streams are constantly being deposited, filling in the basin and eventually obliterating it completely. As already noted, Scotland has over 30,000 standing waters, designated variously lochs, lochans, pools and ponds.

One of the most useful biological classifications of standing waters suggests several major types of open water - oligotrophic (including dystrophic), mesotrophic and eutrophic (including hypertrophic). Oligotrophic lochs are nutrient-poor, usually deep clear lakes which never have oxygen deficiency. Eutrophic lochs are nutrient-rich, usually shallow, turbid lakes which may have an oxygen deficiency in deeper water at some times of the year. Mesotrophic lochs, as the name suggests, are intermediate between oligotrophic and eutrophic lochs. Dystrophic lochs have variable amounts of nutrients but high amounts of humus, making the water brown; they are usually shallow or only moderately deep, when they may show oxygen deficiencies in deeper water. Though this classification does not really cover all types of standing waters, is extremely general and in some ways arbitrary, it has proved its value over a long period.

The great majority of Scotland's lochs are oligotrophic or dystrophic, the former sitting mainly in glaciated rock basins, the latter associated with peatlands. Mesostrophic and eutrophic waters (dealt with below) are far less common - though artificially increasing in number due to cultural eutrophication.

Mesotrophic lochs have a moderate amount of nutrients - more than upland waters, but not such an excess that their ecology is dominated by thick algal blooms. They are therefore productive and typically support a diverse mixture of submerged water plants and their associated animal life. These assemblages are increasingly rare, are under threat in developed countries and are in need of conservation.

Natural mesotrophic standing waters are typically found in the borders of upland areas in the west and north of the UK, their mesotrophic condition being determined by the nature of the catchment, in particular the leaching rate of mineral nutrients. The high conservation value of this habitat is recognised in the designated status of many sites across the UK. These mesotrophic lochs are perceived to be outstanding examples of their type, and have been given the status of Site of Special Scientific Interest and candidate Special Areas of Conservation (SAC) under the EC Habitats Directive. Additionally some mesotrophic lakes may be designated as Special Protection Areas under the Birds Directive. In Scotland the Loch of Menteith

and the Dunkeld-Blairgowrie lochs in Perthshire are some of the best examples of mesotrophic waters.

Eutrophic lochs can be entirely natural or 'culturally eutrophic'. Through the passage of time, fresh waters naturally tend to become silted up and successional changes accelerate through the availability of potential nutrients locked up in deposits, and released when shallowness permits their utilisation, predominantly by macrophytes. This process has been greatly accelerated by eutrophication due to human activities. In lowland areas with fertile soils, for example, intensive arable farming with heavy applications of fertilisers, particularly nitrates, leads to enrichment of ground water and runoff. Eutrophication occurs even in some water bodies in nutrient-poor areas due to nutrients applied during afforestation. Eutrophication also results from domestic sewage and other nutrient-rich effluents as well as agricultural fertilisers. Important nutrients are derived also from fish farming (from fish urine and faeces and from waste feed as it drops to the bottom). Principal among the nutrients concerned are compounds of nitrogen and phosphorus and there have now been many studies of the impact of these on fresh waters.

Thus, whilst there are a number of important natural eutrophic waters, others have changed their status from mesotrophic (in some cases oligotrophic) because of human activities in their catchments. It is important to distinguish between those which are naturally eutrophic (and likely to have a diverse and stable flora and fauna) and those which are culturally eutrophic (and have a fragmented biota).

Many of the nutrients quickly become bound up as part of the benthic biomass and the loch ecosystem in general; this is one of the principal features of the eutrophication process. Instead of a simple system with a constant addition of nutrients, and a constant dilution through the inflow-outflow system of the loch, there is actually considerable accumulation of nutrients within the ecosystem both in its living components (plants, invertebrates and fish) and especially within the bottom deposits. Attempts to reverse the situation are often difficult because of the enormous store which has built up.

Eutrophic lochs have various characteristics, including extensive algal growths, potential deoxygenation of the lower cooler layer of water during stratification in summer and under ice in winter, and a tendency for the fish community to be dominated, not by salmonids as in most oligotrophic lochs, but by coarse fish, especially cyprinids. Cultural eutrophication is likely to be the main cause of the extinction of fish in many lakes, e.g. the Vendace, *Coregonus albula* in the Castle and Mill Lochs in Scotland and the Smelt, *Osmerus eperlanus*, in Rostherne Mere in England.

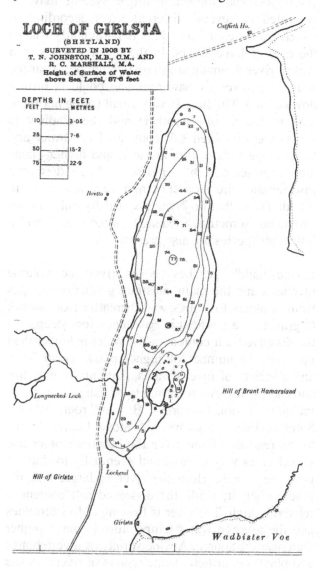

Loch Girlsta on Shetland - a typical glaciated loch (from Murray & Pullar 1910).

Though still encompassing a range of habitats (e.g. shallow stony shores, deep profundal muds, clear open water) lochs are more discrete in their form than rivers. Much of their physical nature has been shaped by past land movements and in particular by the action of ice during the last Ice Age, which

came to an end in Scotland some 10,000 years ago. The grinding action of huge glaciers gouged out many of our loch basins and left terminal moraines which dammed river valleys. Huge blocks of ice left their mark in the form of kettle holes.

One of the features of the lochs and lochans of Scotland is their marvellous variety as habitats for wildlife. Up on the high tops of our mountains are found numerous shallow peat pools. Most are acidic and fishless but have fascinating plants (e.g. the carnivorous Bladderwort) and numerous interesting invertebrates. Lower down in the straths are larger lochs, containing important fish populations, as well as a wide variety of plants and invertebrates. Most of these waters are very pure chemically and naturally slightly acidic, but in areas of limestone (e.g. Durness or Lismore) the waters are alkaline and very clear.

Above and below the surface in these lochs there is a wealth of plants and animals in clean, undisturbed waters. Pondweeds and stoneworts grow in the depths, lilies float at the surface and irises, marsh marigolds, bulrushes and many others line the banks. Among them, bright damsel and dragonflies hover in the sun, catching other insects while whirligig beetles and skaters utilise the surface. Under the water the nymphs of dragonflies, mayflies and stoneflies, water boatmen, shrimps and snails make up the food of fish. The latter, often unseen, can be present in variety, over 15 different species occurring in some of our larger lochs.

A broad range of features characterise a completely natural loch (including the fact that its catchment would also be relatively natural). For example, its basin would not be dammed in any way and the bottom substrates would consist of entirely natural materials like sand, gravels stones and boulders and not concrete, brickwork or gabions. The water quality would be high and reflect the geology and soils of the natural catchment rather than any influence from human activities. The biota too would be characteristic of the geographical area concerned - there would be no introduced alien species nor any loss of native ones. Such systems have two major features - they are sustainable and they have a characteristic, often high, biodiversity.

The shores of many lochs are a major attraction for humans with a wide range of recreation taking place there. Increasing leisure time and mobility in different parts of the world has led to many easily accessible loch shores being used very intensively. Some shores have a naturally high resilience and can absorb considerable pressures from visitor use. However, many shores have a low resilience and can degrade rapidly. It is ironic that the basic quality which attracted people to such shores in the first instance is damaged by their over-use. Fortunately, such deterioration can often be halted and reversed by site management if the process is recognised in time.

Rivers

Many of Scotland's writers have been impressed by the rivers of Scotland; Robert Louis Stevenson was no exception. 'But the streams of Scotland are incomparable in themselves - or I am only the more Scottish to suppose so - and their sound and colour dwell for ever in the memory. How often and willingly do I not look again in fancy on Tummel, or Manor, or the talking Airdle, or Dee swirling in its Lynn; on the bright burn of Kinnaird, or the golden burn that pours and sulks in the den behind Kingussie! I think shame to leave out one of these enchantresses, but the list would grow too long if I remembered all; only I may not forget Allan Water, nor birch-wetting Rogie, nor yet Almond; nor, for all its pollutions, that Water of Leith ...'

Running waters are characterised by a unidirectional current, they rarely stratify, longitudinally, they change gradually from source to mouth, they are always relatively shallow and they have long, often complex, narrow channels. Erosion is characteristic of running waters, and materials so removed may be transported considerable distances. As a further consequence of erosion and deposition, most running waters increase the length of their channels with age, as cutting back to the source and meandering on the flood plain proceed.

Scotland has 950 main river systems, about two thirds of the Great Britain total. The nature of rivers is mainly determined by climate and local topography. Scotland's high rainfall, low evaporation and lack of extreme temperatures create rivers with large flow relative to the size of the catchment, a dense network of rivers and burns and moderate water temperature fluctuations. Scotland's largely mountainous topography results in energetic, steeply flowing, rivers with beds of coarse material - gravels and boulders. Scotland also has slower flowing, more silty, river stretches in the lowlands but these are not typical of the majority of rivers in Scotland.

Although running waters can be classified into these various types, there are often intermediate categories, and indeed a single system may be classified in several different ways according to which part is examined. This is particularly true in the case of large rivers. In the upper reaches of a typical river the main substrate is often peat near the source, followed by bare rock and boulders further downstream. The flow is very small (often not more than trickle in dry weather) and the gradient is steep. The dissolved salt content of the water and the average temperature are low, and higher plants are represented by mosses. The dominant invertebrates are stoneflies and very few fish are found. Those that do occur are usually only Brown Trout, but sometimes Atlantic Salmon, Eels and a few other species are also present.

In the middle reaches of the river the volume increases and the main substrate gradually changes from boulders to stones, and thereafter from stones to gravel. The gradient is now much less steep, and the dissolved salt content of the water is higher than upstream. A number of higher plants occur here, and a variety of invertebrates, particularly mayflies and caddisflies. A wider variety of fish occurs, often including Brook Lampreys, Brown Trout, Atlantic Salmon, Eels, Minnows and Stone Loach. In the lower reaches of the river the substrate of coarse gravel gives way to sand and eventually to fine silt near the mouth. Here the river is large, and the gradient slight, while the dissolved salt content is relatively high. The river is flowing at low altitudes and the average water temperature is much higher than near the source. A wide variety of invertebrates and plants occur here, some typical of rivers, others also found in ponds and lakes. The fish fauna is varied, including cyprinid fish such as Roach and Minnows, as well as Perch, Pike and Eels.

Near the sea the river opens out into a broader estuary and the ecological situation is dominated by the regular presence of salt, or at least brackish, water. Many of the fish found upstream are restricted by this chemistry, but a few are tolerant (e.g. Three-spined Sticklebacks), and many others pass through as they migrate to or from fresh water. Such fish include Sea and River Lampreys, Eels, Atlantic Salmon and Flounders. Typical estuarine species are Allis and Twaite Shad, Sparling, Thick-lipped Grey Mullet and Common Gobies.

Zones characterised by different fish species must not be treated too seriously, for they are obviously dependent on the species concerned being present in the river system. Thus most rivers in the north of Scotland have only two non-migratory freshwater fish species (Brown Trout and Three-spined Sticklebacks), whereas rivers in the south of Europe may have twenty to thirty such species. Schemes of classification developed by ecologists for one area are often transplanted to others with most curious effects.

Brown Trout - the most widespread fish in Scotland (from Wood 1863).

One aspect of running waters which is of major importance to both landscape and fish is waterfalls. There are many spectacular examples of these throughout Scotland. They are important to fish in several ways but in particular where they are of sufficient height to stop the upstream migration of fish. Thus in many rivers, the fish community upstream of a major waterfall can be quite different and much less diverse than the one below the fall.

Much emphasis in recent years has been placed on the need to study the entire catchment of a river in order to understand its ecology. This is no easy task with some of Scotland's largest rivers which have large and complex catchments, with numerous important tributaries. The River Tweed as described by Thomas Stoddart *In Praise of Tweed* is a good example:

*'The lanesome Talla and the Lyne,
An' Manor wi' its mountain rills,
An' Etterick, whose waters twine
Wi' Yarrow frae the forest hills;
An' Gala, too, and Teviot bright,
An' mony a stream o' playfu' speed;
Their kindred valleys a' unite
Amang the braes o' bonny Tweed.'*

Estuaries

Although Scotland has few large estuaries, those of the Solway, Clyde, Tay, Forth and Moray Firths are notable exceptions. In the past, all of these must have been of immense importance ecologically and harboured enormous wildlife resources, including fish stocks. Gradually the riparian reeds were cut down and marshes drained as agriculture and industry developed. Harbours and ports multiplied as the advantages of the firths for shipping were realised and pollution, especially in the Clyde and Forth, reached catastrophic proportions. Many individual fish stocks (especially of migratory and estuarine species) were lost during this period, which stretched from the start of the industrial revolution to the present.

Fortunately, today, the situation is showing a gradual improvement, and indeed some populations are recovering - Atlantic Salmon in the Clyde and Sparling in the Forth being good examples. Nevertheless, the fish communities of all the large estuaries are still substantially impoverished compared to former times and have a long way to go before they can be considered as having recovered fully. For example, of at least 16 previously known populations of Sparling in Scotland, only three survive - in the estuaries of the Cree, Forth and Tay.

Far less is known about the smaller estuaries and inlets around the Scottish coast. It is likely that these have a much greater importance than is realised and some are relatively intact ecologically. The estuary of the Cree is a good example where it appears that, as well as the stock of Sparling mentioned above, there are still spawning stocks of Twaite Shad and possibly even of Allis Shad. Other estuaries such as those of the Morar, Fleet, Tyne and many other Scottish rivers require research to assess their full ecological value, and surveys of their fish stocks must be an important part of such studies. Many other small estuaries are important as the link between fresh and salt water. Seton Gordon describes one: 'In the mist-shrouded Island of Mull ... by the time the stream reaches the seapool its waters are not unworthy of the silvery salmon, and the sea trout, exulting in its strength.'

The River Tay is a fine example of one of Scotland's rivers which has been studied in some detail by the author and his colleagues. Many fine running waters could lay claim to the title of Britain's premier river but the one with most justification is probably the River Tay. With the largest catchment area (5,031 km^2) in Scotland and the greatest flow of any river in Great Britain (twice that of the River Thames), this exciting water is, at one and the same time, a most attractive and valuable resource.

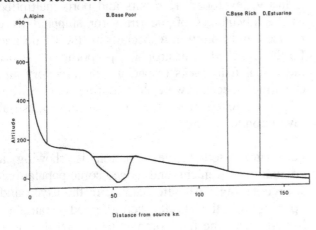

Profile of the River Tay (from Maitland & Smith 1987).

Rising as a small trickle at a height of 625 m in the west of Scotland, near Beinn Laoigh, the source water tumbles down the mountainside to join with other burns and form the River Cononish. Near here, Scotland's latest 'gold rush' started some years ago, and hundreds of tons of rock were mined and processed in the search. As with the barytes mine further downstream, there was concern that this industry would pollute the waters of the river here but this danger has passed with the closure of the mine. Near Tyndrum, the Cononish becomes the River Fillan which passes into Loch Dochart and Loch Iubhair and emerges as the River Dochart; this flows east and eventually passes over the lovely falls at Killin to enter Loch Tay, one of Scotland's many large lochs. Emerging from the loch at Kenmore, and now known as the River Tay, it flows

down a magnificent wooded strath to Ballinluig where it is joined by the River Tummel which doubles its size. Now a major river, it flows steadily towards Perth, joined by the Rivers Braan and Isla. At Perth it becomes tidal as it enters the firth where it is joined by the River Earn before reaching the open sea past Dundee. On its journey from Beinn Laoigh its waters have travelled 148 km.

The river shows many ecological changes along this course. Starting as a montane cascading burn with few nutrients and very little in the way of plants or animals, it slows somewhat to a rocky highland river full of mosses and other weeds as well as many invertebrates and fish. As it enters the strath it slows and meanders, with many deep pools and extensive areas of gravel. This stretch, surrounded by farmland, is very rich and has an abundant aquatic wildlife. Finally it moves into the estuary and mixes with salt water providing yet another varied habitat for aquatic life.

An obvious feature on the river is Loch Tay itself which forms almost one sixth of the total river length and whose deepest point (155 m) is well below sea level. There are many other lesser lochs within the catchment (574 are shown on the 1:50,000 maps) and several of the larger of these have been harnessed for hydro-power. As a result of this, the pattern of flow in the river is somewhat erratic, being related to changes in power demand rather than immediate rainfall. Some water in the River Tummel is transferred from the River Spey.

In Scottish terms, the fish community of the River Tay is a diverse one and, with 20 species, is of considerable conservation interest. No fish occur near the source, but downstream Brown Trout appear and soon Atlantic Salmon and Eels, all of these occurring along the main river to the mouth. Some 13 species are common in the middle reaches, but Arctic Charr is found only in the lochs. As the river changes to estuary saline conditions take control; some fish (e.g. Minnows) disappear but these are replaced by others (e.g. Sparling, Common Gobies and Flounder) which comprise a varied estuarine community here.

Conclusions

Scotland has an abundance of waters of varied types but the demands on this resource are enormous and often conflicting. Each person requires about 100 litres a day. Industry too has huge demands. Wastes from domestic sewage and industry are discharged to rivers. In addition, many adjacent land uses and activities, such as the application of fertilisers, herbicides and pesticides used in agriculture and forestry may affect water courses. Land drainage means that water runs off faster, resulting in higher (and then lower) flows in associated rivers. Fish farming has created new demands. Hydro-electric schemes have harnessed most large mountainous rivers. Finally, water is of major importance for recreation: humans picnic beside it, birdwatch over it, paddle and bathe in it, boat on it and fish in it.

Thus, there are problems, discussed fully in a later chapter, both for humans and for wildlife. Many waters have been so misused that they are unfit for either. Over-enrichment from agricultural fertilisers causes algal blooms on lowland lochs; when these blooms reach a maximum and die, they may use up all the available oxygen, release toxins and fish kills result. Domestic and industrial pollution can eliminate all aquatic plants and animals. Such stretches of extreme pollution act as a barrier to migrating fish which cannot occupy clean stretches upstream. Weirs and dams have the same effect. Aerial pollution has acidified a number of hill lochs, eliminating fish. Other problems are created by the introduction of new plant or fish species.

With so many waters in Scotland there should be enough to go round, but present national planning is inadequate and there are problems and conflicts to be resolved. What is needed is a national framework and policy for the lochs and rivers of Scotland that considers their distribution, quality and value in relation to the needs of local folk. Some are clearly essential to a major need, say, hydro-electricity or water supply; this must be the over-riding factor - though lesser uses may fit in with this. Other waters are of such wildlife importance that conservation needs are paramount. Some of the largest waters are so important nationally (e.g. Loch Lomond) that each requires an individual management plan to reconcile the needs of users to the importance of maintaining its quality. Users must often be prepared to accept compromises and even give up some claims - if not on one water then on another.

A final word on the value and beauty of Scotland's lochs and rivers can be found in the emotive words of George Borrow's *Lavengro* - remembering that he died in 1881 before Berwick on Tweed and its surrounds were administratively incorporated into Northumberland in 1885: 'One morning, I found myself extended on the bank of a river. ... a noble one; the broadest that I had hitherto seen. Its waters, of a greenish tinge, poured with impetuosity beneath the narrow arches ... Several robust fellows were near me, employed in hauling the seine upon the strand. Huge fish were struggling amidst the meshes - princely salmon - their brilliant mail of blue and silver flashing in the morning beam; ... my tears began to trickle. Was it the beauty of the scene which gave rise to these emotions? And as I lay on the bank and wept, there drew nigh to me a man in the habiliments of a fisher. ... "Has any body wrought ye ony harm?" ... "Not that I know of, ... what is the name of this river?" ... "Yon river is called the Tweed; and yonder, over the brig, is Scotland. Did ye never hear of the Tweed my bonny man?" "No," said I ... "I never heard of it; but now that I have seen it, I shall not soon forget it!"'

References

Gordon (1944), Lamond (1911), Maitland (1979, 2001), Maitland & Smith (1987), Maitland *et al.* (1994), McLean (1933), Monro (1549), Murray & Pullar (1910), Robertson (1861), Smith & Lyle (1979), Wood (1863).

'Duray ...The watter of Lasay ther,
the watter of Udergan, the watter of Glongargister,
the waters of Knockbraik, Lindill, Caray, Ananbilley;
all thir waters salmond slaine upon them.'

Donald Monro (1549) *Description of the Western Isles*

*****

A broad classification of the characteristics of the main types of fresh waters in Scotland.

GEOLOGY (rock & soil)	SIZE (area & depth)	ALTITUDE (& latitude)	FAUNA & FLORA	SPECIES DIVERSITY
POOREST	SMALLEST	HIGHEST	PERHAPS NONE	
dystrophic & oligotrophic systems	trickles, upland pools, lochs & rivers	cool waters, often frozen	stenothermic species (e.g. Salmonidae) may dominate	LOW
mesotrophic systems	intermediate waters	intermediate waters	intermediate communities	MEDIUM
eutrophic & hypertrophic systems	lowland lochs, canals & rivers	warmer waters, rarely frozen	eurythermic species (e.g. Cyprinidae) may dominate	HIGH
RICHEST	LARGEST	LOWEST	SOME BRACKISH	

FISH DISTRIBUTION

*'Just ere winter began to take its place among the seasons,
the fish fitted for living in a highly heated medium disappeared.'*

Hugh Miller (1841) *The Old Red Sandstone*

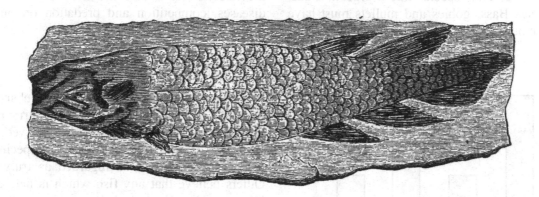

CAN birds distribute fish eggs? It's certainly easy to envisage the ribbon of eggs, up to a metre in length, which Perch lay each May in shallow water, becoming entangled round the legs of a Heron or Coot and then transported to another water body. However, there are other ways in which fish are distributed and these are explored below.

According to Robert Price 'The most important event in the evolution of Scotland's environment over the last 30,000 years was the development of the ice sheet which completely buried both the mainland and surrounding island groups.' This, last, ice age, which came to a close some 10,000 years ago, can certainly be taken as the starting point for any consideration of the present freshwater fish fauna of Scotland. It should not be forgotten, however, that there were many strange fish around in Scotland in the epochs before, some of them memorably recorded as fossils by Hugh Miller in his 1841 classic *The Old Red Sandstone*: 'The old Devonian species of fishes, like those of the present day, seem to have had their favourite haunts and feeding or spawning grounds, and must now be sought for where they corrugated of old. The *Diplacanthus striatus*, for instance, is one of the commonest of the Cromarty Old Red fishes, and the *Dipterus* and *Asterolepis* very rare; whereas at Thurso, Mr. Dick, after years of exploration, never found a single spine of *Diplacanthus*, but not a few noble specimens of *Asterolepis*, and finely-preserved skulls and jaws of *Dipterus*.' Hugh Miller's *Dipterus macrolepidotus* is figured above.

Scotland is rich in fossil fish. Travelling south out of Ayrshire, Alexander mentions '... the Garpel Water ... flowing from the Upper Old Red Sandstone ... down over outcropping Carboniferous rocks. These latter rocks are rich in ... the remains of fishes that swam in the limy or irony or muddy ferny lagoons of the shallow Carboniferous seas.'

Even during the ice age, it must not be thought that there were no fishes associated with Scotland, for the situation was probably similar to that pertaining at present in Greenland and northern Canada, where some fish, notably Arctic Charr, are able to use any running water available and migrate to and from any ice lakes present. Gradually, as the ice cap retreated, just as it is doing today in northern latitudes, more and more rivers and lochs became available for colonisation by those fish which had the capacity for life in both fresh and salt water. The same story is being repeated today in northern Alaska, where scientists have shown that aquatic invertebrates and then salmonids are quickly colonising the streams in Glacier Bay which are appearing as the glaciers retreat from global warming.

Origin

Thus, because of the ice cap, the freshwater fish fauna of Scotland is substantially impoverished compared to the communities found further south in Europe. Nevertheless, 42 out of the 57 species presently found in the British Isles as a whole occur here, and the number is increasing as more species appear from the south. Taking the starting point of the fish communities of Scotland as the closing stages of the last Ice Age, it is clear that euryhaline fishes, many of which come into fresh water to spawn, had no difficulty in invading new waters as the ice receded. Thus species like Sturgeon, shads, Sparling, Sea Bass, gobies and mullets must have occurred in our estuaries for thousands of years.

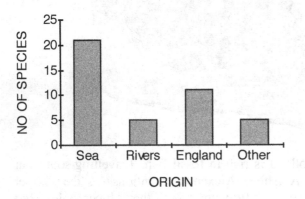

Origins of Scotland's present freshwater fish fauna.

Apart from these mainly estuarine species, the only fish which were able to colonise truly fresh waters as the ice receded were also those with marine affinities and capable of existing in the ice lakes and glacial rivers which prevailed at that time. At most there were then probably only about 12 species, most notable among which were lampreys, Salmon, Trout, Arctic Charr, Powan, Vendace, Eel, sticklebacks and Flounder. These fish probably dominated the Scottish ichthyofauna for many thousands of years. A close study by the author of the early literature has shown that by about 1790 only another seven species had been added to the Scottish fauna - Pike, Carp, Minnow, Roach, Tench, Stone Loach and Perch. Ninety years later (1880) another four species were known to occur in Scotland - Brook Charr (from North America), Grayling, Bream and Chub. By 1970 another seven species (Rainbow Trout, Goldfish, Gudgeon, Rudd, Orfe, Dace and Bullhead) were known to have established viable populations in Scotland, humans appearing to have been responsible for the introduction of all of them. The latest species in this saga are Barbel and Ruffe, discussed below.

Thus the present freshwater fish fauna of Scotland is a mixture of natural immigrants from the sea and from fresh waters in the south, along with many more recent fish which have been brought in by humans from England, continental Europe and even North America. The situation is by no means a stable one even yet and other arrivals can be expected in future years. These will certainly add to the diversity of our fish communities but will inevitably bring with them threats in the form of diseases, competition and predation on our more sensitive indigenous fishes.

Native species

The debate as to which are truly 'native' species in Scotland and which are not is an intriguing one which will never be completely solved. Some scientists believe that only those species with marine affinities can be regarded as truly native. Others believe that any fish which is native to the UK can also be regarded as a native here and therefore, recent arrivals introduced by anglers are natives. Some believe that after a sufficient 'residence time' any species should be regarded as native - the Grayling, for example, although introduced initially, has been established for about 150 years. Some anglers hold an extreme view that there is no reason not to introduce any fish which can establish itself successfully in suitable habitats.

Norman Morrison recorded that '... when on a visit to the Island of Lewis ... I found an eel twelve inches long in a spring of water on the western seaboard of Lewis, about twenty-eight feet from the edge of a precipice ... thirty feet in height ... there is a trickle of water flowing from the pool ... Yet "facts are chiels that winna ding". The presence of the eel in this spring is self-evident proof of its climbing achievements.' He further recorded that 'It is interesting to note that eels are found in the upper reaches of the river Clyde and other city waterways ... how did they get there? That a swarm of elvers could survive the passage through the polluted and foul waters of these rivers ... is improbable, if not impossible. The only other alternative is a cross-country expedition from other neighbouring streams ... surely an object lesson on the amphibian achievements accomplished by these creatures.'

Few people would deny the ability of Eels to move from one water to another yet refuse to accept that other species, in times of flood, in waterspouts or with other natural opportunities, have powers of dispersal. Herbert Maxwell posed the question in

1932: 'I write these lines beside a small burn flowing independently into the sea on the west coast of Scotland. It abounds with loaches, greatly to the profit of certain pink-fleshed trout which inhabit the said water. What I should like to know is: How did the loaches get into this isolated burn?' Some suggestions are given below.

There are a number of ways in which freshwater fish may have arrived in Scotland in the past.

From the sea

Our earliest invaders after the ice age came from the sea; it is not surprising therefore, that some of these fish species are our most widespread species - lampreys, eels, salmonids and sticklebacks. As the climate warmed up, many stocks of these migrants from the sea seem to have lost their marine phase and become 'landlocked', a misnomer really, for in most places there is still ready access to the sea from their resident waters.

Populations of several of these species, such as Brown Trout, Arctic Charr, Vendace and Powan have been isolated from one another for many thousands of years, and have developed distinct and unique characteristics over this period. This created considerable confusion among the early taxonomists who looked at these isolated stocks and originally, several 'species' of Brown Trout and Arctic Charr were described and named. By the beginning of the 20th Century, fifteen different species of Arctic Charr had been described from different parts of the British Isles.

Some species ended up with a very few isolated populations in Scottish lochs, notably the Powan only in Loch Lomond and Loch Eck and the Vendace in the Castle and Mill Lochs at Lochmaben. It is not clear whether these are isolated remnants of more widespread stocks or whether their ancestors were unable to access most catchments because of particular circumstances pertaining during the late glacial period. The author believes that the latter is the case.

Transfer by humans

There is no doubt whatsoever that humans have been a major factor in the dispersal of freshwater fish into and across Scotland, especially over the final 200 years of the last Millennium. It is incontrovertible that foreign species from continental Europe and North America were introduced to Scotland by humans, and the same is true of some other species from England where the initial introductions have been well documented. The Grayling is a notable example of the latter.

The major hydrometric areas of Scotland (from Maitland 2004).

There are many early records of introductions to Scotland. For example, Sir John Clerk of Penicuik wrote in 1742 that 'This year I made the antique cave at Hurley where I had made a large pond, and stocked it with Carp and Tench brought from Corby Castle near Carlyle.' Later, in 1749, he added 'I sometimes draw the Ponds where I have aboundance of Carp and Tench.'

Connections across watersheds

Tate Regan pointed out that 'The Minnow and Loach, probably the only really indigenous freshwater fishes in Scotland north of Loch Lomond and the Firth of Forth, are small species which thrive in little brooks, and are therefore the more likely to spread rapidly, being transferred from one system to another by slight changes in the headwaters ...' It is certainly true that these are the most widely dispersed stream species which have no marine affinity; the author is in agreement with Tate Regan that these two species should be regarded as true natives.

There are a number of places in Scotland where headwater transfer has undoubtedly taken place. Many river systems flowing in different directions actually rise very close to one another and a glance at an Ordnance Survey map will reveal the truth of the traditional old couplet:

'Annan, Tweed and Clyde,
A' rise oot o' ane hillside.'

Proximity of Clyde/Annan/Tweed systems (A: Beattock; B: Little Bog; C: Tweed's Well).

In fact, freshwater fish have only to surmount about ten watersheds to disperse from the Solway to the north coast of Scotland and there is ample evidence that there are a number of watersheds where crossing would be quite easy during periods of flooding. One such place, which the author knows well from electro-fishing studies there, is just beside the M74 at Beattock Summit, where the Evan Water (a tributary of the River Annan) and the Clydes Burn (a tributary of the River Clyde) run close to each other on the same moss before flowing off southwards and northwards respectively. At times of heavy rainfall these burns are connected across

this bog and so Minnows and Stone Loach, which are common in both burns, can pass either way.

An anonymous writer has noted that: 'A similar example is found in the headwaters of the Clyde and Tweed, it being quite a question whether the stream will flow east or west, and in times of spate salmon fry are frequently washed from the headwaters of the Tweed into the upper reaches of the Clyde, above the Falls of Cora.'

Egg transfer by birds

Those who are familiar with the most common 'disputed natives' in Scotland - Pike, Perch and Roach - will be aware that Pike and Roach lay adhesive eggs which attach strongly to aquatic weeds in shallow water, whilst Perch lay long ribbons of eggs which are draped over weed or submerged branches. It is certainly easy to imagine that waterfowl or Herons could readily tangle weed (and eggs) round their legs and then fly to an adjacent water body. This does not prove that it has ever happened! However, the fact that these three species are so widely dispersed in Scotland and their 'spontaneous' appearance in a number of waters does provide circumstantial evidence.

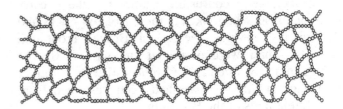

Part of a perch egg ribbon.

Ian Wood records the sudden appearance of several young Perch in his garden pond which was unconnected with, but not far from, the perch population in Tannoch Loch. Ducks regularly commuted between the two waters. The author, during recent work on a remote loch on Islay, remarked to the owner that characteristic ripples which he had seen several times near the edge were probably Pike, something hotly disputed by the owner for, as far as he was concerned, it had never been introduced and this was the last fish he wanted there. However, later that year, it was easy to observe young Pike lazing in the shallows; the conclusion was that they had been brought in as eggs by wildfowl from the only other known pike loch on Islay.

The topic is a controversial one, having been debated by fish biologists over the years – some scientists being against the possibility of transport by birds (e.g. R.A. Jubb and H.M. Smith), others believing there is a good case for such a phenomenon (e.g. A. Gunther and T.R. Roberts). As will be evident from the above, the author is fully in support of the latter and believes that this mechanism has been important in the dispersal of both Pike and Perch in Scotland.

Goldfish survives chimney fall

A GOLDFISH was yesterday recovering with bruised scales after it was dropped down a chimney, bounced off the coals in the fireplace and landed on the hearth in front of an astonished family in Northampton. The fish was apparently dropped by a heron flying over the house.

Newspaper report about a lucky Goldfish.

Perhaps surprisingly, it is true also that live fish can be transported by birds as there are occasional records of them being dropped by Herons. It seems unlikely, however, that such fish would ever be able to initiate new populations - except possibly in the case of Goldfish, which are known to maintain all-female populations.

Transfer in waterspouts

Waterspouts are of relatively infrequent occurrence in Britain and some people, like Tate Regan, would argue that the dispersal of freshwater fish has never been affected by them. Nevertheless, freak instances of fish dispersal occasionally occur: for example in 1916 Meek recorded a shower of sticklebacks falling on Merthyr Tydfil in Wales, which indicated 'that they had been caught up in water spouts'. Although infrequent, it must be remembered that fish have had 10,000 years to disperse in Britain, and the author believes that waterspouts may well have had a part in this. Certainly there are many instances around the world where fish and other aquatic animals have been transferred by storm phenomena.

In 1859 John Lewis, moving timber in a yard in the village of Mountain Ash, South Wales, was peppered by small objects falling out of the sky. They were tiny fish which filled his upturned hat

and were very much alive. On 21 May 1921, thousands of frogs fell on Gibraltar during a thunder-storm - just one among many falls of frogs and fish to have puzzled scientists throughout the century. On 4 March 1998, a shower in Shirley, Croydon, south London, included a large number of dead frogs. Weathermen offered the traditional explanation that they had been sucked up from a pond by a waterspout; however, as has often been objected to in the past, it is a bizarre tornado that can pick up one type of animal from a pool to the exclusion of all others. On 23 October 1947, ichthyologist Dr A.D. Bajkov saw fresh, dead minnows and black bass lying on the street outside the cafe where he was having breakfast in Biloxi, Mississippi, just after a rain shower. His report appeared in *Science* in 1949. More recently, dozens of fish were found in gardens and on roofs in the Borough of Newham, East London, following a thunderstorm on the night of 27-28 May 1984. The London Natural History Museum identified the fish as Flounder and Sparling; there had been no reports of waterspouts in the Thames that night. On 17 May 1996, a fall of more than 20 small fish was witnessed at Hatfield in Hertfordshire. Although they were dead, they seemed to be fresh and, curiously, were warm to the touch.

Further examples are provided by Robert Gibbings: 'It was just then that a squall hit the river ... and ... had drawn up the water into a miniature water spout ... "Rains of fishes" have occurred in many parts of the world ... the records go back to the second century ... A recent fall of fish from the sky occurred about eleven years ago (1930) near Belfast, when, after heavy rain, dozens of tiny red fish were found on the roof of a farmhouse and on the ground round about it. Earlier a storm had occurred at Aberdare ... and the fish came down with the rain in a body like. They covered the ground in a strip about eighty yards long and twelve yards wide, and some of those fish that fell measured as much as five inches in length. Several of them were preserved alive, and were exhibited to the public for some time afterwards.'

Among recent events, *The Glasgow Herald* of June 3, 2000, recorded that 'In October 1996 Scottish housewife Mrs Nellie Straw was driving her family on the A82 Crianlarich-Glencoe road when they were hit by a freak storm. "All of a sudden frogs began to fall on the car - hundreds of them," she said. "They appeared from nowhere." One of the first instances in the new Millennium was in June, 2000 when '... farmers tending their fields in southern Ethiopia ... were pelted with millions of tiny fish which dropped from the air, creating panic among the mostly religious people. Mr Saloto Sodoro, an expert on such matters in the region, attributed the phenomenon to heavy storms in the Indian Ocean which swept up the fish before raining them down on the unsuspecting Ethiopians.'

It is never going to be possible therefore, to reach any definite conclusions as to whether or not some species can be regarded as native or not and the debate will no doubt continue in future years. As Herbert Maxwell put it in 1932: 'Few subjects provide more perplexing conundrums to the reflective loafer than the distribution of certain fresh-water fishes on the surface of the globe.' One fact which does seem clear is that most of the aliens are here to stay, whereas several of our native species are under extreme threat and could disappear in the next few decades.

Tate Regan believed that 'Water-spouts, transference of spawn by birds, and other accidental methods have never, in my opinion, succeeded in establishing a species in a river system from which it was previously absent..' The author, as will be evident from above, disagrees with this assertion; his final classification of the origin of our freshwater species is indicated in the table at the end of this chapter. It can be seen there that the number of established alien species is large.

Charles Darwin's view in his classical *On the Origin of Species* was that 'On the same continent fresh-water fish often range widely, and as if capriciously; for in two adjoining river-systems some of the species may be the same, and some wholly different. It is probable that they are occasionally transported by what may be called accidental means. Thus fishes still alive are not very rarely dropped at distant points by whirlwinds; and it is known that the ova retain their vitality for a considerable time after removal from the water. Their dispersal, however, may be mainly attributed to changes in the level of the land within the recent period, causing rivers to flow into each other. Instances, also, could be given of this having occurred during floods, without any change of level.'

Alien fish

The present situation with regard to controlling the movement of fish by humans into and within Scotland is unsatisfactory. Whereas it was thought until recently that many fish, especially cyprinids, in Britain had already reached their northern limits and were unlikely to disperse much further, this is clearly not the case, as the successful moves made recently by several species (e.g. Gudgeon, Chub, Dace and Ruffe) have resulted in the establishment of thriving new populations north of their previous areas of distribution. For example, in 1970 only native fish species were found in Loch Lomond and its catchment; since then, five alien species have become established there and there is evidence of the introduction of at least three others. All of these new stocks appear to have resulted from deliberate introductions by humans - either intentionally to initiate new populations or accidentally by discarding excess livebait at the end of fishing.

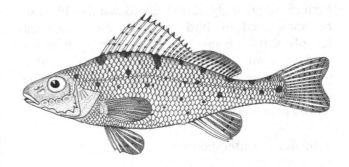

The Ruffe - an aggressive alien species recently established in Loch Lomond (from Maitland 2004).

Some alien species have been established for such a long time that they are accepted as native in some quarters. The Grayling, for example, which is not a Scottish species but was introduced to the River Clyde and others rivers in the 1800s, is the subject of controversy in the River Clyde, because, with the natural return there of the Atlantic Salmon, a native species, anglers regard it and attempts to enhance it as a threat to the Grayling stocks.

Although present controls on the movement of fish are probably adequate in England & Wales, this is not yet the case in Scotland. The Wildlife & Countryside Act 1981 covers exotics but not fish native to Great Britain. If the introduction of fish species (like Ruffe), native to England but not to Scotland is to be controlled then such fish need to be specified under Section 1 of the Import of Live Fish (Scotland) Act 1978. This Act should have included species such as Barbel, Silver Bream, Bleak and Spined Loach. There are, of course, a number of exotic species already established in England (but not in Scotland) and these include Bitterling, Danube Catfish, Largemouth Bass, Pumpkinseed, Rock Bass and Pikeperch. As noted above, movement of these species, theoretically at least, should be controlled by the Wildlife & Countryside Act 1981. The recent legislation brought in by DEFRA prohibits the import of a list of foreign species which could be a threat and the Scottish Executive has followed with a similar list.

A further problem, of course, relates to the movement of freshwater fish within Scotland. The southern species which have become established here relatively recently can be moved around the country and introduced to new catchments without hindrance and these species are much more likely to be successful than native fish in responding to increasing climate change, as they will be favoured by warmer conditions. The recent population explosion of four of these species in Loch Lomond and the River Endrick within a few years of their introduction there shows how quickly they can establish themselves. It is hoped that the new Aquaculture and Fisheries Bill, with the Scottish Parliament at the time of writing, will provide some measure of control on the movement of fish into and within Scotland - alas too late for some waters. Whilst the introduction of alien fish species, accidentally or deliberately, is generally frowned on, some policy decisions are required to confirm which fish species are native and to be given priority, those which are not native but have become established and found to be 'acceptable', and those considered undesirable in Scotland. These

should be eradicated where possible. The listing at the end of this chapter provides the author's view on this question. Certain alien species are regarded as a desirable by some parties, but future policies must regulate their introduction and translocation to ensure that such actions have statutory approval based on sound science.

Conclusion

There is a clear need to maintain a watch on what is happening to freshwater fish communities in Scotland and this can only be achieved by a recording system of some kind. The first comprehensive attempt to record fish distribution in the British Isles was initiated by the author in 1966 and the first maps published in 1972. More recently, the author has established a computer database, SCOTFISH, of the freshwater fish and fish communities in Scotland. This database and its future development in relation to individual fisheries now provide a rational basis for future research, conservation and management of Scotland's native freshwater fishes. Maps of fish distribution by catchment were published in 2004.

References

Anonymous (1907), Adams & Maitland (1991, 2002), Alexander (1925), Darwin (1859), Davies *et al.* (2004), Gibbings (1941), Gray (1892), Gunther (1880), Jubb (1964), Lyle & Maitland (1994), Maitland (1966, 1969, 1970, 1972, 1976, 1977, 1979, 1992, 2000, 2002, 2004), Maxwell (1932), Miller (1841), Morrison (1936), Price (1983), Regan (1911), Roberts (1975), Sinclair (1779), Smith (1945), Sutherland (1934), Wood (1947).

'No stream enters or leaves these mountain tarns, and yet in some there are fish. How did they come there? It may be that the spawn once fell from the beak of a passing sea-bird. I know of no other way.'

Halliday Sutherland (1934) *A Time to Keep*

*****

Scotland's freshwater fish fauna - native and introduced, their origin and status.

COMMON NAME	SCIENTIFIC NAME	ORIGIN	PRESENT STATUS
Native species			
River Lamprey	*Lampetra fluviatilis*	marine	declining
Brook Lamprey	*Lampetra planeri*	marine	declining
Sea Lamprey	*Petromyzon marinus*	marine	declining
Common Sturgeon	*Acipenser sturio*	marine	almost extinct
European Eel	*Anguilla anguilla*	marine	declining
Allis Shad	*Alosa alosa*	marine	declining
Twaite Shad	*Alosa fallax*	marine	declining
Common Minnow	*Phoxinus phoxinus*	fresh water	expanding
Roach	*Rutilus rutilus*	fresh water	expanding
Stone Loach	*Barbatula barbatula*	fresh water	stable/expanding
Pike	*Esox lucius*	fresh water	expanding
Sparling	*Osmerus eperlanus*	marine	declining
Vendace	*Coregonus albula*	marine	extinct/reintroduced
Powan	*Coregonus lavaretus*	marine	stable/expanding
Atlantic Salmon	*Salmo salar*	marine	declining
Brown Trout	*Salmo trutta*	marine	declining
Arctic Charr	*Salvelinus alpinus*	marine	declining
Thick-lipped Grey Mullet	*Chelon labrosus*	marine	stable
Golden Grey Mullet	*Liza aurata*	marine	stable
Thin-lipped Mullet	*Liza ramada*	marine	stable
Three-spined Stickleback	*Gasterosteus aculeatus*	marine	stable
Nine-spined Stickleback	*Pungitius pungitius*	marine	declining
Sea Bass	*Dicentrarchus labrax*	marine	expanding
Perch	*Perca fluviatilis*	fresh water	expanding
Common Goby	*Pomatoschistus microps*	marine	stable
Flounder	*Platichthys flesus*	marine	stable
Introduced species			
Common Bream	*Abramis brama*	angling	expanding
Barbel	*Barbus barbus*	angling	expanding
Goldfish	*Carassius auratus*	aquaria	stable
Crucian Carp	*Carassius carassius*	angling	expanding
Common Carp	*Cyprinus carpio*	angling	expanding
Gudgeon	*Gobio gobio*	angling	expanding
Chub	*Leuciscus cephalus*	angling	expanding
Orfe	*Leuciscus idus*	aquaria	stable
Dace	*Leuciscus leuciscus*	angling	expanding
Rudd	*Scardinius erythrophthalmus*	angling	expanding
Tench	*Tinca tinca*	angling	expanding
Rainbow Trout	*Oncorhynchus mykiss*	angling	expanding
Brook Charr	*Salvelinus fontinalis*	angling	stable
Grayling	*Thymallus thymallus*	angling	expanding
Common Bullhead	*Cottus gobio*	angling	expanding
Ruffe	*Gymnocephalus cernuus*	angling	expanding

FIRST CATCH YOUR FISH

'You see the ways the fishermen doth take
To catch the fish; what engines doth he make?
Behold! how he engageth all his wits;
Also his snare, lines, angles, hooks and nets;
Yet fish there be, that neither hook nor line,
Nor snare, nor net, nor engine, can make thine:
They must be groped for, and be tickled too,
Or they will not be catch'd, whate'er you do.'

John Bunyan (1628-88) *Neither Hook nor Line*

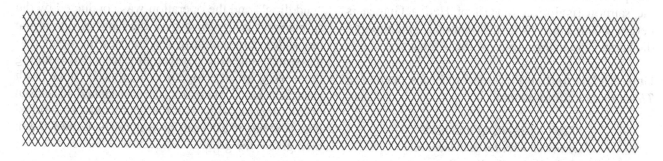

THOUGH fishing has been carried out by humans for many centuries, and several old methods are still in use, many of the most efficient methods of capture are based on relatively recent developments: notable among such are those based on electronics. Regardless of the basic method used to trap the fish, they are usually removed from the water by some form of net. A large number of methods is available for various species and situations; only the most useful general types are described here.

Fish may be caught in a variety of ways, many of which are illegal (e.g. electrofishing) unless local permission and a licence from the Scottish Executive is available - usually only for monitoring or for specimens required for research. Many species, especially their young stages, can be captured in shallow water with a simple hand net from the bank or when wading. Larger fish may be caught by netting - either sweep netting (seining) from the shore or by setting gill nets (hang nets) in open water. The former involves encircling shoals of fish with the net and gradually directing them towards the shore as the net is pulled in. With gill nets, as the name implies, fish become entangled in the mesh of the net, often set overnight.

Almost all nets and traps depend on netting:

formerly this was made of natural fibres, but these have been replaced for the most part by synthetic materials which are lighter, stronger and more durable. Many nets and traps are highly selective; this selectivity depends on the fishes' behaviour, the nature of the catching method, the setting and position of the gear and the size of mesh used. Mesh size is often critical, and clearly must always be small enough to retain the smallest fish which it is hoped to catch. Its importance in relation to gill nets has already been mentioned. Most netting requires a permit of some kind.

Many legal technicalities surround various fishing methods. For example, although many types of fishing by rod and line are legal, the actual baits used vary in their legality. One of the most frowned upon is described by Morris: 'As most trout-fishers are aware, at certain times of the year salmon-roe is as deadly a bait in the hands of these poaching experts as dynamite is in the hands of an anarchist.' Morris also recorded a classic form of poaching which is still commonly practised on many Scottish rivers: '... when I came abreast of the group I saw the poacher take a gaff or "cleek" from his pocket, plunge it into the water, and then bolt up the river-bank with a heavy salmon. The fish was struggling vigorously, blood pouring down its silvery side: an unpleasing sight, but not an uncommon one on this river at the spawning time.'

There are many old accounts of poaching and ways to evade the law. In 1780, Senex wrote that 'On this piscatory excursion our fishing was pretty successful, as each of us had caught a number of flounders and a few eels, but the great prize came to me of a very large salmon fry. I was delighted at my good luck but in great terror, lest the Sheriff ... should get hold of me ... having just issued the following notice: *Glasgow Mercury*, 27th April 1780 - ... the illegal practice of *killing* salmon fry still prevails in this country to the great prejudice of the salmon fishing, a reward of Half a Guinea is hereby offered for each information ... mentioning the names and designations of the offenders ...'

Hugh McDiarmid, in this verse from his poem *The Three Fishes* admitted that:

> '*I am a fisher lad and nane*
> *Can better wield a gad than me,*
> *And a'e day to the burn I've gane*
> *And got me fishes three.*'

Though organised poaching is damaging to fish stocks and distasteful to most folk, in many country districts, poaching is not frowned upon when the objective is just 'one for the pot'. When times were hard in the past, many families depended on what the husband, often unemployed, could manage to bring back from the wild - be it fish, flesh or fowl. The life of the poacher was not an easy one, as this verse from the poem *Fishin'* by Walter Scott indicates:

> '*I try a bit poachin' the keeper comes bye,*
> *An' that's aye the minute my flask has run dry;*
> *It's mebbe I ha'e na seen him for some weeks,*
> *An' then the barbed wire hings me up by the breeks.*
> *When I'm fishin' here, when I'm fishin' here,*
> *There's gey mony jags when I'm fishin' here.*'

At times, in the north and in the south of Scotland, poaching was a matter of survival. Macdonald Robertson recalls the traditional Gaelic saying:

> '*Breac a linne, slat a coille,*
> *is fiadh a fireach,*
> *meirle anns nach do ghab*
> *Gaidheal riamh naire.*'

> '*A fish from the river, a wand from the wood,*
> *And a deer from the forest,*
> *Three thefts no Gael was at any time ashamed of.*'

Hands

The use of hands must be the oldest method of fishing and in Scotland was (and in some places still is) a popular method used by poachers (and small boys!). Some of the author's first fish were caught in this way. Thomas Stebbing, in his life of David Robertson *The Naturalist of Cumbrae* describes how 'He became very expert at catching trout with his hands. The mode of capture was called "guddling", and the operation was conducted by the sportsman's lying down over the bank of one of the moorland burns, putting his hands wide apart into the water under the overhanging turf, and bringing them very slowly together. If this is done with sufficient care, when the fish is touched it seldom offers to move away. But the hands must be worked gently about it till you have it fairly within them; then give it the grip.'

Ratcliffe Barnett, as well as exploring the excitement of splash netting (see below) was also prepared to use his hands when nothing else was available. 'I set out from Glen Tarken, ... before leaving the stream, I stripped and plunged. Then I climbed, refreshed, from one pool to another, and lay down to explore the depths. Yonder in a corner lay a lusty trout, and here lay another. It had been a very dry season, and the big pool had only a mere driblet running from its outflow. So the lusty ones were prisoners. It was not perhaps a very fair game. But instinct is strong, and "Tickled trout are tasty trout, and a weary man must eat." I draw a veil over the chase. ... After a quarter of an hour ... a daigled man with dripping arms sat fondling the two victims on the hot grass. Did ever trout taste sweeter?'

Alexander had had similar experiences: 'Being thirsty I dipped the drink vessel into a pool. A brown-spotted trout darted and dashed out onto the dry shingle, but leapt and wriggled into the next lower pool. I recalled a boyish feast, when hungry after "guddling", on trout fried on the shingle of the "clear winding Devon".'

The author, not having any kind of net as a small boy, was quickly initiated into the art, indeed local culture, of guddling.

Hand nets

Probably the earliest fishing experience of most people was with a simple hand net - usually a plain bamboo cane with a crude mesh net at one end. This, and a jam jar were the stuff that the dreams of many small boys, including the author, were made of. Simple hand nets, perhaps more strongly made than those usually available, are very efficient in small waters, especially burns, and are still used regularly by the author. Hand nets of various types are also widely used in other types of fishing as a method of retrieval of fish, e.g. in electric fishing or purse seining. In angling they may be essential as this incident, related by Lang & Lang, indicates. 'Here, too, befell "the affliction of Bemersyde", when the laird, after a long fight with a monstrous salmon, lost him in the moment of victory. The head of the fish would not go into the landing net, his last wallop freed him; he was picked up dead, by prowlers, - and he weighed seventy pounds. Probably no salmon so great was ever landed by rod from Tweed.'

One of the largest hand nets used legally in Scotland was well described by H.V. Morton: 'Kirkcudbright is the only place I know where a villainous form of salmon poaching is a legal operation! There is a stretch of the Dee belonging to the manor of St. Mary's Isle in which the flinging of a shoulder-net for salmon has been a custom since ancient times. The local legend is that this right - now granted by the Crown - descends from the Franciscan monks of the thirteenth century. ... One night I was invited to accompany the salmon netters on a night's fishing. ... on an island in the Doachs ... We were in the centre of the river. All round us were white tufts where the water cascaded over submerged rocks. ... The net thrower prepared to fish. A grooved wooden implement was strapped to his right shoulder. He took a twenty-foot pole, at the end of which was a big net. He swung it in the air, poised it above him in the manner of a javelin thrower and cast it from him into the torrent. ... out it shot, the net falling neatly as a knife into the water; then, so quickly that I could not follow what happened, it was withdrawn, and inside it were four feet of silver desperation.'

'Poke' nets and 'haff' nets are both really large types of hand net used under licence in the Solway Firth. They are an important local tradition there.

Snares

As well as hand nets, other very simple devices have been used traditionally in some parts of Scotland. Simple snares are one such method used by boys and the method has been related to the author independently by Fred Longrigg and Willie Miller who described it as follows: 'When we were boys, my friends and I used to catch Stone Loach (Beardies) using a noose made from the roots of Ragwort, which was a very common plant in the fields along the Tillan Burn, which flowed past the foot of our garden. We pulled up the plant, selected a long, thin root and stripped off the outer sheath of the root. This left a long, wiry part, at the end of which we tied a running noose. The Beardies were generally to be found under large stones, and if a stone was moved slightly the Beardies would stick their heads out. The noose was gently slipped over the head, just past the gills. A sharp upwards pull usually lifted the fish out of the water. I don't know the origin of this method, and I cannot remember how we came to use it, but it has certainly been passed down the generations. I checked with an 80 year old angler, George Williams, who was brought up in Glassford. He was shown this method by his uncles, and used it in the River Avon and its tributaries. George referred to the plant as "Bunnicles".'

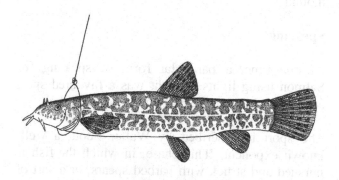

Snaring a Beardie.

Richard Jefferies gives a description of poachers' snares. 'On arriving ... we ... began to fix the wires on the hazel rods. The wire for fish must slip very easily, and the thinner it is, if strong enough, the better because it takes a firmer grip. A single wire will do; but two thinner ones are preferable. Thin copper wire is as flexible as thread. Brass wire is not so good; it is stiffer and too conspicuous in the water. ... At the shank end a stout string is attached in the middle of its length. Then the wire is placed against the rod, lying flat upon it for about six inches. The strings are now wound round tightly in opposite directions, binding it to the stick, so that at the top the ends cross and are in position to tie in the slight notch cut for the purpose. A loop that will allow four fingers to enter together is about large enough, though of course it must be varied according to the size of the jack [Pike] in view.

Heavy jacks are not often wired, and scarcely ever in brooks. ... For jack the loop should be circular; for trout it should be oval;, and considerably larger in proportion to the apparent size of the fish. ... The noose should be about six inches from the top of the rod.'

He then describes the technique. 'After a long look I began to examine the stream near at hand: the rushes and flags had forced the clear sweet current away from the meadow, so that it ran just under the bank. ... Under a large dead bough that had fallen across ... the stream I saw the long slender fish lying a few feet from the bank, motionless save for the gentle curving wave of the tail edges. ... By slow degrees I advanced the hazel rod ... worked the wire up stream, slipped the noose over his tail, and gently got it up to the balance of the fish. ... I gave a sudden jerk upwards, and felt the weight instantly. ... I was not satisfied until I had him up on the mound.'

Spearing

At one time, a particular form of spearing for Salmon using lights at night was a favoured sport among wealthy landowners and poachers alike.

The sport is described by Walter Scott, a well-known exponent: 'This chase, in which the fish is pursued and struck with barbed spears, or a sort of long-shafted trident, called a waster, is much practised at the mouth of the Esk, and in the other salmon rivers of Scotland. The sport is followed by day and night, but most commonly in the latter, when the fish are discovered by means of torches, or fire-grates, filled with blazing fragments of tar barrels, which shed a strong though partial light upon the water. On the present occasion, the principal party were embarked in a crazy boat upon a part of the river which was enlarged and deepened by the restraint of a mill-wear, while others, like the ancient Bacchanals in their gambols, ran along the

banks, brandishing their torches and spears, and pursuing the salmon, some of which endeavoured to escape up the stream, while other, shrouding themselves under roots of trees, attempted to conceal themselves from the researches of the fishermen. These the party in the boat detected by the slightest indications; the twinkling of a fin, the rising of an air-bell, was sufficient to point out to these adroit sportsmen in what direction to use their weapon.'

Leistering for Salmon (from St John 1893).

Lockhart in his *Life of Sir Walter Scott* recorded that 'Sometimes the evening closed with a "burning of the water"; and then the Sheriff, though now not so agile as when he practised that rough sport in the early times of Ashestiel, was sure to be one of the party in the boat - held a torch, or perhaps took the helm - and seemed to enjoy the whole thing as heartily as the youngest of the company.'

'Tis blythe along the midnight tide,
With stalwart arm the boat to guide -
On high the dazzling blaze to rear,
And heedful plunge the barbed spear;
Rock, wood and scaur , emerging bright,
Fling on the stream their ruddy light,
And from the bank our band appears
Like Genii armed with fiery spears.'

Robertson witnessed the same type of fishing by

ordinary Scots: 'I was ... astonished to notice the sparkling of the fairy lights ... amidst the bushes which lined the edge of the stream. ... they were poachers at work! The leader of the party, a tall bearded Highlander ... waded through the water clad in the kilt, holding in one hand a lighted torch ... and in the other a spiked or barbed spear, locally known as "muigheadh" ... I had never before witnessed such a spectacle of blazing torches illuminating the darksome waters, on which were cast moving shadows of a fantastic nature all around ... It was weird! ... Poachers though they were, they were all decent and respectable local men, ... Their lives had so many elements of romance in them, that their infringements of the game-laws were a veritable romance in poaching. ... The bearded leader ... added (1) that it was not their intention to market the spoil for any monetary gain, and (2) that they were only taking part in an old and traditional custom, openly practised in the district, in order to provide food for their wives and children.'

Seine nets

Seine nets function by paying out a wall of netting from a boat so as to surround a certain area of water, enclosing the fish therein; the method is a popular and successful one and is the basis for the traditional, and legal, 'net and coble' fishing for Atlantic Salmon and Sea Trout all around Scotland. In shore seining, the net is set in a semi-circle from a small boat and hauled to the shore. Shore seine nets may, or may not, have a bag in the centre to help to retain fish. Most open-water seines do have such a bag, and the net is set from a boat in a full circle before hauling; two boats are often necessary.

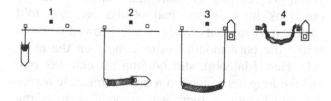

The four stages in operating a seine net (Maitland 1990).

A rather specialised type of seine is that known as a purse seine or ring net; this is set, like the open-water seine net, in a full circle. It is deeper than other seines and along the bottom, running through large rings, is the purse line; as the net is hauled in, this line is tightened so as to close the bottom of the net.

Officially, 'The net and coble is the only method of net fishing which is legal inside estuary limits throughout Scotland, with the exception of the rarely used 'Cruive Trap' held under charter rights. The net is loaded on the coble. Attached to it is a rope held by a fisherman on the shore, who, once the operation begins, must keep the rope in motion by his own exertions. The coble moves across the estuary or the river, shooting the net as it goes. Its course is roughly a semi-circle, finishing on the shore from which the boat started out. The ends of the net are then hauled in and the fish removed.'

Sweep netting is suitable for some rivers and loch shallows but cannot be used everywhere, as a smooth bottom, free of obstructions, is essential for the sole rope to sweep smoothly over the substrate, without becoming snagged. Most species of fish in the area are liable to be caught, providing they are not small enough to escape through the mesh. Occasionally, large and active fish may jump to freedom over the floating head rope. The method is, however, partly selective and has the disadvantage of requiring a boat and a trained crew of netsmen. It may also cause considerable damage to the shallow water habitat over which the net is dragged. However, the fish are collected alive (in contrast to many captured in gill nets) and this is one of the standard ways used by scientists to catch fish for tagging and subsequent release and for other purposes for which live undamaged fish are required.

Gill nets

Gill nets are sheets of netting hung in the water by an appropriate arrangement of floats, leaded lines and anchors. Since such nets depend on fish actually moving into them, they are made of fine materials coloured appropriate to the water in which they are being fished. Many of them are most successful in the dark or where visibility is poor. Gill nets depend on the fish swimming partly through an individual hole in the mesh and then becoming trapped - in many cases simply by the meshes slipping behind the gill covers so that the fish can move neither backwards nor forwards. In other cases the fish may continue to attempt to swim forwards so that its body becomes securely wedged in the meshes. Larger predatory fish may often be caught up by their teeth. The size of the meshes used in each net is critical in relation to the size of fish caught, and it is normal practice to use a range of mesh sizes in any nets fished for survey purposes. Gill nets can be set in any depth of water and in general are much more versatile and less demanding in manpower than seine nets.

Gill nets, however, should only be used by those with experience, for it is essential that they are checked regularly and retrieved after fishing. If a gill net is lost, it will continue to fish as long as it stays in the water and is not too full of fish or otherwise tangled up. The author knows of a case in Wales, where a gill net was lost in a llyn for several weeks. When it was eventually recovered there were many hundreds of fish (mainly Arctic Charr) in its meshes, ranging in condition from fresh and newly caught to weeks' old and stinking.

Shortly after World War II, fine synthetic threads (either monofilament or finely braided or twisted nylon thread) were used for making gill nets. These are much more deadly than the previous nets (of linen or cotton thread) as they are much less visible to the fish and effective even in daylight. Also, synthetic nets have little bulk when not in use, so that a lone fisherman can operate a very long length of netting on his own. Monofilament net is now used almost universally by commercial fishermen and is so efficient that serious overfishing has resulted in many regions. Fish caught in gill nets are very often found dead or are killed during extraction from the net. As just noted, nets which are lost or left in the water for a long time continue to catch fish virtually as long as they remain there.

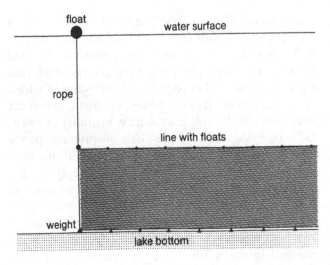

Operating a bottom gill net (Maitland 1990).

In Scotland during 1960 a drift net fishery developed off the east coast. This employed new techniques of suspending curtains of monofilament net about 10 feet deep, often extending for 1,000 yards. The fishery was banned in September 1962. Until very recently, however, such nets were still the basis of the drift net fishery for Atlantic Salmon off the northeast coast of England, an activity which caused much concern in Scotland, as most of the fish caught were on their way to Scottish rivers. After many years of debate and political activity, it was announced in 2000 that the British Government was to buy out the remaining licences in this fishery which were scheduled anyway to finish in a few years. It seems likely that recent high catches of Salmon in the River Tweed may well be due, at least in part, to the demise of this fishery.

Splash nets

Splash nets can be either small seine nets or gill nets which are set in the water for a very short time and fish are actively chased into them. It is a traditional method of coastal poaching in Scotland, especially along the northwest coast. Ratcliffe Barnett describes the procedure well in a chapter entitled *The Splash Net: A Dream of the Western Isles*: 'To all appearances one fishing net is very much like another. And yet there are nets and nets, and a name makes all the difference. Call a herring net a splash net, and the local sheriff may have something to say to you. So I shall relate a dream ... for no man has ever been convicted on the evidence of a thirty-year-old dream. It was in the middle nineties .. the boat was being silently rowed to the shore. ... I ... felt the net heaped up in the bottom of the boat. ... When we reached the burnside Captain Mac got out, took hold of one end of the net, and told Malcolm to pay out slowly, while I rowed the boat across the burn mouth to the shingle on the other side. Here Malcolm, still holding his end, got out, and the long net sagged in a great semicircle across the burn mouth. ... there was a sound of splashing on the other side of the burn. Malcolm bestirred himself and he too began splashing the water. Each man splashed and drew in, splashed and drew in regularly, until the sweep of the horse-shoe net grew less and less across the mouth of the stream. Any fish ... were now secure in the net. ... The net was now a bag from which nothing could escape. Then the real excitement began. The whispered Gaelic went on as the net came in. Was it heavy or light. In my dream I seemed to see a glitter of silver again and again ... Splash and silver. ... Next morning the tinkle of the bell awoke us. ... Henry, the steward, brought in a dish of the most delicious sea trout! ... And the sea trout were as fresh as if they had come straight from a burn-mouth but an hour ago.'

Trawl nets

Trawl nets, though widely used in commercial fishing at sea are rarely used in fresh water in Scotland except for specialised sampling of fish. Trawl nets are essentially open bags of netting hauled through the water at constant speed by a powered boat. The main features of such nets are the shape and length of the bag, and the method used to hold the mouth open. Two main types of trawl are in common use: beam trawls, where one or more rigid poles keep the mouth of the net open, and otter trawls, where planing surfaces known as otter boards keep the net open at either side, while floats and a weighted line keep it open in the middle.

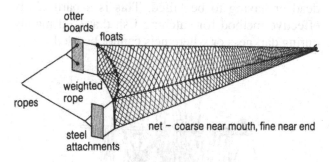

Operating a trawl net (Maitland 1990).

The bags of such nets must be sufficiently long to ensure that fish, having been overtaken by the mouth and moved back into the net, find it difficult to make their way out again.

Traps

Trapping is a method usually used in rivers to catch fish that are migrating (e.g. returning adult Salmon or Sea Trout) as they move upstream to their spawning grounds, or salmon smolts or silver Eels heading downstream to the sea. Trapping is an ancient and traditional method for catching salmonids for food, with its origins lost in the mists of antiquity. Some of the oldest traps were called yairs and examples still exist in some parts of

Scotland. Morton believed that they were very deadly: 'But woe betide the salmon at the river mouths of Scotland! I have sat on a platform above the Solway Dee on the west coast watching men trap salmon in what is called a Yair net.' The method exploits fish when they are at their maximum vulnerability, i.e. they are passing in large numbers through narrow waters. In a few parts of Scotland, Salmon are still harvested commercially by trapping; elsewhere on salmonid rivers trapping may take place to catch ripe fish to provide fertilised eggs for hatcheries or to provide data on stock levels or migration patterns.

The earliest traps were mostly made of stone. Macarthur records that in the 15th and 16th centuries 'Salmon seemed to have been caught mostly in what were called yairs. These were semicircular enclosures of stone built out from the foreshore at suitable places. They had an opening, and fish entered and were caught in them when the tide receded. There was one below Finlaystone. ... In 1678 Thomas Crawford of Cartsdyke, who was a merchant and burgess of Glasgow, constructed one at Gravel Point. The Glasgow Magistrates compleined ... He was fined £100 Scots ...' Marwick records that, about 100 years earlier 'On 29 November, 1581, parliament ratified all acts made by the King (James VI) and his progenitors "anent the destruction of cruvis and yairs, slauchter of reid fische, smolts and the fry of all fisches ...'

Trap nets can be extremely successful in certain situations. Many are relatively mobile and can be moved about easily; others are fixed more or less permanently to the bed of a loch or river. Most traps work on the general principle of fish moving, either by chance or because they are attracted in some way, through a cone or V-shaped entrance into a chamber from which they have difficulty in finding their way out. Many traps have two such chambers connected by a second funnel-shaped partition - this minimises the chances of fish ever finding their way out. Some traps have long sheets of netting which are used as lead nets to deflect fish into them.

A Highlands & Islands Development Board booklet notes that 'Fish moving along the coast are caught in a variety of fixed nets (fixed engines) set out from the shore, while those ascending the rivers are taken principally by net and coble. In Scotland, the most common fixed engines are the 'bag net' and the 'fly net'. The bag net is commonly used on rocky coasts and consists essentially of a trap made of netting to which fish are directed by a "leader" (or line of netting placed along the route the salmon usually follow as they move along the coast). Swimming along the leader, the salmon are led into the mouth of the net, then through a succession of compartments into a final chamber or fish court.'

Legal stake nets, known as fly nets or 'jumpers' depending on their construction, are used on sandy shores. They consist of curtains of netting erected on stakes in the sea-bed and acting as leaders to approaching salmon. Pockets or traps are inserted at intervals to take the fish that are directed along the leader. Unlike bag nets, they are not floating but are fixed to the bottom throughout their length. Fish taken by stake and bag nets are trapped. Macarthur recorded that 'The lower reaches of the Clyde continued to be a good salmon river till the beginning of the nineteenth century. Gradually, as it became contaminated by sewage and chemical effluents, fish life became impossible. In 1811 fishing with stake nets began and replaced the mediaeval yare. They were erected between the West Ferry and Port Glasgow, and on the north side of the river. In one year 2300 salmon and 2000 grilse and other small fish were caught, to the value of nearly £1800.'

Traps, usually in the form of tunnel or fyke nets, can also be used to catch feeding fish such as Eels during their nightly prowling for food. These traps are tethered in shallow water and the fish captured alive, although when large numbers accumulate in the final compartment of the trap net mayhem may arise as the larger fish start to eat their smaller captive brethren. Such trap nets can be modified for research purposes to catch most of the species that come into shallow water. Even tiny elvers are caught in this way in specially designed fyke nets. Recently, Anders Klemetsen has shown in Norway that baited traps set on the bottom in deep water will even take such fish as Arctic Charr.

Rod and line

No outdoor sport has more active participants than angling. The most recent survey indicated that more than 4 million people went fishing at least once a year in Britain alone, and France is said to have 5 million anglers. A huge industry has developed to meet their needs, with much of the tackle being produced in Britain, France and Sweden, and most of the nylon monofilament line being manufactured in Germany. So vast is the business associated in one way or another with angling that if the sport suddenly ceased it would affect the livelihoods of a great many people. Such a disaster would affect not only fishing tackle manufacturers, but also all levels of the fishing industry from the breeder of maggot bait to the owner of a top salmon beat who charges £1,000 or more for a week's sport.

Angling is a slow and usually selective method of sampling fish. It is especially selective for species, size and feeding behaviour, but has its uses in certain special circumstances, where other methods cannot be employed for legal, social or physical reasons. Normal angling methods cause some pain but often not too much damage to the fish and those caught can often be kept alive or marked and released. Barbless hooks can be used for most species, but the use of set lines with many baited barbed hooks, set over the bottom or at any level up to the surface, results in most of the fish being taken dead or having to be killed. This is a particularly effective method for catching fish that feed mainly during darkness or where nets cannot be used.

A good catch of Trout (from Crandall 1914, decoration by Louis Rhead).

The sport of freshwater angling is divided into two basic kinds - coarse fishing and game fishing - both kinds subdividing into various separate interests.

Within coarse fishing there are match anglers who fish against each other purely for the competitive element, or for money or both, so-called 'pleasure anglers' who fish for anything they can catch, and specimen hunters who concentrate upon the capture of exceptionally big fish. Sometimes they will even concentrate on one single species, and there are clubs for specialists in the pursuit of Pike, Carp, Tench and Eel.

Game fishing is rather less fragmented, but there are those who specialise in migratory fish such as Sea Trout and Atlantic Salmon, whilst others find their paradise in river fishing for Brown and Rainbow Trout. The most recent developments in game fishing are trout fishing in reservoirs, and small-water fishing for very big fish - Rainbow Trout reared to record breaking proportions or with special colour forms ('golden trout' or 'blue trout') by fishery and fish farm owners with an eye for publicity.

The boy and the brook (from Crandall 1914, decoration by Louis Rhead).

Angling can be an involved and exciting sport, depending not only on the angler's tackle and how it is used, but on weather, the water and the mood of the fish. It has the capacity to exercise the mind of both adults and children. It is because pleasure can

be derived from angling at so many levels that the sport is so popular. It is claimed, perhaps with some truth, that man goes fishing because he has not entirely forgotten the hunting instincts his ancestors once needed in order to survive. If this is so, then his motive has certainly changed, for very few anglers actually fish for food: coarse fishermen rarely eat what they catch. They retain their fish in nets and release them at the end of the day or after the match. Game fishermen certainly do eat at least some of what they catch, but it cannot be said that they fish simply for food, for this would result in expensive meals due to the often high cost of a day's fishing.

Man is still a hunter, and in those in whom the hunting instinct remains deeply buried, there is still a spark of interest in what the hunter catches. Witness the crowds which gather on the harbour wall when the fishing boats come home. But the anglers' hunting instinct is now converted into a sport, and because of man's penchant for organising almost anything, it is inevitable that angling is organised. Thus there are small clubs, large clubs, national associations and federations, and in turn there are links between the national bodies of various countries and their counterparts in other parts of Europe, North America and the rest of the world. Where relevant and available, Rod Caught Records have been included elsewhere in this book.

One illegal method of fishing which uses a line, but no rod, has been described by Conway: '... Donald discovered, carefully concealed beneath the heather, a well-known contrivance for fishing, belonging no doubt, to some of the shepherds, and yclept, from its deadly effects, the "otter". ... this, one of the most effective weapons of the Highlander for securing his daily meal, ... consists of a small board, fastened to a string, in the same manner and on the same principle as that of the common kite, ... so that when the string is pulled by a person walking along the shore, the board, being previously thrown into the water, darts out to the furthest distance allowed by the string; and as the walker advances along the shore, the "otter" traces out a parallel path through the water. To this board we attached a number of hooks, baited with fly or worm, ... and to this simple contrivance, ... the best and most wary fish frequently become victims. ... This process repeated thrice, we started for home, with rather more than thirty fish ...'

Electric fishing

Electric fishing is a method where, when a current is passed through water between two electrodes, a large proportion of the fish within the electrical field are attracted towards the anode and become stunned so that they can be lifted out with a hand net. A small proportion of the fish (especially the smaller ones) may become completely stunned and remain under stones or entangled among thick vegetation. With equipment working well under optimum conditions, as much as 75% of the fish population can be caught during the first fishing of a length of stream. Much of the remainder can be captured at second and third fishings; the numbers from the three fishings can be used to estimate the original total population in the section of stream being fished.

There are problems with electric fishing, however, and it is usually only useful in small and medium-sized rivers. Very small fish, such as salmon fry, may not be so strongly affected by the current as larger fish and so may escape. If the correct current is used fish will recover completely once the current is switched off and the fish netted and transferred to a container. Although the reaction of fish when within an electric field has been known for a long time, it is only within the last few decades that this collecting technique has been developed and widely used. Originally the electric fishing of a stream required the use of a portable petrol generator, but battery back-pack shocking units have now been developed, allowing the operator great mobility so that isolated waters in rough terrain can now be reached and sampled in this way.

Electric fishing can be one of the least selective methods of capture. Choice of the right type of gear is all important, however, as are appropriate safety precautions, for electric currents used in association with water can, of course, be lethal to humans. Apparatus using alternating currents tends to be easily portable and is useful in clear water where there is little cover for fish. Direct-current equipment, on the other hand, though heavier, is useful in turbid water or where there is a significant amount of cover. If the current used is too powerful, it may kill fish, while if it is too weak it may stimulate but not stun them, thus frightening them away. Weak currents are useful in this respect in forming barriers to lead fish into a trap or fish ladder; recently so-called electric seines have been developed on this principle.

Explosives

The use of explosives is an effective, but highly illegal method of capturing fish. It has on occasion been used by poachers, and supplies of fish have also been available when explosives were accidentally dropped into water, as when bombs hit Loch Lomond in 1941, killing thousands of fish. Of course, in war time the situation was rather different and legal niceties were often ignored! In France, Williams describes the situation he found himself in: '... even coarse fish to men who have not tasted fish for months, become a delicacy, and so, since angling by fair means was unproductive on account of the constant interruptions, a speedier method had to be introduced. The procedure was not one that would recommend itself to the out-and-out Purist, but it must be remembered that the fish were required as food and that the circumstances were exceptional. Neither is there anything new about dynamiting, for it has been carried out by poachers for many a long year. Actually Mills grenades were used ... A sufficient supply being forthcoming a punt was commandeered. ... The sport was by no means unexciting. After a bomb had been thrown, a few seconds elapsed and then a miniature water-spout appeared. Another short interval ... Then slowly a few stunned or dead fish would float upwards to the surface, from whence they were rapidly transferred to the boat. Sometimes thirty or forty fish would be taken at a time ...'

Poisoning

Poisoning is a somewhat drastic but highly effective method of collecting fish. The advantage of this method is that the total stock and all the age groups of a fish population can be counted; invaluable data being obtained on all aspects of the biology of the species concerned - information that can be applied to the management and conservation of stocks elsewhere.

Opportunities to use poison can arise when fishery managers decide to eradicate 'undesirable' species of fish so that new populations of 'desirable' species can be established. For example, the management aim might be to eradicate Pike and perhaps other coarse fish species from a loch so that a trout fishery could be established. If the correct amount of selective fish poison (piscicide) is used all the fish would be killed, but the invertebrates and plant life little affected. After application of the poison, many fish rise to the water surface and if they are netted at this stage and placed in clean water, the majority will recover.

Piscicides (usually Rotenone) have been used in Scotland quite legally on a number of occasions to poison out entire fish populations (usually of Pike or Perch) so that the loch concerned can be stocked with Brown or Rainbow Trout. Regrettably, on only one occasion (by Bill Munro) has the opportunity been used to obtain valuable research data on the fish population concerned.

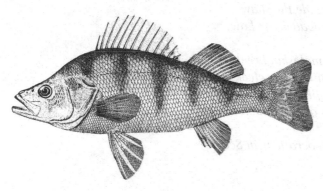

Perch - formerly a common target for poisoning in Scotland (from Maitland 2004).

However, most piscicides also kill all stages of amphibians, so great care has to be taken over the timing of the operation, the conduct of which should be based on a thorough survey conducted beforehand. Some fish poisons such as Rotenone (a derivative of derris) are extremely effective in clearing fish from waters for total population assessments or fishery management purposes. Rotenone acts initially as an anaesthetic and, if fish affected by it are placed immediately into fresh water, they normally recover. The active principle of Rotenone has been known to jungle-dwelling South American Indians for centuries as a means of collecting fish. It is harmless to mammals and birds. Where the law allows, spot applications of Rotenone can be used by fishery scientists to collect samples of fish from small areas of burns and lochs. In Scotland, the use of piscicides is strictly controlled by the Scottish Executive which only issue licences for its application after consulting a number of environmental agencies.

Electronic devices

Though not a method of capture, the use of echo sounders and other electronic devices has recently become important in freshwater fishery work. Many very sensitive echo sounders are now available and are useful for indicating the position of shoals or even individual fish; nets can subsequently be fished in appropriate places to catch these. Occasionally, underwater video cameras may be used.

Echo sounding apparatus is also useful to study fish behaviour. It can be used either from a moving boat to study the numbers and dispersion of fish in a loch at any one time, or from a stationary boat, to study the movement and activity of fish below (see section on fish behaviour using echo sounding in an earlier chapter). One of the main problems in using echo sounding for work of this kind is that of identifying the species concerned; this can often be overcome, however, by netting, underwater photography or explosives.

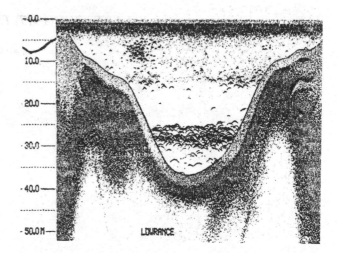

Echo sounding taken by Alex Lyle at Loch Veyatie, showing fish grouping at different levels in the loch.

Hydro-acoustic systems are now available which claim to give accurate information, not only on the numbers of fish present in a loch, but also on the sizes of fish present. Such systems have limitations, however, not only in the accuracy of the count but also in the identification of the fish species concerned. The data must therefore always be treated with caution.

A variety of other methods is available for the study or capture of fish, but most are useful only in specialised situations, and are not of general application. The study of fish behaviour in the field may be carried out by direct observation in shallow clear water, or by the use of sub-aqua techniques, photography and television. Apart from chemicals available to poison fish, a variety of other substances are used regularly in fishery research - for example, where fish require to be anaesthetised during investigations where tags are being fitted. Fish welfare should always be an important consideration regardless of the method or objective of the fisherman.

References

Alexander (1925), Bagenal (1978), Barnett (1927), Conway (1861), Crandall (1914), Holden & Marsden (1964), Jefferies (1879), Klemetsen (1982), Lang & Lang (1923), Lockhart (1871), Macarthur (1932), Maitland (1990, 2004), Marwick (1909), Morrison (1976), Morton (1933), Munro (1957), North (1980), Robertson (1936), Senex (1884), Stebbing (1891), St John (1893), Stuart (1958), Williams (1928), Young *et al.* (1972).

'Twae collier lads frae near Lasswade,
Auld skeely fishers, fand their bed
Ae simmer's nicht aside the shaw
Whaur Manor rins by Cademuir Law,

The pooches o' their moleskin breeks
Contained unlawfu' things like cleeks,
For folk that fish to fill their wame
Are no fasteedious at the game.'

John Buchan (1917) *Theocraticus in Scots*

*****

PART 2

THE FISH

'The loch is fed by two small rills. ... It abounds in a variety of fish ... , including the vendace, the rarest and most elegantly shaped of all the finny tribe, and only found in Britain in this loch, and in the Mill Loch, on the north-west side of the town. Besides the vendace there are pike, perch, two species of loch trout, bream, roach, skelly or chub, greenback, eels, beardie, minnow and banstickle.'

William Graham (1865) *Lochmaben five hundred years ago*

FAMILIES OF FISH

'A branch of one of your antediluvian families,
fellows that the flood could not wash away.'

William Congreve (1670-1729) *Love for Love*

GED OF THAT ILK

MANY well known family names have associations with fish, the Family Geddes, for example, having a long-standing relationship with the Pike, or Ged in Scots. Other fishy associations with family names and with people in Scotland are discussed more fully in a later chapter.

The classification and naming of fish and other living creatures has always appeared something of a mystery to the average layman. Much of the misunderstanding and difficulty has arisen over the fact that biologists use Latin or Greek derivations for the scientific names chosen. This, together with uncertainty over the status of specific, generic, family and other higher rank names, has led to a division between scientists who only recognise a two- or three-worded scientific name for a species wherever it occurs in the world, and laymen who may use widely different names for the same species in different countries, and even in different parts of the same country.

A good example of the complexity of naming fish is found in the Perch, which is widespread across the northern hemisphere in Europe and Asia. Biologists all over the world clearly recognise this fish by the scientific name *Perca fluviatilis*. In Europe alone, however, the following common names are used in different countries (and even within these countries various other local dialect names are found): Bulgaria, *Kostur*; Czechoslovakia, *Okoun ricni*; Denmark, *Aborre*; Germany, *Barsch*; Great Britain, *Perch*; France, *Perche fluviatile*; Yugoslavia, *Grgec*; Italy, *Pece persico*; Hungary, *Suger*; Macedonia, *Ostriz*; Netherlands, *Baars*; Norway, *Abbor*; Poland, *Okon*; Portugal, *Perca*; Romania, *Biban*; Russia, *Okun*; Finland, *Ahven*; Sweden, *Abborre*; Turkey, *Tatlisulevregi baligi*. The value of a single international name for each species is quite clear in this context!

The two-worded scientific name for a species is equivalent to the names given to many individual humans, with the first, generic, name (e.g. *Perca*) equivalent to the family or surname, and the second, specific, name (e.g. *fluviatilis*) equivalent to the Christian or given name. Although of course they are used in reverse here. A species name is often followed by the name of the scientific authority who first described it, and the year the description was published (e.g. *Perca fluviatilis*, Linnaeus, 1758). The scientific name chosen for each species usually refers to some character of shape or habit, but occasionally species are named after particular people or places. Where a subspecies is recognised there is a third, subspecific, name - e.g. *Perca fluviatilis flavescens*.

Different though they are, however, two species which have many features in common may be more closely related to each other than to any other fish, and it is likely that they would be placed in the same genus. Above this, genera are grouped together in families, and then in orders and so on.

Thus, different fish species are classified together in increasingly larger groups, the relationships within each group being believed to be due to common ancestry. It has been felt worth while to give an account of all families of fish occurring in Scotland, but not to concern the reader with groupings above this. As Laurence Sterne wrote in *Tristram Shandy*: 'What is the character of my family ...'

Lamprey Family Petromyzontidae

The lampreys (Family Petromyzontidae - meaning literally 'stone suckers') belong to a small but important group known as Agnatha - literally 'jawless' fishes, the most primitive of all living vertebrate animals. Thus they are quite distinct from all other fish in which have upper jaws fixed closely to the skull and hinged opposing lower jaws. The lampreys, in contrast, have no lower jaws and the mouth is surrounded by a round sucker-like disc within which, in the adults, are strong, horny, rasping teeth. These vary in shape, size, position and number among species, and are an important aid to identification. Lampreys occur in the temperate zones of north and south hemispheres. Fossils are available from the late Silurian and Devonian periods, some 450 million years ago.

Lampreys have several other very characteristic features: they are always eel-like in shape, but have

neither paired fins nor scales. They have no bones - all the skeletal structures being made up of strong, but flexible, cartilage. There is only one nostril, situated on top of the head, just in front of the eyes - the latter rarely being functional or even visible in the young. The gills open directly to each side of the head (i.e. there is no gill cover or operculum) forming a row of seven gill pores behind each eye. Adult lampreys have two dorsal fins which are often continuous with the tail fin.

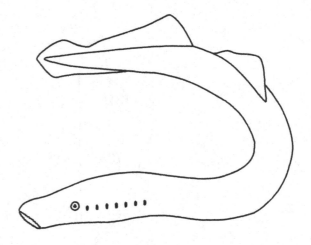

Most species of lamprey have a similar life cycle, which involves the migration of adults upstream into rivers to reach the spawning areas where they spawn in pairs or groups, laying eggs in crude nests. After hatching, the young elongate larvae, known as ammocoetes, swim or are washed downstream by the current to areas of sandy silt in still water where they burrow and spend the next few years in tunnels. They are blind, the sucker is incomplete and the teeth are undeveloped. They feed by creating a current which draws organic particles (coated with bacteria) and minute plants (such as diatoms) into the pharynx. There they become entwined in a slimy mucus string which is swallowed by the larva. The metamorphosis from larva to adult is a dramatic change which takes place in just a few weeks. The rim of the mouth (previously in the form of an oral hood) develops into a full sucker inside which are the rasping teeth; the skin becomes much more silvery and opaque except over the eyes where it clears to give the lamprey proper vision for the first time. The lampreys then migrate, usually downstream away from the nursery areas. Some species of lamprey, such as the Brook Lamprey, never feed as adults - after metamorphosing they spawn and then die - but most are parasitic on various other fish which they attack, either in large freshwater lakes and rivers or in the sea, where most of the adult life is spent.

There are nine species of lamprey in Europe. Three of them occur in Scotland:

River Lamprey *Lampetra fluviatilis* (Linnaeus 1758)
Brook Lamprey *Lampetra planeri* (Bloch 1784)
Sea Lamprey *Petromyzon marinus* Linnaeus 1758

Sturgeon Family Acipenseridae

The sturgeons (Family Acipenseridae) are an extremely interesting and valuable family of primitive large fish which are quite distinct from all other living bony fish. Living sturgeons are similar to fossil ones found in rocks about 100 million years old. They occur only in the northern hemisphere, and the family contains about 25 species. The elongate body has no scales, but has five rows of characteristic bony plates - often an important aid to identification. These plates become smoother with age, in some cases disappearing altogether in old fish. The head is covered with hard bony plates. The spinal column is upturned into the dorsal lobe of the tail fin (a condition known as heterocercal), forming its main support. Internal ossification is incomplete and there is much cartilage instead of bone. The snout is elongate, projecting well in front of the ventral mouth, anterior to which are four fleshy barbels, the exact form of which varies among the different species. The jaws are weak but the mouth itself is unusual in being a protrusible tube, well adapted for its mode of feeding, mainly on bottom invertebrates living in the sediment. The intestine has a characteristic spiral valve.

Many species are mainly marine, but all spawn in fresh water, usually in twos or threes in suitable stretches of large rivers or sometimes in lakes. The adults are often extremely large fish and not particularly efficient swimmers. The eggs are adhesive and stick to rocks, logs, etc. After hatching, the young migrate to nursery areas - usually large lakes, the lower reaches of large rivers or, in many species, the sea. There, growth is very slow; it may be 15 years before maturity is reached and fish migrate back to the spawning areas.

Sturgeons have been valued as commercial species for hundreds, if not thousands, of years. Their large size, ease of capture and tasty flesh and eggs have led to the development of fisheries in all parts of the world where they occur. However, their slow rate of growth has rendered them very susceptible to overfishing; a number of populations have become extinct because of this. Others have disappeared due to major pollution in the lower reaches of rivers which they once frequented. The flesh is eaten fresh or smoked and in some areas is dried. The roe of all species is used as caviar, prepared by removing the ovaries from ripe females, cleaning them carefully and then packing the eggs in brine. Isinglass is produced from a material found in the large swim bladder and is used for water-proofing, preserving and for clearing wine.

There are eight species in Europe. Two of these, the Siberian Sturgeon *Acipenser baeri* and the Sterlet *Acipenser ruthenus*, have been stocked in over 40 waters in England for angling. The Siberian Sturgeon is occasionally caught in the seas around Scotland as an escape from aquaculture in Europe.

There is only one native species, which is now very rare:

Atlantic Sturgeon *Acipenser sturio* Linnaeus 1758

Eel Family Anguillidae

Eels must be among the most familiar of all freshwater fishes with their almost world-wide distribution - they are found in both the northern and southern hemispheres, and from polar waters to the tropics. They occur in the sea, estuaries, continental rivers, isolated ponds, marshes, ditches and large lakes.

Their shape and mode of locomotion is characteristic. Unlike the Muraenidae (e.g. the Moray Eel *Muraena helena*) they possess pectoral fins - but pelvic fins are absent. They have very small scales set deep in the skin which is very slimy. Catadromous in life style, they arrive from the sea into fresh water as small juveniles, often penetrating far upsteam, and spending many years (often 20-30) feeding and growing until mature, then returning to the sea and migrating far out to traditional breeding areas to spawn and die.

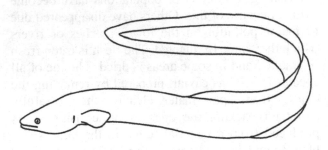

The single genus *Anguilla* consists of 16 species, only one of which occurs in Europe, including Scotland:

European Eel *Anguilla anguilla* (Linnaeus 1758)

Shad Family Clupeidae

The Family Clupeidae, many of which are commonly known as herrings or shads, is a large group of pelagic fishes which are found in seas all over the world except the Antarctic. Most species are marine, but some are anadromous and a few live permanently in fresh waters. There are several genera with a total of about 200 species.

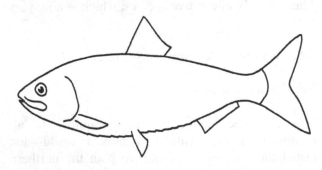

Members of the family are mainly small-to-medium sized fishes with a streamlined laterally compressed body covered by large deciduous, circular, cycloid scales. The ventral edge of the belly has characteristic scute-like scales, forming a toothed edge when viewed in profile. The head has large eyes with fleshy eyelids. The mouth is usually terminal and teeth are either small or absent. The many gill rakers lining the back of the pharynx are long and thin and their number is an important character in identification. There is no lateral line.

Most Clupeidae are pelagic in habit; swimming around in large shoals near the surface. These may include enormous numbers, sometimes millions of individual fish. The principal foods of these shoals are the abundant masses of zooplankton which thrive in the richer parts of the sea. The comb-like gill rakers help to separate the plankton from the water - the former being swallowed, the latter passing out through the gills. The fish themselves form a major source of food for many larger fish and for enormous seabird colonies in some parts of the world. The large size of the shoals, and the ease with which they may be captured, make this one of the most important commercial families of fish in the world. The total world catch is about 30 per cent by weight of all fish caught by humans. The flesh is particularly rich in fats and oils.

European Clupeidae are mainly marine, but a few species are found in fresh water throughout their lives and a number of others enter fresh water either casually or regularly at some time during their lives. There are 11 such species in Europe, two of which occur in Scotland:

Allis Shad *Alosa alosa* (Linnaeus 1758)
Twaite Shad *Alosa fallax* (Lacepede 1803)

Carp Family Cyprinidae

The Family Cyprinidae, which includes the carps, minnows, barbs and similar species, is a large and variable group of fish native to most parts of the world except South America and Australasia. They are all freshwater species and only a few are able to venture occasionally into brackish water (e.g. Roach in the Baltic Sea). This is the largest fish family in the world with some 275 genera and about 2,000 species. It is the dominant family in European fresh waters in terms of numbers of species and is of economic importance in several countries.

Cyprinidae may be small-to-large fishes of quite a range of shapes. The mouth is variable in size and position but never possesses teeth on the jaws. These are replaced functionally by pharyngeal bones with 1-3 rows of teeth which grind against a pair of horny pads on the opposite side of the pharynx. One or two pairs of sensory barbels are often present just beside the corners of the mouth. The body is usually well-covered with cycloid scales and in most fish a lateral line is present. The sexes are often different in appearance during the spawning season, the males becoming brightly coloured with well developed tubercles on the head, body and fins; the shape, position and number of these tubercles are useful aids to species identification.

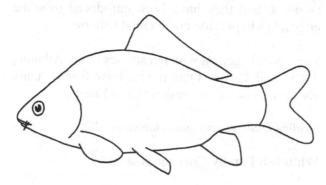

As might be expected from the large number of species, the Cyprinidae show considerable differences in habit and habitat, occupying a variety of ecological niches in fresh waters. Most species feed on invertebrates and only occasionally on fish. A number are omnivorous, feeding on both invertebrates and plants. A few species are entirely herbivorous; such fish have proved to be important to humans in two ways - as efficient producers of fish flesh and as potential biological control agents in waters where algae or higher plants have become a problem.

There are 80 species in Europe; 13 of these occur in Scotland, most of them introduced here:

Common Bream *Abramis brama* (Linnaeus 1758)
Barbel *Barbus barbus* (Linnaeus 1758)
Goldfish *Carassius auratus* (Linnaeus 1758)
Crucian Carp *Carassius carassius* (Linnaeus 1758)
Common Carp *Cyprinus carpio* Linnaeus 1758
Common Gudgeon *Gobio gobio* (Linnaeus 1758)
Chub *Leuciscus cephalus* (Linnaeus 1758)
Orfe *Leuciscus idus* (Linnaeus 1758)
Dace *Leuciscus leuciscus* (Linnaeus 1758)
Common Minnow *Phoxinus phoxinus* (Linnaeus 1758)
Roach *Rutilus rutilus* (Linnaeus 1758)
Rudd *Scardinius erythrophthalmus* (Linnaeus 1758)
Tench *Tinca tinca* (Linnaeus 1758)

Loach Family Balitoridae

The loaches (Family Balitoridae) are a group of bottom-living, small, shy and largely nocturnal fishes only found in Europe, North Africa and Asia.

They live in the shallows of most types of fresh water, both running and standing. The main characteristics of the Balitoridae are a long sinuous cylindrical body, small eyes and a fleshy, ventrally situated mouth surrounded by a number of barbels. Like the closely related carp family, they possess a single dorsal fin and only pharyngeal teeth.

An unusual feature of the family and the closely related Cobitidae is that many of their members can swallow air in stagnant conditions, the oxygen being absorbed through the walls of the gut. A result of this adaptation is that some of the species are particularly sensitive to changes in barometric pressure and become very active and restless on the approach of a storm when atmospheric pressure is falling rapidly. In Europe, one species in particular, the Weather Fish (*Misgurnus fossilis*), used to be kept in glass bowls so that its weather forecasting behaviour could be easily observed. Unlike the drab European loaches, many of the tropical species are very colourful and are popular aquarium fishes.

In Europe, there is only one genus with five species. Only one of these occurs in Scotland:

Stone Loach *Barbatula barbatula* (Linnaeus 1758)

Pike Family Esocidae

The Pike Family, Esocidae, is distributed throughout most of Europe and the temperate regions of Asia and North America, and consists of one genus and five species. They all live in lakes and slow flowing rivers and, in the case of European Pike only, in the low salinity waters of the Baltic and Caspian Seas. The elongate body, large head and huge mouth, filled with long very sharp teeth, coupled with their legendary boldness and ferocity, are features responsible for their sinister image as the 'freshwater shark'. Many exaggerated tales exist concerning their great size as well as their insatiable appetite for other fish and any other animal life that comes their way.

They spawn in the early spring, distributing their adhesive eggs over vegetation in shallow water. A large number of small eggs are produced by each female. A larval stage, possessing an adhesive attachment organ, follows hatching; at this time the small fish are very vulnerable to a range of hazards, both physical (e.g. changing water levels) and biological (e.g. predation).

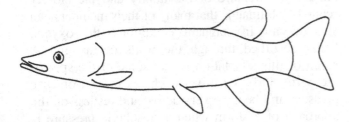

In many parts of their range, their flesh is highly esteemed and priced, while in others they are regarded as pests that consume large quantities of more desirable fish. The largest species, the North American Muskellunge (*Esox masquinongy*) has been recorded weighing over 50 kg and the Pike (*Esox lucius*) at just over half that weight. The latter is the only species found in Europe:

Pike *Esox lucius* Linnaeus 1758

Smelt Family Osmeridae

The Family Osmeridae, or smelts, occur throughout the northern hemisphere, where various species are marine, anadromous or freshwater in habit. All family members are small, silvery, slender fish with long laterally compressed bodies and large mouths. They have well developed teeth and a conspicuous adipose fin indicating affinity with the salmonids.

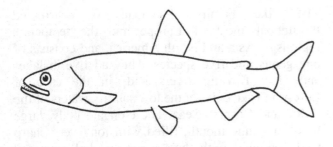

Unlike the salmonids, however, they do not have a pointed pelvic axillary process nor are they spotted or coloured. The scales are thin and cycloid and the lateral line is incomplete. During the breeding season the males (and very rarely, the females) may develop tiny white lumps (tubercles) on the head, body and fins.

Most smelts are popular food fish and have a characteristic odour, similar to fresh cucumber. In some countries they are caught in enormous numbers, and they have been introduced to some large lakes to provide commercial fisheries.

There are six genera with ten species in the Atlantic, Arctic and Pacific Oceans and their basins. Only one species occurs in fresh water in Europe:

Sparling *Osmerus eperlanus* (Linnaeus 1758)

Whitefish Family Coregonidae

The whitefishes are a temperate, northern hemisphere family found in cool, clean waters in northern Europe, Asia and North America. Their classification is complicated and often controversial but currently there are thought to be three genera and about 20 to 30 species.

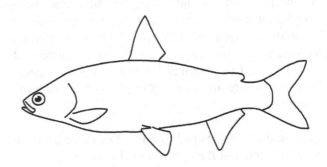

In appearance, coregonids are typically salmonid in form and background coloration, but they have very large scales and do not have any colour, patterning or spots. They also resemble salmonids in ecological requirements and behaviour in that they need a cool, unpolluted environment and include wholly freshwater and anadromous populations, the latter particularly in northern parts of their range.

In large lakes, where they feed mostly on pelagic and/or benthic invertebrates, they usually occur in large shoals and can be the basis of very important commercial fisheries. Their white flesh is excellent in texture and in flavour. In North America, several species provide anglers with sport and have been widely redistributed for this purpose and also for commercial exploitation in the USSR. Spawning takes place in the winter when eggs are laid over clean gravel and stones in shallow water, to hatch in the spring. In one genus, *Stenodus* - the 'Inconnu' of rivers flowing to the Arctic Sea - individuals grow, atypically, to a very large size (up to 130 cm), have a large underhung mouth and are fish eaters.

Because of the great variety of form found in the members of this family, their identification is often a matter of some difficulty. The situation is made even more difficult in some places, such as the Alpine lakes of Europe, where new species have been introduced, hybridising with the local species to such an extent that both the original types have disappeared, leaving a single new 'race' in possession of the habitat. The few populations of whitefishes in Scotland are thought to have originated, like those of Brown Trout and Arctic Charr, from anadromous ancestors which migrated into fresh waters after the last glaciation.

Eight species are found in Europe; two of these are native to Scotland:

Vendace *Coregonus albula* (Linnaeus 1758)
Powan *Coregonus lavaretus* (Linnaeus 1758)

Salmon Family Salmonidae

The Salmonidae is one of the world's best known families of fish. Although originating in the northern temperate zones of Europe, Asia and North America, their palatability, sporting value and adaptability have resulted in the establishment of members of the family in highland areas of tropical countries where the rivers are sufficiently cool, and in temperate lands in the southern hemisphere. The life histories of both Atlantic (*Salmo*) and Pacific (*Oncorhynchus*) Salmon have caught the popular imagination - the struggle up-river in the face of all manner of predators and physical barriers, the negotiation of high waterfalls to spawn in the natal stream and finally, in the great majority of cases, death due to exhaustion from their efforts. Following this sacrifice, the return migration of the vulnerable offspring downstream to the ocean begins the cycle all over again.

One of the diagnostic physical features of the Salmonidae is the small fleshy adipose fin situated between the dorsal fin and the tail (though a few other families of fish, such as the Osmeridae, also have such a fin). Various spots, large and small,

black, red or yellow, are also a characteristic feature, while juvenile fish usually bear 'parr marks' in the form of regular dark blotches ('thumb prints') along their sides.

All the Salmonidae spawn in fresh water, so no member of the family is entirely marine, although a number of species and races spend most of their lives at sea. Some species consist of both anadromous and wholly freshwater (so-called 'land-locked') races.

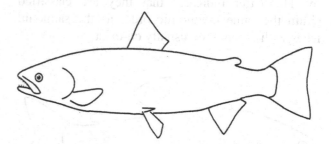

All species spawn during the coolest period of the year, usually in running water but sometimes in relatively still water in lakes; in either case they require a clean silt-free bottom for the survival of their eggs and larval stages (alevins). In most stream spawners, a nest - known as a redd - is hollowed out in the gravel by the female into which her eggs are laid and simultaneously fertilised by the male. The female then fills in the redd, covering the eggs. A feature of the anadromous Salmonidae are 'precocious' male juveniles. These small fish mature before they have reached their sea-going stage and are capable of fertilising the eggs of returned fully grown adult females.

Of the 12 species which are found in Europe (several of them alien), five occur in Scotland:

Rainbow Trout *Oncorhynchus mykiss* (Walbaum 1792)
Atlantic Salmon *Salmo salar* Linnaeus 1758
Brown Trout *Salmo trutta* Linnaeus 1758
Arctic Charr *Salvelinus alpinus* (Linnaeus 1758)
Brook Charr *Salvelinus fontinalis* (Mitchill 1814)

Grayling Family Thymallidae

The grayling family Thymallidae consists of only one genus, *Thymallus*, of six species whose distribution is confined to the temperate and arctic zones of Europe, Asia and North America. They occur in most systems that drain to the Arctic

Ocean. Their habitat is cool swift flowing streams and deep cold lakes.

A characteristic of the family is the faint smell from their palatable flesh (first commented on by the ancient Greeks and which inspired St Ambrose of Milan to call this 'the flower of fish') which has been likened to that of the herb thyme, hence the origin of the scientific name of the family and the genus. Other characteristics are the high and colourful dorsal fin and the presence of an adipose fin. The latter indicates that they are classified within the same taxonomic order as the salmonid fishes with whom they usually co-exist.

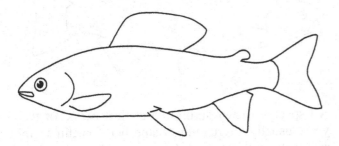

Grayling tend to occur in shoals and on occasions large numbers can be caught by angling. In Arctic regions, grayling were at one time cropped in large numbers to provide food for dogs.

The single European species has been introduced to Scotland:

Grayling *Thymallus thymallus* (Linnaeus 1758)

Grey Mullet Family Mugilidae

The mullets (Family Mugilidae) are very common fish in most oceans, particularly in shallow inshore waters. Many species penetrate brackish and fresh waters, but they usually enter estuaries, lagoons and the lower reaches of rivers for short periods only.

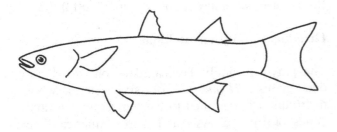

The mullets are elongate but sturdy fish, whose bodies are only slightly compressed laterally. The mouth is terminal and large, but the teeth are either very small or absent. In some species fleshy eyelids are prominent. The body is covered by large cycloid scales which extend partially on to the head. There is no lateral line. The gill rakers are long and slender.

Mullets are fast swimming shoaling fish which often come into shallow waters in large numbers to feed, their backs breaking the surface. Much of their food consists of filamentous algae, but invertebrates are also eaten in some numbers. The gut is remarkably long with a powerful muscular gizzard anteriorly which crushes and breaks up food, subsequently digested in the intestine which may be spiral in form. Spawning takes place in the sea, usually during spring in inshore waters, but relatively little is known about their breeding biology. Mullet fry are exceedingly common along some shores and enter streams in these areas. The adults run into larger rivers from time to time.

There are several genera, with a total of about 100 species found in tropical and temperate waters. Three species occur around Scotland and may enter fresh water from time to time:

Thick-lipped Grey Mullet *Chelon labrosus* Risso 1826
Golden Grey Mullet *Liza aurata* (Risso 1810)
Thin-lipped Grey Mullet *Liza ramada* (Risso 1826)

Stickleback Family Gasterosteidae

The sticklebacks (Family Gasterosteidae) belong to a group of small prolific fish consisting of five genera and seven species, which are found in a great variety of habitats in the temperate and arctic zones of Europe, Asia and North America. They occur in the sea and in fresh water. One genus, consisting of only one species, is wholly marine, while other species are equally at home in salt or fresh water. Some races are anadromous. Although of little direct economic value they act as useful prey fish for larger species, converting small organisms into food for exploited predatory species.

Sticklebacks are small laterally compressed fish with well-developed dorsal and pelvic spines, the characteristic of the family. The mouth is small and adapted for crushing their principally invertebrate diet. The body may be naked or covered by a variable number of bony plates. All species show a characteristic behaviour at spawning time involving colouring up and nest building by the males, and the ritualised behaviour to attract females and persuade them to lay their eggs in the nest. This is followed by a period when the male closely guards the nest and fry.

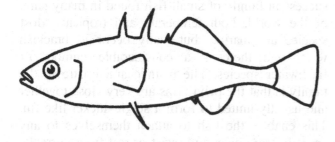

Their rigid behaviour patterns and adaptability to captivity have made them one of the most studied of all laboratory animals, while in the field, the wide range of morphological and meristic variation makes them ideal material for study by evolutionary biologists. The family is represented in Scotland by two species:

Three-spined Stickleback *Gasterosteus aculeatus* Linnaeus 1758
Nine-spined Stickleback. *Pungitius pungitius* (Linnaeus 1758)

Sculpin Family Cottidae

The sculpins (Family Cottidae) are predominantly marine fishes found in northern and arctic regions, where they mostly inhabit shallow coastal waters. A few species are found in fresh water.

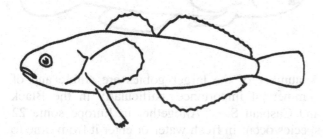

They are unusual looking little fish with a squat appearance, large cavernous mouth, opercular spines or 'horns' and mottled skin. Structurally, cottids are very advanced and they are most successful fish in the habitats which they occupy.

They are relatively poor swimmers and normally bottom-dwelling, living on invertebrates, fish and fish eggs. Most species are known to guard their eggs in some kind of nest, often under a stone or some other cover.

The family have no sporting and almost no commercial value and are sometimes considered a pest by anglers. Only a few of the 300 or so species live in fresh water. Three are found in Europe but in Scotland, there is only one representative:

Common Bullhead *Cottus gobio* Linnaeus 1758

Sea Bass Family Serranidae

The Family Serranidae - the sea basses (sometimes called sea perch) - occur in the coastal waters of most seas of the world, both temperate and tropical. Though the majority of the species are marine, a few are found in brackish and sometimes in fresh waters. Some are anadromous. In North America, one anadromous species, the Striped Bass (*Dicentrarchus saxatilis*) has been successfully established in fresh waters as a popular game fish.

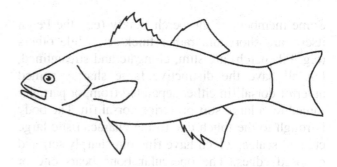

They are characteristically well built, streamlined, laterally compressed, deep bodied fish with sharp spiny anterior dorsal fins and ctenoid scales. The mouth has numerous pointed teeth usually arranged in bands. The preopercular bone is serrated. Both the dorsal and the anal fins have sharp spiny rays anteriorly and soft rays posteriorly.

All species are very predatory, feeding on a variety of bottom invertebrates and fish. They are important commercially and to sport fishermen in many countries. The family contains over 400 species. Only two species enter fresh water regularly in Europe and only one is found around Scotland:

Sea Bass *Dicentrarchus labrax* (Linnaeus 1758)

Perch Family Percidae

The perch family (Percidae) is distributed widely throughout the northern hemisphere in Europe, Asia and North America. The European Perch *Perca fluviatilis* has been successfully introduced into Australia and New Zealand. Most of the larger members of the family are familiar to sport and commercial fishermen, particularly the Perch (or Yellow Perch *Perca flavescens* in North America) and Pikeperch *Sander lucioperca*, with its close North American relative the Walleye *Stizostedium vitreum*, and are much sought after. Their flesh is firm, white, flaky and highly palatable.

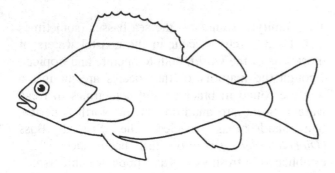

Some members of the perch family (e.g. the Perch itself) are short and rather thick-set, while others (e.g. Pikeperch) are slim, elongate and streamlined, but all have the distinctive large sharply spined anterior dorsal fin either separated from, or partially joined to, a large soft posterior dorsal fin. The body is rough to the touch due to the characteristic large ctenoid scales, which have fine but sharply serrated exposed edges. The opercular bone bears one or more sharp, rear-pointing spiny outgrowths. The mouth is unusually large for the size of the fish and has several rows of teeth. Fishermen are always advised to handle struggling percids with some caution!

The perches are highly predatory in their feeding habits and the larger members of the family are most adept at working as a group, rounding up shoals of small fish which they then catch and eat. Spawning takes place during the spring and summer. In some species, including the Perch, the eggs are laid in long ribbons draped over stones and vegetation.

The family consists of six genera and 16 species, 12 of which are found in Europe. Only two members of the family are found in Scotland, one of these relatively recently introduced:

Ruffe *Gymnocephalus cernuus* (Linnaeus 1758)
Perch *Perca fluviatilis* Linnaeus 1758

Goby Family Gobiidae

The Gobiidae, commonly called gobies, is a successful family of small fish found in many parts of the world, both temperate and tropical. Most species are marine, but many occur in brackish water and there is a considerable number of freshwater species. The main characteristic of the family is that the pelvic fins are very close together and usually united to form a single sucker-like fin. This enables the fish to attach themselves to any smooth firm object and resist or rest from currents, etc. The body is normally elongate, though often broad and squat anteriorly. The head is large and the lips and cheeks well developed. The lateral line is either incomplete or absent.

Often very abundant fish, gobies are usually benthic in habit and common in shallow coastal areas of the sea. A number of species are pelagic. At spawning time, the males become much darker (some species turn pure black), their fins elongate, and the shape of the head alters. They build simple nests - usually in the shelter of shells, stones or weed - and guard the adhesive eggs until the larvae emerge and swim away. Many species are short-lived, dying within a year of hatching.

A number of the larger gobies are edible and of commercial importance, particularly in the Black and Caspian Seas. Altogether in Europe some 22 species occur in fresh water or enter it from time to time during their life histories. Many species occur round the shores of Scotland and some come into estuaries, but only one of these is regularly found in fresh water:

Common Goby *Pomatoschistus microps* (Kroyer 1838)

Flatfish Family Pleuronectidae

The Family Pleuronectidae, commonly known as flatfishes, is a large group of mainly marine fish found in most seas of the world. As their common name implies, these fish are strongly compressed, have both eyes on one side and long, well-developed, soft rayed dorsal and anal fins.

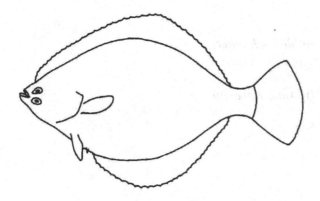

This is one of the most distinct families of fishes owing to their flattening and asymmetry. They commence life swimming about normally, but soon, instead of swimming in the normal position, they lie and swim on one side, skimming over the bottom. The eye on the lower side gradually moves to the upper side, involving modification of the head.

Interestingly, the eyes can be raised and moved independently, thus increasing the field of vision. The upper side is pigmented while the lower side is usually white. They are well known for their rapid adaptation to the colour and patterning of they substrate on which they lie. Flatfishes are carnivorous, feeding on a variety of other animals. Mainly bottom-living fishes in continental inshore waters, they are widely distributed in tropical and temperate seas, with a few species in arctic waters.

There are many genera, but only one species regularly enters fresh waters in Scotland.

Flounder *Platichthys flesus* (Linnaeus 1758)

Unscientific 'families'

Most anglers recognise two major 'families' covering the fish they are trying to catch - 'game' fish and 'coarse' fish. It is very important that these terms are defined clearly, especially in relation to any legislation which may deal with them. If the terms are to stand legal scrutiny they must have a firm foundation in science and, ideally, acceptance by the angling community.

Game fish are those freshwater fish which are generally regarded as 'salmonids'. Coarse fish are 'all the others'. These are not really adequate definitions.

Salmonid fish, in scientific terms, belong to the Order Salmoniformes. This order is treated in two different ways by various taxonomists, as either:

(a) One Family - Salmonidae - with three Subfamilies: Salmoninae, Thymallinae and Coregoninae.

Or

(b) Three Families - Salmonidae, Thymallidae and Coregonidae.

Either way, the following eight fish species (two of which are not natives) in the Order Salmoniformes are represented in Scotland:

(i) Salmonidae (Salmoninae)

    Rainbow Trout *Oncorhynchus mykiss*
    Atlantic Salmon *Salmo salar*
    Brown Trout *Salmo trutta*
    Arctic Charr *Salvelinus alpinus*
    Brook Charr *Salvelinus fontinalis*

(ii) Thymallidae (Thymallinae)

    Grayling *Thymallus thymallus*

(iii) Coregonidae (Coregoninae)

    Vendace *Coregonus albula*
    Powan *Coregonus lavaretus*

There appear therefore to be two options in defining game fish - either:

(1) Include all eight of these salmonid fish (i.e. Salmoniformes) within the definition of game fish.

Or

(2) Use the narrower definition of salmonid fish (i.e. Salmonidae / Salmoninae) which would include only five species.

The author's personal preference is for the first of these to be regarded as 'game' fish, as all these species have a great deal in common and are quite distinct from all other species which would therefore be regarded as 'coarse' fish. Presumably the latter would include even small species such as Common Minnows and Common Bullhead, for even these are angled by some fishermen.

References

Houghton (1879), Kottelat (1997), Maitland (1972, 2000, 2004), Maitland & Campbell (1992), Nelson (1994), Regan (1911).

*'All happy families resemble each other,
each unhappy family is unhappy in its own way.'*

Leo Tolstoy (1828-1910) *Anna Karenina*

*****

River Lamprey *Lampetra fluviatilis* Creathall-na-h-aibhne

*'The Lampern, or River Lamprey... is a curious little fish,
though by no means beautiful'*

Gregory Bateman (1890) *Freshwater Aquaria*

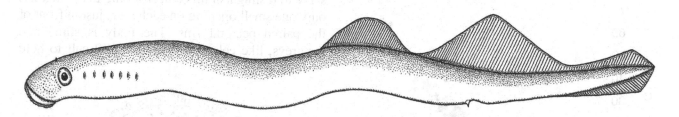

THOUGH not uncommon in many Scottish rivers, the River Lamprey is rarely seen by the public, for the larvae are hidden from view in silt beds and the adults hide during the day whilst on their spawning migration from the estuary where they have been feeding.

The notable population in Loch Lomond is quite different in appearance and behaviour from the normal form. No research had previously been carried out on this unusual population and the author recalls the excitement of coming back from North America in 1979 with experience gained there and plans to trap lampreys in the River Endrick. In October these traps were immediately successful in catching many specimens of the unusual small, but very dark, form of River Lamprey in the Endrick - the first time they had ever been seen in significant numbers. Subsequently, thanks to the care taken by Ken East, some of these lampreys were persuaded to spawn in a laboratory tank in Edinburgh. Later, after many attempts to catch them during the feeding phase in Loch Lomond, success was achieved when the first live, feeding specimen was hauled aboard a small rubber dinghy, attached to a Powan in a gill net set near Tarbet.

The River Lamprey has a variety of common names which include Juneba, Lamper Eel, Lampern, Nine-eyed Eel, Nine Eyes, Seven Eyes and Stone Grig. In the Borders, George Bolam noted that 'Day gives *Barling*, *Cunning*, and *Spanker Eel* as Northumbrian names for this species, but I have only noticed the use of the last.' John Carter recalled the name *Ramper* being used in the Borders. A

literal translation of the scientific name is 'River Rocklicker'.

John Gay, in his poem *To a Young Lady with some Lampreys*, written in 1720, clearly believed that they had special properties:

*'With lovers 'twas of old the fashion
By presents to convey their passion:
No matter what the gift they sent,
The lady saw that love was meant.*

*Why then send lampreys? Fie for shame!
'Twill set a virgin's blood on flame.
This to fifteen a proper gift!
It might lend sixty-five a lift.*

*I know your maiden aunt will scold,
And think my present somewhat bold.
I see her lift her hands and eyes.*

*"What eat it, niece; eat Spanish flies!
Lamprey's a most immodest diet:
You'll neither wake nor sleep in quiet.".'*

Size

Of the three British lampreys, the River Lamprey is intermediate in size between the large Sea Lamprey and the small Brook Lamprey. The average adult length is around 30 cm with a corresponding weight of some 60 gm, but large specimens over 40 cm can be found. The unusual race in Loch Lomond is small - often less than 20 cm. Specimens of this race collected by the author in the River Endrick ranged from 170 to 243 mm. There are no angling records for this species in Scotland, or indeed anywhere else, as it cannot be caught by rod and line other than accidentally. The largest specimen recorded by the author in Scotland was from the River Teith - it measured 362 mm in length and weighed 76.7 gm.

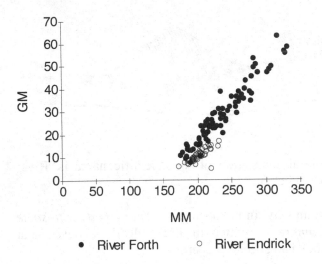

Length / weight relationship in River Lampreys from the Rivers Forth and Endrick (Maitland et al. 1994).

Appearance

Young River Lampreys (known as ammocoetes), prior to their sudden change (metamorphosis) to the adult form, are usually a dull translucent grey-brown in colour and indistinguishable from the ammocoetes of Brook Lampreys. However, they metamorphose at a smaller size (90-130 mm on average as opposed to 110-170 mm in Brook Lampreys) when they become very silvery along the sides and belly, darkening to grey on the back. During their feeding stage in the sea, they retain this silvery colour but as the time to return to fresh water to spawn draws near, they lose the silvery sheen and become darker all over. The back (which is not mottled like the Sea Lamprey) is a uniform dark olive to dark grey; this changes to a brownish yellow on the sides gradually lightening ventrally. The fins are mainly dark brown. The large eye has a golden iris, flecked with brown. The mouth is a sucker, circular when open and feeding, a longitudinal furrow when closed and the lamprey is swimming.

The River Lamprey is easily distinguished from the Sea Lamprey on the basis of size and colouration, and the patterning of the teeth inside the mouth is

also quite different. However the general body shape of the two species is similar. Thus the River Lamprey has a long streamlined eel-like body with two dorsal fins which are separate from each other, though the second is continuous with the tail fin. There are no paired fins and the seven gill openings on each side of the head are obvious and easily serve to distinguish this fish from the Eel which has only one small opening on each side, just in front of the paired pectoral fins. The body is slimy and lampreys, like eels, are extremely difficult to hold when alive.

Mouth of River Lamprey (from Maitland 1972).

The unusual race in Loch Lomond is quite distinct from the normal form. Apart from being much smaller, it is markedly darker, sometimes almost black in colour. The sucker is relatively very much larger, as is the head and the eye. These differences are not clearly understood, but may be to do with the fact that the Lomond form appears to spawn earlier and at lower temperatures than the normal River Lamprey. In spite of substantial effort by the author and his colleagues, none have ever been seen spawning, which may perhaps take place at night, or very early in the morning.

Distribution

River Lampreys are found only in western Europe where they range from southern Norway to the western Mediterranean in coastal waters and estuaries and in accessible rivers. The species usually feeds and grows in the sea but spawns in fresh water (a life cycle known as anadromous) but there are a few land-locked non-migratory populations isolated from the sea in Finland, Russia and, of course, in Loch Lomond. George Bolam, in his review of fish in the Borders, noted that it is

'Less frequent than the last [Sea Lamprey], but taken in Tweed ... and probably also elsewhere along the coast.'

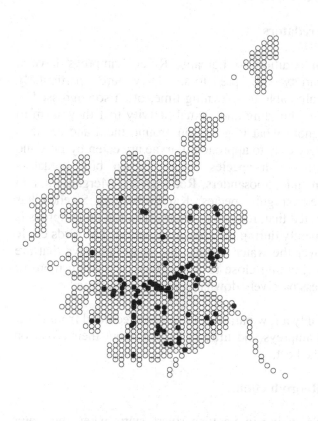

Recorded distribution of River Lamprey in Scotland.

The ammocoete larvae occur in silt beds in many of the larger rivers in Scotland. Occasionally they are found in suitable silts in large lochs. They are absent from a number of rivers because of pollution or obstacles which the adults cannot surmount during the spawning migration - these may be natural waterfalls or artificial dams, etc. River Lampreys often occur in association with the other two British lampreys but occasionally (e.g. as in one small stream in the English Lake District: Coylton Beck) they may, for reasons unknown, occur as pure populations.

Once the type of silt bed in which the larvae are likely to occur has been recognised it is usually rather easy to determine whether or not ammocoetes are present. It is particularly simple for the fishery biologist who, armed with an electric fishing machine, can simply attract them from their burrows by turning on the current - once in the water near the electrodes the larvae are stunned and easily netted. However, the observant naturalist can simply lift handfuls of silt on to the river bank and, if present, the larvae, which may be from 1-12 cm, will soon be seen wriggling out of the silt and back into the river. The author, as a boy, found this method by chance to be very successful for Brook Lampreys in the River Allander. Needless to say having watched them all return to the river and burrow quickly there, the silt on the river bank should be returned also.

Ecology

The River Lamprey is a migratory species which grows to maturity in estuaries in Scotland and then moves into fresh water, usually between August and November, to spawn in clean rivers and burns. The larvae spend several years in silt beds before metamorphosing and migrating downstream to estuaries. After metamorphosis, the young River Lampreys can still burrow but their main purpose seems to be to descend downstream to the sea, usually during spring. In the estuaries of our major rivers they can be found in some numbers and they spend 1-2 years there. Studies by the author in 1996 indicated that River Lampreys are common in the Firth of Forth and are entrained at Longannet Power Station at a rate of over 46,000 per year. Most of the lampreys taken there during summer are silvery with full guts, but in the autumn, most animals are much darker in colour, have stopped feeding, are mature and ready to migrate upstream. They have been taken in traps as far upstream as Deanston Weir on the River Teith and the Water of Mye at Buchlyvie on the River Forth. Spawning has been observed by the author in the Water of Mye and in the River Devon at Tillicoultry.

The exception to this cycle is the unusual population found in the Loch Lomond catchment. Here, most of the population spawns in the River Endrick, between Drymen Bridge and the Pot of Gartness, in early April, but after metamorphosis, instead of migrating down to the Clyde estuary, as might be expected, almost all the population stay in fresh water in Loch Lomond where they feed mainly on Powan. Both the downstream transformers and the feeding adults have proved difficult to catch, but, since about one-third of the adult Powan in Loch Lomond bear scars from attacks by River Lampreys, the population of the latter must be quite a large one. They have not been seen in any of the other rivers entering Loch Lomond.

William Brereton, writing in 1636, noted that Loch Lomond was famous for three wonders 'Waves without wind, fish without fins, and a floating island.' It has been suggested that the fish concerned were the Lomond River Lampreys, which are found throughout the loch.

Food

The diet of larvae and adults is completely different.

The main food of the larvae is fine particulate matter, mainly micro-organisms such as desmids and diatoms which they draw down into their silt burrows. They have a very sophisticated feeding mechanism. A current with food is drawn into the burrow by the pumping action of the gills. Particles are trapped in the mouth on to a slimy mucus spiral which is regularly secreted by the larva and then swallowed.

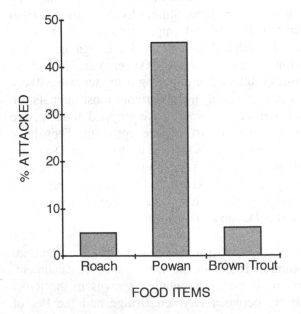

Percentage occurrence of lamprey wounds on fish in Loch Lomond.

During the feeding phase of adults in estuaries they prey on a variety of estuarine fish, but particularly Herring, Sprat and Flounders; these are the main host species whose scales occur in the guts of lampreys from the Firth of Forth. They often inflict extensive damage on these hosts by rasping away large amounts of flesh from the back. The lampreys

themselves are not particularly attractive at this time for, though very silvery, they have a very bloated appearance due to the entire gut being full of blood and fish flesh.

Predators

In addition to humans, River Lampreys have a variety of predators. They are particularly vulnerable at spawning time, often so engrossed in nest building and sexual activity that they seem to ignore what is going on around them and are thus very easy to approach. Larvae are eaten by Eels and other fish species, while known bird predators include Goosanders, Red-breasted Mergansers, and several gull species. Kaj Sjoberg in Sweden has noted that, for gulls, 'Their predation on lampreys is mainly during the spawning period. The birds circle over the water surface and snap lampreys that are swimming close to the surface or drifting more or less passively downstream after spawning.'

Only a few parasites have been recorded from River Lampreys and little is known about their effect on the host.

Reproduction

Spawning in Scottish rivers starts when the water temperature reaches 10-11°C, usually in March and April. The spawning grounds are areas of small stones and gravel in flowing water. The nest, which may be constructed by up to a dozen or more adults, is an oval depression, 30-70 cm across and 2-10 cm deep. As mentioned above, there is a high mortality at this time, for various piscivorous birds (e.g. Herons and Herring Gulls) and mammals (e.g. Otters and Mink) quickly learn that easy meals are available at lamprey spawning grounds in the spring. All lampreys die after spawning. After hatching, young larvae drift downstream, burrow in silt beds and start to feed.

Growth

After metamorphosis (July-September), at 4-5 years of age, the young adults move from life in their dark tunnels in silt in fresh water, migrating downstream during darkness at lengths of 90-130 mm to estuaries (e.g. the Firth of Forth), where they spend 1-2 years of pelagic existence in open water feeding on a variety of estuarine fish. On their return journey to the spawning areas they have normally reached lengths of 250-350 mm.

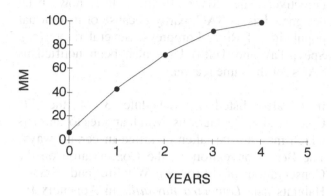

Growth of larval River Lampreys in the River Teme, England (Hardisty & Huggins 1970).

As indicated above, the Loch Lomond population remains entirely in fresh water during its life and, though full details of its life cycle there are not yet known, most metamorphosed animals seem to migrate down to the loch at about 120 mm in length. The fully grown adults return to the River Endrick to spawn at lengths of only 180-210 mm. They have never yet been observed spawning anywhere in the river.

Value

The ancient Romans greatly esteemed lampreys, but because they have no scales they were forbidden to Jews, along with other scaleless fish. River Lampreys are of considerable commercial value in parts of Europe; in both Sweden and Finland there are major fisheries for them, using mainly traditional basket traps during the upstream spawning migration. After capture the lampreys are heavily salted to remove the slimy mucus and then grilled or smoked and taken to the fish markets, where they are bought as a delicacy. In some places in Finland, they are so popular that there are grill houses which sell nothing but hot grilled River Lampreys, more or less the equivalent of fish and chip shops in Scotland! The price in the Helsinki fish market in 2000 was some £32 per kilo! Substantial fisheries did at one time exist for this species on some large British rivers (e.g. the Severn), and indeed it was from a surfeit of this species (possibly in the form of the lamprey pie that the citizens of Gloucester annually presented to the sovereign) that both King Henry 1st and King John

were supposed to have died! It is, however, no longer of any commercial importance in the British Isles other than as bait for anglers.

Herbert Maxwell recorded in 1904 that 'They were esteemed a high delicacy in old times, and are still to be seen in the shops of London fishmongers in spring, but it is to be presumed that they have lost their ancient repute, as I have never seen them served at table. It is probably one of the facts in history that stick most firmly in the memory of schoolboys, that Henry I of England died after eating too freely of a dish of lampreys.' Maxwell, however, also noted that 'Lamperns have never been so highly esteemed for the table as [sea] lampreys, albeit Buckland declared that "there is no finer dish" and that when he was fishery inspector the people of Worcester used often to complain to him that they could not get lamperns to stew, because, although multitudes were caught in the Severn, they were all packed off wholesale to be used as bait for cod by the North Sea fishermen. ... it has been stated before a Royal Commission that a single fishermen has taken as many as 120,000 in a season, ...'

William Houghton noted in 1879 that: 'In Pennant's time Lamperns ... were potted with the larger species, being preferred by some people to it, as being of milder flavour. ... great numbers are caught in wicker weels from October to March .. In order to get rid of the mucus the fish are placed in hot water in a vessel and whisked about with a bunch of straw; they are then put in cold water for a short time, and are ready for stewing or potting, under either of which happy conditions they are most delicious food, even rivalling the white-fleshed Eel in richness of flavour. The ordinary price for fresh Lamperns is about ten shillings a hundred.' In the Borders, George Bolam, writing in 1919, noted that 'It occasionally finds its way into our shops, but there is no local demand for it.'

The author, on a visit to Finland in 1985, remembers eating River Lampreys there. Having seen them caught in considerable numbers in extensive basket trap fisheries in northern Finland, he then moved to Helsinki, which is the main market for them. Here, they are deslimed in salt or salt water, roasted and then sold hot from special stalls - rolled in a piece of paper; you eat the lamprey from head to tail, skin and all, and since they have no bones, there are no problems. They were especially nice with a pint of Finnish beer, listening to the music of Sibelius!

Tate Regan recorded that 'the Irish naturalist, Thompson, once observed, on a warm summer's day, a number of Lamperns attached by their mouths to the under surface of the leaves of water lilies in a pond near Belfast, the wriggling of their dangling bodies producing a strange effect.'

Although W. Furneaux felt that 'We can hardly recommend the Lamprey as an ornament to the aquarium, nor can any hopes be entertained of keeping the fish alive for any length of time in a small tank.', River Lampreys are actually relatively easy to keep in well aerated aquaria, where the larvae will bury themselves in silt on the bottom - possibly never to be seen again for years!. After metamorphosis in the spring, the adults will spawn quite readily in captivity. Gregory Bateman commented that 'The Lampern, or River Lamprey... is a curious little fish, though by no means beautiful, and has much attraction for the aquarium keeper, for it has a remarkable structure, it gives no trouble, and is very graceful while in motion. ... As he can easily climb by the help of his mouth, the tank in which he is confined must be covered.' The adults, after entering fresh water, when they stop feeding, will live for long periods (i.e. over a year in cool conditions) in suitable aquaria and can be induced to spawn there. During our work on the Lomond population, Ken East successfully induced the black adults to spawn in tanks in Edinburgh.

Conservation

The River Lamprey has declined in Scotland over the last hundred years due to pollution, river barriers and habitat destruction and, though not yet distinctly threatened, is in need of general conservation measures to restore populations to their former status. It is given no special protection under British legislation but is regarded by the author as Vulnerable in Scotland. However it is listed in Annexes IIa and V of the EC Habitats Directive (1992) and this new conservation legislation obliges member states to (a) designate sites to form part of the 'Natura 2000' network comprising Special Areas of Conservation (SACs), (b) protect such sites from deterioration or disturbance with a significant effect on the nature

conservation interest (and take steps to conserve that interest), and (c) protect the species of Community interest listed in the Annexes to the Directive. The River Endrick has now been designated as an SAC, partly because of its unusual population of River Lampreys. Several rivers (e.g. Spey, Tay and Teith) have also been notified as SACs for the same reason.

It is also listed in Schedule 3 of the EC Conservation Regulations, which applies to 'animals which may not be taken or killed in certain ways'. The Bern Convention on the Conservation on the Conservation of European Wildlife and Natural Habitats lists *Lampetra fluviatilis* in Appendix III, which permits some exploitation of its population.

The dwarf population in Loch Lomond and the River Endrick deserves special conservation measures - especially as there are some signs that it, and its host the Powan, may be in decline. Martin Hardisty indicated in a letter that he 'would much prefer to regard your dwarf lamprey as a landlocked variant whose special features may be attributed to a relatively long period of genetic isolation.'

A future conservation programme for the River Lamprey in Scotland should include:

(a) Regular monitoring of its distribution throughout Scotland.
(b) Further research on its biology in the Loch Lomond catchment.
(c) Regular monitoring of adults in the Rivers Endrick (Drymen) and Forth (Longannet).
(d) Regular monitoring of larvae in the SAC Rivers Endrick, Spey and Teith.
(e) The feasibility of extending the nursery area in the River Endrick by adult translocation upstream.
(f) The feasibility of creating a 'safeguard' population in the Sloy or Carron catchments.

References

Relatively little research has been carried out on the River Lamprey in Scotland and the following short list includes a number of relevant valuable studies carried out elsewhere in Europe.

Bateman (1890), Bird & Potter (1979), Bolam (1919), Brereton (1636), Buckland (1880), Day (1880), Furneaux (1896), Hardisty (1961b),

Hardisty & Huggins (1970), Hardisty & Potter (1971), Houghton (1879), Huggins & Thompson (1970), Kainua & Valtonen (1980), Lamond (1922), MacDonald (1959), Maitland (1972, 1980, 1997), Maitland *et al.* (1983, 1984, 1994), Maitland & Campbell (1992), Maitland & Lyle (2003), Maxwell (1904), Morris (1989), Morris & Maitland (1987), Pennant (1761), Regan (1911), Robertson (1875), Schoonoord & Maitland (1983), Sjoberg (1980), Stevenson *et al.* (1995), Thompson (1849), Tuunainen *et al.* (1980), Valtonen (1980).

*'It was not the fault of the Lampreys that King Henry died,
it was his own fault for eating too much.'*

William Houghton (1879) *British Fresh-water Fishes*

*****

Brook Lamprey *Lampetra planeri* Creathall-an-uilt

'usually found in the smaller streams ... and never goes down to the sea.'

George Bolam (1919) *The Fishes of Northumberland ...*

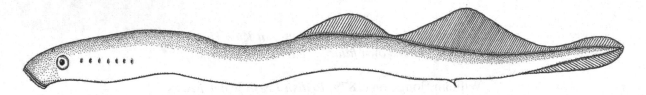

AS a small boy, trying to catch Stone Loach one day in the River Allander near Milngavie, the author was astonished to see what appeared to be a small black eel swim up to a sand bank at the edge of the river and disappear instantly into the sand. Curiosity aroused, Beardies were forgotten and an attempt was made to catch this unknown creature. Not having anything with which to dig into the sand, the only recourse was to scoop masses of sand by hand onto the adjacent bank and there, soon enough, was a wriggling 'fish' that was then placed in a standard jeely jar of the day. Indeed was it a 'fish' for it had no eyes, nor paired fins and a row of holes down each side of the head! When some sand was put into the jar, it immediately disappeared from view into the substrate. This was my first view of the ammocoete larva of the Brook Lamprey - the only lamprey in the Allander at that time, because pollution from the local paper mills downstream blocked off all access from the Clyde Estuary of the other, migratory, species. Now that the Allander is clean again, perhaps River and Sea Lampreys will appear there too.

Some of the first work on lampreys in Scotland was initiated by Harry Slack at the University of Glasgow, and Ted MacDonald, his PhD student in the late 1950s, was a kenspeckled figure cycling around Loch Lomondside in a towel tricycle with a tank for lampreys (instead of a basket) on the front. Ted had very poor eyesight and could not drive, so Harry Slack thought up the idea of a towel tricycle to make him mobile.

The Brook Lamprey is also known as Mud Lamprey and Pride. George Bolam notes that in the Borders it is 'Provincially known as *Lampern*, *Nine-eyes*, and *Blind Eel*.' A literal translation of the scientific name is 'Planer's Rocklicker'.

Size

The Brook Lamprey is the smallest of the three Scottish lampreys, usually maturing at a length of some 130-150 mm. Some populations are known where the adults may be much smaller than this, e.g. on Skye adults spawning in the small burns there may be less than 100 mm. However, at some sites they may be larger - for instance in the River Endrick (and its tributary the River Blane) which flows into Loch Lomond adults may reach 160 mm and occasionally even longer; the longest recorded there so far was 170 mm in length.

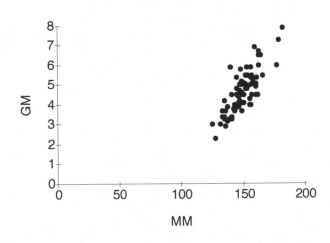

Length / weight relationship of Brook Lampreys in the River Endrick (Maitland et al. 1994).

Brook Lampreys are at their largest as soon as they metamorphose. Thereafter, in contrast to the other two lamprey species which feed and grow as adults, since Brook Lampreys do not feed after maturing, they lose weight (and length) and so the adults arriving on the spawning grounds are actually smaller than they were as metamorphosing larvae in the nursery beds.

Appearance

Edward Boulenger pointed out that 'The larvae of lampreys or prides, as the infant creatures are called, are very different from their parents when they first hatch from the eggs.' The ammocoetes are semi-translucent and mainly dull grey-brown in colour, though a 'golden' form does appear from time to time, with very much reduced pigment. After metamorphosing to the adult form they are much more silvery, especially along the sides and belly. The back remains dark grey-brown. The larvae of this species are virtually indistinguishable from those of the River Lamprey except when nearing metamorphosis.

The adults are small eel-like fishes with two dorsal fins, usually joining both each other and the tail fin. There are no paired fins. The eyes, which are covered by opaque skin in the larvae, are large and bright in the adults. Characteristically the teeth are blunt and much less developed than those of the predacious River and Sea Lampreys. The maxillary plate is wide but there are no lower labial teeth; the mandibular plate has 5-9 blunt teeth.

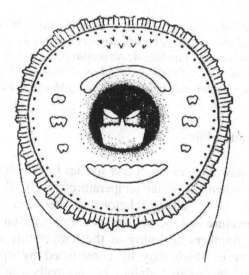

Mouth of Brook Lamprey (from Maitland 1972).

The author has recorded a golden form in the River Endrick and Ross Gardiner, working on the River Spey near Brae on 26 August 1994, during an electrofishing survey for ammocoete larvae, found a complete albino. The absence of melanin resulted in it being a remarkable (and beautiful) golden colour and good photographs were obtained. This was the only specimen like this out of 3,795 ammocoetes examined by Ross from Scottish rivers.

Tate Regan reminded us that 'Lamprey larvae, formerly called prides, are now more commonly called ammocoetes. They were originally thought to be a distinct species *Ammocoetes branchialis*, and it was in this species that the transformation from larva to adult was first witnessed by A. Muller in 1856. He described how he had watched Planer's Lampreys spawning in a brook near Berlin and had seen the eggs hatch out in about three weeks and develop into Prides, previously known as *Ammocoetes branchialis.*'

Distribution

Recorded distribution of Brook Lamprey in Scotland.

The Brook Lamprey is a purely freshwater species occurring in streams and occasionally in lakes in north-west Europe, especially in basins associated with the North and Baltic Seas. It occurs over much of the British Isles, but is absent from much of Scotland north of the Great Glen, including the Northern and all but a few of the Western Isles. Henry Lamond found lampreys spawning in the lower reaches of the Luss Water and they are common in most of the burns flowing into larger lochs south of the Great Glen. Ammocoetes are quite common in some of these lochs.

The ammocoete larvae, like those of other lampreys, occur in suitable silt beds, mainly in running water but sometimes in large numbers in silt banks in lochs. Larvae were found by the author in considerable numbers in suitable silt beds at the edges of Lochs Doine, Voil and Lubnaig during a survey of the River Teith. In the 1980s, a salmon farmer at Loch Ness took the author to a silt bank adjacent to the effluent from his smolt farm near Foyers where larvae were abundant, due, no doubt, to the rich organic waste coming from the farm. Some highland burns are virtually devoid of the silts which are essential habitat for the larvae; it is only where such sediments are found in associated lochs that populations can survive. Many of the burns they use in these situations are quite small - the author remembers seeing them dying in tiny burns along Loch Lomondside during a drought, when they were trapped in drying up pools. The Brook Lamprey is the most abundant and widespread of the Scottish lampreys, often found in the absence of the other two species, for example above a pollution or physical barrier which prevents the other species reaching that part of the river from the sea.

Ecology

This species does not feed as an adult and so other fish evoke no response. The larvae, having light sensitive cells in the skin, always actively move away from light, for the most part remaining sedentary within their burrows. However, if disturbed they will swim around rapidly until they find suitable silt into which to burrow. They are capable of completely disappearing into sand in just a few seconds and larvae placed in a glass jar with water and a few centimetres of sand will quickly vanish from sight! As spawning time approaches the metamorphosed adults move out from the silts and start to migrate upstream (often in large numbers) till they reach suitable spawning grounds. These are areas of small stones and gravel in flowing water where the current is present but not too strong. Very characteristically they spawn at the lower ends of pools just where the water is starting to break up into a riffle.

Food

The larvae feed, like those of other lampreys, by filtering fine organic particles, especially diatoms and other algae as well as protozoans and detritus,

from the surface of the silt around the mouths of the burrows in which they spend virtually all their larval years. The ciliary mechanism and the mucus threads involved in the collection of this food form a complex, but very efficient feeding mechanism. Water is drawn down into the burrow by the pumping action of the pharynx as the larva breathes and food particles in the water are sieved off by the mucus and then swallowed. The adults do not feed after metamorphosis.

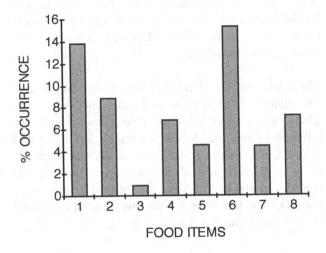

Food of larval Brook Lampreys in Highland Water, England. All the items are algae (1: Achnanthes; 2: Cymbella; 3: Eunotia; 4: Navicula; 5: Nitzschia; 6: Pinnularia; 7: Synedra; 8: Chlamydomonas), though some detritus was recorded as well. (Moore & Potter 1976)

Reproduction

The spawning season of this species in British rivers starts when the water temperatures reach 10-11°C. There is a clear relationship between water temperature and the number of animals at spawning sites, numbers declining as the temperature drops. The nest, which may be constructed by up to a dozen or more adults, is normally an oval depression about 20-40 cm across and 2-10 cm deep. The actual spawning act is similar to that of other lampreys though the Brook Lamprey on account of its small size is less fecund, producing only about 1,500 eggs per female. After hatching, the young larvae leave the nest and distribute themselves by drifting downstream and burrowing in suitable areas of silty sand. By this time all the adults are dead, for none seem to survive long after spawning.

Predation

Lampreys are rarely seen by the general public except at spawning time when they become very obvious. Then, the otherwise cryptic and nocturnal creatures seem to 'throw caution to the winds', moving into shallow clear water in broad daylight to start their complex and fascinating communal nest building activities. Of the three Scottish species, Brook Lampreys are those most often seen, and in April and May they can be readily observed in many thousands of streams in different parts of the country. At this time they are extremely vulnerable, being eaten in considerable numbers by Herons, gulls and sawbill ducks, especially Goosanders as the research by Mick Marquiss and his colleagues has shown. Lampreys, probably mostly Brook Lampreys, formed an important part of the diet of this bird in the Rivers Tay, Tweed and Esk (Borders). They are also collected by anglers to use as bait.

Frank Buckland noted that the Brook Lamprey 'has two enemies, the eel and the fisherman; the former devours the sand-pride, the latter seeks it for bait for pollock.'

Very little is known about their parasites.

Growth

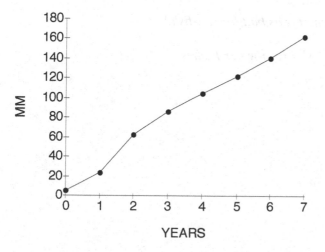

Growth of Brook Lamprey larvae in the River Usk (Hardisty & Huggins 1970).

Larval life seems to vary considerably in different parts of Europe; in the Scotland it is probably about

6 or 7 years. The larvae are some 3-5 mm on hatching and about 12-15 cm at metamorphosis which takes place between July and September, usually simultaneously (i.e. within 3-4 weeks) in any one population.

The adults migrate upstream after metamorphosis, continuing to burrow in silts beds like ammocoetes or hiding under stones during the day. Since they no longer feed, they lose weight (and length) from metamorphosis up to spawning time, when the females suddenly become heavier as the eggs take up water prior to spawning.

Trapping of Brook Lampreys by the author at Drymen Bridge on the River Endrick showed that the mean length of migrating adults changed from 160 mm in January to less than 119 mm in May of 1984.

Value

This species is rarely considered to be of any commercial value. Occasionally anglers use the larvae and sometimes the adults as bait. William Houghton recorded that 'The fishermen use these fish as a bait in whiffing for Pollacks . ... Excepting as bait or food for other fish it has no value.' Herbert Maxwell confirms this view: 'When sought after by fishermen, it is as bait for sea-fish.' - as does George Bolam: 'A Pike, it may be remarked, will often take a Lamprey greedily when it refuses to look at any other bait.'

Although W. Furneaux noted in 1896 that 'We can hardly recommend the Lamprey as an ornament to the aquarium, nor can any hopes be entertained of keeping the fish alive for any length of time in a small tank.', the larvae will actually live well in aquaria and though they are not often seen there, since they spend all their time burrowed in the substrate, they can be induced to metamorphose and even spawn if the conditions are right. In this connection the species is a useful one for school laboratory demonstrations. The author has kept the larvae in aquaria with other fish, when they were not seen for months, living buried in the sandy substrate on the bottom. Suddenly they produce a surprise when they metamorphose and are seen hanging on to the sides of the tank - an admirable time to make a close inspection of their complex suckers and blunt teeth. Thereafter, as Gregory

Bateman rightly points out 'The Brook Lamprey will breed in an aquarium, making a sort of nest in a gravelly part of the bottom, ...'

Conservation

The Brook Lamprey is given no special protection under British legislation and is regarded by the author as of Lower Risk in Scotland. It is, however, listed in Annex IIa of the EC Habitats Directive (1992) and this new conservation obligation obliges member states to (a) designate sites to form part of the 'Natura 2000' network comprising Special Areas of Conservation (SACs), (b) protect such sites from deterioration or disturbance with a significant effect on the nature conservation interest (and take steps to conserve that interest), and (c) protect the species of Community interest listed in the Annexes to the Directive. The River Endrick has now been designated as an SAC, partly because of the Brook Lamprey population there. The Rivers Spey, Tay and Teith have also been notified as SACs because of their lamprey populations.

There are many other populations of Brook Lamprey in Scotland some, coincidentally, as Alex Lyle and the author have shown, in National Nature Reserves and Sites of Special Scientific Interest, thus having some legislative and practical protection in these areas.

A future programme of research on the Brook Lamprey in Scotland should include:

(a) Regular monitoring of its distribution throughout Scotland.
(b) Further research on its biology in the Loch Lomond catchment.
(c) Regular monitoring of larvae in SAC Rivers (e.g. Endrick, Spey and Teith).
(d) Research on the apparent dwarf races on Skye and elsewhere.
(e) The role of loch sediments as a habitat for ammocoetes.

References

Relatively little research has been carried out on the Brook Lamprey in Scotland; the following short list, however, includes a number of relevant valuable studies carried out elsewhere in Europe.

Bateman (1890), Bolam (1919), Boulenger (1946), Buckland (1873), Furneaux (1896), Hardisty (1944, 1961), Hardisty & Huggins (1970), Hardisty & Potter (1971), Houghton (1879), Huggins & Thompson (1970), Lamond (1931), Lyle & Maitland (1992), MacDonald (1959), Maitland (1972, 1980), Maitland et al. (1994), Maitland & Lyle (2003), Marquiss & Carss (1991), Maxwell (1904), Moore & Potter (1976), Morris & Maitland (1987), Regan (1911).

'When sought after by fishermen, it is as bait for sea-fish.'

Herbert Maxwell (1904) *British Fresh-water Fishes*

*****

Sea Lamprey *Petromyzon marinus* Creathall-na-mara

'The sucking disc is covered with conical teeth arranged in oblique radiating series'

Herbert Maxwell (1904) *British Fresh-water Fishes*

HUMAN attitudes to the Sea Lamprey vary greatly across its range of distribution and so it is an animal with very mixed fortunes. Though the Great Lakes Fishery Commission in North America spends some 10 million dollars annually trying to eradicate Sea Lampreys in the Great Lakes, the species is a delicacy in Portugal, currently fetching up to £40.00 per specimen there. Females are particularly valuable because of their roe. The species was once popular as food in England; indeed William Houghton contended that 'The species now under consideration is the one which brought King Henry the First to an untimely end.' J.M.D. Meiklejohn confirms this: 'Henry I died in Normandy in 1135. He was very fond of lampreys and after a day spent in hunting he ate very heartily of them. His body was brought to England and buried in Reading Minster which he had himself built.'

The Sea Lamprey is also known as Marine Lamprey and Lamprey Eel. In the Borders, George Bolam noted that 'Our fishermen call these fish *Lamperns* or *Sookers*...' A literal translation of the scientific name is 'Sea Stonesucker'.

Size

The Sea Lamprey is by far the largest of the Scottish lampreys, sometimes reaching a length of 100 cm and a weight of 2.5 kg. The normal adult length is around 50 cm. There are no angling records, but Henry Lamond recorded in 1931 that 'I expect that Mr. Andrew Colquhoun, Luss, is the only angler who has ever caught a lamprey with a fly. In striking a rising sea-trout on one occasion he missed the sea-trout but hooked and landed its attendant lamprey!' George Bolam records that 'One I examined at new Water Haugh on 26th March

1896 was over 3 feet long, and 8 or 9 inches in circumference, which is about the maximum size to which the species attains.'

Larvae normally metamorphose at about 140-160 mm, but individuals as long as 180 mm have been recorded by Pat Manion in North America.

Appearance

The body is very long and cylindrical, except at the tail where it is laterally compressed. The overhung (inferior) mouth has the form of a large circular sucker frilled with extensions of the skin known as fimbriae. When closed, the mouth has the form of a slit, but when open for attachment it forms an oval sucking disc whose diameter is greater than that of the head or pharynx behind.

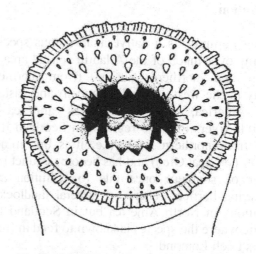

Mouth of Sea Lamprey (from Maitland 1972).

Inside the mouth are numerous hard sharp teeth arranged in concentric rows. Above and around the tongue the teeth are especially large and bicuspid;

below the tongue is a huge multiple transverse tooth with up to 10 cusps. The tongue itself has several large complex teeth.

The eyes are of moderate size and positioned on each side of the head just behind the single nostril and in front of the seven pairs of gill openings. The two dorsal fins are distinctly separate in the young, but much closer in the adults. The first of these (the lower) originates just behind the middle of the body; the second (which is slightly higher) terminates just in front of the small tail fin. There are no paired fins. The skin is smooth and scaleless. There is no lateral line and no vertebrae, the entire skeleton being cartilaginous. At spawning time, the males develop a distinct ridge along the back, whilst the females have a pronounced fold of skin behind the vent.

Colour varies greatly with age. The ammocoete larvae are dark greyish brown above and a light grey below. Occasionally specimens are found which are lacking in pigment - these have a yellowish gold appearance. Newly metamorphosed animals are a slaty grey-blue above, changing gradually to a metallic bluish on the sides, and to a pale white on the belly. In adults, the main body colour is brownish grey (paler ventrally) with extensive black mottling. The body colour lightens to a golden brown (almost orange sometimes) at spawning time.

Distribution

The Sea Lamprey is a native anadromous species occurring over much of the Atlantic coastal area of western and northern Europe (from northern Norway to the western Mediterranean) and eastern North America, and in estuaries and easily accessible rivers in these regions. In Scotland it is absent from northern rivers (i.e. it is uncommon north of the Great Glen) and has become extinct in a number of southern ones due to pollution and engineering barriers. There are several landlocked populations in North America but in Scotland the only site where the species is known to feed in fresh water is Loch Lomond.

In the Borders area, George Bolam noted that this species was 'Not uncommon and has been recorded from time to time from the estuaries of most of our rivers. ... on the Tweed it has been taken at least as

high up as Kelso... About the mouth of the Tweed one or two are generally taken every March or April, but I have seen it there as late as 28th June, at which season it is usually found higher up the river.'

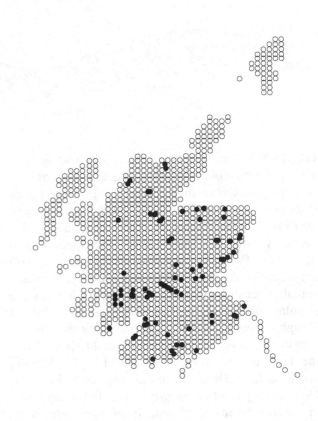

Recorded distribution of Sea Lamprey in Scotland.

The larvae (known as ammocoetes) are usually found in silty sands in running water, though in some places they may occur in silt and gravel beds in lochs (e.g. Loch Lomond). The habitat occupied by the larvae of all species seems to be very similar; indeed in Scotland, Sea, River and Brook Lampreys may often be found together at the same sites. Where suitable substrates are present they are found in streams and rivers upstream as far as the adults are able to migrate; they are stopped by high waterfalls or weirs, dams and severe pollution. In a study of the distribution of lampreys in the River Teith, the author found that Sea Lamprey larvae were present up to the Falls of Leny, but not above.

David Dunkley recalls having problems with the fish counter on the North Esk at Logie which he traced to a large Sea Lamprey stuck on the downstream face of the counter weir. This was sufficient to upset the electrical balance of the counter trying to compensate for the presence of the lamprey, which has a different conductivity from the volume of water its displaces.

Ecology

Several research workers have measured the conditions at places occupied by the ammocoetes in an attempt to define their habitat precisely. The optimum particle size of the beds of sediment in which they occur is said to be 0.18-0.38 mm, and to include clay, silt and sand fractions. Shade (which appears to be related to the types of micro-organisms on the surface) and water velocity appear to be important factors connected to the suitability of sites. Normally, suitable sites are found only in some parts of each river system and in some rivers there may be none at all. Within the stretches of suitable gradient, adequate sites are often found in conditions of slowing current, where deposition of sand and silt occurs (e.g. in eddies, backwaters, behind obstructions or at the edges of streams).

Relatively little is known about the precise habitats occupied by adult Sea Lampreys. Though adults are sometimes caught at sea, the precise conditions in which they occur have not been described, nor is it certain which fish are the main host species. Most adults found in fresh water are either migrating upstream to spawn or dying after spawning. Habitat seems only to be important in relation to their ability to get to the spawning beds. Just before spawning they may be found in calmer water above the spawning areas or just below protecting obstructions, etc. The nests are normally built in areas of flowing shallow water among sands and gravels of varying particle size.

Food

There is little evidence for any differences in the food or feeding habits of the ammocoete stage of the three British species of lamprey. All appear to feed from within their burrows on fine particulate matter, mainly micro-organisms, desmids and diatoms in particular. In addition, various unicellular animals including ciliates, euglenoids and rhizopods have been found in ammocoete guts in some numbers. The role of detritus as food is uncertain, but large amounts appear to be eaten during the summer months. Most of the food taken in by the larvae comes from the superficial

sediments in the vicinity of the larval burrows. The system of ciliated tracts in the pharynx, used as a means of transporting food on strands of mucus towards the intestine, is complex.

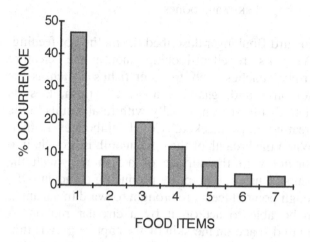

Food of larval Sea Lampreys in Snyder-Deadhorse Creek, Michigan (1: Synedra; 2: Fragilaria; 3: Navicula; 4: Gomphonema; 5: Meridion; 6: Cocconeis; 7: Other algae) (Manion 1976).

After metamorphosis and the downstream migration to the sea, the adults feed on fish there, but detailed evidence on their feeding habits is fragmentary (except in the specialised case of the purely freshwater populations in North America which have been intensively studied). They seem to feed on a wide variety of marine and anadromous fishes, including Sturgeon, Herring, Salmon, Cod and Haddock. Salmon and Sea Trout entering rivers often bear fresh scars attributable to attacks by this species. Recently, the author has received information and photographs of young Sea Lampreys feeding on Sea Bass off Guernsey. Tate Regan recorded that '... it is on record that large numbers attached themselves to a Basking Shark, not leaving it until it was dead.' However, it is unclear if such a large fish can be killed in this way and Denis Fairfax suggests that the Sea Lamprey 'utilises the skin of the shark even though it does not seem that lampreys can actually bite into the flesh as the skin is too tough'. Peter Malloch, who has obviously seen many, recorded that 'The salmon appears to be a favourite with the lamprey, judging by the number that have come under my notice marked in this fashion. ... Occasionally they are brought ashore by the nets, firmly fixed to a salmon, and only desist when they see they run the risk of being caught. I feel sure that many salmon are killed

by the gnawing wounds inflicted by this creature.'
George Bolam noted that 'Our fishermen ...
frequently find them attached to Cods or other fish
upon their lines, more than one ... sometimes
sticking to the same fish, which is quickly reduced
to a bag of skin and bones.'

Edward Boulenger described the method of feeding:
'With its repellent-looking mouth the creature
firmly attaches itself to other fishes - such as the
salmon - and, gnawing away the flesh, taps its
victim's life stream, usually with fatal results to the
creature thus attacked.' Couch elaborates further:
'When the breadth of the open mouth is brought into
contact with the surface of a fish on which the
lamprey has laid hold, by producing a vacuum, the
rough pointed teeth are brought forward in a manner
to be able to act on it by a circular motion. A
limited space on the skin of the captive prey is thus
rasped into a pulp and swallowed, and a hole is thus
made which may penetrate to the bone.' The adults
do not feed after they have started on their
spawning migration. Ross Gardiner recalls, while
diving in the River Tay at Logierait about 1980,
touching a large Sea Lamprey that was at rest
attached to a stone. 'It got such a fright that it swam
straight to the edge and beached itself (temporarily)
on the shore.'

Unusually, in Loch Lomond, just as in the Great
Lakes of North America, some Sea Lampreys feed
in fresh water. The author has seen large Sea Trout
leaping there with Sea Lampreys attached and has
in his fish collection a specimen taken from an
Atlantic Salmon angled in the loch. In the past,
others have made similar observations. Henry
Lamond recorded that in August 1919 he 'hooked a
salmon of 19 lb. ... While the fish was swerving to
and fro alongside the boat we both observed that a
sea lamprey was fixed to its shoulder. ... The
lamprey was fully eighteen inches in length and
about five in girth.' Lamond further notes Atlantic
Salmon 'leaping with the lamprey attached to it, as
one so frequently sees sea-trout leaping. I have, as
many anglers also have done, caught sea-trout and
netted them with the lamprey still adhering.'

Reproduction

The Sea Lamprey usually spawns in June in
Scottish rivers, when the water temperature reaches
at least 15°C. Normally, males appear on the nesting

sites first and are apparently highly attractive to
females, possibly by the secretion of an olfactory
sex attractant. The numbers of eggs produced by the
females in some populations have been estimated
by research workers and average about 172,000 per
female. The eggs are small (0.80-1.25 mm in
diameter) and an opaque white colour when laid.

William Jardine noted that 'They ascend our rivers
to breed about the end of June, and remain until the
beginning of August. ... their sucker-like mouth, by
which they individually remove each stone. Their
power is immense. Stones of a very large size are
transported, and a large furrow is soon formed. The
P. marinus remain in pairs; two on each spawning
place; and while thus employed, they retain
themselves affixed by the mouth to a large stone.'
Ross Gardiner, when diving in the River Spey on 12
July 1983, with the water temperature at 22°C,
recalls watching Sea Lampreys digging out a nest
and then spawning. Ross and his colleague Mike
Donaghy were able to watch them under water,
taking still photographs and cine film.

After hatching, larvae leave the nest and drift
downstream, distributing themselves among
suitable silt beds. The duration of larval life varies
but averages about five years. Larvae do move
about from time to time, especially when disturbed.
Pat Manion found that it is common for the larger
larvae to move at night into deeper water. The age
of larvae has mainly been calculated from length
frequency analyses, for there is no known method
of ageing them. Metamorphosis to the adult form
takes place between July and September, the
process usually taking a few weeks. The time of the
main migration downstream seems to vary from
river to river and relatively little is known about
them after they reach the sea, where they have been
found in coastal and off-shore waters. The spawning
migration in Europe takes place in April and May
when adults migrate back into fresh water.

Predators

During spawning, the newly laid eggs may be eaten
by Minnows, Eels and Bullheads, which can be seen
moving in and out of the redds. Newly hatched
larvae are also eaten as they drift downstream out of
the redd - mortality seems to be very high at this
stage. Later, the mortality rates in ammocoete
populations are probably rather low and consistent

throughout the larval period. Apart from the effect of fluctuating physical factors, especially during the embryonic period, it is known that the larvae are eaten by Eels, sticklebacks and other fish as well as several different birds (e.g. Herons). A high mortality probably occurs at metamorphosis; the adults, however, appear to be less vulnerable to predation at spawning time, probably because of their large size and the fact that they generally spawn in deeper and faster flowing water than either River or Brook Lampreys.

Only a few parasites have been recorded from lampreys and nothing is known about their effect on the host.

Growth

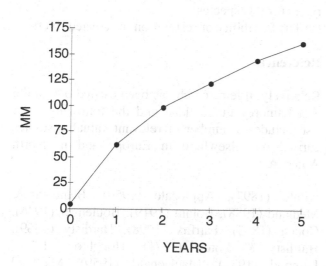

Growth of larval Sea Lampreys in Lynde Creek, Ontario (Beamish & Medland 1988).

An amazing experiment by Pat Manion and Bernie Smith in the Big Garlic River in Michigan revealed exciting new facts about the growth of young Sea Lampreys. In 1960, adults were introduced and spawned in the river above waterfalls where there were no other lampreys. As expected, the young from this spawning started to metamorphose and migrate downstream in 1966 - but continued to do so for at least 19 years thereafter! Thus the growth and age at metamorphosis of this species can be very variable - apparently very dependent on local variations in temperature and nutrition, and perhaps other factors as well .

After metamorphosis (July-September), usually at about 5 years of age, the young adults, at a length of some 15 mm, migrate during darkness to estuaries and eventually to the sea where they spend 1-2 years feeding on a variety of fish. On their return journey to the spawning areas they have reached lengths of 45-55 cm, sometimes larger. As indicated above, some individuals feed, not in the sea, but in Loch Lomond but full details of the life cycle there are not yet known.

Value

In Scotland, there never seems to have been a tradition of eating, or even marketing, Sea Lampreys. According to W. Parnell, 'the fishermen in the Forth, above Alloa, when they accidentally take Lampreys in their nets, invariably return them again to the water having a prejudice against them. Consequently they are never under any circumstances seen in the Edinburgh markets.'

Further south, however, Sea Lampreys were important commercially: Tate Regan records that 'The City of Gloucester used to present the reigning monarch with a dish of lampreys every Christmas.' William Houghton admitted that 'I have never tasted the Sea Lamprey, but if the quality of its flesh is at all similar to that of the River Lampern, it must be most delicious eating.' F.G. Aflalo agrees: 'This lamprey is much esteemed as food, and is caught in wicker baskets specially constructed and placed in the mud.'

Lampreys are still popular in some countries today, as indicated by this e-mail sent home by David Henty to his mother on 21 February, 2001: 'Just thought I'd say hello from Portugal! Having a great time - the weather is fantastic and so is the food. I actually ate lamprey last night - I think it's something of a local delicacy and apparently they're "in season". I hadn't realised how big they were! They had a tank of them in the window, and they were about a yard long and as thick as my forearm with suckers that must have been up to two inches across. You could see as they stuck to the glass that they were full of hundreds of tiny teeth.'

In a letter to the author from Portugal, fisheries scientist Alberto Matos enclosed two recipes for Sea Lampreys and noted that 'Usually they are caught in the rivers in the months of January and February and they are much appreciated here by local people.'

The larvae live well in aquaria provided they have a suitable substrate in which to burrow and are fed regularly on fine particulate foods (e.g. a suspension of yeast). However, because of the burrowing habit they are rarely seen. Eventually they metamorphose and start to swim freely about the tank and are easily visible. None of the adults caught in our rivers will feed, for they are all sexually mature and intent on spawning. However, they have very substantial food reserves at this time and will live for many months under suitable cool water conditions. They can easily be induced to spawn in large aquaria if they are given suitable gravel and a reasonable artificial current.

Conservation

The author recalls seeing, in 1979, thousands of Sea Lampreys being poisoned in a river in Michigan as part of the control programme there using a specially designed lampricide. This programme has been running for several decades now at an annual cost of some 10 million dollars. Meantime, in Europe, this species was declining fast and has been declared Endangered in some countries. Apart from problems of pollution and river barriers, there has always been a threat to this species from anglers, who regard it as a pest and, up until recently, have killed spawning adults at every opportunity.

Though the Sea Lamprey is given no special protection under British legislation it is the least common of the three lamprey species in Scotland and regarded by the author as Vulnerable. Fortunately it is listed in Annex IIa of the EC Habitats Directive (1992); this new conservation obligation commits member states to (a) designate sites to form part of the 'Natura 2000' network comprising Special Areas of Conservation (SACs),

(b) protect such sites from deterioration or disturbance with a significant effect on the nature conservation interest (and take steps to conserve that interest), and (c) protect the species of Community interest listed in the Annexes to the Directive. Several rivers (e.g. Spey and Teith) have now been designated as SACs, partly because of the populations of Sea Lampreys there.

A future programme of research on the Sea Lamprey in Scotland should include:

(a) A regular review of its distribution throughout Scotland.
(b) Further research on its biology in Loch Lomond.
(c) Regular monitoring of larvae and adults in the River Leven (Lomond).
(d) Regular monitoring of larvae and adults in SAC rivers for this species.
(e) The feasibility of restoration in selected rivers.

References

Relatively little research has been carried out on the Sea Lamprey in Scotland and the following short list includes a number of relevant valuable studies carried out elsewhere in Europe and in North America.

Aflalo (1897), Applegate (1950), Beamish & Medland (1988), Bolam (1919), Boulenger (1946), Couch (1865), Fairfax (1998), Hardisty (1969), Hardisty & Potter (1971), Houghton (1879), Lamond (1931), MacDonald (1959), Maitland (1972, 1980, 1984, 1994, 2004), Maitland *et al.* (1994), Maitland & Lyle (2003), Malloch (1910), Manion (1976), Manion & Smith (1978), Maxwell (1904), Meiklejohn (1903), Newth (1930), Parnell (1838), Regan (1911), Smith (1957).

'I feel sure that many salmon are killed
by the gnawing wounds inflicted by this creature.'

Peter Malloch (1910) *Life History of the Salmon*

*****

Common Sturgeon *Acipenser sturio* Stirean

'It is a very large fish, attaining a length of 18 feet'

Tate Regan (1911) *British Freshwater Fishes*

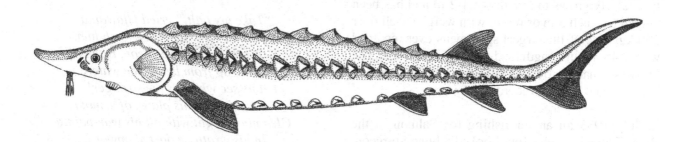

SOMETIMES caught accidentally by commercial fishermen, William Yarrell recorded that '... the Sturgeon is considered a royal fish; the term being intended to imply that it ought to be sent to the king, and it is said that this fish was exclusively reserved for the table of Henry the First of England.' Fred Buller comments that 'Everyone in Britain associates the sturgeon with Royalty and most believe that all landed sturgeons become the Monarch's property. It all started with legislation (Prerogative Regis Crown Fisheries and Wreck of the Sea) enacted at the time of Edward II in 1324, giving the King the right to all English but not Scottish "fishes royal (whales, dolphins, porpoises and sturgeons) stranded on the shore or caught in inshore waters.' A postscript continues 'In 1970 a Bill was presented to the House of Lords to abolish the Crown's rights to royal fishes and other prerogatives. This was amended to retain the Monarch's right to royal fishes including sturgeons. ... The wisdom of the Lords demonstrates that a law passed 600 years ago can still have great utility, not least because it enables the Palace kitchens, aided by their deep freezers, to continue to provide, on notable occasions, an historic dish - Royal Sturgeon.'

There is, however, some considerable doubt as to the reality of this, for, in 1966 when the author was initiating a scheme to map the distribution of freshwater fish in the British Isles, he wrote to the Comptroller of the Queen's Household - thinking that there might be a wealth of records of fish offered and received there. In fact, the reply indicated that advantage was never taken of such offers and that there was therefore no information

available. However, reports still highlight this connection with Royalty, for example *The Daily Telegraph*, 8th December, 1986 records: 'Sturgeon Gift. A 15-lb. 3 ft. 6 in. sturgeon - the fish whose roe makes caviar - has been caught off the Sussex coast by Mr Richard Goodsell, 32, of Rye. It has been offered to the Queen, according to ancient tradition. It is worth about £150 and is the first caught in the area for 36 years.'

In Scotland, the capture of a Sturgeon is usually noteworthy. In the late 19th century, James Fraser recorded such an event: 'The 7 of August a sturgion fish was taken in the Yarr of Drumchardeny, within our paroch of Wardlaw. It was 12 foot in length, a monstous creatur. In all my travels I never saw so big. Severall English came out from Inverness, who had not seen such another. They bought it at a very great rate, to preserve it pickled, the fish being meat and medecin; they barrelled it, and sent it to London, wher it will sell very deare. The report is that such fishes coming ashore is ominous, and presages the death of some eminent person.' The best example of an eminent person Fraser could provide was the brother of the Laird of Cawdor, who accidentally shot himself some weeks later.

Though sometimes described as a vagrant to Scottish waters this was certainly not the case originally. It may have entered fresh water relatively infrequently, but it is clear that Scottish coastal waters and parts of the North Sea were a major feeding ground for this species.

An alternative name for this species is Baltic Sturgeon. A literal translation of the scientific name is the 'Stirring Sturgeon'.

Size

This enormous fish is now only a rare vagrant to fresh waters in Scotland since it never breeds here. It regularly grows to lengths of 1-2 m and has been known to reach 3 m or more, with weights well over 200 kg. One of the largest specimens ever recorded was 3.45 m in length and weighed 320 kg. The females usually grow to a larger size than the males. There are no angling records.

In July 1933 an angler fishing for Salmon in the River Towy in Wales foul-hooked a huge Sturgeon; the fish eventually beached itself. It was 2.79 m in length, 1.49 m in girth and weighed 196 kg. It was a female and contained 36.3 kg of caviar - but it was not accepted by the Royal Household. George Bolam recorded several large Sturgeon in the Borders: 7 ft and 140 lb. at the mouth of Tweed on September 1853; 8.5 ft and 217 lb. at Yarrow Haugh on 3 August 1872; and 7 ft 10" and 160 lb. at Sandstell on 31 August 1909.

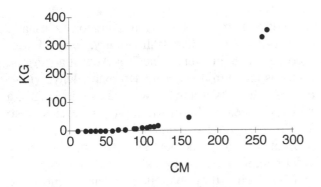

Length/weight relationship in the closely related Acipenser oxyrhynchus (Magnin 1963).

Historically, William Houghton noted that 'One of the largest British specimens on record is mentioned by Pennant; it was caught in the Esk, and weighed four hundred and sixty pounds.' The Leeds Mercury of February 1st, 1879 recorded that 'The largest Sturgeon ever delivered at the port was brought into Grimsby on Wednesday morning by the smack 'The Kitty' ...The Sturgeon was four feet eleven inches in circumference, eleven feet nine inches in length, and weighed forty-four stones and a half. It was sold to Mr. A. Clifton, fish merchant.' Tate Regan notes that 'It is a very large fish, attaining a length of 18 feet ... records of British examples of 7 or 8

feet are too numerous to mention.' Unfortunately, large specimens like these are now all too rare.

The size of sturgeon obviously impressed Henry Longfellow who included this passage in *The Song of Hiawatha* in 1855:

> '"Take my bait," cried Hiawatha,
> Down into the depths beneath him,
> "Take my bait, O Sturgeon, Nahma!
> Come up from below the water,
> Let us see which is the stronger! ...
> ... Clashing all his plates of armour,
> Gleaming bright with all his war-paint;
> In his wrath he darted upward,
> Flashing leaped into the sunshine,
> Opened his great jaws, and swallowed
> Both canoe and Hiawatha.'

Appearance

This fish was at one time virtually unmistakable for any other species found in Scottish waters. One of the most characteristic features is the five rows of hard bony plates along the back and sides of the body, the row on the back having some 9-15 of these plates, the lateral rows about 24-36 plates and the rows on either side of the belly about 9-13 plates. The snout is long and pointed, bearing two pairs of barbels hanging down midway between the anterior tip and the mouth. The latter can be partially projected during feeding as a short tube. The ventral fin carries a strong spine.

The body surface is soft and scaleless, protection being given by the rows of hard bony plates. The colour is usually a dark bluish brown on the back lightening down the sides to a pale whitish yellow ventrally. Tate Regan noted that, formerly, 'It has been assumed that specimens with a large number of lateral scutes, as, for example, one said to be from the Tay, with thirty-eight on one side and forty on the other, belong to an American species (*A. maculosus*) which has been supposed to cross the Atlantic occasionally.' However, Regan believed they are all *Acipenser sturio*.

More recently, a complication has arisen from the appearance of the Siberian Sturgeon *Acipenser baeri*, which has been introduced to the catchments draining to the Baltic Sea. These fish are straying into the North Sea and are being caught occasionally around the coast of Scotland. In addition, specimens of the freshwater Sterlet *Acipenser ruthenus* are imported as aquarium and pond fish and may escape or be released to local waters. Further, both these alien species are now being released in lakes in England to provide fisheries for sturgeon.

Distribution

The original distribution of this magnificent fish was much wider than at present. In the sea it occurred along the entire coastline of Europe from the North Cape to the Baltic, Mediterranean and Black Seas. A similar species occurs along the Atlantic coast of North America. The Common Sturgeon is no longer common and indeed is nearing extinction, occurring only occasionally in our seas and breeding in only a few European rivers, notably the Gironde. One or two are still caught each year off British coasts by commercial netsmen, but only very occasionally does one ever venture into fresh water here. As noted above, some of the specimens caught in the last few years have not been the Common Sturgeon, but actually the Siberian Sturgeon, numbers of which have been released into the Baltic Sea.

The adults favour the lower reaches of large rivers and the young remain there for some time before descending to the sea. Here many of them apparently stay in the nearby coastal waters, though others obviously do move about for considerable distances to be caught many hundreds of miles away. They appear to live mainly over soft sandy or muddy bottoms to which their feeding mechanism is adapted.

W.J. Gordon notes that 'It is a solitary fish that appears in our rivers in the spring, and takes to the sea in winter ...' Records of Sturgeon caught by commercial fishermen around Scotland have included the following sites: Brodick, Musselburgh, Queensferry, St Andrews, Carradale, Loch Fyne, River Nith, Scrabster, Eyemouth, the Moray Firth, Buckie and Banff. G. Low noted in 1774 that he had heard of a specimen which came ashore in one of the north isles of Orkney. J.A. Harvie-Brown & T.E. Buckley recorded that 'At Barvas a Sturgeon was taken some years ago' and 'A Sturgeon was found on the beach of North Uist in 1887, as I am informed by the late Mr. John Macdonald of

Newton. I have some of the scales he sent me by which it was identified'. Rintoul & Baxter noted that the Sturgeon were 'Taken occasionally, usually about Musselburgh or Queensferry, but has occurred as far up as Alloa.' George Bolam noted in 1919 that 'Dr Johnston was able to rank the Sturgeon as of almost annual occurrence at the mouth of the Tweed, in 1837; at the present day only an occasional wanderer visits us, and as these are usually captured, it is not difficult to understand the reason if the species has become less common than formerly. ... Perhaps about a dozen have been taken in the neighbourhood of the Tweed during the last thirty years.' A note in *Barrow's Worcester Journal* in July, 1833 recorded that 'There was caught in a stake net, near Findhorn, Scotland, a sturgeon, eight feet six inches long, three in width and weighing two hundred and three pounds.'

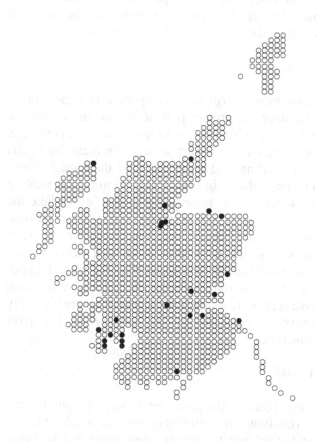

Recorded distribution of Common Sturgeon in Scotland.

Though now only a rare vagrant to Scottish rivers, the question of whether this magnificent fish ever spawned here is an interesting one. Certainly at one time it was tolerably common around the shores of

Scotland and came well up the estuaries of all the large rivers. Service recorded in 1907 that 'The shad or rock herring was also another of the fish that came into fresh water to spawn. So also was the sturgeon, huge examples of which were annually caught when running riverwards in early summer.'

South of the Border, Fred Buller noted that 'In the 12th Century, if we give credence to an account written by Fitzstephen [in Couch], suggesting that a dish of sturgeon could be obtained readily at a hostelry in the City of London ... then sturgeon must have been caught regularly in the Thames and other large unobstructed rivers.' Buller continues: 'Andrew Kones ... has found sturgeon scutes ... from sites of Saxon occupation at Ipswich and York.' 'Doncaster Museum has a record of 90 sturgeon from the area..' Houghton recorded that 'This fish is occasionally taken in Salmon nets on various parts of our coasts; generally speaking, in the estuaries, or not far up the rivers; sometimes, however, it ascends rivers to a considerable distance.'

Ecology

Like most sturgeon, this species is anadromous, spending the greater part of its life in the sea but returning to fresh water to spawn when it is mature. After at least 10 years at sea, the maturing adults stop feeding and move back to their natal rivers, entering these in the early spring and moving upstream to the spawning grounds, which take the form of pools where the water is deep (several metres) and flows over a gravel bed. After spawning, the adult fish drop back down to salt water and start to feed again, many of them dying or caught by predators during the journey. The young fish stay in fresh water for a year or so but gradually move downstream to the estuary of the river concerned.

Food

The young Sturgeon feed mainly on bottom invertebrates in their rivers, especially midge larvae and other insects, worms, crustaceans and molluscs. The fish feeds by rooting about on the bottom with its snout, detecting the presence of invertebrates by means of its barbels, sucking them in through its tubelike mouth - often with a great deal of mud. In the sea they feed in the same manner but mainly on the larger worms (including polychaetes),

crustaceans and molluscs there, also taking some fish - especially sandeels and gobies.

Because adults do not feed during the spawning migration they are in poor condition after spawning. However, they do start feeding in fresh water as soon as spawning is over and, after they reach the sea they feed heavily and are able to regain condition again fairly rapidly.

Reproduction

Little is known about the details of the actual spawning act which takes place in early summer when the dark coloured sticky eggs, about 2-2.5 mm in diameter, are laid over gravel and vegetation, to which they adhere. Females can produce from 800,000-2,400,000 eggs each, depending on size. These hatch into larvae about 11 mm long, in about 3-6 days at water temperatures of some 12-18°C and the yolk sac lasts them a further six days before they need to start feeding.

Predators

Relatively little is known about the predators of this large fish. It is probably eaten by piscivorous birds and mammals when young, but adults seem to have few predators being well protected by their bony scutes. Deaths from attacks by Sea Lampreys have been recorded in the Atlantic Ocean. The main predator over the 20th century has undoubtedly been man.

Growth

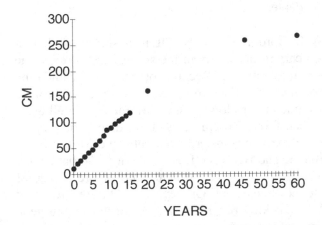

Growth of the related Acipenser oxyrhynchus in the St Lawrence River, Canada (Magnin 1963).

Growth is fairly rapid after hatching though it does depend to some extent on local conditions, especially food and temperature. Most fish reach at least 10 cm by the end of their first year; many may reach twice this length. Some start to move down to the sea after this time but most others stay on for another one, sometimes two, years. In the sea, growth is steady and fish are usually at least a metre in length when they first start to mature at about 8-12 years of age. Thereafter growth is rather slow, depending on the frequency of spawning among various individuals.

Value

Though Marian McNeill records that 'Malcolm IV granted the half of the fat of the royal fishes which might come into the Forth on either shore', the Sturgeon is of little value in Scotland today. They used to appear occasionally on fishmonger's slabs for curiosity value. Tate Regan records that 'According to Macpherson, a Sturgeon recently captured may be a dangerous companion, and one has been know to cut a man to the bone with a blow of its tail, the dorsal spines and the sharp edged lateral scutes making the latter a formidable weapon of offence.' Thomas Shirley noted that 'The common way of killing sturgeon is with a harping-iron for they take no bait; and when they feed they rout in the mud with their snouts like hogs.'

■ *THE Herald* reported that a 51lb royal sturgeon, caught in the Firth of Clyde by Mr Andrew McCrindle, skipper of the *Golden Venture*, was offered to the royal household by Ayr Lord Provost Mr Alex Handyside. It had been bought by an Ayr businessman for £23. The sturgeon was male and "contained no caviare".

Captures of Sturgeon are so rare that they are reported by the media.

As far as the flesh is concerned, Yarrell believed that 'it is generally stewed with rich gravy, and the flavour is considered to be like that of veal.' Charles Dickens wrote in 1880 that 'The flesh of the sturgeon is looked upon with suspicion little short of aversion by some persons ... but ... the great chefs, Francatelli and Ude, used to aver there were one hundred different ways of rendering sturgeon fit for an Emperor.' F.G. Aflalo, however, was not enthusiastic: 'The flesh has a faint pink tinge, and there is a good deal of fat. It is not bad eating, but rather coarse, and rarely fetches anything more than a very low price in the market.' Though the roe is very valuable as caviar and widely sought after, William Houghton was of a different opinion: 'This Sturgeon's roe is the caviar of commerce, a thing, in my opinion, disgusting in appearance, offensive to the smell, and horrible to the taste.'

On the continent of Europe, however, it is another matter - the flesh of the Sturgeon is highly prized there and even more so its roe which is processed to make caviar, an extremely valuable commodity supplied to high class restaurants all over the world. The pursuit of this and other species of sturgeon for their valuable roes has been one of the main reasons for the decline in the numbers of all species in recent years. However, the development of Sturgeon farms in France and elsewhere, though primarily for the commercial production of caviar, may help in the conservation of this species if the farms are managed wisely. Such farms may help in two ways: firstly by reducing the need to catch many fish from the wild; secondly, by releasing part of the young stock they obtain from wild parents back into the wild again. The author remembers eating Sturgeon (probably farmed *Acipenser baeri*) - but not its roe (!) - every day during a visit to Moscow and St Petersburg in 1986. Caviar was available at meal times but it was orange in colour and made from the roe of Chum Salmon.

Small sturgeon are available in aquarium shops in Scotland from time to time, but these are the young of another European species, the Sterlet *Acipenser ruthenus*. This is a small freshwater species which is very attractive to watch in an aquarium, but which is difficult to feed and keep over a long period.

Conservation

Though listed in Schedule 5 of the Wildlife & Countryside Act 1981, Annex IIa and IVa of the Habitats Directive 1992, and Schedule 2 of the Conservation Regulations 1994 such legislation is very late in the day as far as the Sturgeon is concerned; the only hope of survival for this species is the practical conservation action which is being taken in Europe, especially in France, where it is hoped that a captive breeding programme may help to save this valuable species from extinction. The author recalls visiting the research facility involved beside the River Dordogne in France in 1995, where attempts are being made to captive breed this Sturgeon and save it from extinction.

Scotland does not, but could, contribute to the conservation of Sturgeon by adopting one or more of the following:

(a) An action plan whereby any Sturgeon caught around Scotland are kept alive and transferred to one of the breeding programmes on mainland Europe.

(b) Alternatively, a facility where young fish could be reared from eggs and sperm obtained from adults caught accidentally in Scottish waters. These could eventually be released to the North Sea in order to help to support the highly threatened European stock.

(c) A voluntary ban on caviar by both the public and commerce, unless obtained from farmed fish.

References

Virtually no research has been carried out on the Sturgeon in Scotland, other than the recording of those caught by commercial fisherman at sea. Most of the available literature concerns work done elsewhere in Europe.

Aflalo (1897), Bolam (1919), Buller (1989), Couch (1865), Dickens (1880), Fraser (1833), Gerasimov (1988), Gordon (1902), Harkness & Dymond (1961), Harvie-Brown & Buckley (1889), Houghton (1879), Low (1774), Magnin (1963), Maitland (1966), McNeill (1929), Regan (1911), Rintoul & Baxter (1935), Service (1907), Shirley (1784), Yarrell (1836).

*'This Sturgeon's roe is the caviar of commerce,
a thing, in my opinion, disgusting in appearance,
offensive to the smell, and horrible to the taste.'*

William Houghton (1879) *British Fresh-water Fishes*

*****

European Eel *Anguilla anguilla* Easgann

*'The waters of Scotland abound in eels,
but you will never find a Scot who does not treat
with shuddering or contempt the idea of eating one.'*

Herbert Maxwell (1904) *British Fresh-water Fishes*

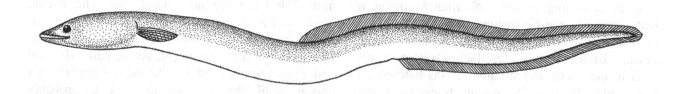

THE ability of the European Eel to move out of the water and travel overland is legendary, but rarely observed. Frank Buckland noted that 'The eel, as is well known, will live a long time out of water. This habit is of the greatest service to him, as sometimes it is necessary for him to migrate from place to place by an overland route.' One of the few proofs of this was recorded by George Bolam '... in 1878 Thomas Sligh, so long fisherman at Carham, ... had settled the question to his own satisfaction by catching an Eel in a common steel rabbit-trap. Having observed sundry slimy tracks leading from the river to the remains of a Salmon lying some distance away upon the bank, he suspected that Eels might be the cause of them, and placing some traps in the "runs", was not surprised to find an eel caught in one of them the next morning.' W.J. Gordon recorded that 'Eels will travel overland wherever it is fairly moist, and have even been reported by a gardener for eating his peas. They are quick of hearing, and in captivity will come to be fed when called.'

Joscelyn Lane is one of the few people ever to have observed migrating Eels on land: 'There had been a few showers, and not far from the town the car was held up by a wide column of eels crossing the road. ... they were coming along a deep ditch .. from a spot about twenty yards higher up, where they could be seen wriggling through the grass of an adjoining field on their way down to the ditch. ... the eels were still crossing the road and streaming along the bottom of a ditch on the opposite side for some thirty yards, where I lost sight of them as the ditch diverged from the road. ... There was no water in either of these ditches, along which several hundred

eels must have travelled when I was there.'

An even less common observation was recorded by George Bolam in the Borders: 'The extraordinary habit which young Eels have of occasionally leaving the water, in large numbers and without apparent reason, to crowd together amongst the grass at the river's edge, or even to attach themselves to overhanging branches of trees, has frequently been commented upon in print, usually to be received as only a fisherman's yarn. I have, however, more than once had ocular demonstration of the habit ... on the Till at Castle Heaton ... I was wading in order to reach some Trout that were rising under the willows on the opposite bank, when I noticed that some branches overhanging and dipping into the water seemed to be literally alive with Eels. ... the twigs ... had developed rootlets ... None of the eels exceeded six inches in length, most of them being an inch or two shorter, and there were something like a hundred of them quite clear of the water, besides many more partially out of it, or swimming round the branches seeking to effect a landing. Some of them had wormed their way up the twigs at least a foot above the water... When I shook their branch they dropped helter-skelter into the water, but when undisturbed they paid no attention to me, although I stood within a yard of them and took many of them in my hands for examination.'

Until the close of the last century, the complicated life history of the freshwater Eel was a mystery, subject to various mythical and magical explanations, e.g. the long hairs from the tail of a stallion were said to turn into Eels when dropped into water, or alternatively that they arose by spontaneous generation. David Cairncross, in his book The Origins of the Silver Eel claimed that 'Eels were bred from beetles'. However, during the first three decades of the 20th century, the true story was gradually revealed - one of the great pieces of natural history detection. Even to this day however, no-one has ever seen a spawning Eel or knows exactly where they spawn, or how they navigate to reach the spawning grounds. Norman Morrison, a great admirer of the Eel commented that 'It is a far cry from a mountain tarn in mid Scotland to the Leeward Islands in the West Indies, yet every eel found in our lochs and streams ... was hatched on these banks.' It is 'A fascinating drama from real life, to be eternally repeated.'

The Eel has a variety of other common names, including Astan, Broad-nosed Eel, Bulldog Eel, Frog-mouth, Glut, Gorb Eel, Grig, Silver Eel and Yellow Eel. In the Borders, John Carter recalled Sapey, whilst in Caithness Harvie-Brown & Buckley say it is called *Tammy yaa* or *Yaa*. A literal translation of the scientific name is 'The Little Snake'.

Size

Male Eels are usually smaller than females. Fully grown mature male Eels, on their return migration to the sea, are seldom longer than 47 cm whereas the females are seldom less. Individuals of around 40-90 cm are sizes normally encountered by anglers and commercial fishermen. The Scottish rod-caught records stands at 2.495 kg (5 lb 8 oz) for an Eel caught in Loch Ochiltree in 1987 by T. May. The Scottish sea records stand at 1 lb 13 oz 7 dm (boat), caught in the Gareloch, and 3 lb (shore), caught in Ayr Harbour. The British rod-caught record stands at 5.046 kg (11 lb 2 oz) for an Eel caught in 1978 at Kingfisher Lake, Hampshire.

Peter Malloch noted that 'In Loch Leven they often weigh 6 or 7 lbs, and I have known them to exceed this weight in the River Tay.' Elsewhere in the British Isles, Tate Regan has recorded some very large specimens: 'R Lubbock's record of one of more than 20 lbs., taken near Norwich in 1839, can scarcely be questioned, whilst W. Yarrell saw at Cambridge the skins of two said to have weighed 23 and 27 lbs., which were taken on draining a fen-dyke at Wisbech.'

Appearance

The European Eel with its serpentine form and slimy body is unmistakable. The body is cylindrical anteriorly becoming rather flattened posteriorly. The gill openings are only small slits and the eyes too are small and covered by a thick translucent skin. The tubular nostrils protrude just below the eyes. There is a pair of pectoral fins, but unlike most fish there are no ventral fins. The dorsal, caudal and anal fins are joined into one continuous fin which has over 1,000 rays. The scales of the European Eel are transparent, small, narrowly oval and deeply embedded in the skin. They do not appear until the Eel is 15-18 cm, probably regardless of age. They are laid down as concentric rings of small bony particles and are not ridged by circuli like most other teleost scales. The lateral line is clearly defined.

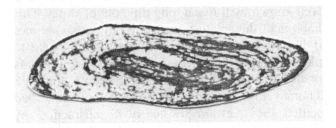

Scale from European Eel.

Eels have teeth, and the author well remembers his colleague Andy Rosie's amusement when he (the author) was bitten by an Eel beside Loch Lomond during routine measuring for length (not an easy matter with live Eels!).

European Eels, during their freshwater or estuarine life, are known as 'yellow eels', but actually vary considerably in colour; when ready for returning to the sea they are called 'silver eels', a fairly accurate description. Eels in fresh water are usually dark brown above the lateral line and yellowish below, but individuals which are pale yellow-amber to dull brown may occur apparently in response to habitat - the yellow individuals occurring where the water is shallow and the bottom sandy. On maturation, Eels assume a grey or silver-grey colouration and their eyes become much enlarged, in preparation for life in the depths of the oceans.

The shape of the head was once said to vary with feeding habits, and this gave rise to the recognition of two distinct forms: the sharp or narrow-nosed eel (with the smaller mouth) and the broad-nosed eel, which was said to feed on large organisms. The latter was once referred to as a Grig, Glut or Frog-mouthed Eel and given specific status as *Anguilla latirostris*. These variations were allegedly associated with feeding behaviour only and not with geographic location, sex, age or habitat - though the broad-nosed variety appeared to be more common in estuaries. It has now been shown, however, that the great majority of European Eels fall into an intermediate category, with continuous variation between the two extremes.

Distribution

In Europe, Eels are found in all type of fresh waters which are accessible from the sea, and in estuaries and shallow coastal waters. As well as occurring as far north as Iceland, the species also occurs around the Mediterranean coasts of North Africa and Asia Minor, and in the Black Sea. Female Eels tend to remain longer in fresh water than males and continue to migrate further upstream; thus eventually attaining a greater size and age than most males. This tendency can also give rise to the impression that in some waters, most Eels are female and larger, age for age.

During the day, Eels usually lie buried under stones or weed, or in mud or whatever substrate is available to them. St John recorded on May 3rd, 1848, that 'When fishing near the river I found great numbers of young eels under the gravel, my attention being first called to them by a Skye-terrier, which accompanied me, and who employed himself in turning over the stones and eating the eels which he found under them.' George Bolam noted in 1919 that 'The young, formerly regarded as a distinct fish under the name of Beardless *Ophidium*, ascend our rivers as Elvers,' Although, as discussed above, Eels have a reputation for being able to travel overland in wet weather and thus reach virtually all fresh waters; in fact their powers are more restricted. Alastair Stephen was unable to find any Eels during electrofishing in the catchment of the River Dee, apparently due to the hydro-electric dam at Tongland near the mouth of the river acting as an impassable barrier. Other fish (e.g. Sparling) also seem to have been affected by this barrier.

Yet Eels can certainly negotiate other difficult obstacles as noted by MacIntyre: '... there are no

elvers a few miles from the sea on the Kintyre rivers, but thousands of the small eels they quickly grow into. I saw the rock at the lower Glen Lussa fall black with little eels at the great July floods, all still. They come out on the rock only on days of flood; but the eels negotiate the Upper Fall of Glen Lussa, which is more than a salmon can do.'

Recorded distribution of European Eel in Scotland.

Eels also have a great tendency to slip into any orifice under water. David Carss records that in the summer of 1984, 'while living in a small cottage on a West Highland estate, our water supply failed. A plumber was called and the supply pipes checked, evidence pointed to a blockage rather than a burst pipe. The header tank was examined first - it held water draining from a small, peaty, low-lying lochan. The first section of the pipe to be removed revealed the problem - an eel stuck head-first down the pipe.' Donald Mitchell has a similar story: 'In January 2000, I was called to the Edrachilles Hotel near Scourie to examine a large Eel caught up in a water pump. It was 72 cm in length and weighed 1.8 kg. The same thing had happened a few years previously when an even bigger Eel was stuck so firmly it had to be cut out of the pump.'

Ecology

From late winter to spring, vast numbers of eel larvae, known as elvers, arrive all along the coasts of Scotland (and much of the rest of northern Europe), migrating upstream into all kinds of fresh waters from large rivers and lakes to damp flushes and tiny ponds. At this stage the elvers are around 5-7 cm in length and almost transparent, but with conspicuous black eyes showing. As they make their way upstream, often in dense columns, following the edges of a stream or river they are preyed upon by various fish (including larger Eels) and avian predators. Thus their seasonal arrival is often advertised by the presence of gulls and other birds lining the waterside and by Trout caught bulging with them. All manner of seemingly insurmountable barriers are overcome by these advancing elver hordes, often at great cost to themselves. Dam walls are scaled wherever there are damp areas - though few elvers may actually survive to surmount the crest and continue upstream. On being confronted by a waterfall, they can wriggle around the edges through damp moss and scale wet, almost vertical, rocks. Water intake systems may be entered - elvers sometimes emerging from cold water taps in houses, causing domestic concern.

There are many accounts of the elver invasion: Herbert Maxwell records that 'elvers ... appear in such prodigious numbers that I have seen a Scottish trout-stream slate-coloured from bank to bank with the throng for a distance of twenty or thirty yards. It is a display of the prodigality of Nature, ...' Peter Malloch notes that 'The eel fry come up all our rivers in countless thousands, distributing themselves in every stream and loch in Great Britain. They appear in the Tay about the 20th of April, when both sides of the river to the width of about 10 yards are crowded with them. In their movement upwards on a quiet day they make a ripple like that caused by wind blowing on the surface of the water. At this time they are 3 to 4 inches long.'

The timing of this annual elver invasion varies from year to year and area to area, but in Scotland it usually occurs from March to June with a peak around May; some late arrivals can still be seen as late as August, however. The numbers of elvers arriving at the coast also varies considerably from year to year. Obviously they do not all attempt to find their way to the headwaters of each river system but gradually become distributed among most of the suitable waters. In Scotland, however, small elvers can commonly be found trying to scale the dam walls of hydro-electric schemes at least 48 km above the estuary.

One aspect of eel behaviour that is rarely commented on was noted by J. Medcof in his work on the closely related American Eel. He firstly recorded 'eels lolling on weed ... Many eels lay motionless, looped in inverted "U"s over stems of these water plants ... By the end of July great numbers of these lolling fish were regularly seen ...' Secondly: 'Eel balls on bottom ... On August 17 three spherical clumps of eels (diameters about 0.5m) were sighted. ... The fish were knotted together in tight, complicated, motionless masses clustered about the bases of rushes at a depth of about 1.5 m.' Finally: 'Free-floating balls. Several experienced fishermen stated that they had seen "eel balls" ... One was about 2m in diameter - "big enough to fill two barrels". Flipping tails on the outside of the ball apparently caused slow rolling.'

In Europe, Fries and his colleagues had previously noted that 'Frequently the migrating eels knot themselves together in bunches, and large bundles, often a fathom in circumference, are seen lying in the lakes or trundling down the streams .... Thus it appears that both North American and European Eels form on-bottom resting balls and free-floating balls.'

Medcof further recorded fascinating local attitudes in Canada to the so-called eel balls: 'Another fisherman ventured the opinion that free-floating balls at the surface might be the basis for the traditional "great beast of Lake Ainslie." This monster was known to the Gaelic-speaking of the community as "beathach mor Loch Anaslaigh." Tales of it were common among the older folk.' Unknown to Medcof, in Scotland more than half a century earlier, Morrison had noted that: 'In the Island of Lewis there is an interesting legend about a monster fish living in some of the deep lochs ...The fish is known as "Seilich Uisge" (The Water Brute). I am quite certain that this allegory had its origin in an eel or eels ... which grew to a gigantic size. About eighteen years ago an eel measuring five feet in length was caught in a loch in Lewis. ...

The Loch Ness monster may be one of these huge eels. I have put forward this theory in an article in *The Glasgow Herald* on 6th December, 1933.'

Food

Edward Boulenger believed that 'The growing eel is voracity personified, and having exhausted the resources of a given lake or stream, travels overland under cover of night, or in flood time, and so discovers new worlds to conquer.' Foraging for food takes place mainly after dusk. Their sense of smell is very acute, a feature that is exploited by Eel fishermen whose baited traps can attract Eels from a large area by day or night. The diet of Eels is catholic, but the relative proportions of invertebrates to fish taken varies with the size of the individual and from population to population. Small Eels live mainly on invertebrates including midge larvae, crustaceans and molluscs. Larger Eels take an increasingly larger proportion of fish. In other situations, stream-living Eels over 20 cm take significant numbers of fish whilst in yet other waters large Eels rely mainly on invertebrates, though they will take large animals such as frogs if they get the chance.

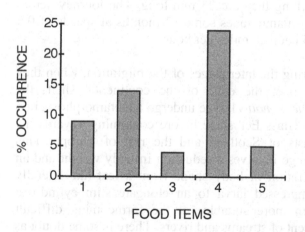

Food of Eels in the Dhub Lochan (1: mayfly larvae; 2: alderfly larvae; 3: fly larvae; 4: fish; 5: newts) (Shafi & Maitland 1972).

In Loch Tummel, Eels were found by Niall Campbell to be feeding extensively on small aquatic worms which live under stones or buried in the substrate - organisms hardly ever found in the stomachs of Trout and Perch which co-existed with the Eels there. Other items in the diet of Eels there included a wide range of invertebrates such as freshwater slaters and leeches. While diving in the River Dee at Glen Tanar, on 1st September, 1993, Ross Gardiner saw an Eel (ca 35-40 cm) actually take - but not swallow - a salmon parr from under a rock. The Eel then swam around with the parr in its mouth and was so engrossed in trying to swallow it that Ross was able to catch it in his hand and hold it for a short while. Even then, it did not release the parr!

On warm still summer nights, Eels will push their heads out of the water along suitable shores to seize invertebrates such as emerging caddis flies. In doing so they make a distinct sucking sound as the prey is taken in. This behaviour has also been recorded in the American Eel under similar conditions and the sound described as 'chirping'. During winter, as the water temperatures fall, Eels become progressively less active and pass the season relatively moribund, concealed somewhere on the bottom and probably not feeding.

Norman Morrison recorded that '... on the west side of the Island of Lewis, I witnessed a remarkable phenomenon. There is a stream infested with eels passing through this township where housewives are in the habit of gutting fresh herrings.. One day while watching this operation, a swarm of eels gathered in the pool and began to feed voraciously on the entrails of the fish. These creatures were so bold that they actually seized the herring in the woman's hand. In order to test their boldness ... I picked up a length of gut and walked down to the other end of the pool, ten yards from where we were standing. I laid down my bait four feet from the edge of the water ... Exactly three minutes afterwards, a large eel came out of the pool, wriggled up the slope towards the offal, and in a flash seized it and slid back into the stream ... This is surely a singular illustration, not only of the extraordinary smell of the fish, but of their daring and courage when foraging for food.'

Eels are considered by some to be very damaging to salmonids and Peter Malloch went so far as to state 'I consider the eel by far the greatest enemy that salmon and trout have.' Though Eels do take small Salmon and Trout and their eggs, these are not an important item of their diet. John Colquhoun found that 'Fish ... are also apt to prey upon their own species. I had ocular proof in the summer of 1859 that the fresh-water eel is as great a cannibal as the pike. One of my children set a line, baited with an earth-worm, in the burn that runs past Scalastal, in Mull. A small eel took the bait, and was itself swallowed by a big one. Although untouched by the hook, the large eel was pulled ashore before disgorging its tiny neighbour. It was nearly as thick as a man's wrist, and weighed five pounds.' The author has noted that Eels seem particularly fond of both Powan and Arctic Charr. When either of these species is caught up in gill nets during summer they may be attacked by Eels, which eat first the soft parts and then strip off the muscle. Sometimes only a head remains and the only trace of Eel is mucus in the net meshes. Brown Trout, Pike or Perch in the same nets are usually undamaged.

Predators

Eels, particularly small ones, have many predators, being a common prey of Herons, Goosanders and Red-breasted Mergansers as well as other fish such as large Eels, Pike and occasionally Trout. Eels are often the main food of Otters and they are also eaten by Mink. J.A. Harvie Brown & T.E. Buckley noted that 'It is amusing to see a cormorant with a large eel, the head of which has disappeared down its throat, but whose tail is tightly curled round the bird's neck.'

They are fairly tolerant of pollution of different kinds and are often one of the last fish to disappear as a river deteriorates. In those areas of upland Scotland affected by acid deposition, they were found by the author to be the only fish left after the others have been eliminated. Eventually, of course, they too disappear with the increasing acidity.

Eels have various of internal parasites. Those noted by Winifred Frost in Cumbria include roundworms (Nematoda: *Raphidascaris cristata*), tapeworms (Cestoda: *Bothriocephalus claviceps)*, flukes (Trematoda: *Diplostomum* sp.) and anchorworms (Acanthocephala: *Echinorhynchus truttae*).

Reproduction

European Eels, of course, are the only freshwater fish that leave the continent to spawn elsewhere. The return migration of silver eels to the sea takes place after the maturing Eels have been in fresh water for a varying number of years. From the lochs and rivers of Scotland, Eels appear to make their

way steadily south-west across the Atlantic to the Sargasso Sea, somewhere to the east of the Bahamas and south-west of Bermuda, where they spawn, it is said, at depths of 100-200 m, and die. Spawning is thought to take place from February until May. Each female produces several million eggs which are about 1 mm in diameter and said to be buoyant, floating about in the sea until they hatch. However, no spawning European Eel, or one about to spawn, has ever been caught nor has any pelagic egg ever been found.

The smallest European Eel larvae found so far, caught at depths of 90-270 m, measured less than 6 mm and were still bearing the remains of their yolk sacs. The young, known as *Leptocephalus* larvae, are elongate and flattened ventrally, with small pointed heads and relatively large mouths with many teeth. They were originally thought to be a separate species of fish. Dick Shelton describes a single specimen caught on 9 June, 1996, close to the surface above the 1,000 metre on the edge of the Scottish continental shelf south of the Wyville-Thomson Ridge. It was 'virtually transparent apart from the thread-like gut along the ventral edge and the eye.' At this stage they feed mainly on small planktonic animals, and within two months of hatching they are 25 mm long. The journey across the Atlantic takes some 12 months at speeds of 0.5 to 5 nautical miles per hour.

During the later stages of the migration, when they are over the edge of the continental shelf, the *Leptocephalus* larvae undergo a metamorphosis into the 'Glass Eel' stage before continuing towards the coasts of Scotland and the rest of Europe. This change involves a reduction in body weight and an alteration in body shape from a leaf-like laterally compressed larva to an elongate slim cylindrical form more suitable for the forthcoming difficult ascent of streams and rivers. There is some doubt as to how the young fish move in the sea and Gordon Williamson poses the question: 'Do glass eels and leptocephali normally drift in vertical positions?' As the larvae approach coastal waters, pigmentation develops and they become 'elvers'. After a period of adjustment in brackish water they swim against the current and set off on their invasion of fresh water.

Growth

The age of European Eels is usually determined by

an examination of their otoliths or 'ear-bones', which are leaf-like in shape with a conspicuous shoulder on one side, but in many cases these are difficult to interpret and sometimes only a proportion of the fish in one sample can be aged. Scales have sometimes been used too but this method has even more shortcomings. Scales first appear when the Eel has reached a length of 16-18 cm, after anything from 2-6 years in fresh water, appearing in succession in different areas of the body. Considerable skill and experience is required to use either method.

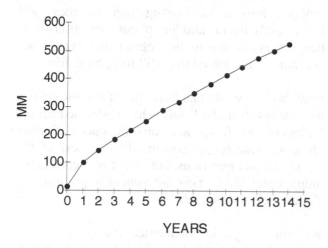

Growth of Eels in the River Barrow, Ireland (Moriarty 1984).

Eels start to mature by 8 years of age but some may still be feeding in fresh water after 36 years. Some Scottish yellow Eels have been aged at over 50 years of age. However, on average, in Scotland, silver Eel males generally have spent some 7-12 years in fresh water and average about 36 cm in length, while silver Eel females have spent 9-19 years and average around 46 cm.

Value

It is during the migration of silver Eels that the species is most vulnerable to harvesting by humans, and great numbers, weighing thousands of tonnes, are taken annually across Europe in traps and nets set to catch them during their annual rapid passage downstream. Peak movement takes place on dark moonless nights when river levels are high. Commercially, Eels are one of the most important freshwater fish in Europe and they are caught in enormous numbers in fyke nets and traps of various kinds. All sizes may be cropped by humans. Elvers

are trapped in huge numbers, formerly for direct human consumption (as 'elver cakes') but now mostly for stocking ponds and polder lakes for growing on.

In 1873, Frank Buckland recorded an extract from his official report to the Home Office on Scotch Salmon Fisheries: "I was much surprised, during my inspection of salmon rivers in Scotland, not to find any apparatus at work for catching eels. I venture, therefore, to state my opinion that the eel fisheries of Scotland should at once be worked. If this were done in a practical manner, I feel convinced that a considerable revenue from eels might be derived by individuals, and the public at the same time benefitted by an increased supply of good food, now neglected! It cannot be otherwise than clear that thousands of tons of eels are allowed to escape down into the sea during the autumn months from the rivers and lakes of Scotland! ... Why should all these tons of eels be lost to the public?'

'I trust therefore, that Scotch salmon fishery proprietors will see the importance of utilising the eels which they now allow to pass them. Eel fisheries, if properly and scientifically carried out, will not interfere with the salmon fisheries.'

Richardson recorded, for an English river, that 'Immense quantities of silver eels are taken in August, September and October, in nets at Eel-Stank, about half-a-mile down the River Eamont. In five or six hours, eight or ten horse loads have been caught; but such large quantities only on the darkest and stormiest nights. The largest eels commonly go last; some have weighed upwards of 9 lb.' George Bolam was less enthusiastic, mentioning '... professional fishermen at the mouth of the Tweed, where one or two men still eke out a precarious living by catching Eels for market by means of wooden or wire box-traps sunk in the river.'

Harvie-Brown & Buckley noted that in Wick, where Eels were abundant in the harbour, 'The inhabitants disregard them, but they are a prize to the Norwegian sailors who frequent the port. These latter ... come prepared with a peculiar instrument, made of a long pole ... Into one of the edges numerous straight pieces of barbed iron are driven so close together as to hold any unfortunate eel they may come into contact with, ... Nearly a barrelful of eels is sometimes caught thus in a single night. As they cannot all be consumed fresh, they are salted, and taken to Norway, where they are considered a great delicacy.'

Having caught an Eel, there is yet a problem in handling it, as Piers of Fulham noted in 1400:

> *'What fisshe is slipperer than an ele?*
> *ffor whan thow hym grippist and wenest wele*
> *Too haue him siker right as the list,*
> *Than faylist thou off hym, he is owte of thy fyste.'*

James O'Gorman describes 'a practice adopted for killing eels ... that of spearing for them ... so extensively practised, that when the rushes grow up, there is a regular flotilla of spearmen. A man stands on his bundle, poking before him with his long-handled spear. When he takes an eel, he bites its head between his teeth, and then strings it up with a needle on a long cord. Anything so hideous as the appearance of these fellows, their faces begrimed with blood and dirt, can hardly be imagined.'

European Eels are commonly caught by sport fishermen, but they are not universally popular, for though some anglers specialise in eel fishing, many more abhor them for their snake-like appearance and slimyness and difficulty in handling. When hooked they often twist and turn in such a way as to completely tangle up the line. They are subsequently difficult to unhook and only to be killed with some difficulty. Their powers of survival are legendary among fishermen. Wilfred Morris recorded the following amusing incident in 1929, at the expense of his angling partner. 'I waded cautiously in to the fish, but coming within reach of it with the net, an amused laugh escaped me. "Man, ... it's a muckle eel!". He forthwith hauled, and soon there was wriggling on the gravel a very fine eel indeed, of between three and four pounds at least. The revulsion of feeling that angler experienced was more than he could bear, and he wept for chagrin. ... This happened on a Scottish river, where few anglers prize the eel, alive or dead, and where the majority shun the slimy, wriggling things as they would a venomous snake.'

Norman Morrison, though an eel enthusiast, noted that '... they were repugnant to the Egyptians. Their slimy, snake-like appearance and their supposed affinity with the serpent species have been chiefly the cause of this prejudice. To the same prejudice may be traced the reason why this excellent fish in Scotland is entirely rejected as an article of food.'

Their flesh is white, firm and with a distinctive flavour, and it has the highest fat content of all our fishes, being over 30 per cent of body weight in silver eels. Their calorific value is considerably higher than that of fresh-run Salmon. They may be eaten fresh, pickled or smoked. The latter are a delicacy in parts of Europe, especially The Netherlands, where the author has eaten them on several occasions. Fresh Eels should never be dropped into a hot frying pan as they will immediately uncurl and 'jump' out - an observation that has given rise to the legend that even when decapitated and gutted they still hang on to life.

Izaak Walton wrote that 'It is agreed that the eel is a most dainty dish; the Romans have esteemed her the Helena of their feasts, and some the queen of palate pleasure.' Tate Regan concurred: 'The flesh of the Eel is rich and nutritious, but sometimes considered indigestible; Silver Eels are generally preferred to Yellow Eels for the table.'

William Houghton commented that 'The Jews - excellent cooks and judges of what is good - refuse to eat the Eel at this very day, though they are perfectly aware that it has scales. Amongst the Scotch there is great antipathy to Eels; whence derived one cannot say, unless from an objection to their snake-like form. ... It would appear from Partington's British Cyclopedia, that the Scottish objection to Eels as an article of food, is mainly due to their supposed unwholesomeness.' In 1809 John Carr noted that 'The natives of the Highlands and the Hebrides still continue their dislike to eels, as an article of food, which they never touch.' Herbert Maxwell continued this belief, saying that 'Yet, although there is no more intelligent and practical peasantry in the world than the Scots, there exists among them a strong and universal prejudice against eels. The waters of Scotland abound in eels, but you will never find a Scot who does not treat with shuddering or contempt the idea of eating one. ... However, this is all nonsense of course; the eel is excellent, nutritious food, whereof the supply is in no danger of running short in the British Isles. On the contrary, the resources of our waters in the matter of eels is well nigh inexhaustible, and it is to be regretted that they are not more generally developed, ...' In a later article (1932) he wrote that 'Some day the Scottish peasantry may wake up to the partiality of their English compatriots for eels, and turn the abundance in their own waters to profitable account.'

For once, Herbert Maxwell was wrong in his historical information. In fact, Eels were at one time regarded as a very valuable species in Scotland and large numbers were both caught and eaten in some places. For example, discussing Eels in the Outer Hebrides in 1888, Harvie-Brown mentions a burn on Berneray where 'In a north wind, which produces a spate in the burn, these eels are carried down and stranded upon the sand below tide-mark, and are there collected and relished as a great delicacy.' In Fife, Rintoul & Baxter record that 'So abundant were they in the old days, that the lands of Strathendrie, in the parish of Leslie, were, before the Reformation, subject to an annual tax of some thousands of eels to the Abbey of Inchcolm.' In 1880, John Colquhoun recalled that 'Several years ago a Thames fisherman offered to take the eel-fishing of Loch Lomond for a good rent from my late brother. Though assured that he would make nothing of it, the man of enterprise stood to his proposal, and at the time of the yearly eel-migration set his eel-pots in the Leven, the effluent of the loch. He was so well repaid that he was anxious to renew the lease, sending an enormous specimen as proof of his success to Rossdhu. From his experience of the Thames, the eel-merchant knew that the only road to the sea for the migrating fish was by the river Leven, and that he could secure them on their seaward voyage.'

Henry Lamond recorded in 1931 that: 'Many years ago they were fished for in the loch by an Irish crew, who made a moonlight flitting of it and left neither cash nor information behind them.' These Lough Neagh Irishmen apparently used long lines. The fish are retained in submerged wooden boxes until a sufficient number is got to make a "draft". Six shillings a stone seems to be the usual price fetched ... the draft containing three stones weight. ...' In 1928 'They killed and sent to the London market ... 84 drafts of eels which realised £87.'

The eel fisheries of Loch Leven were particularly productive at one time. An excellent account of them has been given by Burns-Begg who recorded 'the take of eels in recent years' in stones as: 1865 -

500; 1866 - 300; 1867 - 300; 1868 - 400; 1869 - 360; 1870 - 411; 1871 - 565; 1872 - 400.' Burns-Begg further recorded that '... an annual feu-duty or feu-mail, composed partly of a stated quantity of eels, was ... payable by the proprietor of Lochleven to the Archbishop of St Andrews. .. The original letter ... dated 18th February 1573 ... is addressed "To the Right Honorable WILLIAM DOUGLAS, Off Lochleven. "Right Honorable Sir, - Eftir maist hartlie commendatione please wit I writt to you befor desyring you to haif send to me the feu mailes of the lands of Bischopschyre and also so many barrellis of elis as ye had wont to pay to my predecessouris, ... Commend me hartlie to the Lady your spouse, quhom with you I committ to the protection of the Eternall. ... Youris at power, J. Sanctandrois.' Subsequently, William Earl of Morton had also to pay the feu-duty: '... dated 4th November 1606, the eels forming part of the feu-duty are thus referred to: "8 cadi sive barellae anguillarum salsarum".

Andrew Grant, in The Old Statistical Account of Scotland (1793) recorded that 'The most remarkable fishing in this river [Leven] is that of eels. In the month of September they begin to go down from the loch in great numbers .... but only at night. When this season arrives the fishers place their nets in the river, which they draw every two hours during the night, and frequently find them full.'

The importance of these eel fisheries to local people is indicated even earlier in the Kinross Kirk Session Records: 'ACT ANENT YE EEL-FISHING' "Kinross 27 July 1701. - Sess. : mett - after prayer. The Session taking ye eel-fishing into yr consideraton finds yt's it is a manifest breach of ye Sabbath forasmuch as ye fishers travel three myles to it on Saturdays night and fishes till Sabbaths morning near sun rising, comes back al ye three myles carring yr eele eyr on yr own or horse back or oyr ways delivers ym. (to such persons as hes ym taken at a set price) at ye place qr they are taken, and again after divine service immediately returns to ye same eel-fishing travelling ye sd three myles, sets yr netts on ye Sabbath afternoon towards night, and therefor dows unanimously conclude (according to and as far as ye word of God gives light) that it ought to be put a stop to as a manifest and palpable breach of the Lord's day.'

Frank Buckland noted in The Field of June 6, 1863:

'Some time since I received, through the kindness of a friend, four cakes composed entirely of young Eels. ... These Eels, or Eelvers, ... are about four inches long ...The women catch them at night ... They are thrown into a tub of salt, which cleanses them; they are then boiled, and pressed into cakes, which are cut into slices and fried, making the most delicious food.'

Eels are curious and consequently of interest to aquarists. Ross Gardiner records that on the River Tummel, between Faskally and Loch Tummel, on 10 August 1979, he and Peter Geddes were surprised to watch an Eel of about 80 cm repeatedly come out of cover to attack the end of a glass fibre pole which was being used for depth measurements. The water temperature was 17°C.

Elvers sometimes turn up in strange places! Whilst in the "Gents" at Scrabster Pier, Niall Campbell 'was horrified at seeing elvers swimming in the urinal trough. My first thought was that I had passed them!'

Eels can live for a long time in captivity. Aalfred, a pet Eel which has lived in the bathtub of a German family for 34 years was caught in 1969 by Paul Richter. The angler took him home for supper, but his children refused to let him kill the Eel which has lived with the family ever since.

Gregory Bateman suggests that 'There are few fish, with the exception, perhaps, of the lampern, which will give less trouble in an aquarium than the Sharp-nosed or Common Eel ... Its movements when swimming are exceedingly graceful, and its wants are few and supplied without difficulty. ... The Eel should be procured when very small - about 5 in. long is a suitable size for an aquarium.' The author has kept Eels indoors on several occasions, the main requirements being regular live, or at least fresh, food - and an eel-proof cover on the aquarium.

Conservation

The Eel has no special protection in Scotland and many people would scoff at the idea that a fish which arrives all along the shores of northern Europe in such abundance would need special conservation measures. However, there is clear evidence that stocks of Eels have declined in recent years and it would be wise to consider conservation measures now so that fisheries can be managed on a sustainable basis. It is recommended that:

(a) Elvers are monitored at several rivers on both west and east coasts.
(b) Yellow Eels are monitored at several sites, possibly through commercial fishermen.
(c) Silver Eels are monitored at several sites, again possibly through commercial fishermen.

References

Relatively little research has been carried out on the European Eel in Scotland, but there is an extensive literature from other parts of Europe.

Bateman (1890), Bertin (1956), Bolam (1919), Boulenger (1946), Buckland (1873, 1881), Burns-Begg (1874), Cairncross (1862), Carr (1809), Colombo et al. (1984), Colquhoun (1878), Frost (1946), Gordon (1902), Grant (1793), Harvie-Brown & Buckley (1889), Houghton (1879), Jones (1959), Kennedy (1984), Lamond (1931), Lane (1955), Lowe (1951, 1952), Maitland (1997), Maitland & Campbell (1992), Maitland et al. (1987), Malloch (1910), Maxwell (1903, 1904), Medcof (1966), Moore & Moore (1976), Moriarty (1973, 1974, 1975, 1978, 1983, 1984), Morris (1929), Morrison (1936), O'Gorman (1845), Regan (1911), Rintoul & Baxter (1935), Sadler (1979), Schmidt (1922), Shafi & Maitland (1972), Sinha & Jones (1966, 1975), St John (1891), Tesch (1982), Tucker (1959), Walton (1653), Williamson (1987, 1995), Williamson et al. (1999), Yarrell (1836).

'It is agreed that the eel is a most dainty dish;
the Romans have esteemed her the Helena of their feasts,
and some the queen of palate pleasure.'

Izaak Walton (1653) *The Compleat Angler*

*****

Allis Shad *Alosa alosa* Gobhlachan

'a monster Herring had been taken in the fresh-water part of the Tay,
and that it weighed no less than five pounds and a half.'

Frank Buckland (1873) *A familiar history of British fishes*

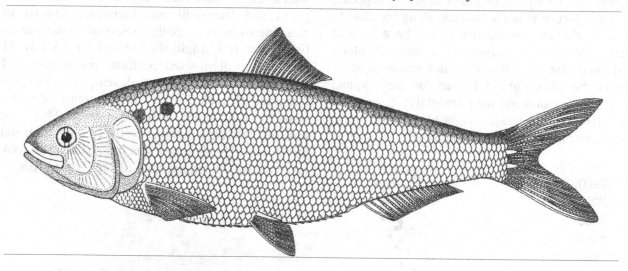

THE Allis Shad and the Twaite Shad are often confused with each other; indeed the two species are very similar - so close in fact that hybrids between the two are not uncommon. In a study of the two species in the Solway area, the author found that only a few percent of the combined population consisted of hybrids, but some recent work in Portugal has indicated that hybrids in some rivers there may account for as much as 30 percent of the population. This is believed to be a recent phenomenon caused by the severe restrictions on the spawning areas available due to engineering works and pollution in the rivers concerned, and bodes ill for the separate identity of these two species in the future.

The Allis Shad has various common names, including Ale Wife, Chad, King of the Herring, May Fish, and Shad. Its main distinction from the Twaite Shad is that the latter always has less than 90 gill rakers on the first gill arch whilst the Allis Shad always has more than 90. In the Borders, Bolam noted that 'Curiously enough, this species is often called a Twait by our fishermen, and so becomes confused with the next. It is also known locally as *Rock Herring, Herring Hake*, and *Damit-* or *Daming-Herring*. The last, it has been suggested, may be a corruption of *Dame of the Herrings*; but fishermen whom I have interrogated have always

maintained that, in whichever tense it was used, the word came from the verb "to damn", though the origin of its application they could never explain. *Queen of the Herrings* is another pet name of this Shad.' Frank Buckland noted that 'The Allice, or Allis, the Alewisse Shad ... sometimes visits the Tweed.' A literal translation of the scientific name is '*The* Shad'.

The Allis Shad is now regarded as so rare in Scottish waters that individual fish taken at sea by commercial fishermen are now recorded on a database operated by the FRS Marine Laboratory, Aberdeen.

Size

This is the larger of the two British shad, commonly growing to a length of about 30-50 cm, exceptionally to some 70 cm. The present British rod-caught record weighed 2166 gm and was caught off Chesil Beach in Dorset in 1977. There is no Scottish rod-caught record. The largest Allis Shad collected by the author from salmon netsmen in the Solway area was a female of 515mm and 2183 gm' taken at Innerwell on 29 May, 1991. The largest male was 496 mm and 1619 gm' also taken at Innerwell on 29 May, 1991. Frank Buckland records that 'In June, 1872, I received a telegram from Captain Hill of Perth, to say that a monster herring had been taken in the fresh-water part of the Tay, and that it weighed no less than five pounds and a-half. ... The basket with the fish arrived ... upon clearing out the straw, I could not help laughing to find that the herring turned out to be nothing but a jolly old common shad ... I suppose, from this fact, it is not a common thing for shad to come up the Tay, or the fish would have been at once recognised.' However, Peter Malloch confirmed that such fish were not uncommon: 'A fair number are caught by the nets on the Tay, and those I have examined have invariably been filled with spawn. The average weight of those that have come under my notice was about 5 lbs.'

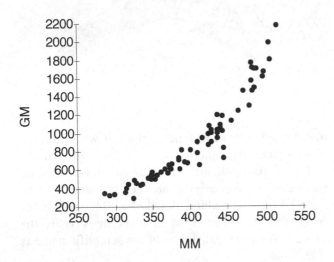

Length / weight relationship of Allis Shad caught in salmon nets in the Solway area (Maitland & Lyle 2005).

Appearance

Allis Shad, like most other members of the family, are deep bodied, laterally compressed streamlined fishes. Characteristically they have a well-developed median notch in the upper jaw (into which the lower jaw fits neatly), no teeth on the vomer bone on the roof of the mouth and more than 70 scales laterally. There is no lateral line. The jaws are equal in length and the gill covers have characteristic radiating striations on them. The eye has a curious membrane across the front and rear portions - a feature highlighted by Herbert Maxwell: 'The shad is remarkable as being the only British fish of fresh water that possesses eyelids. These are merely folds of transparent skin, opening with a vertical slit, instead of a horizontal one, as in birds and mammals.'

The scales are large and fragile, readily coming away from the fish during handling and so fish often appear damaged. The author worked on shad during the 1980s and remembers the beauty of the large (2+ kg) fresh Allis Shad taken in salmon stake nets on the Solway and their wonderful shining scales. These scales are very characteristic of the genus and the radii are transverse instead of radiating from some central focus as in most other fish. There is a single short dorsal fin with 18-21 rays and medium-sized pectoral, pelvic and anal fins. The caudal fin is well developed and deeply cleft, and characteristically some groups of scales from the caudal area extend on to it. There are normally 90-120 gill rakers on the first gill arch and this is probably the most specific character to look for, especially when trying to identify young fish.

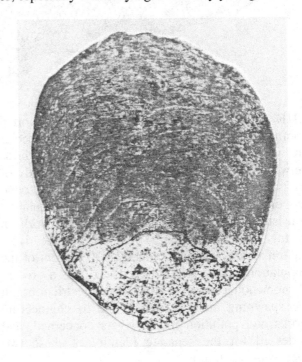

Scale from Allis Shad.

Hybrids with Twaite Shad have been identified by the author among shad in the Solway area. In 1993 and 1994, six specimens with an intermediate number of gill rakers (60-90) were found among 93 shad examined (22 Allis Shad and 65 Twaite Shad).

Distribution

This fish is found along the coasts of western Europe from southern Norway to Spain and in the Mediterranean eastwards to northern Italy. It occurs mainly in shallow coastal waters and estuaries but during the spawning migration it may migrate many kilometres up some of the larger rivers. It has suffered considerably from pollution, overfishing and river obstructions and is now a rare fish over most of its range. There are still, however, large populations in a few continental rivers; the author remembers his pleasure at seeing thousands of adult Allis Shad pass through the observation chamber of a fish lift on the River Garonne in France. Here, electronics rule, and all the fish are automatically counted and identified by shape, through computerised visual imagery.

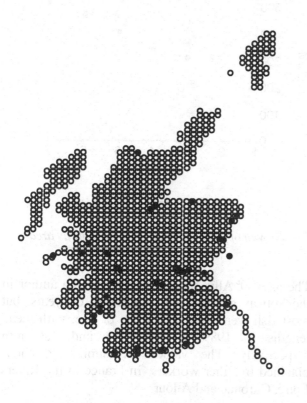

Recorded distribution of Allis Shad in Scotland.

Though William Houghton believed it to be: 'In some seasons abounding in certain parts of the Severn and the Wye.' this species is rare around the British Isles now and no spawning sites are known. In Scotland, Parnell recorded that 'it is of rare occurrence in the Firth of Forth .. it is frequently reported that Herrings of a large size, measuring from twenty to twenty-four inches in length, are occasionally taken off the Dunbar and Berwick-shire coasts, and which fishermen name the Queen of the Herrings, but that it is probable that the fish they allude to is the Allis Shad.' Herbert Maxwell

noted that 'Shad occasionally ascend the Tay, the Tweed and other rivers on the east coast whilst George Bolam writes that it was 'Frequently taken at the mouth of the Tweed and along the coast on both sides of the Border. I have had specimens from Shields, Amble, Boulmer, Holy Island, and Eyemouth, all caught during September, which is the usual month of its appearance with us.'

In fact, the species still occurs around the Scottish coasts and is taken from time to time by commercial fishermen. It seems to be most common, however, in the Solway area where a study by the author used specimens taken as a bycatch in the salmon nets.

Ecology

W. Aflalo notes that 'Connected by many honest fishmongers with the salmon ... the Shad certainly resembles that fish in its anadromous tendencies ...' In those few rivers on mainland Europe where Allis Shad still spawn, mature fish run up from the estuaries into rivers during late spring, thus giving it the name of May Fish in some areas. Dennys recalls it as 'The Shad, that in the spring time commeth in ...' In some of the larger European rivers it has been known to ascend upstream for several hundred kilometres. Shoals of fish accumulate in suitable pools and spawning takes place there at night. Afterwards the spent adults drop downstream to the sea again, many of them to die there.

Food

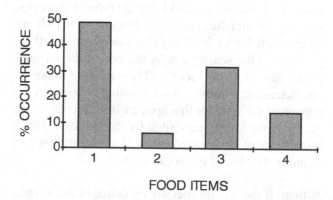

Food of Allis Shad in the Solway area (1: copepods; 2: mysids; 3: fish; 4: vegetable fragments) (Maitland & Lyle 2005).

The food of the young fish is mainly bottom-living river invertebrates, especially midge larvae and crustaceans. The food of adults in the Solway area, feeding in salt water, consisted largely of invertebrates, especially planktonic crustaceans (e.g. calanoids and euphausids), but also to some extent of larger crustaceans (e.g. mysid shrimps) and small fish such as Sprat and Herring. Fine vegetable fragments were common in the stomachs, attributed to the filtering method of feeding.

The major item of diet was clearly zooplankton, especially *Calanus*, and at times the stomachs of some fish were distended with many thousands of these small crustaceans. In one Allis Shad, taken in July, 1993, the stomach contents represented over 10% of the weight of the fish.

Reproduction

The clear eggs (about 4.4 mm in diameter) fall to the bottom, remaining there in crevices until they hatch some 4-8 days later. The fry are about 10 mm on hatching but rapidly grow to 8-14 cm after one year. By this time many of them have descended to the sea and the remainder follow during their second year. The adults mature after 3-4 years at about 30-40 cm when they start the spawning migration again.

Tate Regan gave a good account of the their spawning behaviour: '... they usually breed in May or a little later (hence the German name *Maifisch*), and make a considerable commotion, swimming rapidly at the surface, and pressing together and thrashing the water with their tails during the process of deposition and impregnation of the eggs, which sink and lie free at the bottom. The fish are so exhausted after breeding that many perish; others get back to the sea, so that by the end of the summer none are to be seen.' The adults make a characteristic noise at spawning, and in a conservation area for this species in France at Agen on the River Garonne, visited by the author in 1994, recordings of this sound taken at night are used to monitor the size of the population.

Although there is no specific evidence of Allis Shad ever spawning in Scotland, R. Service, who was a reliable observer of detail in the Solway area, noted that 'The shad or rock herring was also another of the fish that came into fresh water to spawn.' Further work is needed on this topic in Scotland, especially in the Solway area.

Predators

Various parasites have been recorded from Allis Shad. In the Solway, fish were commonly found to have both nematodes and trematodes in their stomachs. Externally, the crustacean parasites *Caligus elongatus* and *Lepeophtheirus salmonis* were found attached to the skin.

Growth

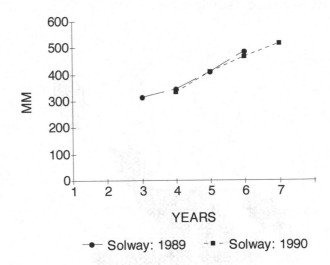

Growth of Allis Shad caught in the Solway area
(Maitland & Lyle 2005).

The ages of Allis Shad examined by the author in the Solway area ranged from 2+ to 6+ years, but most fish were 3, 4 or 5 years of age, with mean lengths in 1989 of 345, 412 and 486 mm respectively. The values are similar to those obtained by other workers in France in the Rivers Loire, Garonne and Adour.

Value

Though this fish is highly esteemed for the table in several countries there are relatively few fisheries for it nowadays, largely because it is so rare. Some of the larger French rivers still have local fisheries using special lift nets. However, opinions on its culinary virtues are not universally high; it has been described as "a plebian fish excluded from all reputable banquets", and (in Northern Ireland) as "the bony horseman". William Houghton had a higher opinion: 'The flesh, as an article of diet, is fair, and though far inferior, in my opinion, to that of a Herring, it is nevertheless good food.'

In 2006, a number of shad were seen by Alex Lyle for sale in a fish shop in Port Seton. Later the same year, several shad were caught by anglers fishing in the River Forth at Stirling. One of these, caught by Rab Brown on 10 August, was an Allis Shad. Alan Ayre reports that 'They have been caught on a variety of lures including salmon flies, toby spinners, and one on a worm.'

Conservation

Like many large anadromous species, it is particularly vulnerable during the spawning migration both to local fisheries, and to the hazards of pollution and obstructions so common now in the large European rivers in which the species was once abundant. Such problems have been known for over a century – William Houghton pointed out in 1879 that 'The Severn navigation weirs prevent the ascent of Shad and Flounders beyond certain parts of the river;' Many of the spawning grounds have been destroyed by river engineering and bad management and so the author was delighted to see the nature reserve which has been created on the River Garonne, solely to protect the spawning grounds of Allis Shad there.

It is unclear why the Solway Firth should be so favoured by Allis Shad if they do not spawn in the area. Perhaps the reason is similar to that postulated by Henry Thoreau for a similar situation in North America: 'Still patiently, almost pathetically, with instinct not to be discouraged, not to be reasoned with, revisiting their old haunts, as if their stern fates would relent, and still met by the Corporation with its dam. Poor shad! Where is thy redress? When nature gave thee instinct, gave she the heart to bear thy fate? Still wandering the sea in thy scaly armour to inquire humbly at the mouths of rivers if man has perchance left them free for thee to enter.'

The Allis Shad is finally being given special protection in several countries in Europe, and hopefully may recover in at least parts of its former range. In Scotland, the Allis Shad is given limited protection under Section 9(1) (Intentionally killing, injuring or taking it) and Section 9(4)(a) (damage or destruction to any structure or place they are using for shelter or protection) of Schedule 5 of the Wildlife & Countryside Act 1981. It is also listed in Annexes IIa and Va of the EC Habitats Directive and Schedule 3 of the Conservation Regulations 1994.

Conservation action for this species is needed in Scotland and the following actions are suggested as an important part of any programme:

(a) Intensify research on this species in the Solway area to clarify its status there.
(b) Monitor all catches around Scotland by commercial fishermen.
(c) Collaborate with European geneticists to identify the origin of stocks around Scotland.

References

Apart from the work carried out in the Solway area by the author and Alex Lyle, little research has been carried out on Allis Shad in Scotland and we are dependent for other information on work carried out elsewhere in Europe.

Aflalo (1897), Bolam (1919), Buckland (1873), Dennys (1613), Gordon (1920), Houghton (1879), Maitland (1977, 1989), Maitland & Hatton-Ellis (2003), Maitland & Lyle (2005), Malloch (1910), Maxwell (1904), Parnell (1838), Regan (1911), Service (1907), Thoreau (1889), Wheeler *et al.* (1975).

'... and it is as quick of hearing and as fond of music as the Twait Shad.'

W.J. Gordon (1920) *Our Country's Fishes*

*****

Twaite Shad *Alosa fallax* Gobhlachan

*'It is quick of hearing, frightened at thunder, and so fond of music that
the Germans attach little bells to their shad nets,
which ring under water, and not only attract the fish,
but keep them lost in admiration as the nets are drawn in!'*

W.J. Gordon (1920) *Our Country's Fishes*

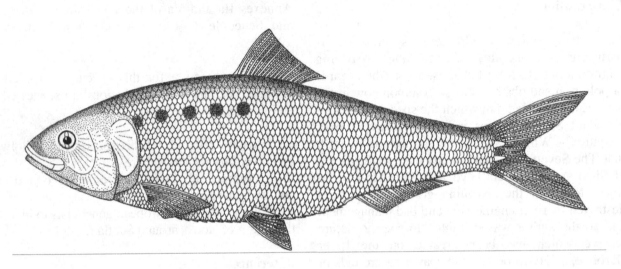

ONE of the problems with old records of shad is to know which species was involved. As William Houghton noted many years ago: 'The two British species of Shad are ... so similar ... in general form and appearance, that they have, both in ancient and modern times been frequently confounded.' As recorded in connection with the Allis Shad, this similarity is emphasised by the fact that the two species regularly hybridise in the wild. The author remembers his surprise at finding several hybrids of Twaite and Allis Shad among specimens taken from salmon stake nets on the Solway during the 1990s; this was the case too with fish taken in 2004. In addition, there can be the problem of confusing local names or misidentification by fishermen. For example, a number of years ago, one fisherman in the Solway area produced a batch of 'shad' for the author's research which turned out to be Scad *Trachurus trachurus*, a totally unrelated species.

Like the Allis Shad, the Twaite Shad has a variety of common names, including Bony Horseman, Chad, Goureen, Herring Shad, Killarney Shad, May Fish, Queen of the Herring, and Shad. Its main distinction from the Allis Shad is that the latter always has more than 90 gill rakers on the first gill arch whilst the Twaite Shad always has less than 90. Many accounts indicate that Twaite Shad can be distinguished from Allis Shad by the number of spots on the sides, the latter having only one or two. The author has found that this is not the case and agrees with George Bolam that 'The spots are, however, very inconsistent in either species.' Bolam goes on to point out that, in the Borders, 'Our fishermen do not distinguish between the species; but the most spotted fish are usually called *Twaites* or *Herring Hakes*, those with only single spots on the shoulders *Shads* or *Daming Herrings*.' A literal translation of the scientific name is 'Deceptive Shad'.

The Twaite Shad is now regarded as so rare in Scottish waters that individual fish taken at sea by commercial fishermen are now recorded on a database operated by the FRS Marine Laboratory, Aberdeen.

Size

Though both are large herring-like fishes, the Twaite Shad is normally rather smaller than the Allis Shad, adult fish usually averaging some 25-40 cm with a maximum of about 55 cm. The present Scottish shore record is for a fish of 0.765 kg (1 lb 11 oz) caught by J. Morrison at Garlieston in 1972. The British rod-caught record stands at 1.417 kg for two fish - one caught in 1949 near Deal in Kent and the other in 1954 near Torbay in Devon. The most recent notable catch, which appears not to have been ratified, was a fish of 1.247 kg caught in 1978 at Garlieston in the Solway Firth.

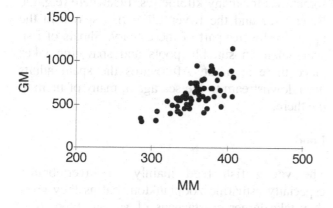

Length / weight relationship of Twaite Shad caught by salmon netsmen in the Solway (Maitland & Lyle 2005).

The largest Twaite Shad caught in the Solway by salmon netsmen and made available to the author for study was a female 429 mm in length and 1.032 kg in weight, taken at Carsluith on 23 July, 1993. The largest male was 403 mm and 0.75 kg, taken at Portling on 26 May 1994.

Appearance

Like the Allis Shad, the Twaite Shad has a transparent membrane across the front and rear parts of the eye, no teeth on the vomer bone on the roof of the mouth and a marked median notch in the upper jaw into which the lower jaw fits, but there are only about 40-60 rakers on the first gill arch and less than 70 lateral scales. The gill covers have a characteristic radial sculpture and, like most other members of the family, the body is rather flattened laterally. There is no lateral line.

The ventral scales are characteristically keeled - especially in the region behind the pelvic fins. In colour, the Twaite Shad is usually a dark steely blue along the back which grades into a silver yellow on the sides and then into a silvery white on the belly. There are usually about 5-10 round dark spots on

each side of the back, starting just behind the head and decreasing in size posteriorly.

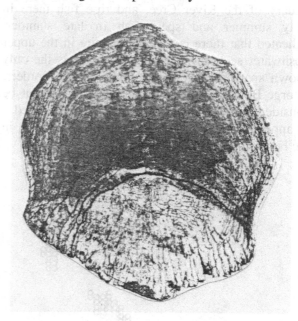

Scale from Twaite Shad.

Hybrids with Allis Shad have been identified among shad in the Solway area. In 1993 and 1994, six specimens with an intermediate number of gill rakers (60-90) were identified among 93 shad examined (65 Twaite Shad and 22 Allis Shad).

Distribution

The Twaite Shad occurs along most of the west coast of Europe, from southern Norway to the eastern Mediterranean Sea and in the lower reaches of the large accessible rivers along these coasts. With the exception of isolated populations in large lakes, the normal habitat of this species is the sea - especially coastal waters off the southwest coast of northern Europe. It has been found at depths down to 100 m but it normally occurs in water much shallower than this. The eggs and young are found in the lower reaches of large slow-flowing unpolluted rivers where there is easy access from the sea. In general, Twaite Shad do not migrate upstream as far as Allis Shad, often spawning in the freshwater tidal section of estuaries.

In Scotland, Twaite Shad seem to occur all round the coast and are caught occasionally by commercial fishermen. Adults come into most estuaries less frequently than previously, but are still moderately common in the Solway where the author and Alex Lyle carried out a study of the two shad species. Twaite Shad were commonest in the estuary of the River Cree, and ripe fish there in early summer and spent fish in late summer, indicated that there is a spawning site in the upper freshwater section of the estuary. This is the only known spawning site in Scotland. In the Borders, George Bolam recorded that it 'Must at present be considered rare in the district. ... Parnell recorded it as appearing in July and disappearing in August in the Forth; '

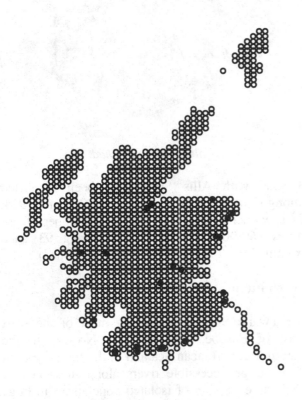

Recorded distribution of Twaite Shad in Scotland.

There are a number of extremely interesting non-migratory populations of this fish in a few of the larger European lakes such as Como, Garda, Iseo, Lugano and Maggiore. In the British Isles, Lough Leane in Killarney has such a population which appears to have been isolated here for thousands of years. It is known as the Goureen and was given subspecific recognition by Tate Regan in 1916 as *Alosa fallax killarnensis*. This fish is rather smaller than its marine based relatives, rarely growing longer than about 25 cm. When the author visited Lough Leane in the late 1990s he was concerned to learn of major threats to this unique population -

pollution, eutrophication and general lack of catchment management.

Ecology

In those rivers where Twaite Shad still spawn, mature fish run up from the estuaries during late spring, thus - as with the Allis Shad - giving it the name of May Fish in some areas. In a few rivers (e.g. the River Wye) it ascends well upstream, sometimes for many kilometres. Elsewhere (e.g. the River Cree and the River Elbe) fish spawn in the upper freshwater part of the estuary. Shoals of fish accumulate in suitable pools and spawning takes place there at night. Afterwards the spent adults drop downstream to the sea again, many of them to die there.

Food

The young fish feed mainly on invertebrates, especially estuarine zooplankton, but as they grow they take larger crustaceans of various types (e.g. shrimps and mysids) and also small fish. The latter was the main food item found in the stomachs of Twaite Shad in the Solway Estuary, where adults feed to an appreciable extent on other fish, especially the young of other members of their own family such as Sprat and Herring. Plant debris, though present in some stomachs was an unimportant part of the items identified.

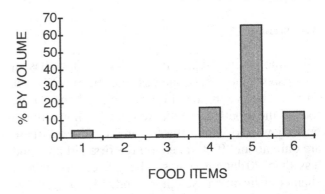

Food of Twaite Shad in the Solway area (1: molluscs; 2: copepods; 3: ostracods; 4: mysids; 5: fish; 6: other food items) (Maitland & Lyle 2005).

Reproduction

At maturity, the adult fish stop feeding and gather in the estuaries of rivers in early summer (April and May), thereafter moving upstream to spawn in mid-

June in the stretches above the influence of high tide. Usually the males move upstream first, but they are soon joined by the females and spawning takes place in flowing water over stones and gravel among which the eggs sink. Twaite Shad are fairly obvious, and vulnerable to anglers, when on the spawning grounds; the author has watched them in the River Usk chasing each other with much splashing. Frank Buckland noted that 'In the language of fishermen, the shad are said to thrash the water with their tails, and on a calm still evening, or night, the noise they make may be heard at some distance.' The females, depending on their individual size, each produce about 75,000-200,000 eggs. The eggs take about 4-6 days to hatch, and the young drop quickly downstream in the current to the quieter waters of the lower river and upper estuary where they start to feed and grow. The non-migratory populations seem to spawn in, or in the vicinity of, large rivers entering the lakes and the young soon find their way into the lake.

Though the author and Alex Lyle spent many hours looking for juvenile Twaite Shad in the Solway area we were unsuccessful. Juvenile fish were, however, taken by the author in the River Usk in Wales; hundreds of both Twaite and Allis Shad juveniles were subsequently collected during a trip with a shrimp boat in the Gironde Estuary in France.

Predators

Various parasites have been recorded from Twaite Shad. In the Solway, fish were commonly found to have nematodes and trematodes in their stomachs, while externally, the crustacean parasites *Caligus elongatus* and *Lepeophtheirus salmonis* were found attached to the skin.

Growth

Growth in the first year is fairly rapid: they can reach some 5 cm in 6 months and 10-15 cm after one year. Thereafter, growth is steady, most fish reaching about 20-25 cm after 2 years and 25-30 cm after 3 years. The males start to mature after three years and are therefore spawning with older and larger females at first. The females themselves do not start to mature until they are about 5 years old. Full size is usually reached at 8-10 years of age and some fish may live (and grow slowly) for several years beyond this.

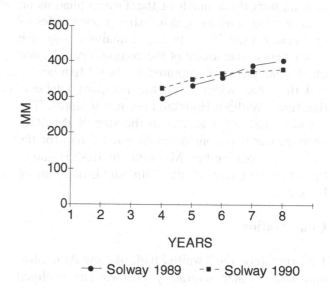

Growth of Twaite Shad caught in the Solway area (Maitland & Lyle 2005).

In the Solway area, the ages of Twaite shad collected from salmon nets ranged from 3+ to 7+, most fish being 4, 5, or 6 years old, with mean lengths in 1989 of 332, 357 and 390 mm respectively. These values agree with studies elsewhere in the British Isles and Europe. No juvenile fish have ever been seen by the author in Scotland, in spite of the strong evidence that adults do spawn in the estuary of the River Cree.

Value

Though it is now much less common than formerly, the species is still fished commercially in some parts of Europe with various types of nets, especially in estuaries during the start of the spawning migration. Some of the catch is smoked before sale. The species is also angled seasonally in some rivers which still have a reasonable run of fish (e.g. the Rivers Wye and Severn) where it is usually taken on a small metal spoon. A fish of 450 g is considered a good specimen. The smaller Goureen is sometimes taken by rod and line in Lough Leane in Killarney.

In 2006, a number of shad were seen by Alex Lyle for sale in a fish shop in Port Seton. Later the same year, several shad were caught by anglers fishing in the River Forth at Stirling. Alan Ayre reports that 'They have been caught on a variety of lures including salmon flies, toby spinners, and one on a worm.'

Few authors think much of the Twaite Shad as an item of food. Ausonius, discussing poor articles of diet commented 'Who is not acquainted with the green Tench, the solace of the common people, and the Bleak, the spoil captured by boys' fish-hooks, and the Shad which is in the fire-pans, plebeian fish-fare?' William Houghton seemed to agree: 'The Twaite Shad never attains to the size of the other species, nor is it considered as good a fish for the table.' - as does Herbert Maxwell: 'Its flesh is said to be inferior to that of the Allis shad, and full of bones, ...'

Conservation

Unfortunately, the Twaite Shad, like the Allis Shad and many other migratory species has declined dramatically in Europe. Across the Atlantic, a similar position exists, as indicated by Henry Thoreau many years ago: 'Salmon, Shad, and Alewives were formerly abundant here, and taken in weirs by the Indians, who taught this method to the whites, by whom they were used as food and as manure, until the dam, ... put an end to their migration hitherward.'

The Twaite Shad is given limited protection in Scotland under Section 9(4)(a) (damage or destruction to any structure or place they are using for shelter or protection) of Schedule 5 of the Wildlife & Countryside Act 1981. It is also listed in Annexes IIa and Va of the Habitats Directive and Schedule 3 of the Conservation Regulations 1994.

In Scotland, more action is required than just legislation, and in particular, further research is needed on the population in the Cree Estuary in order to identify the spawning areas and develop a management plan for their conservation.

Conservation action for this species is needed in Scotland; the following actions are suggested as an important part of any programme:

(a) Intensify research on this species in the Solway area to clarify its status there.
(b) Identify spawning grounds in the Cree Estuary.
(c) Monitor all catches around Scotland by commercial fishermen.
(d) Collaborate with European geneticists to identify the origin of stocks around Scotland.

References

Apart from the work carried out in the Solway area by the author and Alex Lyle, little research has been carried out on Twaite Shad in Scotland and we are dependent for information on work carried out in other parts of Europe.

Aprahaimian (1981, 1985, 1988), Bolam (1919), Buckland (1873), Claridge & Gardner (1978), Gordon (1920), Houghton (1879), Kennedy (1981), Maitland (1989), Maitland & Hatton-Ellis (2003), Maitland & Lyle (2005), Maxwell (1904), O'Maoileidigh *et al.* (1988), Parnell (1838), Thoreau (1889), Trewavas (1938).

'... *and the Shad which is in the fire-pans, plebeian fish-fare?*'

Ausonius (Undated) *Carmen de Mosella*

*****

Common Bream *Abramis brama* Briantach

'the Breme is a noble fysshe and a deynteous.'

Juliana Berners (1496) *A treatyse of fysshynge wyth an angle*

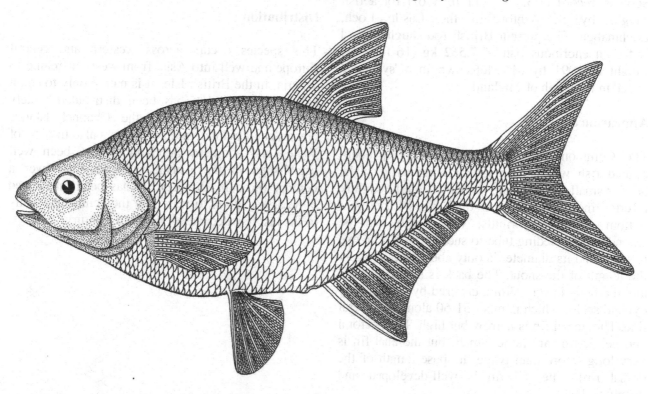

DAME Julia Berners, writing in 1496, obviously admired the Common Bream and it is certainly true that this elegant fish - the deepest bodied of any freshwater fish in Scotland - has a certain dignity and charm. The author's first experience of this species was at the Castle Loch in Lochmaben in 1964, when many of these enormous 'dinner plates' (John Flannagan's expression!) were netted during a search for Vendace there. Some of them were well over 2 kg. In the same catch there were a number of Roach x Bream hybrids – the first record of this occurrence in Scotland. Several authors have been fascinated by the Bream, including George Orwell, who wrote in his 1939 volume *Coming Up for Air*: The pool was swarming with Bream, small ones, about four to six inches long. Every now and then you'd see one of them turn half over and gleam reddy-brown under the water.'

Alternative common names, occasionally used for this species are Bellows Bream, Bronze Bream and Carp Bream. A literal translation of the scientific name is 'River Bream'.

Size

This is one of the larger members of the carp family in the British Isles, adult fish commonly reaching lengths of 30-50 cm and weights of 2-3 kg.

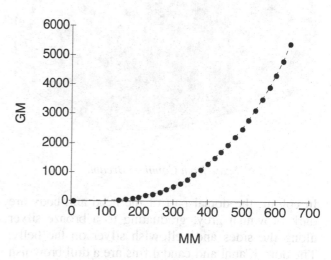

Length / weight relationship in Bream from an Irish lough (Kennedy & Fitzmaurice 1968).

Exceptionally they may reach over 80 cm in length and 9 kg in weight. Fred Buller & Hugh Falkus record a large specimen: 'Britain's sometime largest Bream: 13 lb. 12 oz found dead in Startops End Reservoir, Tring, on 19 November 1931.' The Scottish record is 5.1 kg (11 lb 4 oz) for a fish caught by S. Ashton in the Castle Loch, Lochmaben. The present British rod-caught record is for an enormous fish of 7.512 kg (16 lb 9 oz) caught in 1991 by M. McKeown in a 'syndicate water' in the south of England.

Appearance

The Common Bream is an exceptionally deep-bodied fish with strong lateral compression. The head is small and there are no barbels on the slightly inferior mouth, which has thick lips. Like other bottom feeding cyprinids, the mouth can be extended as a feeding tube to suck in mud. The eye is small and its diameter is only about two-thirds of the length of the snout. The back is rather humped and the body is very slimy, covered by large strong cycloid scales which number 51-60 along the lateral line. The dorsal fin is narrow but high. The pectoral and pelvic fins are rather small, but the anal fin is very long (more than twice the base length of the dorsal fin). The tail fin is well-developed and deeply forked.

Scale from Common Bream.

In colour, the dorsal part of the head and body are dark brownish grey, graduating to a bronze silver along the sides and yellowish silver on the belly. The dorsal, anal and caudal fins are a dull brownish grey as are the anterior edges of the pectoral and pelvic fins. Small Common and Silver Bream (not

yet recorded in Scotland) can be confused with each other; fin ray and lateral line scale counts may be necessary for accurate identification. Also, the eye of the Common Bream is set further back in the head than the Silver Bream and is relatively smaller.

Distribution

This species occurs across western and central Europe and well into Asia, from western Ireland to Siberia. In the British Isles it is native only to south eastern England but has been distributed widely across England, including the Channel Islands (Jersey) and Ireland, and now occurs also in parts of Wales and southern Scotland. It has been well established in the Annan catchment for over a century (probably longer) but has been introduced by coarse anglers to waters in the central belt only in the last few decades of the 20th Century.

Recorded distribution of Common Bream in Scotland.

Ecology

The Common Bream is a fish of lowland areas, preferring rich, muddy, weedy lakes, reservoirs and slow flowing rivers and associated canals - also the favoured habitats of Silver Bream, Carp and Tench. W.J. Gordon says 'The Bream is gregarious, so that where there is one there are generally more.' Its ability to withstand very low levels of oxygen means that it can survive in poor conditions for considerable periods. In some parts of Europe it occurs in slightly brackish waters (in the Baltic Sea) but always migrates back into fresh water to spawn.

During the winter, like Silver Bream, Carp and Tench, it seeks deeper water and may occur in densely packed shoals sheltering in suitable places, often under ice cover, existing on the fat reserves built up during the summer. Successful recruitment in Scotland depends on warm summers. No growth takes place during the winter.

Food

At first the young feed mainly on small plankton, especially rotifers, protozoans and small crustaceans, but they tend to move to a more benthic diet after a few weeks, for which their extrusible mouth is especially adapted. When feeding the fish work their way along the bottom with a characteristic oblique head-down posture pushing the tubular mouth into the bottom to detect and expose suitable food items.

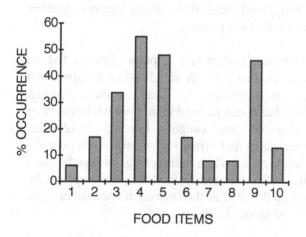

Food of Bream in an Irish Lough (1: Cladocera; 2: Asellus; 3: Gammarus; 4: midge larvae; 5: caddis larvae; 6: mayfly larvae; 7: corixids; 8: beetles; 9: molluscs; 10: others) (Kennedy & Fitzmaurice 1968).

Molluscs are a particularly important food of large Bream, along with freshwater shrimps and slaters. This behaviour often leaves characteristic 'Bream pits' on the bottom where an individual fish has stayed for a time feeding on some particularly favoured food items, exposing a hole up to 10 cm in diameter and 5 cm or so in depth in so doing. Their general diet includes worms, molluscs, crustaceans and various benthic insects (e.g. caddisfly, mayfly and midge larvae). A small amount of vegetable matter is eaten in the form of filamentous algae and the leaves of higher plants.

Reproduction

Spawning takes place when water temperatures move above 15°C in May and June (sometimes later in cold years), when the mature fish move into shallow warm weedy areas. In Ireland, Kennedy and Fitzmaurice noted that 'the period during which confirmed bream spawnings have been recorded in Ireland is May 15 to June 9 - a period which coincides with the blossoming of the hawthorn ...' Water temperatures were between 19.5-23.0°C. The males develop numerous large whitish or yellowish nuptial tubercles on the head and body at this time and appear to be partly territorial, selecting suitable small territories and defending them against other males. Spawning itself takes place mainly during late evening and the night when mature males pursue the ripe females into the weed beds, where the eggs are laid and fertilised amidst much splashing. The yellowish eggs, which are about 1.5-2.0 mm in diameter, are adhesive and stick to the weeds down to a depth of 3 m. Each female may spawn several times over a week or so until all her eggs are laid. The actual number varies with the size of the female, but can range from 90,000 to 340,000. The young hatch in about 5-10 days depending on water temperature (9-10 days at 12-13°C).

Bream hybridise with a number of other cyprinids. Bream x Roach are probably the commonest of all cyprinid hybrids found in Great Britain, though Bream x Rudd hybrids are very frequent in Ireland. Bream x Roach hybrids from the Castle Loch near Lochmaben, were examined by the author and found to be almost intermediate in character between the two parent species.

Predators

Little has been published concerning the predators of Bream but it can be assumed that, as well as coarse fishermen, many of the 'usual suspects' (e.g. Pike, Herons, Otters) are involved.

Growth

The fry, which are about 6.0-6.5 mm on hatching stay attached to plants until the yolk sac is fully absorbed in about 7-10 days. They are then just over 8 mm in length and start to shoal and feed. Growth seems to be very variable according to local conditions, but some fish reach 7 cm in length after one year and 12 cm after two years. Growth is steady thereafter and they may reach typical adult sizes of 25-35 cm after six or seven years. They start to mature earlier than this, however, usually in the third or fourth year. In Ireland, Bream seem to mature rather later, females at 7-10 years of age at a fork length of 28 cm in slow growing populations and at 36-37 cm in fast growing situations. Males tend to mature earlier and at a smaller size. It is quite a long-lived species and fish often live for 10-15 years and sometimes for 20 years or more. A Bream of 23 years has been recorded in Ireland.

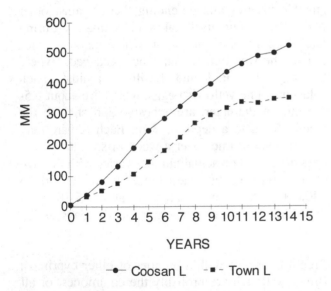

Growth of Common Bream in Coosan and Town Loughs in Ireland (Kennedy & Fitzmaurice 1968).

Value

In central and eastern Europe this is an important commercial species, caught in large numbers in seine nets and traps. In Britain, William Dugdale recorded '... that about the year 1419 a single Bream was valued at twenty pence, when the day's labour of a mason or master carpenter was less than

sixpence.' Evidently they were not held in such high esteem a few centuries later, for F.G. Aflalo noted that 'These large Norfolk Bream are much used as bait for the crab-pots on the coast.' In Sweden during the middle of the 18th Century, early fish culture techniques were developed in which ripe male and female Bream (and Perch) were induced to spawn in containers, where the eggs were hatched in large numbers, the fry being subsequently reared on in ponds.

The flesh is thought by some to be very tasty, though small specimens tend to be rather bony. Others think that the flesh is soft and muddy. Izaak Walton noted that 'Though some do not, yet the French esteem this fish highly, and to that end have a proverb: "He that hath breams in his pond is able to bid his friend welcome.",' but adds that Bream are not to everyone's taste. William Houghton agrees 'The flesh of the Bream is generally soft, insipid, and full of bones, and in little estimation for the table.' but notes that '... the quality of the flesh of various fish depends to a considerable extent on the character of the water inhabited by them, and on the season of the year.' Herbert Maxwell thought that 'the flesh of bream is of rather less than moderate repute.' The quality of flesh in Bream for culinary purposes, as in many cyprinids, varies according to the habitat in which it has recently lived. Those from slow moving muddy waters tend to taste accordingly, whereas those from cleaner, fresher waters are far better eating.

The Common Bream is a popular sporting fish in many areas, being much sought after by specimen anglers in particular. By heavy ground baiting, sizeable shoals can be held in an area while many of their number are caught. Edward Boulenger, however, notes that 'Bream have earned a reputation for cunning. ... when attempting to evade their enemies, they frequently facilitate their escape by stirring up the mud at the bottom of the lake or river in order to cloud the water.'

Gregory Bateman noted that 'The Carp-Bream ... is rather a good-looking fish, but it is not so easy to keep in health in an aquarium as any of the fish already mentioned. It should be obtained when young, and placed in a tank which presents a large surface of water in proportion to its size.' This is actually true of any of the larger cyprinids which are to be kept in an aquarium.

Conservation

The Common Bream has no special protection in Scotland and, as an introduced and potentially damaging species, does not warrant any particular measures.

References

At the moment, the geographic range of Bream in Scotland is restricted and virtually no research has been carried out here. Useful studies are, however, available from elsewhere in Europe.

Aflalo (1897), Bateman (1890), Berners (1496), Boulenger (1946), Bucke (1974), Buller & Falkus (1988), Goldspink (1981), Goldspink & Banks (1981), Gordon (1944), Hartley (1947), Houghton (1879), Kennedy & Fitzmaurice (1968), Maitland (1966), Maxwell (1904), Orwell (1939), Svardson (1950), Walton (1653), Wood & Jordan (1987).

'Ful many a fat partrich had he in mewe,
And many a brem, and many a luce in stew.'

Chaucer (1328-1400) *Canterbury Tales*

*****

Barbel *Barbus barbus* Breac-theusagach

'It is very quick of hearing,
and often makes a noise when caught,
and growls under the water.'

W.J. Gordon (1920) *Our Country's Fishes*

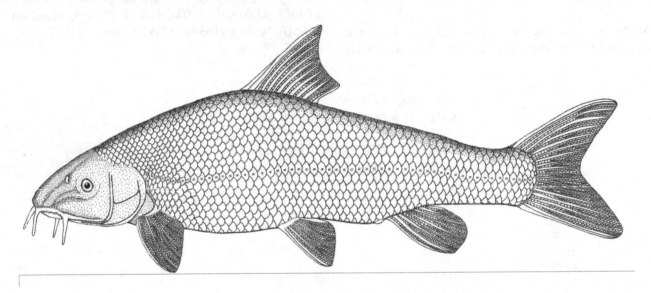

THE Barbel belongs to a very successful group of fish, found in many countries, its species seen variously as trophy fish, in tropical aquaria and on dinner plates. Herbert Maxwell notes that 'The *Barbus* genus is the largest in the carp family, comprising nearly 200 species in Africa Asia and Europe. One of its best known species is the mighty Mahaseer *Barbus mosal* of the mountain rivers of India and neighbouring countries, which can reach weights up to 150 lbs and has scales as large as the palm of ones hand.'

Frank Buckland records that 'The Barbel, called the Bearded Fish, from the barbs or wattles at its mouth, though unknown in Scotland, is frequent in England, and generally throughout Europe.' A literal translation of the scientific name is 'The Bearded Fish'.

Size

Adult Barbel are usually some 40-60 cm in length and 1-2 kg in weight, but the species can grow much larger - up to 100 cm in length and reach a weight of over 8 kg in very favourable waters. The present Scottish record is 3.969 kg (8 lb 12 oz), for a fish caught in the River Clyde in 1997 by A.G.

McLellan. Larger fish have, however, been caught since, and some of these will no doubt be accepted as records. The present British rod-caught record is for a fish of 8.55 kg (19 lb) caught in 2001 in the River Great Ouse by Tony Gibson, though again other larger specimens have actually been caught in England but not ratified.

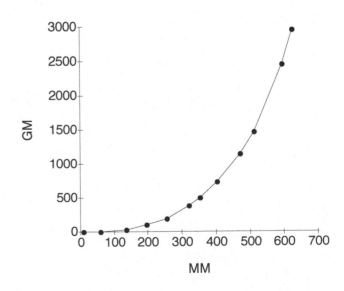

Length / weight relationship in Barbel from the River Avon, England (Hunt & Jones 1975).

Appearance

The Barbel is characterised, as its name suggests, by the possession of two pairs of long fleshy barbels on the equally fleshy upper lip, one pair (the smaller), just in front of the snout, the other at the rear angles of the mouth. The mouth is placed ventrally back from the snout. The pointed pharyngeal teeth end in short hooks and are in three rows, 5+3+2 on each side. There are usually about 17 gill rakers on the first arch. The body, evolved for active swimming, is long and rounded with very little lateral flattening, although it is rather flat along the belly, as befits a bottom-feeding fish. The body is covered in medium-sized cycloid scales of which there are 55-65 along the lateral line. The exposed edge of the scales, which are normally rather deeply embedded, is typically rather pointed.

Scale from Barbel.

The dorsal fin is high and convex on its free edge, originating just behind the mid-point of the body and with the strongest of its anterior 3-4 spines serrated behind, followed by 7-9 soft branched rays. The anal fin is long and when laid flat against the body almost reaches the origin of the tail fin. It has 3 spiny rays anteriorly followed by 5 soft rays.

In colour, the Barbel is usually a brownish-green on the head and back which grades to a golden-brown on the sides and then to creamy white on the belly. The fins are a dull yellowish-green, sometimes with an orange tinge.

Distribution

The Barbel does not occur in northern Europe but is found from western France across central Europe to the Black Sea. In the British Isles, it was, as a native species, formerly confined to the east and south-east of England but has been gradually re-distributed in England by angling interests to several southern river systems such as the Medway, the Severn and the Bristol Avon.

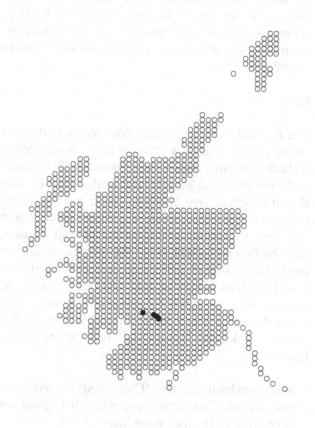

Recorded distribution of Barbel in Scotland.

It is absent from Ireland and was, too, from Scotland until very recently, when, following the successful translocations in England, introductions of adult Barbel were made by coarse fishermen towards the end of the 20th century. These were found by Willie Miller and the author to be successfully breeding in the River Clyde, as proven by juvenile fish which were starting to be caught in the year 2000. Since then, numbers of fry have been collected by Willie Yeomans and Willie Miller by netting in shallow water at the edge of the river.

Ecology

The Barbel is a bottom-living fish which occurs usually in the lowland reaches of large clean rivers, where there are stretches of clean gravel and weed beds. As Edward Boulenger described it 'Unlike carp and tench, barbel often favour fast-running streams, and during the breeding season repair to shallows where the water is clear and the bottom is stony or gravelly.' It is a very good swimmer and may also be found in other favourable parts of rivers, especially where deep pools offer some protection during the day. Fast water, by weirs and bridges, are among its favourite haunts. It tends to occur in considerable numbers in such habitats, often forming schools during the day-time resting period.

Food

The fry start to feed on small crustaceans and insect larvae changing as they grow to larger invertebrates including worms, crustaceans, molluscs (both snails and mussels), and mayfly and midge larvae. Tate Regan observed that '... they appear not to disdain any sort of animal or even vegetable matter, which they find by rooting about on the bottom ... using their barbels as feelers to aid them in the search for food.' Large Barbel will take small fish when they can. They feed mainly at dusk and during the night, moving back at dawn to sheltered resting places under banks or weed beds. During winter, Barbel move to deeper water and become relatively inactive.

Frank Buckland records 'That most observant of poets, Flood, has aptly recorded his grubbing propensities, in the following lines:'

> 'On they went, and hunted about,
> Open-mouthed, like chub or trout,
> And some with upper lip thrust out,
> Like that fish for routing – a barbel.'

Reproduction

Spawning takes place in May and June or even into July when adult fish move upstream and congregate in large numbers near the spawning grounds, which are over clean gravel and among open weed beds in flowing water. The males have by this time developed prominent facial tubercles which extend in rows on to their backs and they chase the females until mating is achieved. Some authorities say that a rough redd is prepared.

Each female can carry from 3,000-30,000 eggs depending on her size. The eggs are liberated into the water and fertilised immediately by the sperm which the males (often more than one male may accompany a female) release at the same time. The eggs are yellow, some 2-2.5 mm in diameter, sticky, and adhere to weed and to the gravel. The incubation period varies with water temperature, but normally takes some 10-15 days.

Predators

Little has been published concerning the parasites and predators of Barbel, but is can be assumed that the 'usual suspects' (e.g. Pike, Herons, Mergansers, Otters) are involved, as well, of course, as anglers.

Growth

After hatching, the young fish, which are about 7-8 mm in length, drift downstream into quiet shallow stretches of water where they begin feeding after the yolk sac has been absorbed.

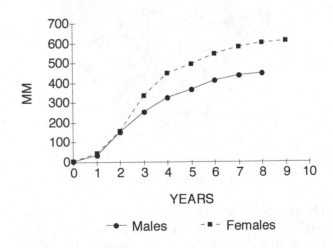

Growth of Barbel in the River Avon, England (Hunt & Jones 1975).

In good habitats the young fish may reach some 10 cm after one year and 15-20 cm after two years. They mature normally at about 4-5 years of age, the males usually about a year earlier than the females and correspondingly often smaller.

Value

In a few parts of Europe the Barbel is caught commercially, especially during the spawning migration when it may be taken in large numbers by traps and seines. The flesh is highly regarded in these areas, though bony, but the roe (and possibly even the flesh during the spawning period) is poisonous, causing severe stomach disorders. Juliana Berners in 1486 gave one of the earliest warnings about eating Barbel: 'The barbyll is a swete fysshe, but it is a quasy mete, and peryllous for mannys bodye.' Evidence is given by John Hawkins who in notes to his edition of *The Compleat Angler* which he published in 1760 'that one of his servants, who had eaten part of a Barbel, but not the roe, was seized with such a violent purging and vomiting as had like to have cost him his life.'

Tate Regan continued with further warnings '... opinions differ as to its value as food, the flesh being white and firm, but rather coarse; the eggs are more or less poisonous, sometimes inducing violent purging and vomiting, and also weakening the heart so much that fainting may result; the poisonous secretion is sometimes absorbed by the flesh of the lower part of the fish, which may thus produce similar effects, and to be safe it is best to eat Barbel only in the late summer and autumn, and to remove the roe as soon as possible after the fish is caught.' The safest option is offered by Herbert Maxwell: 'Well, and what are you to do with your Barbel when you have got him? That is just the least satisfactory part of the performance. Were barbel a culinary prize, like salmon, the sport would be a noble one; but most people account the fish fit for nothing better than to feed pigs withal.'

However, the Barbel is primarily an angler's fish and is respected for its cunning and valued for its fighting qualities in Great Britain, where it is of no commercial importance. Isaac Walton, insisting upon its cunning and the consequent difficulty of catching it, advocated 'amongst other baits, the use of cheese that had been soaked in clarified honey.' Herbert Maxwell noted that 'Altogether this fish is well equipped with organs of propulsion, which enable him to make a grand fight when hooked.' Tate Regan agreed: 'The Barbel is a strong and active, yet wary, fish, and affords fine sport to the angler.'

Conservation

The Barbel has no special protection in Scotland and, as a recently introduced species which is spreading in England, does not warrant any special measures.

References

No research has yet been carried out on Barbel in Scotland, so most of our information is available only from research carried out in other parts of Europe.

Berners (1496), Boulenger (1946), Buckland (1873), Gordon (1920), Hancock *et al.* (1976), Hunt & Jones (1974, 1975), Maitland & Miller (2002), Maxwell (1904), Regan (1911), Smythe (1959), Walton (1653).

'The barbyll is a swete fysshe,
but it is a quasy mete,
and peryllous for mannys bodye.'

Juliana Berners (1496) *A treatyse of fysshynge wyth an angle*

*****

Goldfish *Carassius auratus* Iasg-oir

'The Golden Chi at dawn is the beauty of fishes.'

Su Tzu-mei (ca 1050) *Lines in reply to Tzu-lu*

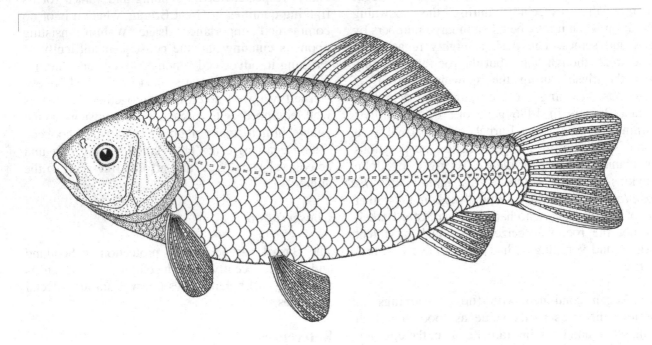

THE Goldfish is one of the few fishes which have been cultivated by humans for many hundreds of years - as William Houghton has noted: 'The Goldfish has been domesticated in China for ages ... perhaps no other fish is more subject to variation both in shape and colour, and none more brilliantly beautiful than the vivid golden varieties.' Chinese literature has many mentions of this species, for example Yo K'o (1173-1240): 'At the present time there are fish breeders in Chung-tu [Peking] who can change the colour of fish to gold. The Chi [goldfish] is the most prized for keeping, and the Li [carp] comes next.'

Hundreds of years later, Goldfish were brought into the British Isles. Christopher Lever records that 'Goldfish were introduced into England about the year 1691, but remained exceedingly scarce till 1728, when a great number were brought over, and presented to Sir Matthew Decker, by whom they were usually distributed around London.' Subsequently Goldfish became widely available, this species remaining one of the most popular of aquarium and pond fishes. Part of this popularity is undoubtedly due to the hardiness of this species.

Frank Buckland once placed a dozen Goldfish in wet grass and then in a cloth in his carpet bag. 'There they remained all night'. On his arrival home in the afternoon, six of them were still alive.

It has many alternative common names including Gibel Carp, Comet, Shubunkin, Veiltail and numerous other varietal names. A number of varieties have extreme distortions of body shape, finnage, eyes and head outgrowths (e.g. Bubbleyes, Bramblehead, etc.) which the author finds grotesque. A literal translation of the scientific name is 'Golden Form of the Dark Fish'.

Size

Most Goldfish, especially those kept in aquaria, do not grow to a large size and lengths of 10-15 cm are common in adults. However, with good conditions and space in large ornamental ponds or even in the wild, the species can grow up to 30 cm in length and 1 kg in weight. Specimens over 40 cm in length have been reported from Asia; Tate Regan recorded that 'Some of the largest specimens I have seen, certainly several pounds in weight, were in the ponds at Hampton Court.' The Scottish Rod-caught record stands at 0.709 kg (1 lb 9 oz), for a fish caught by B. Stevenson in the Forth and Clyde Canal in 1978. The British rod-caught record is a fish of 2.594 kg (5 lb 11 oz) caught in 1994 by D.Lewis at a Stillwater Pond in Surrey.

Appearance

In shape, the body can be very variable but is usually rather deep and laterally compressed. It is normally less deep than the Crucian Carp, with which it is often confused, but in the Goldfish the strongest anterior spines in the dorsal and anal fins are always deeply serrated (only slightly so in the Crucian Carp) and the anal fin usually has only 5 soft rays (usually 7 in the Crucian Carp). There are no barbels at the mouth and the body is covered with large cycloid scales of which there are 28-33 along the lateral line. The dorsal fin is usually concave along its free edge and the caudal fin is more deeply notched than in the Crucian Carp.

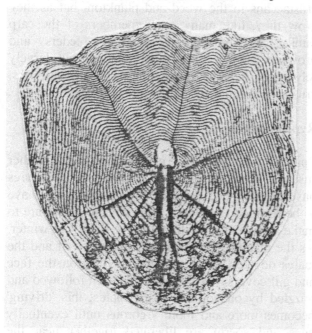

Lateral line scale from Goldfish.

All young Goldfish are similar in colour, which is mainly a dull olive greenish-brown, sometimes with a yellowish tinge. As they grow, however, remarkable colour changes can take place. In the wild, most fish darken to be olive green on the upper head and back, lightening somewhat on the sides to a dull yellowish green on the belly. In domesticated aquarium and pond bred stocks

however, the fish may have been line bred for any one of a number of colours, including white, yellow, gold, red, blue or black or any combination of these with a variety of patterns. It is largely due to these lovely colours that the fish is so popular as a pet. Gordon believed '... the colour seeming to vary with the temperature, the warmer the water the more golden being the fish, the handsomest and healthiest examples being found in the pools in which waste steam is condensed, where the average temperature is 80 degrees.'

Distribution

This species was originally native to Asia and parts of eastern Europe. It has however, been introduced and has established itself in many parts of the world, now occurring widely throughout the warmer parts of Europe. In the British Isles, where it first appeared in the late 17th Century, it is common in most parts of England but there appear to be only a few isolated populations established in suitable ponds in the Scotland.

Recorded distribution of Goldfish in Scotland.

At one time, Goldfish were abundant in the Forth and Clyde Canal at Clydebank. Here, a heated effluent from the Singers factory created ideal conditions for this species, as well as others, including the tropical aquatic plant *Cabomba caroliniana*, which is normally grown in Scotland only indoors in heated aquaria! It was reported to the author one summer that Goldfish were particularly abundant and, on arrival there one Saturday, he was met with the surprising sight of dozens of small boys industriously engaged in catching Goldfish, either for pets or to sell to local pet shops. The author's own efforts were less successful than some of the boys, so he ended up buying some of the better specimens to take back for study! Unfortunately, the factory has now closed down and, though there are still some Goldfish in the canal, they are very few and far between.

Ecology

Goldfish prefer rich weedy ponds and canals, where there is plenty of cover and the bottom is muddy. They occur in rivers only in very slow flowing lowland situations and rarely have large populations there. They are frequently found in artificial situations, especially ornamental ponds in parks and on large estates, where their ancestors were no doubt introduced for their attractive colours. In such situations the entire population may have reverted to the 'wild' colour, for the golden and other light coloured forms are very vulnerable to Herons and other predators. Goldfish are among the most temperature tolerant of all fishes - they can survive low temperatures of around 0°C and high temperatures of up to 40°C.

One of the reasons for the success of the Goldfish as an aquarium inhabitant is its hardiness. As Cleeland Bean noted '... endurance likewise belongs to the goldfish, and a surprising instance of such endurance must be the specimen which was bought from a Glasgow pet shop in 1954, and was alive twenty years later in 1974. The fish ... was kept in a large bowl on the sideboard. Flick, as the goldfish was aptly called had survived various accidents and on two notable occasions he was given up for dead, yet came alive in fine style under the influence of an unusual stimulant' - sometimes known as *the water of life*. This medicine confirms the report by Gregor Bateman that 'When Gold-fish, by accident, have been allowed to remain a long time out of water,

they may frequently be revived by administering a little brandy, even if they are apparently quite dead.' Another example of its hardiness is reported in this newspaper item: 'A Goldfish was yesterday recovering with bruised scales after it was dropped down a chimney, bounced off the coals in the fireplace and landed on the hearth in front of an astonished family in Nottingham. The fish was apparently dropped by a Heron flying over the house'.

However, others disagree and one anonymous writer in *The Aquarist* of May, 1962, wrote 'Great care must be taken of goldfish, as they are very susceptible; and hence a loud noise, strong smell, violent or even slight shaking of the vessel, will often destroy them.'

Food

The young fish feed almost entirely on protozoans (the 'infusoria' cultures of aquarists) and small crustaceans in the weed and plankton, but as they grow they, like many other members of the carp family, become mainly bottom feeders and browsers, eating some plant material but largely relying on benthic invertebrates such as molluscs, crustaceans and insects (e.g. midge larvae).

Reproduction

Spawning only takes place in warm weather (usually June and July) when water temperatures have reached 15-20°C, and these temperatures have to be retained for several weeks if the young are to make sufficient growth to survive their first winter. As the mature fish ripen the females swell and the males develop white nuptial tubercles on the face and gill covers. Ripe females are each followed and nuzzled by one or more ripe males; this 'driving' becomes more and more vigorous until eventually eggs and sperm are liberated together near the surface among weed when the spawning act reaches its climax. This performance is repeated many times during the day until the female is spent. Fecundity varies very much according to size and can range from 6,000-380,000 eggs per female. The eggs, which are a clear yellow colour some 1.5 mm in diameter, stick to the vegetation, mostly in warm positions near the surface. Development takes 3-8 days depending on temperature. After hatching, the fry keep to the shelter of vegetation.

Their dependence on a warm climate is emphasised by Frank Buckland. 'They thrive best in a temperature somewhat higher than the mean; and hence breed very freely in ponds where the water from steam engines, thrown off to be cooled, has an average temperature of eighty degrees.'

The spawning act is well described by Arthur Boarder, a well known breeder and exhibitor of Goldfish: 'When in breeding condition the male fish will usually shows small, white, raised tubercles on the gill-plates and perhaps on the front rays of the pectoral fins The female should look fatter in the body when viewed form above. ...There should be no doubt as to whether the fish are spawning as the males chase and nudge the females towards the bunches of plants ... then with a sudden and violent thrashing of tails, the whole bunch of fish force through the fine-leaved plants and eggs are released and fertilised.' The author will never forget his excitement when, as a schoolboy, his Goldfish first spawned! The sun had warmed the aquarium which was thick with weed and the male Goldfish chased the female all over the tank, occasionally driving her up to the surface where the eggs which were sprayed out, immediately attaching to the weed.

One very notable feature about the reproduction of this species is the fact that in a few places the entire population can consist of females. For instance in lowland ox-bow ponds in Russia, samples of as many as 600 fish all proved to be female. However, the scientists studying them demonstrated that, in the absence of males, these fish spawned readily in the wild with other members of the carp family with which they co-existed (e.g. Common Carp and Crucian Carp) whose sperm, though unable to fertilise a Goldfish egg, stimulated it to start dividing so that it developed into a young fish - genetically identical to its mother, and of course another female. This type of reproduction, producing clones of identical genetic make up is called gynogenesis and is very rare among fish species.

Predators

Herons are the most notorious predators of Goldfish and the author, like many hundreds of other people with garden ponds, has lost fish to this daring bird. Gulls too can be a pest especially in shallow ponds. No doubt also many of the 'usual suspects' (e.g. Pike

and Otters) also feast on them when they get the chance.

Growth

The young fish are some 4-6 mm on hatching and subsequent growth is very dependent on water temperatures. In northern situations they may reach only 2-3 cm by the winter and rarely survive, but further south they may be 10 cm after one year and 15 cm after two.

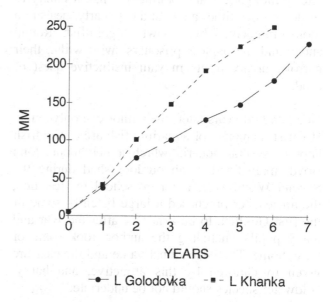

Growth of Goldfish in two Russian Lakes (Berg 1965).

They are normally mature in their third or fourth years and can live for up to 10 years in the wild, often much longer in captivity, as noted below.

Value

Goldfish are of little commercial importance as food and only in a few places are they of interest to anglers. They are, however, a major species for aquaria and ponds and there is an enormous international trade in this fish, with many hundreds of thousands being moved annually from warmer areas in the south of both North America and Europe, where the species is easy to breed and rear, to the cooler areas further north. There is also a bewildering variety of exotic, often highly distorted forms, originally developed in China and Japan, with spherical bodies, telescope and bubble eyes, lion heads, double veil and fan tails and so on. Superimposed on these shapes are many colour patterns and the keeping and breeding these forms is popular in many parts of the world. The author has to confess to a dislike of these largely crippled forms which have been bred to remove all the elegance and efficiency of the wild shape of this lovely fish.

Traditionally many people have kept their fish in 'goldfish bowls', something Herbert Maxwell obviously disapproved of: 'Now I do not suppose that it would be possible to devise a more heartless proceeding - that is, it would be heartless were it not utterly thoughtless and brainless - than a family to sit down three times a day and eat hearty meals in a room ornamented by a bowl of gold-fish. Round and round the hapless prisoners swim within their narrow, dreary limits in vain instinctive quest of food.'

One hundred years later, the author can only agree. His first memories of aquarium fish are of Goldfish kept in several aquaria which a neighbour, John Boyd, maintained in an outdoor shed during the Second World War. Later, at school in Bearsden, the art teacher purchased a large batch of some of the first Goldfish to be available after the War and many pupils, including the author, took some of these home. The educational value and the pleasure given to children by this attractive, and hardy, coldwater species should not be underrated!

There are many jokes at the expense of Goldfish. A recent newspaper report noted that 'According to *The Journal of Aquatic Animal Health*, rainbow trout are falling victim to whirling disease, which causes them to swim round in circle,s. It'll be goldfish next.'

Sometimes Goldfish can live for many years in quite small tanks. *The Independent* (14 October 2005) reported the death of Goldie at the astonishing age of 45, a pet kept by two generations of the Hallett family in Devon. This beat the previous record for longevity held by Tish which died in 1999 in Yorkshire aged 43!

The Goldfish still remains the most popular coldwater aquarium and pond fish in Scotland and many other countries. It has done so for many decades, as K.L. Brown writes concerning the Scottish Aquarium Show for that year: 'Looking at a show catalogue for 1936, for instance, the biggest section, by far, was the coldwater section with 10 Goldfish classes.' This confirms the much earlier view by Frank Buckland. 'The extreme elegance of their appearance, the splendour of their colouring, the ease and agility of their movements, and the facility with which they are preserved in small vessels, have made these little fish great favourites.'

Conservation

The Goldfish is not a native of Scotland and, though a popular aquarium pet, is not in any way threatened as a species, so no special conservation measures are needed.

References

No field research has been done on this species in Scotland so ecological information is available only from other European countries where Goldfish have been studied.

Allen (1987), Anonymous (1978), Bateman (1890), Bean (1978), Berg (1965), Boarder (1978), Brown (1977), Buckland (1873), Davies (1963), Gordon (1902), Hervey & Hems (1968), Houghton (1879), Lever (1977), Maitland (1971), Maitland & Campbell (1992), Maxwell (1904), Regan (1911).

*'Round and round the hapless prisoners swim
within their narrow, dreary limits in vain instinctive quest of food.'*

Herbert Maxwell (1904) *British Fresh-water Fishes*

*****

Crucian Carp *Carassius carassius* Carp di-farf

'... the Crucian is eminently retentive of life, and will live for a long time either without water or in water whose impurities would poison other fish.'

W. Houghton (1879) *British Fresh-water Fishes*

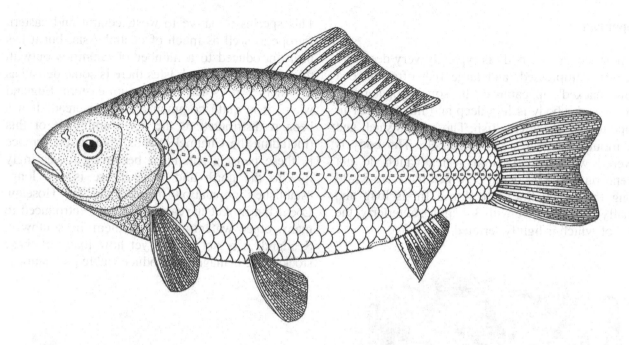

THE Crucian Carp is very similar in appearance to the Goldfish - indeed the species are closely related and, at one time believed by some to be the same species. It is an exceedingly hardy species - Yarrell notes that 'Pallas, in his *Zoogeographia*, mentions the Crucian Carp as inhabiting all the lakes and marshes of Russia and Siberia, and as yielding the inhabitants of those countries a delicious food. It is taken with nets in the vicinity of Jakutsk in the winter time and is then quite torpid. The fishermen select the largest and return the others into the water, where they revive on the advent of spring. Even when the lakes freeze to the bottom, it survives, and when a thaw comes issues from the mud, in which it had buried itself.'

This species occurs further north than any other member of the carp family in Europe; the author has visited the most northerly population in the world, which is near Tromso in northern Norway. Crucian Carp are remarkably resistant to the poor conditions which occur in its northern habitats during winter and fundamental research by Ismo Holopainen in Finland has shown that it is able to withstand, not only long periods of near freezing, but also many weeks of total deoxygenation. No other fish in Europe is so tolerant and it suggests that, although this species has a relatively restricted distribution in Scotland at the moment, it will be able to survive in a wide variety of waters once it finds its way there.

Other common names include Crowger, Gibel Carp, Prussian Carp. A literal translation of the scientific name is 'The Dark Fish'.

Size

This deep-bodied fish is superficially very similar to the 'uncoloured' Goldfish, but whereas in the latter the spines in the dorsal and anal fins are both deeply serrated, in the Crucian Carp these spines are only very lightly serrated. The Crucian Carp varies considerably in its size at maturity. Gordon notes that '... it thrives in water that would kill any other, and is generally small, rarely exceeding 7 inches in length.' In some populations the majority of the adult individuals are only 5-10 cm in length, whereas elsewhere it may commonly grow to 20-30 cm. Under ideal conditions it has been known to grow to 50 cm in length and over 2 kg in weight. There is no Scottish record at the time of writing; the qualifying weight is 0.9 kg (2 lb). The British rod-caught record stands at 1.956 kg (4 lb 5 oz) for a specimen caught in 1999 at Surrey Pit by Adrian Eves.

Appearance

In most stocks, the body is typically very deep and laterally compressed, and large fish often have a 'hump-backed' appearance. In some populations, however, the body is less deep and much closer in shape to that of the Goldfish. The head is small and the mouth terminal with no barbels. The body is covered with large cycloid scales which do not extend on to the head and of which there are 31-36 along the lateral line. The dorsal fin is long and usually convex above with 3-4 spines anteriorly, the third of which is lightly serrated.

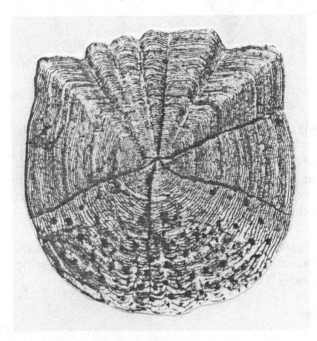

Scale from Crucian Carp.

In colour, the Crucian Carp is mainly an olive green to golden green on the upper head and body, grading to a lighter brassy green on the sides and dull brown on the belly. Frank Buckland records that 'its back has a metallic lustre, and the eye is of golden yellow.' The dorsal and caudal fins are brownish, whilst the pectoral, pelvic and anal fins are a dull reddish brown. Like young Tench, young fish may have a dark mark at the base of the tail.

Crucian Carp will hybridise with Common Carp where they occur together in the wild. The hybrids, which are sterile, are intermediate in form between the parent species and have small barbels, usually only one pair.

Distribution

This species is native to west, central and eastern Europe as well as much of central Asia, but it has been introduced to a number of countries outwith this area. In the British Isles there is some debate as to whether it is native to south-eastern England where it is quite common and widespread. It has been re-distributed to the north and west of this (including the Channel Islands), but its occurrence in these parts is patchy, it becoming increasingly rare to the north and west. There is one long-established population in Scotland (in Hoselaw Loch) but in recent years it has been introduced to many other waters, most of them in southwest Scotland. It is not known yet how many of these stocks have managed to produce viable populations.

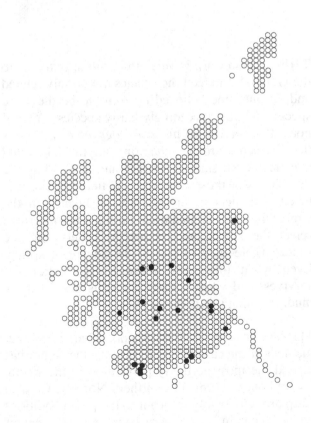

Recorded distribution of Crucian Carp in Scotland.

An anonymous article, which appeared in 1909 in the History of the Berwickshire Naturalists' Club, mentioned that in Hoselaw Loch: 'A few carp have been introduced lately, but sufficient time has not elapsed to determine whether they will survive their translation to peaty water.' It seems likely that these fish may have been Crucian Carp and that these were the progenitors of the stock which is well established there now. In 1995, David Fraser caught Crucian Carp in Loch Rannoch (an unlikely habitat) - individual fish had puncture marks which looked very like hook marks from live baiting. However, subsequently Crucian Carp were found to be common in a small lochan in the Rannoch catchment, along with several other cyprinid species which had been introduced there during the 1990s.

Ecology

The Crucian Carp favours small rich ponds and lakes in lowland areas where there is abundant weed growth, though it may occasionally be found in slow flowing rivers and canals. It requires reasonable summer water temperatures (in excess of 20°C for several weeks) for successful spawning and growth of the fry, but is more tolerant of stagnation and poor oxygen conditions in both summer and winter than most other species. So much so that William Houghton records '... the Crucian is eminently retentive of life, and will live for a long time either without water or in water whose impurities would poison other fish.' Edward Boulenger recorded that 'Numbers have been sent alive to this country from the Continent, either packed in damp weed or frozen up in blocks of ice, the water composing which having been richly charged with oxygen prior to freezing.' Ismo Holopainen's recent research in Finland confirms that this is an extremely hardy species, well able to tolerate very low temperature and oxygen conditions. It is likely to do well in all parts of Scotland.

Food

After absorbing the yolk sac (this may take 1-2 days) the fry feed on protozoans and small crustaceans in the plankton and among the weeds. As they mature they may eat some plants but rely mainly on benthic invertebrates, especially worms, molluscs, crustaceans and various insects (midge larvae especially).

Reproduction

Adult fish become sexually active during early summer, when the females fatten and the males develop nuptial tubercles. Ripe females are actively pursued by one or more males and spawning takes place near the surface in thick weed beds, the eggs and sperm being released simultaneously during orgasmic shuddering when the pair are usually just at the surface. Fecundity of the females depends very much on size and may range from 10,000-300,000 eggs per fish. The eggs, which are a clear pale yellow and 1.4-1.7 mm in diameter, are adhesive and stick readily to the vegetation. They hatch in 5-10 days at water temperatures of 15-25°C, requiring approximately 100 day-degrees for development.

Predators

Little has been published concerning the parasites or predators of Crucian Carp but it can be assumed that many of the 'usual suspects' (e.g. Herons, Grebes, Otters) are involved – as well, of course, as anglers.

Growth

The young are about 4 mm on hatching and growth is very variable depending on temperature and habitat. However, under favourable conditions they can reach 5 cm after one year and 10-15 cm after two years.

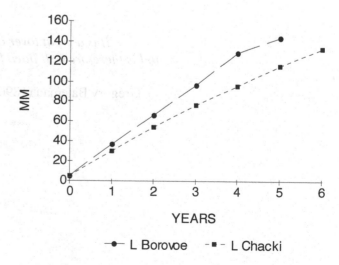

Growth of Crucian Carp in two Russian lakes (Berg 1965).

They usually mature after 3-4 years and may live for up to 10 years on average, though specimens older than this are not uncommon and in captivity the species may live for several decades.

Value

The Crucian Carp is of considerable value in parts of eastern Europe where it is cultured in ponds and cropped in the wild using traps and seines. It is especially valuable in situations where mild pollution or eutrophication create low oxygen conditions during summer or winter, which would be inimical to most other species. Though it is less favoured as a food fish than the Common Carp, hybrids of the two species perform well in some situations and are used from time to time.

It is also a popular sport fish in some areas proving, like the Common Carp, a wily fish which presents a challenge to the angler.

William Houghton was not over-enthusiastic about eating Crucian Carp '... and although its flesh cannot be considered dainty food, yet good well-nourished specimens are by no means to be despised.' Frank Buckland said that the flesh of the Crucian Carp has not the muddy taint which naturally pertains to the common Carp. Gregory Bateman agreed and was more positive about it as a food item 'The Gibel Carp is very good to eat when properly cooked, and it does not taste, it is said, of the water in which it lives.'

It is an attractive aquarium and pond fish and becomes tame rapidly. Gregory Bateman noted: 'The Prussian Carp ... is if possible hardier than the Common Carp. It is a great lover of stagnant water, and is therefore well fitted for a life in the aquarium.' For several years, the author had several attractive specimens in one of his aquaria, which were caught in the lochan near Loch Rannoch. They become tame very quickly and were an interesting fish to keep and watch. However, like other alien species, they should not be released to the wild.

Conservation

The Crucian Carp is given no special protection in Scotland, and as an introduced species here, which is expanding its range rapidly through translocations by coarse fishermen, no conservation measures are required. Indeed, there is a strong case for its removal from some waters where it has become established.

References

Virtually no research has been carried out on the Crucian Carp in Scotland so most available information is from research elsewhere in Europe.

Anonymous (1907), Bateman, (1890), Berg (1965), Boulenger (1946), Buckland (1873, 1880), Gordon (1902), Holopainen *et al.* (1986, 1997), Houghton (1879), Maitland (1977), Marlborough (1966), Maxwell (1904), Penttinen & Holopainen (1992), Tonn *et al.* (1994), Yarrell (1836).

'It is a great lover of stagnant water,
and is therefore well fitted for a life in the aquarium.'

Gregory Bateman (1890) *Freshwater Aquaria*

*****

Common Carp *Cyprinus carpio* Carbh

*'Carp will not thrive in Scotland until some means be discovered
for meliorating the climate and giving soft quality to the water.'*

Anonymous (1843) *New Naturalist's Library*

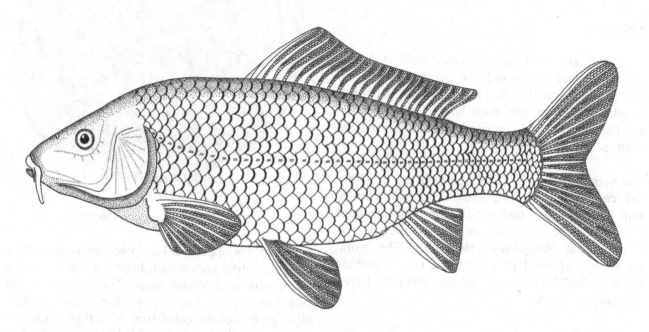

DAME Juliana Berners in her book on angling, published in 1486, writes of the Carp: 'a daynteous fysshe, but there ben but few in Englonde, and thereforre I wryte the lesse of hym.' A similar comment was made by Mascall in 1590: 'The Carpe also is a straunge and daintie fish to take, his baites are not well knowne, for he hath not long been in this realme ... and because not knowing well his cheefe baites in each moneth, I will write the lesse of him.' However, as the species became commoner, it became popular and Isaak Walton regarded it as '... the queen of rivers; a stately, a good, and a very subtle fish.'

Its origin is described by Hilda Allen, who records that 'It is generally recognised that the common form of carp originated in central Asia and became useful as food fish. They eventually became spread across Asia and Europe into the warmer areas where they could live naturally. The early Chinese writings of around 300 A.D. refer to the cultivation of carp and the earliest record of carp in Japan dates back to around the year 700 A.D. but it is not known if these were coloured carp.' Since then, Carp have become widely distributed in England

and have also been introduced to a number of watersheds in Scotland where they have established populations in suitable waters. Many anglers regard this as the most wily of fish, as indicated by the comments of Sheringham: 'You cannot, of course, fish for big carp in half a day. It takes a month. So subtle are these fishes that you have to proceed with the utmost caution.'

There are many strange stories and some superstitions regarding Carp. Tate Regan (quoting Smitt) records that 'Carp are often troubled by the male frogs, which under the influence of sexual excitement attach themselves firmly to the head of the sluggish Carp, and with their forefeet press the eyes of the fish so hard as to produce blindness.' Edward Boulenger observed that 'In warm weather carp come to the surface ... and produce noises which may aptly be described as "smacking of the lips".' Frank Buckland mentions 'a lake in Ireland on which the natives believe fairies dance by the light of the moon.' On investigation he found 'no fairies, but... great carp smacking their lips as they took in gulps of air.' The author's tame Carp are certainly very noisy when eating at the surface.

There are several cultivated varieties of the Common Carp, including Leather Carp, Mirror Carp and Koi. Keeping of the latter in special ponds is now a very popular hobby and good specimens can fetch thousands of pounds. A literal translation of the scientific name is 'Carp of Carps'.

Size

This well-known fish has been cultivated and moved around the world for many hundreds of years. Adults usually reach lengths of 40-50 cm and weights of 2-3 kg, but much larger fish can occur, and fish of up to 30 kg at over 40 years of age have been recorded from Europe.

The Scottish record is 13.23 kg (29 lb 2.5 oz) for a fish caught in the Castle Loch, Lochmaben by S. Ball. The present British rod-caught record stands at 25.571 kg (56 lb 6 oz) caught in 1998 by K. Cummins at Wraysbury, Berkshire. The world record was claimed in 1985 for a large Common Carp of 26.22 kg caught in the Potomac River, Washington, USA.

Appearance

The Common Carp is a sturdy fish with a laterally compressed stream-lined body which is usually well covered by large cycloid scales, of which there are some 35-39 along the lateral line. Old fish may become very deep and much less streamlined. Characteristically, the head carries two pairs of barbels on the upper lip, the posterior pair, in the angles of the mouth, being the longer. The pharyngeal teeth are in three rows and rather flattened with slight grooves.

The scaling of this species, like its colour, shows some variation and several varieties have been bred with different scale patterns. These range from the standard wild form, illustrated above, which is completely scaled ('King Carp'), through a form which has only a single row of very large plate-like shiny scales along the lateral line, and perhaps a few elsewhere ('Mirror Carp') to a completely scaleless form with a smooth skin ('Leather Carp'). All of these forms and variations of them can now be found in the wild in the British Isles. The dorsal fin is long and bears anteriorly, as one of 3-4 spines, a serrated spine.

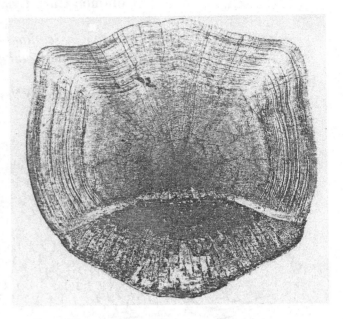

Scale from Common Carp.

In colour, the upper head and back are usually olive brown to olive grey which becomes lighter on the sides, with a yellowish tinge. The belly is a dull yellowish brown. The dorsal fin is usually a dark olive green but the other fins are a lighter greeny-brown, sometimes with a reddish tinge. Small Carp may be confused with Crucian Carp, wild Goldfish or Common Bream. In domesticated forms a complete range of colour varieties, mixtures and patterns has been bred for ornamental (pond) purposes. These are Koi Carp, ranging from silver ('ghost') and pure white individuals through yellow, orange, red and brown to blue and black and an amazing range of mixtures of these. Under ultra-violet light, these fish may be strongly iridescent.

Distribution

The Common Carp was originally an Asian fish found only in the warm temperate band from Manchuria to the Black Sea. Because of its importance as a food fish it has been widely introduced, and is now found in all the major continents and most parts of the world where the summers are warm enough to induce spawning and allow adequate growth of the young. It is widespread throughout central and southern Europe and is found in most parts of the southern half of the British Isles including Ireland and the Channel Islands, where it arrived during the 15th or 16th Centuries.

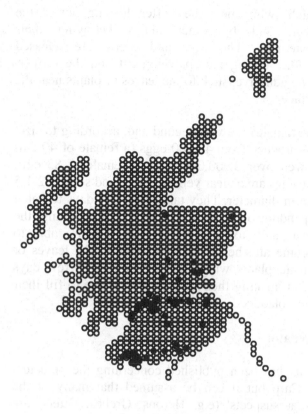

Recorded distribution of Common Carp in Scotland.

In Scotland a number of populations do occur but it is likely that the species is reaching its northern limit here and, as in Ireland, can only breed successfully during warmer than average summers. However, Carp are very hardy fish and transport well, having a low oxygen demand, a high temperature tolerance and the ability to survive out of water for a considerable time. These factors have no doubt been helpful to its wide dispersion, especially in ancient times. Rintoul & Baxter recorded that Carp 'Were introduced into Duddingston Loch by the Earl of Abercorn before 1795 to stock his canals and ponds, but were eaten by pike or washed away by the flood.' In recent years, the author has seen specimens from well established populations in the Lomond (Culcreuch Loch) and Forth (Danskine Loch) catchments.

Ecology

The preferred habitats of the Common Carp are rich ponds and lakes, canals and slow-flowing rivers where there is an abundance of aquatic vegetation. It can sometimes be found in large numbers and though the young move around together it is not really a shoaling fish. Its tolerance of low oxygen concentrations allows it to inhabit very rich waters where the decomposition of organic matter is rapid. The higher the water temperature, the more active Carp are. Gordon records them '... spending the winter in the mud, and rising to the surface in summer among the weeds, when it can be heard grunting like a pig. It is very quick of hearing and discriminating in attacking its food, but it can be tamed so as to feed from the hand, and will come for its dinner when the bell rings.'

Food

Carp are truly omnivorous feeders and will eat a wide variety of food items, digestion of these being considerably helped by the action of the pharyngeal teeth which grind up tough or solid items. Carp do not have a stomach, the oesophagus merely emptying into the expanded upper intestine, which acts as a 'crop'. The young feed mainly on small crustaceans of different kinds which occur in the plankton and among weeds, but as they grow they start to feed more and more on benthic invertebrates, especially midge larvae, crustaceans and molluscs. They often do this by sucking in quantities of mud from the bottom and then expelling it, subsequently taking in the invertebrates they have disturbed by this action.

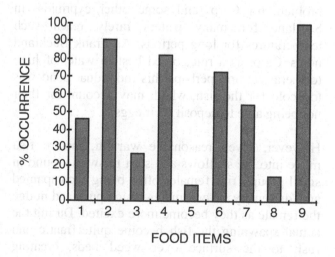

Food of Common Carp in Garcine Pond, Camargue, France (1: algae; 2: seeds; 3: molluscs; 4: crustaceans; 5: mayfly larvae; 6: beetles; 7: midges; 8: others; 9: detritus) (Crivelli 1981).

The extensible tubular mouth is capable of penetrating some distance into bottom muds and silts which are stirred up as a consequence, often increasing water turbidity. As a result, they have often been responsible for considerable alteration of habitat, clear weedy waters becoming turbid and weedless as the mud is regularly stirred up and light for the growing plants is eliminated. For this reason, anglers and conservationists have regretted their introduction to many areas.

Richard Wydoski and Richard Whitney record that, in North America 'In the early days ... The carp was billed as the greatest food fish ever, and every homesteader wanted carp in his ponds lakes and streams. ... Today money is being spent to remove carp ... In 1955 an estimated 1.5 million carp were killed with a fish toxicant at Malheur Lake, Oregon, because of the damage they were causing ...'

Standing quietly by the edge of the water, the author has spent many hours watching Carp in Danskine Loch and other waters including some in France. On bright days and if the water is clear (often it is not in carp waters) they can be seen swimming lazily round, stopping occasionally to feed and sometimes sprinting off if disturbed slightly.

Reproduction

Spawning takes place during summer when the water temperature reaches at least 18°C. This is a problem for Carp (and some other cyprinids) in Scotland for many waters rarely reach such temperatures for long periods. As Frank Buckland notes 'Carp, as a rule, breed best in water of high temperature, and perhaps this individual pond was too cold for the fish, which may account for their not being able to deposit their eggs.'

However, given reasonable warmth, mature fish move into the shallows and start to swim around in small groups, ripe females often being accompanied and 'driven' by several males which push and nudge the female as they become more excited. During the actual spawning the fish become quite frantic and rush to the surface over weed beds, creating considerable disturbance and noise. The species, which is otherwise very wary and often difficult to observe, becomes very obvious to the onlooker at this time. Tate Regan describes the process as follows: 'The Carp spawn on quiet shallows among the weeds, to which the eggs adhere. Sometimes a female is attended by two or even three males,

which swim above her, often leaping out of the water, and by other antics betraying their excitement.' The eggs and sperm are released together during the spawning act and the former, being adhesive, stick to the leaves of plants near the surface.

The females are very fecund and, according to size, can produce from 40,000 eggs (a female of 40 cm) to well over 2,000,000 eggs (a female of 85 cm). The eggs are a clear yellow colour and some 1.2-1.5 mm in diameter. They take about 3-7 days to hatch depending on water temperature. On hatching, the larvae sink towards the bottom, but are able to become attached by their mouths to the leaves of aquatic plants where they remain for a few days before making their way to the surface to fill their swim bladders.

Predators

Little has been published concerning the predators of Carp but it can be assumed that many of the 'usual suspects' (e.g. Herons, Grebes, Otters) are involved. They suffer from several diseases.

Growth

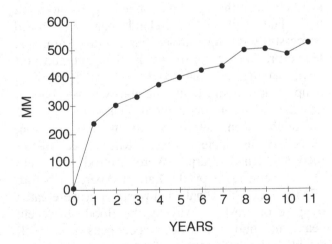

Growth of Common Carp in the marshlands of the Camargue, France (Crivelli 1981).

The young are quite small (4-6 mm) when they hatch but soon grow rapidly if temperatures are high so that they may be 10-15 cm at the end of one year and 20-25 cm by the time they are two years old. They start to mature during their third or fourth years when they have reached lengths of 30-40 cm and may live, on average, for 10-15 years - often much longer in captivity. One British rod-caught record Carp was 15 years old, but lived a further 20 years in captivity. They can be aged by their scales.

During the winter, Carp retire to deep water where they remain in a torpid state. For it to overwinter safely it has been shown that a Carp must build up a significant fat reserve: a fish at the beginning of winter with a condition factor of less than 2.9 may not survive. Overwinter mortality of carp has been observed in Britain as a result of a long, cold winter. The acute sensitivity of Carp to even a small rise in the ambient water temperature was dramatically demonstrated during the mild winter of 1987-88 in England when water temperatures in ponds and lakes rose to over 7°C. Many specimen Carp were taken by anglers - a previously unheard of winter event - and the angling column of a national newspaper reported that about a dozen Common Carp of 13.6 kg or over had been caught during January. It would seem that, roused from their winter torpor, the Carp had to start looking for food to meet an increase in their metabolic demand.

Value

The Common Carp has been cultivated for over two thousand years and is of considerable importance commercially in many countries, especially those far from the sea, where sea fish are unobtainable. In early times, the eggs of Carp were used as a substitute for sturgeon caviar by Jews as, according to the Old Testament, Sturgeon (and other scaleless fish) were unclean and thus genuine caviar was not 'kosher'. It is caught commercially in the wild but is also reared in enormous numbers in special ponds. Daniel recorded that 'It is common practice in Holland to keep *carp* alive for three weeks or a month, by placing them in a net, well wrapped up in wet moss, hanging them up in a cellar or cool place, and feeding them with *bread* and *milk*.'

Edward Boulenger noted that 'Until the close of the seventeenth century very few country seats of any consequence were without their stews or fish ponds, and the carp figured very largely in the "stock" of such installations. Old works on fish culture give elaborate instructions for the feeding, etc., of the carp, one such work advocating the suspension of a horse's carcase over the pond, in order that the maggots accruing therefrom might descend in a gentle rain upon the piscine population below.'

It is also a very important sport fish: its wily nature, potentially large size and fighting qualities when hooked have made it a sought after species by specimen carp anglers who will spend long hours patiently fishing for it throughout the night. Izaak Walton called the Carp the 'river fox' - a tribute to its wilyness. Samuel's comment is 'The best bait that ever I did know for the killing of Carpe, is, a quantitie of sufference, with a good deale of patience, and as much silence as may be possible, all these mingled together.'

Izaak Walton obviously valued Carp: 'I will tell you how to make this carp, that is so curious to be caught, so curious a dish of meat as shall make him worth all your labour and patience.' However, William Houghton did not think much of Walton's recommendations: '... the culinary art as practised and recommended by Izaak Walton is not requisite. The Carp should be merely boiled; a little melted butter and walnut pickle is a better condiment for a fountain-feed Carp than old Izaak's sweet marjoram, thyme, parsley, savory, rosemary, onions, pickled oysters, anchovies, cloves, mace, orange and lemon-rinds, and claret wine, etc.' Edward Boulenger was less enthusiastic: 'The carp, if submitted to proper treatment, is edible.' Herbert Maxwell, however, felt that he knew the secret of final rearing: 'The monks of old knew far better how to treat the Creator's gifts than to drag carp out of a miry pond and deliver them straightway upon the table, with all their impurities upon them, imparting a disgusting flavour of mud to the flesh. ... they bestowed them in stews constantly replenished with pure water, and fed them up on boiled grain or other fattening material, calculated to sweeten and enrich the flesh.'

Its popularity as an aquarium and especially a pond fish has also given it a considerable commercial importance in this field, especially with the enormous recent popularity of the keeping of Koi Carp. These are now widely available in a large range of sizes, colours and prices (which can vary from one or two pounds to several thousand for a single fish!). The origin of ornamental varieties has been described by Hilda Allen: 'Carp were raised primarily as food but they were often regarded as pets and so it was natural that any unusual mutation of colour would not go unnoticed. ... with the appearance of coloured carp the rich and noble families began to keep them for their ornamental beauty alone .. Carp easily become tame household pets and they provided enjoyment and recreation during the winter months. ... It would be quite impossible to describe the enormous number of Koi varieties available today. The highly-skilled techniques of Japanese breeders over many years, and which continue today, offer us a tremendous choice of colours, patterns and sealings.' 'Koi are coloured carp and variations derived from the Common Carp (*Cyprinus carpio*) are known in Japan as Ma-goi.'

Conservation

Carp are given no special protection in Scotland, nor do they need it as an alien species which can be quite damaging to sensitive aquatic habitats.

References

Little research has been carried out on the Common Carp in Scotland; most available data have been gathered elsewhere in the world.

Allen, (1977), Anonymous (1845), Berners (1496), Beukema & de Vos (1974), Boulenger (1946), Buckland (1873, 1879), Crivelli (1981), Daniel (1812), Fitzmaurice (1983), Gordon (1902), Houghton (1879), Hruska (1961), Lubbock (1845), Maitland (1964), Mascall (1590), Maxwell (1904), Regan (1911), Rintoul & Baxter (1935), Samuel (1577), Sheringham (1925), Stein & Kitchell (1975), Walton (1653).

'.. no fish baffle the fishermen so completely as large carp...
fen-men regard these fish with mysterious awe ...
and look upon him as what the Scotch call "No cannie".'

R. Lubbock (1845) *The fauna of Norfolk*

*****

Gudgeon *Gobio gobio* Guda

'It is not mentioned as occurring in Scotland.
Is this merely negative evidence?
Is not the Gudgeon an inhabitant of any of the waters of Scotland?'

William Houghton (1879) *British Fresh-water Fishes*

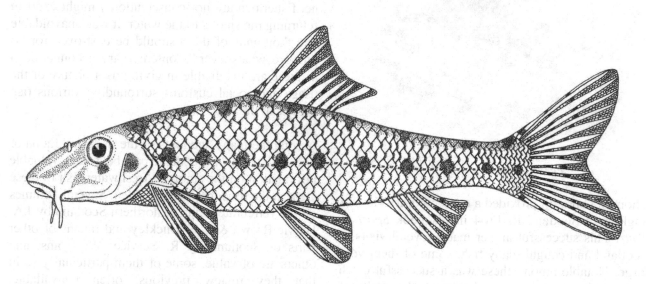

IT is not very clear if the Gudgeon was already established in Scotland when William Houghton asked his questions, but if not, it was probably introduced soon afterwards. Perhaps his disbelief as to its occurrence north of the Border arose from the fact that it was not only abundant in the waters in England which he knew, but that he found it so easy to catch. Hence Houghton's belief that 'The expression "to gudgeon a man", i.e. to deceive him, may have originated from the ease with which a Gudgeon is taken by a bait.' In contrast, Cholmondeley-Pennell noted that 'Gudgeon fishing! ... I maintain to be, *par excellence*, the sport of the poet and the philosopher.' Probably few coarse fishermen would agree with him today.

Edmund Waller, however, had a warning to anglers in his poem *Beware of lady gudgeon-fishers*

'The Ladies angling in the chrystal lake,
Feast on the waters with the prey they take;
At once victorious, with their lines and eyes,
They make the fishes and the men their prize.'

A literal translation of the scientific name is 'The Goby-like Gudgeon'.

Size

The Gudgeon is a small fish, usually maturing in Scotland at about 12-15 cm, although fish of up to 20 cm and over 220 gm have been recorded elsewhere. There is no Scottish record at the time of writing and the qualifying weight is 4 oz. The British rod-caught record is for a fish of 0.141 kg (5 oz) caught in 1990 by D.H. Hull in the River Nadder, Wiltshire.

Appearance

This is a distinctive fish, unlikely to be mistaken for any other except when very small when it could be confused with small Barbel or Stone Loach. The body is elongate and rounded with only slight ventral flattening and some lateral compression towards the tail. There is a single pair of long barbels, one situated in the posterior corner of each side of the ventral mouth, whose lips are thick. The body is well covered in medium-sized cycloid scales of which there are 38-44 along the well-developed lateral line. The sensory organs of the lateral line are exposed and very sensitive. The scales can be used for age determination.

feed upon salmon more than thrice a week.' In spite of Norman Simmond's deplorable comments about Richard Franck's writing skills, he did provide useful information about fish in Scotland during this period.

Perch (from Wood 1863).

Thomas Thornton provided a contemporary view of angling in Scotland in 1804 in his book *Sporting Tour*. This successful angler made several visits to Scotland and caught many fish, some of them very large. Notable among these was a successful catch in Loch Lomond. '... passed through the straits of Loch Lomond, a most likely place for pike: still we had no rise, when, in turning gently round a dark bay, I felt a fish strike ... After much trouble he was secured in the landing-net, and proved to be a perch of ... seven pounds three ounces, or thereabouts. He was very thick about the shoulders, and I regret I did not measure him, as I never saw a fish so well fed.'

Naturalists

Although most accounts of the natural history of parts of Scotland tended to concentrate on the three most popular fffs (fur, feathers and flowers), some authors did mention a fourth - fish. Notable among these was Herbert Maxwell, who had an especial interest in fish (and indeed published a notable text on British freshwater fish), and was a copious writer. Within a series of fascinating volumes (each containing many essays on different aspects of the countryside), published at the turn of the 20th century under the general title *Memories of the Months*, are several useful articles on fish in Scotland. For example, an extract from 1922: 'In the Cree, another of the Solway rivers frequented by sparlings, the members of an angling association which rented the salmon-fishing became aware of

the destruction wrought by the small-meshed nets of the sparling-fishers among the smolts of salmon and salmon-trout descending to the sea in April and May, this being the very period when the sparlings seek the top of the tide to spawn. Consequently, gravid smelts and migrating smolts were hauled out indiscriminately, and however much care conscientious fishermen (and it is rumoured that all net fishermen are not conscientious) might exert in returning the smolts to the water, it was unavoidable that thousands of them should be destroyed for no purpose whatsoever.' Contemporary accounts such as these are invaluable in giving us a picture of the usage and social customs surrounding various fish species at the time.

Among the many accounts of the flora and fauna of different parts of Scotland can be found valuable information on the status of freshwater fish in these areas at the time they were written. Several volumes on the vertebrate fauna of northern Scotland by J.A. Harvie-Brown & T.E. Buckley and faunas of other parts of Scotland by R. Service, W. Evans, and others are of value, some of them particularly so in that they review previous, often unavailable, literature.

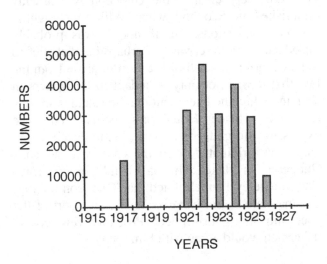

The numbers of Powan caught in Loch Lomond at the end of the First World War (data from Lamond 1931).

There are a few classical accounts of individual Scottish waters which deserve special mention and one such is Henry Lamond's *Loch Lomond* published in 1931. This is an exceptionally fine account of the fish and fisheries of Scotland's largest loch, now the focus of our first National Park, and his study is referred to several times in the

John Berry's estate in Fife for several decades. A number of years ago, the author and Ken East were able to show John that the population was still viable during electrofishing there and several large specimens were captured. It also occurs now in Loch Lomond.

Ecology

The Gudgeon is thus largely a fish of running water. Aflalo noted that 'Thriving equally in still or running water, with a preference perhaps for the latter, this is *the* fish of the Seine.' It prefers fairly fast running water with a clean bottom of sand and gravel, with some weed beds to provide cover. Here, large schools may build up during the summer months. Its oxygen requirements are similar to those of Grayling and Bullhead. It also occurs in slower flowing lowland systems in lesser numbers where it tends to move into deeper waters during the winter when feeding activity declines.

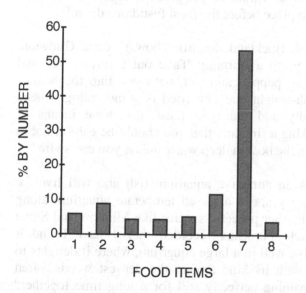

Food of Gudgeon in the River Cam, England (1: molluscs; 2: Gammarus; 3: cladocerans; 4: copepods; 5: mayfly larvae; 6: blackfly larvae; 7: midge larvae; 8: others) (Hartley 1948).

Edward Boulenger states that 'The German word of "groundling" well describes its habitat since it delights in gravelly river-beds where it often congregates in considerable numbers, hiding for preference beneath banks or bridges.' Brian Stott and his colleagues found that 'gudgeon are capable of homing after being displaced both upstream and downstream and ... to a remarkable degree of precision.'

Food

The fry of Gudgeon feed mainly on small crustaceans in open water or among weed. As they grow they start to feed more and more on the bottom but still rely largely on invertebrates there, especially oligochaete worms, crustaceans, molluscs and a variety of insect larvae. Filamentous algae and fish eggs are also sometimes eaten. At this time they often occur in the same habitat as Stone Loach.

Frank Buckland records that 'They swim in shoals, and are assembled by raking the bed of the river, a process called by the fishermen "scratching their backs", when they immediately crowd to the spot in search of food.'

Reproduction

Reproduction takes place in May and June when the water temperature is around 14°C. The males develop tubercles on the head and front of the body. Spawning takes place in shallow flowing water among plants and gravel to which the opaque, greyish-yellow eggs, some 1.4-1.8 mm in diameter, stick in small groups after they have been laid and fertilised. Edward Boulenger comments that 'The transparent eggs have a clear blue or yellowish "bloom" and are adhesive, one to three thousand being deposited in small clumps on stones, etc., the process of spawning being spread over a lengthy period.' Each female spawns several times in succession, often with several different males until she is spent, having laid about 1,000-3,000 eggs in total. The eggs hatch in 10-20 days depending on local water temperatures. After hatching, the larvae drift downstream to slightly quieter water, usually along the edge of the river among vegetation, where they absorb the yolk sac and start to feed within 5-6 days.

Predators

Little has been published concerning the predators of Gudgeon but many of the usual piscivores (e.g. Herons, Mergansers, Otters, and Mink) are involved – as well, of course as coarse fishermen.

Growth

The larvae are only about 4-5 mm on hatching and by the end of their first summer may have grown to a length of 5 cm. After two years they may have reached 10 cm and by their third year they are mature at 12 cm or more. By this time, most females are slightly larger than males of the same age.

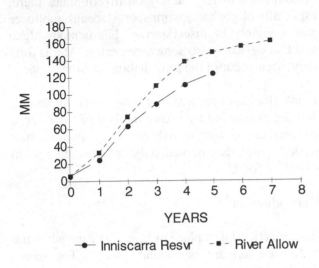

Growth of Gudgeon in two Irish waters (Kennedy & Fitzmaurice 1972).

Most fish live for 4-5 years some reaching 7 or 8 years of age. A proportion of both sexes matures at 2+ years of age, around 7-8 cm in length. No size difference between males and females was noted.

Value

The Gudgeon is of little importance commercially, although it is esteemed in some parts of France for the table. Elsewhere, including Scotland, it is only occasionally sought after by anglers, but is used mainly as bait for other fish. In some waters, Gudgeon are an important food for large predators such as Trout and Pike, hence their value as live bait. The recent extensions of its distribution in Scotland, including the Loch Lomond catchment, are partly as a result of Gudgeon being carried about by anglers for use as live bait and then tipped into the water at the end of the day's fishing. William Houghton noted that '... they are gregarious in their nature, and readily taken with a worm, affording capital sport in their small way, to those anglers who care more for numbers than size, and who can appreciate a most excellent food.'

William Houghton goes on say that 'In point of flavour the Gudgeon approaches that of the Smelt or

Sparling, and in my opinion is one of the best of fresh-water fish we possess. It is a pity, from a gastronomic point of view, that the Gudgeon does not exceed the length of six or seven inches.' Cholmondeley-Pennell agreed 'In a gastronomic point of view, *gobio* gives precedence to none; a fry of fat gudgeon, eaten piping hot, with a squeeze of lemon juice, is a dish "to set before a king".'

Writing a few years later, J.J. Manley was equally enthusiastic 'Let me say that, in my humble opinion, however mean a fish the gudgeon may be thought whereon to exercise the angler's skill, he is worthy of all commendation as a fish for the angler's table, and indeed the board of the most fastidious *gourmet.* There are few freshwater fish worth the salt with which they must be eaten, if eaten at all, but oesophagistically I am enthusiastic about our *Gobio fluviatilis.*' Edward Boulenger agreed: 'It is very good eating when fried and is appreciated as a delicacy on the Continent, especially in France, where a "friture de gougons" is regarded as a dish fit to place before the most fastidious diner.'

Frank Buckland describes how to cook Gudgeon, but gives a warning. 'Take out a frying pan, and butter, pepper, and salt, with you into the boat, a fresh-caught gudgeon fried is grand eating. I need hardly add that you must also have means of making a fire, and that you should be careful not to burn the boat in deep water unless you can swim.'

It is an attractive aquarium fish and will live for many years in a mixed temperate aquarium along with other peaceful species like Minnow and Stone Loach. Furneaux recorded that 'I have found it thrive well in a large aquarium, where it delights to lie with its kind among the densest weeds, often remaining perfectly still for a long time together.' Gregory Bateman has the final word about its value to humans: 'The Gudgeon ... is a general favourite. It is a favourite with the aquarium keeper, for it is hardy, handsome, and easily tamed.; it is a favourite with the fisherman, for it is a bold biter and provides excellent sport; and it is a favourite with the epicure, for when freshly caught and properly cooked it is most delicious to eat.'

Conservation

The Gudgeon has no special protection in Scotland and, as an introduced species which is common in places and continuing to disperse across the country with the help of coarse fishermen, needs no particular measures.

References

Some work has been carried out on this species in Scotland and there is also a substantial body of research from other parts of Europe.

Aflalo (1897), Bateman (1890), Bean & Winfield (1989, 1992), Bolam (1919), Boulenger (1946), Buckland (1873), Cholmondeley-Pennell (1866), Furneaux (1896), Harman (1950), Hartley (1948), Houghton (1879), Kennedy & Fitzmaurice (1972), Ladich (1988), Manley (1877), Mann (1980), Stott (1967), Stott *et al.* (1963).

'What gudgeons are we men,
Every woman's easy prey!
Though we've felt the hook, again
We bite and they betray!'

John Gay (1688-1732), *The Beggar's Opera*

*****

Chub *Leuciscus cephalus* Pluicean

'Oh , sir! a chub is the worst fish that swims;
I hoped for a trout for my dinner.'

Izaak Walton (1653) *The Compleat Angler*

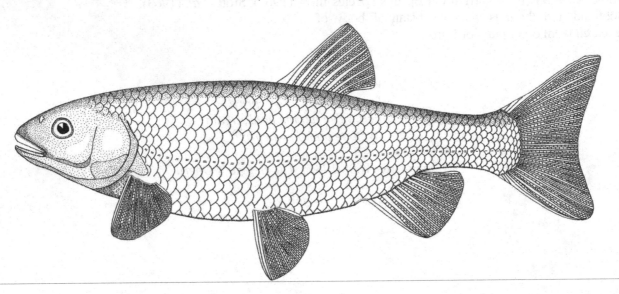

UNLIKE Izaak Walton, William Houghton appeared to think highly of the Chub and favoured it being dispersed more widely than it was in 1879: 'However, it is probable that the Chub may be artificially increased by its introduction into rivers where it is not at present found ...' Others, like Bridgett, describing a fish he had just made contact with, seemed to agree with Walton: 'All my hopes returned ... The fight aroused suspicions. After a kick or two the fish allowed itself to be dragged unceremoniously upstream and into the net. It was a half-pound chub, the only one I have ever captured, and I have no wish whatever to add to the number. These experiences were humiliating.'

Herbert Maxwell, however, had a solution for those who did not like this species: 'Now dace and chub are chiefly known in the southern and Midland counties; but it is in the north that wise men have devised an insidious way of ridding themselves of the pest. ... here salmon and trout anglers avail themselves of the process known by the delightful title of "fuddling chub". ... Boil 1.5 lb. of rice until rather soft (not so soft as for the table), let it cool, then add 1.5 lb. of flour, one ounce of *Cocculus Indicus*, and crumble up with the whole a threepenny loaf of stale bread. Mix all together with the hands, and throw into the haunts of the chub in pieces about the size of a pea. The chub eat it, presently float on the top incapably drunk, and may be ladled out.'

Chub have a range of other common names including Bottling, Chavender, Chevin, Loggerhead, Lob, Poll, Pollard and Skelly. The last name should not be confused with Schelly - the Cumbrian name for *Coregonus lavaretus* in that part of England. Travis Jenkins noted that 'The name "Chub" is of doubtful origin. ... At any rate, the names of this fish in different languages have some reference to the shape of the head. In Germany the fish is known as *Dickkopf*, thick-head; in French *chevin* or *chevenden*, from *chef*, a head.' A literal translation of the scientific name is 'Big-headed White Fish'.

Size

The Chub is one of the larger cyprinids, with an average adult length of some 30-40 cm and a corresponding weight of 500-900 gm. Much larger specimens can occur, however, and fish of over 60 cm and 4 kg have been recorded from central Europe.

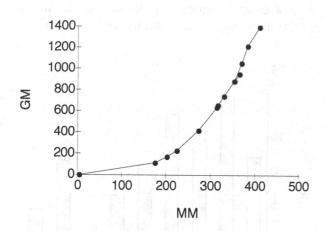

Length/weight relationship in Chub (from Hellawell 1971).

The Scottish record is 2.835 kg (6 lb 2 oz), for a fish caught in the River Annan in 1999 by L. Breckell. Although larger fish have been reported unofficially, the present British rod-caught record is for a fish of 3.912 kg (8 lb 10 oz) caught in 1994 by P. Smith in the River Tees, near Blackwell, County Durham.

Appearance

The body is elongate, but sturdily built and with only slight lateral compression. The head is rather small but very blunt when seen from above and with a wide forehead. The mouth is large and terminal, its angle reaching back to the level of the eye.

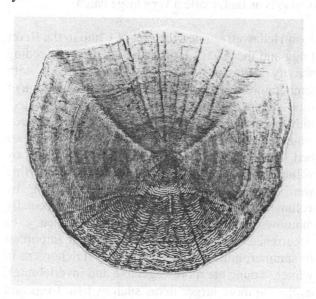

Scale from Chub.

The body is covered with thick cycloid scales, which number 44-46 along the lateral line. These can be used for age and growth studies, though opercular bones are also of use in this context. The origin of the dorsal fin is behind the base of the pelvic fins, and its free edge is characteristically squared. The paired pectoral and pelvic fins are rather small. The anal fin is distinctly convex along its free (rear) edge. The tail fin is large and clearly forked.

The basic colour on the head and back is a dark olive greenish grey. This grades to a greenish silver on the sides and eventually to a dull yellowish white on the belly. However, most of the scales on the body have a dark edge which gives the whole body a reticulate appearance when seen close up. The fins are mainly a dull grey in colour, but the pelvic and anal fins are usually faintly reddish.

Distribution

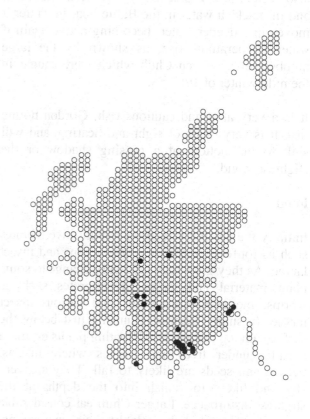

Recorded distribution of Chub in Scotland.

The Chub occurs across much of central and southern Europe, from Portugal in the west to the

Caspian Sea in the east. In the British Isles it is native to the south east of England but has been dispersed across much of the rest of England and into parts of Wales. It is absent from Ireland and formerly occurred in Scotland only in the Rivers Sark and Annan in the south west. However, it has recently been introduced to the Clyde catchment and is now common in the River Endrick, where the author first recorded shoals of young fish there in 1986. Though normally a river fish it does occur in some lochs and has been common in the Castle Loch at Lochmaben (where the author first netted it in 1964) for many decades.

Ecology

The Chub is mainly a river fish and shoals occur in the lower and middle reaches of lowland rivers. Large individuals tend to be more solitary in their habits. Chub are typical of the stretches of mixed habitat found in the middle reaches, where lengths of fast water with stones and gravel alternate with slower pools containing some weed and silt. It has also been found in some lochs (e.g. the Castle Loch) and in brackish water in the Baltic Sea. In winter it moves into deeper water, becoming active again if water temperatures rise, as shown by the large numbers of specimen Chub which were caught in the mild winter of 1987-88.

It is a very alert and cautious fish, Gordon noting that 'It is very quick of sight and hearing, and will sink to the bottom at a passing shadow or the slightest sound.'

Food

Initially the small fry feed on minute invertebrates such as rotifers, protozoans, crustaceans and insect larvae. As they grow they gradually change to some plant material and larger invertebrates such as worms, molluscs, crustaceans and various insect larvae. In sunny weather, Chub lie just below the surface among the leaves of floating plants or, more usually, under trees and bushes where insects, berries and seeds are likely to fall. They are very shy and likely to vanish into the depths at the slightest disturbance. Larger Chub eat considerable numbers of small fish, notably their own young, Eels, Dace, Roach, Gudgeon and Minnows. They also eat other quite large organisms such as crayfish, frogs, voles and young water birds as well

as some plant material. Boulenger noted that 'In parts of central Europe it is known as the 'mouse eater' from its supposed partiality for water-rats.'

Food of Chub in the River Lugg, England (1: diatoms; 2: filamentous algae; 3: macrophytes; 4: Eel; 5: other fish; 6: crayfish; 7: frogs; 8: insects; 9: others)(Hellawell 1971).

The most common plant food of Chub in the River Stour in Dorset was the Water Buttercup. The fruits of Mulberry, Elderberry, Bramble and Cherry are traditional baits for Chub and they certainly feed on these naturally as Wheeley describes: 'A large damson tree overhung the river near Pangbourne. One windy day when the fruit was ripe, an angler happened that way, and noticed a fine company of chub greedily gulping the damsons blown off the tree into the water. He took the hint, and, using damsons as bait, made a very large catch.'

John Hellawell studied the diet of Chub in the River Lugg and the Afon Llynfi: 'The intensity of feeding activity was significantly correlated with temperature.' 'The diet ... embraced a wide range of food items but vegetable material predominated. Substrate material was frequently encountered which suggested considerable feeding on the river bed. Fish were the most important animal food, by volume.' No salmonids were found, but Grayling were eaten. 'The most important prey fish, by volume, was the eel but cyprinids, especially minnows, were most important by frequency of occurrence.' Vegetable foods were most important in summer, animal foods, especially Trichoptera in winter. Young ate more vegetable and invertebrates, adults ate more larger items such as fish, frogs and crayfish.

Their voraciousness in feeding was clearly known to Thomas Hughes when he wrote *Tom Brown's School Days*: 'It isn't often that great chub, or any other coarse fish, are in earnest about anything, but just then they were thoroughly bent on feeding, and in half-an-hour Master Tom had deposited three thumping fellows at the foot of the giant willow.' Derek Mills recalls watching Chub in the Border Esk in June 1975, 'rounding up a shoal of Minnows in shallow water - just like placid sheep dogs.'

Reproduction

Spawning takes place from May to June when water temperatures rise above 12°C. The males develop small spawning tubercles on the head and upper body and drive ripe females over suitable areas (usually weed beds) where spawning takes place. Young fish tend to spawn in shoals, whereas there is a tendency for the older, larger, fish to spawn in twos and threes. The fecundity of females ranges from 25,000-100,000 eggs depending on size. The eggs are pale yellow, some 0.7-1.2 mm in diameter and stick to weeds, stones and gravel. Incubation takes 5-9 days depending on local water temperatures and the young drift into shallow slow flowing water after hatching.

The Chub, like several other cyprinids, hybridises with other members of the family and crosses between it and Bleak, Roach and Rudd are not uncommon.

Predators

Little has been published concerning the predators of Chub but it can be assumed that many of the 'usual suspects' (e.g. Herons, Mergansers, Otters, Mink) are involved – including, of course, coarse anglers.

Growth

The fry are some 7-8 mm on hatching and initial growth is slow, since they only reach 3-4 cm at the end of their first year. Thereafter growth is more rapid: they are usually some 10 cm at the end of the second year and 15-18 cm at the end of the third year. Males start to mature after 3 or 4 years (the majority are mature by 5 years) and females after 4 or 5 years (the majority by 7 years). By this time they are 20-30 cm in length and growth continues

after maturity. They normally live for 10-12 years. Females tend to grow faster and live longer than males, although males appear to outnumber females in early age groups.

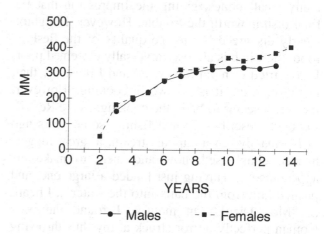

Growth of Chub in the River Lugg, England (Hellawell 1971).

John Hellawell in his extensive study of Chub in the River Lugg and the Afon Llynfi found that 'The maximum age observed was 15 years. ... Dominant year classes were observed. The strongest was 1959, which was attributed to exceptionally-warm, dry, sunny weather during the period May-September. ... Female chub grew more rapidly and attained greater ultimate size than the males.' The most rapid growth occurred after check formation (of scales) in May or June and continued until September.'

Value

The Chub is regarded as a very bony fish and is of practically no commercial importance anywhere in Europe. It is, however, a popular sporting species and is angled for in many of the countries in which it is found. It has the advantage, from the angler's point of view, of taking artificial spinning lures and particularly wet and dry flies like Trout. Manley recorded that 'the Chub, as a fish for sport, is by no means to be despised, though he is not so strong, plucky and determined as some others when hooked.' Williams emphasised the ambivalent attitude to Chub among anglers: 'In rivers where trout abound, chub are quite rightly classed as vermin. In other waters they are well worth fishing for'. Judging by the extent to which it has been introduced to new waters in Scotland in recent years, it must be a popular fish with coarse anglers.

Izaak Walton and several later writers have given involved recipes for cooking Chub in spite of its many small bones, giving the impression that the final dish is worth the trouble. However, Houghton clearly disagreed 'As to the quality of the flesh as food, the Chub is almost universally esteemed poor; I have tried it in various ways, and I must say that, in my opinion, it is not worth cooking.' However, the secret seems to be in the cooking, as the Rev. R. Lubbock describes. 'When fishing some years ago in Normandy, on a capital stream, a 'great logger-headed chub' used, now and then, to make his appearance. ... Having just landed a large one, and pushed him from the bank into the water ... I heard ... "Mon Dieu!" On inquiry, I found the poor woman perfectly horror-struck at my thus throwing pearls away. I thought she was going to weep, as she explained that ... she would have made of it "quelque chose magnifique"... In the evening I left (her) three ... and on the following day ... received a most hearty invitation to dine off my own chub. ... It was, as she promised, very good ... and altogether it might be regarded as a real victory achieved by the cuisiniere over stubborn materials.'

W. St Leger's opinion of its edibility is not recorded, but it was clearly a fish he liked to look at when he was eating as he indicates in *A Gallop of False Analogies*:

'There is a fine stuffed chavender,
A chavender or chub,
That decks the rural pavender,
The pavender or pub,

Wherein I eat my gravender,
My gravender or grub.

Farewell to peaceful pavender,
My river-dreaming pub,
To bed as sweet as lavender,
To homely wholesome gravender,
And you, inspiring chavender,
Stuffed chavender or chub.'

The Chub does well in aquaria and ponds and the author has kept it in tanks with Dace (both from the River Endrick) and other species, for long periods. Gregory Bateman agrees: 'The Chub ... is a hardy and handsome fish, and will live for a long time in an aquarium, under judicious management. He is rather a large eater for his size and seems to prefer insect food.'

Conservation

The Chub has no special protection in Scotland and requires none for it is an introduced species and its distribution is expanding, not declining.

References

Little work has been done on the Chub in Scotland as it is a relative newcomer in most places except the southwest.

Bateman (1890), Boulenger (1946), Bridgett (1929), Buckland (1873), Cragg-Hine & Jones (1969), Fitzgibbon (1847), Gordon (1902), Hellawell (1971), Hickley & Bailey (1982), Houghton (1879), Hughes (1898), Mann (1976b), Maxwell (1904), Walton (1653), Wheeler & Easton (1978).

'His burnished gold outside hides a miserable interior.
He is neither fish, flesh, nor good red herring.'

Edward Fitzgibbon (1847) *A Handbook of Angling*

*****

Orfe *Leuciscus idus* Orf

'It is a surface frequenter,
that is after the early morning chill has gone off the water.'

Jack Hems (1977) *A garden pond*

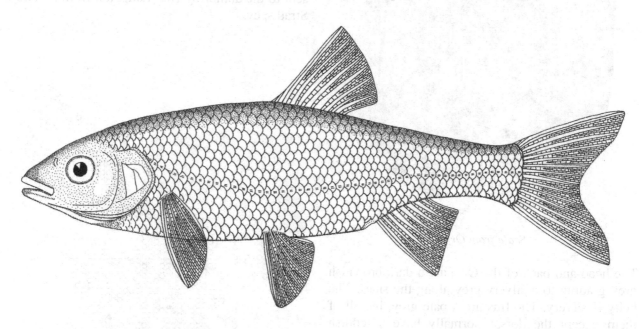

JACK Hems was enthusiastic about the qualities of this species as a fish for the ornamental pond: 'The golden orfe grows to about the size of a common carp, but is much more streamlined in shape. It is pinkish to orange yellow in colour and sometimes marked with a few brownish to blackish spots on the head or body or both. It is a surface frequenter, that is after the early morning chill has gone off the water.' Unlike many other coarse species, which have been introduced to Scotland by anglers, most populations of Orfe here seem to have been introduced to ornamental ponds by estate owners and keen fishkeepers.

Ide is an alternative common name for the species. A literal translation of the scientific name is 'White Ide'.

Size

Like many other cyprinids the 'average' size of the Orfe depends very much upon its habitat, and fish in a small garden pond, for instance, may mature and spawn at 15-20 cm and not grow much beyond that. In the wild, however, they can grow larger, commonly maturing at 30-40 cm in length and 700- 1,200 gm in weight. Exceptionally they may grow much larger than this; fish of up to 100 cm and 8 kg have been recorded from some parts of eastern Europe. The present Scottish rod-caught record stands at 3.604 kg (7 lb 15.5 oz) for an Orfe caught at Culcreuch Castle Loch, Fintry. The present British rod-caught record is for a fish of 8 lb 5 oz 4 dm caught in 2000 by Michael Wilkinson at Lymm Vale, Cheshire.

Appearance

Though the young fish are very slim and laterally compressed, the adults are rather thick set in proportions, with a slight humped back behind the head. The head is quite large and the mouth is broad, terminal and set obliquely. The body is covered with large cycloid scales, which number 56-61 along the lateral line. The dorsal and anal fins are rather short and the former is situated behind the level of the origin of the pelvic fins. The free edge of the dorsal fin is convex whilst that of the anal fin is straight or even slightly concave - a character which distinguishes it from its relative the Chub, where the anal fin has a convex free edge. The anal fin has a longer base than the dorsal fin.

Scale from Orfe.

The head and back of the Orfe are a dark brownish grey grading to a silvery grey along the sides. The belly is silvery. The fins are a pale grey, but all of them except the dorsal normally have a reddish tinge. A golden variety has been bred selectively and is a popular and lively fish for aquariums and ponds. More recently a 'blue' variety has appeared on the market and seems to be quite popular with pondkeepers.

Distribution

The Orfe is native to much of Europe and Asia occurring naturally from eastern France to Siberia, north of the Pyrenees and the Alps. It also occurs in brackish water in the Baltic. It has been introduced much more widely, however, and though never native to the British Isles occurs at a number of scattered sites, mainly in southern England and typically in artificial ponds in large gardens and estates where it was originally introduced as an ornamental species.

In the north of England a typical example is described by George Bolam: ' ...was introduced by Mr C.J. Leyland at Haggerston Castle, soon after the new lake was formed there in 1892, and apparently did remarkably well, as during the succeeding 10 years I used frequently to see large examples swimming near the surface of the water.'

It is established in only a few waters in Scotland, most of them ornamental ponds. However, an apparently viable population has recently been discovered in a small lochan which connects to Loch Rannoch; two large specimens were recently sent to the author by Bob Laughton from a pond in Strathspey.

Recorded distribution of Orfe in Scotland.

Ecology

The favoured habitats of this species are clean slow flowing rivers but it also occurs in some lakes and in the slightly brackish water in the Baltic Sea. It is a fast swimming fish, often feeding in shoals near the surface in shallow water, though in winter, like many other cyprinids, it seems to retreat to deeper waters.

Food

The young fry feed mainly on small invertebrates such as protozoans, rotifers and crustacean larvae. Later they feed on larger invertebrates, for example molluscs, crustaceans, and insect larvae.

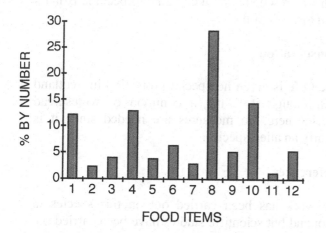

Food of Orfe in the River Kavlingean, Sweden (1: plants; 2: molluscs; 3: worms; 4: cladocerans; 5: copepods; 6: Asellus; 7: beetles; 8: midge larvae; 9: blackfly larvae; 10: others; 11: fish; 12: detritus) (Cala 1971).

Large Orfe may become piscivorous, eating other cyprinids such as Roach. As much of its time is spent swimming near the surface, aerial food, such as insects which fall on to the water, can be important at certain times of year.

Reproduction

Spawning takes place mainly in May and June when the water temperature rises above 10°C. The males develop tubercles on the head and anterior body and start to shoal in the spawning areas in streams or along the stony shores of lakes a few days before the females. When the ripe females appear the males drive them over suitable stones and weed beds and spawning occurs accompanied by much splashing in the shallow water. The pale yellow eggs, ca 1.6-2.2 mm in diameter, are slightly adhesive and stick to stones and weed. Each female may lay from 39,000 to 114,000 eggs, depending on her size. In streams, the shoals of adults drop back downstream again after spawning. The fry hatch in about 8-16 days depending on local water temperatures.

The Orfe will hybridise with other cyprinids in some habitats to produce infertile hybrids. None of these seem ever to have been recorded in Scotland.

Predators

Little has been published concerning the predators of Orfe but it can be assumed that, as well as anglers) many of the 'usual suspects' (e.g. Pike, Herons, Grebes, Otters) are involved.

Growth

The young are only 8-10 mm at hatching but after absorbing their yolk sacs soon start to grow and may reach 10-12 cm after one year and 12-15 cm after two years.

They mature in their third or fourth years and may live for up to 20 years.

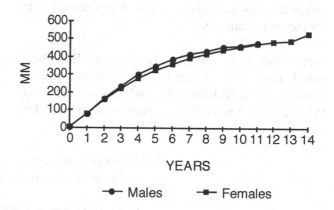

Growth of Orfe in the River Kavlingean, Sweden (Cala 1971).

Value

The Orfe is still fished commercially in a few places in eastern Europe, where it is taken in nets and traps, particularly during the spawning migration. It has the reputation of being a tasty fish to eat.

It is a popular angling species in many countries where it may be taken with both worm and fly, and sometimes by spinner.

David Carss recalls that in the summer 1974 'our next-door neighbours were on holiday and my friend Grant and I devised a novel fishing trip. Tackling-up my split cane fly rod with a small dry fly, I cast over the garden fence into the neighbour's fish pond. An instant take ! Keeping the line taught, we scrambled over the fence armed with the landing net. After a disappointing fight I landed a beautiful orfe of about 20cm. This was one of our more successful fishing trips!'

The golden variety is a popular aquarium and pond fish. Gregory Bateman observed that 'The Golden

Orfe ... is a very beautiful and interesting little fish, and is in every way suitable for the aquarium.' Ian Harman is even more enthusiastic: 'The Golden Orfe is one of the most valued of fish by the aquarist, being perhaps the most attractive of all golden-hued fishes.'

Arthur Boarder comments that 'This fish thrives well in a large pond and does not mind any cold. It can soon grow to eighteen inches long and as it is a surface swimmer and feeder so is usually seen. It also shoals well and so is very attractive.' The author has kept Golden Orfe in both aquaria and ponds and has to confess that the latter habitat is by far the more suitable.

Electrofishing in the grossly polluted Red Burn, near Cumbernauld, Brian Clelland 'was surprised to come across a Golden Orfe - which was presumed to have been flushed down a drain somewhere.' The fish was rescued and taken home to become Brian's daughter's first pet!

Conservation

The Orfe is given no special protection in Scotland and, though it is not a common or widespread species here, no measures are needed since it is clearly an alien species.

References

No work has been carried out on this species in Scotland but scientific studies have been carried out elsewhere in Europe.

Bateman (1890), Boarder (1978), Bolam (1919), Cala (1971), Harman (1950), Hems (1977), Lever (1977), Maitland (1977).

'The Golden Orfe ... is a very beautiful and interesting little fish, and is in every way suitable for the aquarium.'

Gregory Bateman (1890) *Freshwater Aquaria*

*****

Dace *Leuciscus leuciscus* Seorsa de Iasg Aibhne

'It is of a beautiful silvery whiteness, and exceedingly quick in its movements, darting like an arrow away from any person or object that may alarm it.'

William Houghton (1879) *British Fresh-water Fishes*

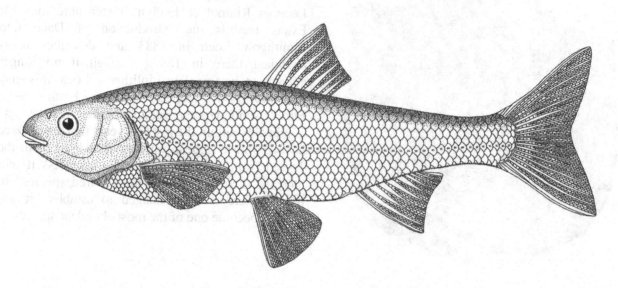

THE first catchment in which the Dace appeared in Scotland was the Tweed; Derek Mills recalls that 'Sweep netting in the River Tweed near Norham in 1972 yielded several hundred Dace in one haul. They were very good eating.' Few people in England would think of eating Dace, but about 100 years ago they were praised by Charles Marson in 1906: 'Dace are by no means to be scorned as table fish. How could they fail to be sweet eating, when their own tastes are so pellucid and aerial? A fine dish of them fried in delicate fats, as soon as the happy angler comes home, makes a royal banquet, with a suspicion of vinegar, a little cress, and some brown bread and butter. You could not do better, believe me, if they were troutlets which hissed and spat in the pan. A little white wine - Chablis shall we say - goes well with them.'

One of the problems with Dace in Scotland is that they do well in our rivers and have similar behaviour and feeding habits to young Atlantic Salmon and Brown Trout, as W.J. Gordon observed: '... often swimming near the surface and leaping out of the water. ... It is a fast swimmer, quick of sight and hearing, and will dart for some distance out of danger when alarmed.' Thus they may well compete with native salmonids and can certainly be a nuisance to anglers.

Alternative common names for this species are Dare, Dart, and Graining. A literal translation of the scientific name is 'The White Fish'.

Size

In most habitats, Dace grow to an adult size of 20-25 cm and a weight of 150-200 gm. Exceptionally, they may reach lengths over 30 cm. The present Scottish rod-caught record is for a fish of 0.553 kg kg (1 lb 3 oz 8 dm) caught by G. Keech in the River Tweed at Coldstream in 1979. The present British rod-caught record is for a fish of 0.574 kg (1 lb 4 oz 4 dm) caught by J.L.Gasson in 1960 in the Little Ouse, near Thetford in Norfolk. However, larger fish have been reported but not substantiated.

Appearance

The body is fairly slim and streamlined with some lateral flattening. The head is narrow and small, as is the mouth which is inferior, its posterior angle not reaching back as far as the level of the eye. The body is covered in large cycloid scales which number 45-55 along the lateral line. Scales are used for age and growth determination. The origin of the narrow-based dorsal fin is above the base of the pelvic fins. Its free edge is concave, as is that of the anal fin. The caudal fin is quite long and deeply forked. Young Dace and Chub can be distinguished by the shape of their anal fins - concave in the Dace but convex in the Chub.

Scale from Dace.

In colour, the Dace appears mainly to be a silvery fish, but the head and upper body are an olive greenish grey grading into silvery grey along the sides and silvery white on the belly. The iris of the eye is distinctly yellowish, often with small dark spots. The fins are a pale grey, but the pectoral, pelvic and anal fins usually have a yellowish orange tinge. William Houghton's description is that 'The Dace is one of the most elegant-shaped fish which we have; it is of a beautiful silvery whiteness, and exceedingly quick in its movements, darting like an arrow away from any person or object that may alarm it.'

Distribution

The Dace is a widespread species across much of Europe and Asia from the Atlantic rivers of southern England and France in the west to Arctic Ocean rivers of Siberia in the east.

In the British Isles the species is native to southeast England but it has been redistributed, mainly by live bait anglers, since the last ice age and is now found over much of England and Wales and the border rivers of southern Scotland, notably the Tweed, as mentioned above. It has recently been moved further north by anglers and is now established in the Rivers Clyde, Endrick (where the first specimens were found by the author in 1986) and Forth , and most recently in Loch Awe - a further major extension of its distribution.

Leonora Rintoul & Evelyn Baxter note that 'Mr Evans records the introduction of Dace into Linlithgow Loch in 1883 and describes it as abundant there in 1892.' Though it no longer appears to be found in Linlithgow Loch, this may have been how it gained entry to the Forth system. Further south, George Bolam recorded in 1919 that ' ... it is only within the last few years that small Dace have begun to appear amongst Roach taken in the Tweed from Twizel downwards: doubtless it will increase there along with the kindred species.' In rivers where it has managed to establish it has quickly become one of the most abundant species.

Recorded distribution of Dace in Scotland.

In the River Forth during the summer of 2000, the author and Alex Lyle were astonished at the abundance of Dace, shoals of which made a loud 'swishing' noise - disturbing the surface as they swam swiftly around in shallow water.

Ecology

The principal habitat of the Dace, which is a shoaling species, is the middle reaches of clean, fast flowing rivers and small streams, largely in lowland areas, but it also occurs in the lower slow flowing reaches of some rivers and in a few lakes.

Food

The young fry feed initially on microscopic food such as diatoms, rotifers, protozoans and crustacean larvae. As they grow they eat larger invertebrates such as tubificid worms, molluscs, crustaceans and a range of insect larvae, but during warm weather they are also very active surface feeders, a considerable proportion of their diet consisting of terrestrial insects and adult aquatic insects which have fallen or alighted on the water surface. They also eat algae and higher plant fragments. They continue to feed during the winter, when molluscs and caddis larvae form important elements in their diet.

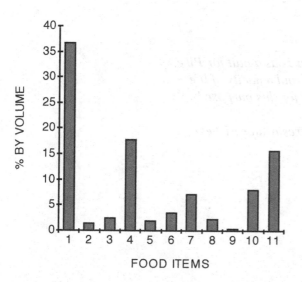

Food of Dace in the River Lugg, England (1: algae; 2: molluscs; 3: crustaceans; 4: caddis larvae; 5: stonefly larvae; 6: mayfly larvae; 7: midge larvae; 8: beetles; 9: others; 10: aerial; 11: detritus) (Hellawell 1974).

Reproduction

Spawning occurs from February until April and the Dace is our earliest breeding cyprinid. The males develop tubercles all over the head and much of the body, and the maturing fish move upstream to suitable spawning areas - usually shallows below stretches of riffle - the males normally a few days ahead of the females. Spawning takes place over gravelly or stony shallows, often where there is some weed growth, and the yellowish eggs, which are about 1.5-2 mm in diameter, are deposited in large numbers during the frantic spawning chases of females by the males. The number of eggs laid by each female ranges from 3,000-27,000. Incubation takes from 2-3 weeks at 12-15°C.

Predators

Little has been published concerning the predators of Dace but it can be assumed that, as well as coarse fishermen, many of the 'usual suspects' (e.g. Pike, Herons, Otters) are involved.

Growth

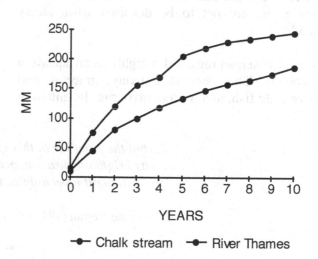

Growth of Dace in the River Thames, England (Mathews & Williams 1972).

The young hatch at a length of 7.5-8.0 mm and grow fairly rapidly, reaching about 6-8 cm at the end of their first year and 10-15 cm at the end of their second year. A few of the larger fish mature in their first and second years, but most mature in their third and fourth years, though they keep growing, rather more slowly thereafter.

They rarely live longer than 7 or 8 years, with males tending to grow faster than females. Dace are known to hybridise with Bleak, Rudd and Bream.

Value

Dace are of little value as a commercial species and are not taken anywhere except as a bycatch when other species are being sought. They are a useful bait species to fishermen in some areas and some anglers fish specifically for them in suitable rivers, where they provide considerable sport for fly fishing as they are even faster risers than Trout. On the other hand, many salmonid anglers consider them a serious pest as they may compete with young salmonids in the river, and they are a nuisance when they are persistently caught during angling. Tate Regan noted that '... the chief use of this species is as a bait for Pike, its bright colouration, activity and tenacity of life making it especially suitable for this purpose.'

William Houghton, though not over-enthusiastic, believed that 'though not held in great repute in a gastronomic point of view, being rather soft in flesh and full of small bones, fine specimens out of our clear rivers are not to be despised when nicely fried.'

Gregory Bateman regarded it highly as an aquarium inhabitant: 'The Dace is a lively, graceful, and active little fish, and a great favourite. Its glittering scales are very conspicuous in the aquarium. The Dace is hardy and soon becomes quite tame.' One Dace he kept 'lived altogether about eight years in captivity, and during this time grew from 2 in. to 6 in. in length.'

Conservation

The Dace has no special protection in Scotland and needs none for it is an introduced species which is still increasing its area of distribution through the activities of anglers.

References

Some research has been carried out on Dace in the River Tweed, but most studies have been carried out in other parts of Europe.

Bateman (1890), Bolam (1919), Cragg-Hine & Jones (1969), Gordon 1902), Hellawell (1974), Hickley & Bailey (1982), Houghton (1879), Kennedy (1969), Kennedy & Hine (1969), Mann (1974), Mann & Mills (1986), Marson (1906), Mathews & Williams (1972), Mills (1982), Mills *et al.* (1985), Regan (1911), Rintoul & Baxter (1935), Starkey (1976), Williams (1965).

'... *but the chief use of this species is as a bait for Pike,*
its bright colouration, activity and tenacity of life
making it especially suitable for this purpose.'

Tate Regan (1911) *British Freshwater Fishes*

*****

Common Minnow *Phoxinus phoxinus* Sgildaimhne

'There is no more sociable fish than the minnow'

Herbert Maxwell (1904) *British Fresh-water Fishes*

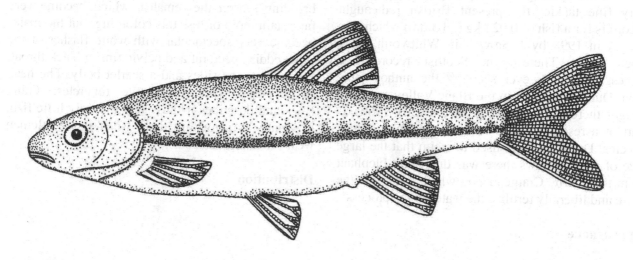

ONE of the earliest fishy experiences of the author, living just to the north of Glasgow, was the catching of Minnows in Craigton Burn, near Milngavie, and taking them home to keep in aquaria - they are excellent aquarium fish and can live for at least five years given the right sorts of food and an appropriate aquarium habitat.

The situation appears to have been no different in the Edinburgh area more than 50 years before, as endearingly recalled by Alasdair Alpin MacGregor in his autobiography *Auld Reekie*: 'One of the commonest diversions to which the children of our neighbourhood had recourse was fishing the steamy edges of the Union Canal for minnows, or 'mennans', as Edinburgh bairns call them. Of a Saturday, one might see scores of barefoot, weather-bronzed children from the poorer quarters, angling with great diligence for these tiny fish ... The tackle appropriate to minnow-fishing was humble enough. It consisted of a glass jar carried by a loop of string, a thin bamboo cane and a tiny detachable net. As often as not, the net was home made. Auld Reekie mothers fashioned them out of bits of muslin, as a means of keeping their offspring out of the way when busy with domestic problems of a Saturday afternoon. ...Youngsters would topple into the slimy fringes of the canal when engaged in this piscatorial pursuit. When the wee fish were more numerous than usual, and congregated close to the bank, the more deft could scoop them out with cupped hands. The more ambitious frequently slipped in ... and it was no uncommon occurrence to see and hear a greeting laddie making for home in his dripping breeks, without tackle or catch.'

One of the most curious of fishes, several fish biologists, including the author, Alex Lyle and Ross Gardiner, have, at various times, had the ticklish experience of Minnows nibbling their toes and legs whilst working barefoot in various Scottish burns and rivers. Perhaps this species has a future with beauticians as an expert in exfoliation!

The Minnow has a great variety of local names, including Baggie, Baggit, Banny, Jack Barrel, Jack Sharp, Meaker, Mennet, Mennon, Mennot, Menon, Minim, Minnin, Peer, Penk, Pink, Shadbrid and Streamer. The tiny fry are known as Pinheads. George Bolam recalls that in the Borders the species is 'Provincially known as Mennam. Males, in May, when the belly becomes red, are called Streamers by boys, perhaps on account of the manner in which they crowd up all side streams, often where there is scarcely enough water to cover them' In Hawick, someone who is unwell is described as being 'sick as a baggie'. A literal translation of the scientific name is 'The Small River Fish'.

Size

This, the smallest member of the carp family in Scotland, matures at lengths of 6-9 cm, but in some favourable habitats can grow up to 12 cm. Minnows are not infrequently taken by anglers fishing with very fine tackle; the present British rod-caught record is for a fish of 0.024 kg (13.5 dm) which was caught in 1998 by J. Sawyer in Whitworth Lake, Spennymoor. There is no Scottish record. The largest specimens ever seen by the author came from Dumbrock Loch in the Blane Valley and were caught there by David Scott in the early 1960s as part of a research study on the physiology of this species. David jokingly used to claim that the large size of the Minnows there was due to an Elephant from the nearby Craigend Zoo which used to bathe there and liberally fertilise the water in the process!

Appearance

The body is slim and rather round in cross-section with only slight lateral flattening. The snout is blunt and the small mouth is subterminal and has no barbels. The body has rows of very small rather embedded scales along the sides - of which there are 85-100 along the lateral series. There are no scales on the belly. The lateral line itself is incomplete posteriorly. All the fins are of medium size and rounded. The dorsal and anal fins have much the same size and shape. The origin of the anal fin is just below the last rays of the dorsal.

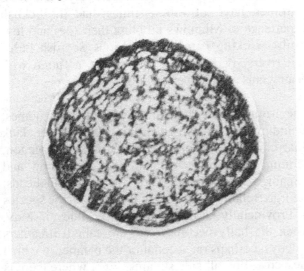

Scale from Minnow.

The colour of this species is very variable, both inside and outside the breeding season. The normal colouration is an olive greeny-brown on the upper head and back and a whitish grey on the belly. These are separated along each side by a line of colour which can vary from a series of indeterminate dark and gold blotches to a complete black stripe from head to tail fin. During the breeding season the females, which become very fat, retain more or less this colouring, but the males become very spectacular with white flashes at the fins, reddish pectoral and pelvic fins, a black throat, green along the sides and a scarlet belly. The head is peppered with tiny white tubercles. Frank Buckland describes it thus. 'It is a lovely little fish, especially in spawning time, when the gentlemen wear red and green coats.'

Distribution

Recorded distribution of Minnow in Scotland.

This species is very widespread in fresh waters across virtually the whole of Europe and Asia, from the Atlantic coast of Ireland to the Pacific coast of China. Niall Campbell recalls catching them in Lake Baikal. It occurs in brackish water in some parts of the Baltic Sea. In the British Isles, though presumably originally found only in the south-east of England, it is now widespread in England, Wales, Ireland and Scotland - with the exception of the Northern and Western Isles. Though Harvie-Brown and Buckley noted in 1887 that 'In twenty years we have never once met with the minnow in Sutherlandshire.'; this is no longer the case, as it now occurs in several catchments there.

There is considerable argument as to whether the Minnow originally arrived in Scotland naturally or was introduced by humans. Though there is no doubt that Minnows have been moved around by anglers, it is perfectly possible, and the author believes likely, that this small stream fish arrived here by its own efforts. It is found in the upper reaches of many streams and is quite capable of crossing watersheds at times of flood. An excellent example of where it can do so is just beside the M74 at Beattock Summit, where the head waters of the Evan Water (a tributary of the River Annan, flowing south) rise in a bog which also has the source of the Clydes Burn (a tributary of the River Clyde, flowing north). Minnows are abundant here.

Ecology

The Minnow occurs in a wide range of habitats, from small fairly fast flowing streams to large lowland rivers, and from small upland lochs to large lochs and reservoirs. It must have access to clean gravelly areas for spawning - either in running water or on the wave washed shores of lochs. This is a shoaling species which can often occur in enormous numbers, especially in summer when shoals, usually comprising only one year class, move into very shallow water at the edges of pools (often only a centimetre or so deep) to feed and sunbathe. It is probable that temperature conditions at these sites are optimal for growth, but they are vulnerable at this time and, as noted by George Bolam, even '... Grey Wagtails have been noticed to catch them.' During winter, fish move into quieter water and often spend long periods under stones or in thick vegetation in a semi-comatose condition.

They are very inquisitive fish, and after being frightened away initially by the appearance of a human, will come back to have a better look. If the bottom is disturbed by a stone or net, Minnows quickly come forward to seek any food items that have become exposed. As mentioned above, on several occasions when working barefoot in the River Endrick the author has had shoals of Minnows nibbling at his feet and legs! This curiosity often leads to their downfall for a small transparent plastic trap, baited with a piece of bread or other attractant, can catch large numbers of Minnows. Once one fish is inside the others seem to be attracted in even more.

During winter, Minnows shelter under stones or large weed clumps. At Loch Walton, in the early 1960s, the author remembers having to break up ice about 5 cm thick, in order to obtain samples of Minnows from under the stones there.

Minnows also appear to be curious of large fish which could be potential predators and two or three may break away from a shoal to investigate. They can then communicate their fear or apprehension to the rest of the shoal which may immediately flee. Frightened or injured Minnows give off a 'fear' substance from their skins which quickly warns the rest of the shoal of impending danger. Alarm substances are emitted by many members of the carp family. They seem to be rather non-specific in their action so that an 'alarm' raised by one species can be communicated to other cyprinid species in the vicinity. As one author aptly puts it 'the situation is analogous to the use of similar alarm calls by different species of small birds'.

Food

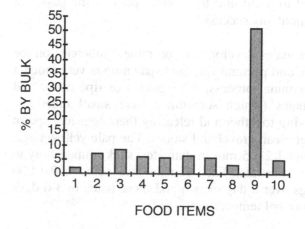

Food of Minnow in the Altquhur Burn (1: worms; 2: crustaceans; 3: mayfly larvae; 4: stonefly larvae; 5: caddis larvae; 6: midge larvae; 7: beetles; 8: others; 9: algae; 10. surface insects) (Maitland 1965).

The young start to feed on minute foods such as rotifers, protozoans and small crustaceans as well as plant material like diatoms. As they grow they continue to eat a varied diet, taking many crustaceans and bottom invertebrates such as worms and midge larvae, as well as quite a high proportion of plant material - mostly algae. They are truly omnivorous in their feeding behaviour.

Apart from plant material, the food of Minnows is much the same as that of Trout, and where they occur in large numbers it is likely that they must compete with Trout for the available food supply. Shoals can often be seen rising to surface food.

Reproduction

Reproduction takes place from April to June when water temperatures reach 10°C and more. As they come into breeding condition and the males colour up there is a migration to suitable spawning beds. These may be nearby on a clean wave-washed loch shore or where an underground spring upwells through the gravel, or at some distance upstream. Shoals of adult fish often migrate considerable distances at this time, from a loch into an inflow or from a large slow-flowing river or canal into a swifter side tributary. Niall Campbell has seen them wriggling upstream through wet Willow Moss, with their backs out of the water, when they could be scooped up by the handful. David Dunkley recalls 'seeing Minnows trying to leap into the fish pass at Philiphaugh on the River Ettrick. They were watched for some considerable time and they tried hard to jump into the bottom pool of the pass, but without any success.'

The males develop very prominent tubercles on the face and pectoral fins, and spawning is very much a communal process, with groups of ripe males and females (which sometimes have small tubercles) driving together and releasing their eggs and sperm over clean gravel and stones. The pale yellow eggs, about 1.2-1.5 mm in diameter, stick immediately to this substrate. Each female lays some 200-1,000 eggs over a day or two and these hatch in 3-6 days at normal temperatures.

Predators

As well as the 'usual suspects' (e.g. Pike, Herons, Otters), a variety of other, smaller, predators prey

on this common fish. In 1900 Herbert Maxwell recorded that 'On the opposite bank a hill burn flows into the river, rippling over gravelly shallows, a favourite gathering-ground for minnows. My friend the dipper was just as busy fishing as I was, and with better luck. He dashed repeatedly through the shoal ... At last, after many failures, he caught a minnow, went ashore with it, ... pinched the wriggle out of the wretch, winged his flight to a cascade ... where, no doubt, a little hungry circle of Masters and Misses Dippers were awaiting their fish course.'

Growth

The young are only 4-5 mm on hatching (and known colloquially in Scotland as 'pinheads') but take a few days to absorb their yolk sac and start feeding, by which time they are about 7-8 mm. Growth is very variable according to local conditions, but after one year most fish are 3-4 cm, after two years some 5-6 cm and after three years about 6-8 cm. They may live, and continue growing at a slower rate, for up to 7 or 8 years. Maturity in both sexes is usually reached in the second year, though fast growing individuals may be mature after one year. Mature females are normally larger than males.

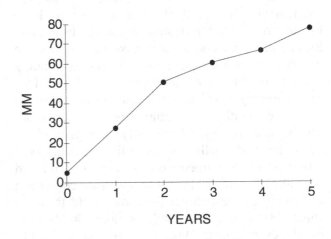

Growth of Minnows (Mills & Eloranta 1985).

Although the main spawning season is spring and early summer, very small 0+ Minnows can be found in the late autumn, suggesting a more protracted spawning season in some places. In nearly all habitats, numbers of small immature Minnows can be found at any time of the year. Probably as an adaptation to faster growth, on sunny days small

Minnows can frequently be found crowding into the shallowest water at the edges of lochs and rivers. As an astute observer of nature, John Keats picks this up in his poem *On a Summer's Day*:

'Where swarms of minnows show their little heads,
Staying their wavy bodies 'gainst the streams,
To taste the luxury of sunny beams
Temper'd with coolness. How they ever wrestle
With their own sweet delight, and ever nestle
Their silver bellies on the pebbly sand!
If you but scantily hold out the hand,
That very instant not one will remain;
But turn your eye, and they are there again.'

Value

Though caught in enormous numbers in seine nets and traps in a few parts of northern Russia for use as fertiliser or animal food, the Minnow is really of no commercial importance in Scotland or anywhere else in Europe other than as a bait fish where it is used both as live bait and dead bait. Its use as live bait has resulted in a widespread distribution.

Minnows are angled for occasionally. Taylor records that 'Though so diminutive in size, the minnow may be compared, for the excellency of its taste, to some of the most famed fish. ... on hot days, they will very eagerly bite all day long, and afford great sport to youths and others that like to angle for them.'

They are reputed to be palatable when fried like whitebait; Izaak Walton gives directions on 'how to make a minnow tansy, wherein the fish being gutted, but not washed, must be fried with yolk of eggs, cowslips, primrose, and tansy.' Herbert Maxwell recorded that 'William of Wykeham gave a great banquet at Winchester to Richard II and Queen Anne, and among the fish served were seven gallons of minnows, costing 11s 8d.' J.J. Manley noted that 'small Minnows are an excellent substitute for real Whitebait.' and Gregory Bateman agrees: 'Man has also found that they make delicious food when properly cooked.'

In his angling diary, Patrick Smythe recorded '16 June Coming back from Church caught a minnow with my hands. I ate minnow.' Frank Buckland noted that 'When at Winchester, I used to pickle minnows with vinegar and spice, and keep them in pickle bottles. They were capital eating, especially when, as a "junior", I was not over-fed and had to look out for what extra "grub" I could get hold of.'

Few people eat Minnows nowadays, but Dick Shelton recalls that 'My own experience of eating Minnows (72 taken by long-trotting a single maggot on a miniature rod) is that they are excellent headed, gutted and then fried in a batter of egg and seasoned flour. One problem in some areas is that Minnows may pick up an earthy taint - leaving the headed and gutted bodies overnight in salted water (to which a little vinegar has been added) is effective in dealing with this problem.' Another contemporary account is that of David Carss 'In the summer of 1975, abandoned all day beside Fenwick Water from about 08.00h by my friend's dad as he commuted to Glasgow, our 'upstream fly-fishing in rough burns' project was plainly going to be a disaster. Within an hour, I'd fallen in and was soaked from head to foot and we'd eaten all our lunch. Only 8 hours to go. After following the burn downstream for some distance, it cascaded over impressive falls into a deeper pool. Abandoning our scruples, we dug around for some worms and were soon casting out. We soon hit a shoal - of Minnows. After some time we'd amassed quite a haul. By now extremely hungry, we cooked each fish individually over Grant's cigarette lighter with a bit of butter and the tin foil left from our sandwiches. We ate the minnows 'white-bait' fashion, but I can't remember what they tasted like.'

It is widely used by anglers as a bait species (hence its wide redistribution throughout the British Isles) and is an important forage fish for larger fish species such as Pike and Perch and for many aquatic birds like Herons and Kingfishers. It is easily caught for bait; Aflalo comments that 'They are also endowed with a fatal curiosity that prompts them to congregate over a net in which are tied fragments of red wool, ...' Derek Mills remembers when he was a boy and needed live bait, he used to wade out into the shallows of a stream and placed a jar over his toes. The Minnows came to nibble his toes and swam into the jar and were soon transferred to a bait can. Trout normally appear to have some difficulty in catching Minnows (compared to the apparent ease with which they catch sticklebacks) - except at spawning time when large numbers of Minnows are eaten at the spawning beds.

The value of this species to fisheries is uncertain. Though promoted as beneficial by Frank Buckland 'Wherever there are minnows the trout are always very good: minnows should be introduced into trout streams.', there is evidence from research in Scandinavia that Minnows are actually detrimental to populations of Brown Trout.

Being easy to obtain and keep, it has also been used widely in laboratory experiments, and the early scientific work on fish hearing used this species.

It is an interesting, peaceful and attractive aquarium fish - very tame individuals have been kept for several years in aquaria by the author. Gregory Bateman agrees: 'The Minnow ... is a beautiful, hardy, and as a schoolboy would say, "cheeky" little fish. Though Minnows are generally found in clear and running water, yet there are no fish which are ready to take more kindly to life in confinement than they. ... Mr Jones, of Woolwich, put a few of these fish in the same aquarium with the silver carp ... and thirteen years passed before the last of them died.' Herbert Maxwell's affection for this little fish is unmistakable: 'There is no more sociable fish than the minnow; he is never to be found save in company with his kind. No edible substance comes amiss to his unflagging appetite; if the morsel is too large for a single minnow to tackle, his companions gather thickly round, and soon demolish it among them.' Henry Lamond too was fond of it: 'I have an affectionate regard for this little fish. For more than two years at Luss I kept a pet minnow in a vacant goldfish bowl, and we became the best of friends. ...

When I entered the room he would jump from the water and turn endless "cart-wheels" till I got him a fly to eat. The yellow kind of cake which has an occasional sultana raisin in it became his favourite dish, and he grew to be very fond of any rice pudding made with eggs...' Harman sums it up: 'There are few more delightful sights to a fish-lover than an aquarium containing a shoal of Minnows.'

Conservation

The Minnow receives no special protection in Scotland and indeed none is necessary, for this is a widespread and successful species which is still expanding its distribution in the northern part of Scotland, partly due to transfers by anglers who use them as bait. Like other small fish, it deserves to be spared from the despicable practice of livebaiting.

References

There have been several studies of the Minnow in Scotland and also in other parts of Europe.

Aflalo (1897), Bateman (1890), Bibby (1972), Bolam (1919), Buckland (1873), Frost (1943), Harman (1950), Harvie-Brown & Buckley (1887), Houghton (1879), Lamond (1931), Lein (1981), Levesley & Magurran (1988), Maitland (1965, 1966), Manley (1877), Maxwell (1900, 1904), Mills & Eloranta (1985), Pitcher *et al.* (1986), Rasottos *et al.* (1987), Scott (1963), Smythe (1956), Stott & Buckley (1979), Taylor (1800), Walton (1653), Wootton & Mills (1979).

'Though so diminutive in size, the minnow may be compared, for the excellency of its taste, to some of the most famed fish.'

Samuel Taylor (1800) *Angling in all its branches*

*****

Roach *Rutilus rutilus* Roisteach

'He is rather a handsome fish,
and by no means the "fresh-water sheep"
which some people think him to be.'

Gregory Bateman (1890) *Freshwater Aquaria*

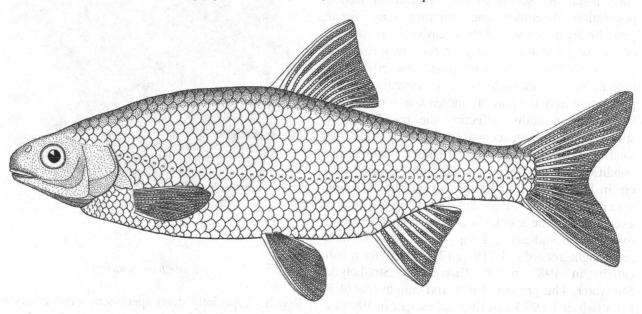

THE Roach is regarded by the author as a native species which has moved north naturally on the west coast as far as Loch Lomond. However, there appears little doubt that it has been introduced by some anglers to many catchments on the east coast, where it has not been made welcome by other fishermen. For example, Herbert Maxwell notes that 'Mr Harvie-Brown has recorded the introduction of this fish to the district of Moray in 1886 or 1887, when a stock was turned into the Loch of Spynie. Persons who own or manage valuable waters ought resolutely to refuse to establish roach among them. The roach in the Castle Kennedy lochs above referred to have multiplied to such an extent that the trout-fishing, which was formerly excellent, has very much deteriorated. The roach are now netted out in cartloads, and applied as manure to the land.'

Further south in Scotland, George Bolam writes of the Roach as 'Another alien on the Borders, where it was wholly unknown till about 1898, when quite small ones began to appear amongst minnows in the Tweed at Twizel. They increased rapidly, both in numbers and size, examples of half a pound in weight being pretty frequently found in the salmon-

nets there by 1903.' 'During the netting operations carried out in 1913, for the destruction of coarse fish, the Tweed bailiffs destroyed no fewer than 1436 Roach between 7th April and 9th August, ...' 'On 12th December 1889 the late Mr George Gray informed me that, for a good many years previously, he had kept Roach in a pond at Milfield, for the purpose of keeping himself supplied with fresh bait for winter Pike-fishing. They had bred freely in the pond, and increased rapidly, and there was no obstruction to the descent of small ones to the Till at any time.'

These incidents and many others indicate just how fast some alien species can establish themselves and also that it is virtually impossible to get rid of them once established.

The local name of Braise was well known to people in the west of Scotland, for example to John Colquhoun of Luss: 'One day a boy brought me a basket of five very broad-shaped fish with red fins, like the bream [actually Roach] or "braise" of Loch Lomond.' A literal translation of the scientific name is 'The Red Fish'.

Size

The size of Roach in different populations is very variable and dependent on local conditions. In some waters there are huge populations of small stunted fish, whereas in other waters where their abundance is less, they grow much larger individually. Year class instability seems to be a feature of Roach population dynamics and stunting may occur periodically as a result of the occasional appearance of an exceptionally strong year class as a result of particularly favourable conditions. Conversely, the growth rate of individuals in a normally stunted population may temporarily increase as a result of a large scale mortality affecting the population. A strong year class can account for as much as half the total population present. Under reasonable conditions, adult Roach are normally some 20-25 cm in length and 200-300 gm in weight. In a few waters fish of up to 35 cm and 1 kg occur and exceptionally the species has been known to reach over 50 cm and about 2 kg. The present Scottish rod-caught record is 1.219 kg (2 lb 11 oz) for a fish caught in 1987 by P. Russell in Strathclyde Stillwater. The present British rod-caught record is for a fish of 1.899 kg (4 lb 3 oz) caught in 1990 by R.N. Clarke in the River Stour in Dorset.

Appearance

Though relatively slim when young, the adult fish is rather deep-bodied and laterally compressed with a small head. In stunted fish, the body tends to remain rather slim and does not deepen. The mouth is terminal and slightly oblique and has no barbels. The body is well covered by large cycloid scales which number 42-45 along the lateral line and characteristically have a number of furrows which radiate from the centre of the scale over both the exposed and embedded areas. Scales are used for age determination, as are the gill covers (opercular bones). The origin of the high dorsal fin is directly above the base of the pelvic fins. The paired pectoral and pelvic fins are rather small, but the anal fin is well-developed. The tail is well developed and deeply forked.

The upper head and back of the Roach is a dark olive greenish brown grading to a silvery brown on the sides and pale white on the belly. The iris of the eye ranges from yellow in small fish to a strong red in adults. The dorsal and tail fins are a dull brownish, but the remaining fins have a reddish tinge, especially the pelvic and anal fins which may be a dark red in some specimens.

Scale from Roach.

Roach - especially dead specimens - can easily be mistaken for Rudd, while the quite common Roach x Rudd hybrids may also provide identification difficulties. Roach also hybridise with Common Bream, Silver Bream and Bleak.

Distribution

The Roach is a common and widely distributed species throughout much of Europe and Asia from Ireland in the west to Siberia in the east. It is very common in many canal systems, occurring also in slightly brackish water in mainland Europe, in the Baltic, Black and Aral Seas. Several subspecies are recognised in different parts of its range.

In the British Isles it is common throughout most of England, including the Channel Islands (Jersey) and parts of Wales. In Scotland it occurs on the west coast as far north as Loch Lomond and Loch Ascog on Bute, and on the east coast from the Tweed to Morayshire, including the catchments of the Rivers Forth, Tay and North Esk. It is abundant in the Forth and Clyde Canal and in many ponds and lochs in the central belt. During the 1990s Roach and several other coarse species were introduced to numerous waters in southwest Scotland.

worms, molluscs, crustaceans and many species of insect larvae, as well as much detritus, filamentous algae and higher plants.

Although much more of a bottom feeder than Rudd, Roach will take adult insects from the surface of lochs and rivers, especially during warm sunny weather.

It has been suggested that the more animal food that there is in the diet, the faster and larger Roach will grow. The most stunted populations are found where detritus is the main source of nutrition. These stunted fish, when moved to richer waters where more animal food is available, will respond by increasing their rate of growth markedly. Roach will certainly grow more rapidly where they can feed on molluscs, but do not seem to exploit this food source until they have reached about 15 cm in length. There is evidence that while young Roach indiscriminately eat much detritus, older fish are more selective, rejecting material that cannot be digested.

Recorded distribution of Roach in Scotland.

Ecology

The Roach is a shoaling species which is tolerant of a wide range of temperature conditions and even some mild pollution. It occurs in a variety of habitats, mainly in lowland areas, ranging from small weedy pools to large clear lochs (e.g. Loch Lomond) and from small streams to the largest lowland rivers (e.g. the River Tay) and often their upper estuaries. In winter, large shoals move into deeper water where they feed relatively little until the water starts to warm again in spring. One of their sporting virtues, however, is that they are more readily caught than Rudd during the winter (hence one of the motives for spreading them in Ireland - to attract the tourist 'off season'). During the mild winter of 1987-88 specimen Roach were taken throughout the winter.

Food

The young feed on very small invertebrates (e.g. rotifers, protozoans and crustacean larvae) at first along with some vegetable matter (e.g. diatoms). As they grow they move to larger invertebrates, especially planktonic crustaceans and eventually to a diet of a wide variety of bottom animals such as

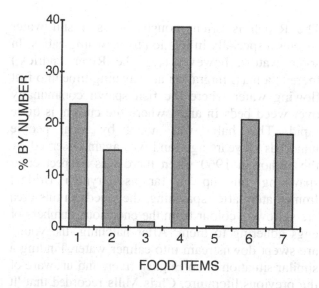

Food of Roach in the River Frome, England (1: mayfly larvae; 2: midge larvae; 3: caddis larvae; 4: molluscs; 5: crustaceans; 6: others; 7: algae) (Mann 1973).

Being largely mud swallowers and therefore having to ingest and process large amounts of indigestible material, Roach typically do not have a stomach (or pyloric caecae) but the intestine is very long - around 15 times the length of the fish. With their well-adapted mouths, Roach are able to penetrate some 5 cm into the bottom mud, about the same as Bream but less than Tench or Carp.

Reproduction

Spawning takes place in April, May and June when the water temperature rises above 12°C. The mature males develop numerous small white tubercles over the head, anterior fins and much of the body, and they start to chase ripe females, some of which may also have tubercles. Eventually the fish, often in small groups of one or more females and several males, dash into weed beds where the eggs and sperm are released together, usually near the surface amid much splashing. Tate Regan states that during the breeding season 'they press together in a dense mass and by their movements against each other produce a hissing noise.' The eggs, which are pale yellow in colour and 1.0-1.5 mm in diameter, are adhesive and stick to the weeds, often in enormous numbers. Each female may lay 5,000-200,000 eggs according to size. The eggs hatch in 5-10 days depending on local temperatures, the new fry remaining attached to weed for a further few days until the yolk sac is absorbed and they can start to swim freely. In some waters, an intersex condition has been recorded in Roach.

The Roach is often thought of as a still water species, especially in relation to spawning habits. In some waters, however (e.g. the River Endrick,) there is a mass migration at spawning time into fast flowing water where the fish spawn communally over weed beds in areas where the current is quite rapid. This habit was noted by local people hundreds of years ago, and was again observed by the author in 1960 when there was a spectacular spawning run up as far as Drymen Bridge. Immediately after spawning, the weeds in this area are yellow in colour from the enormous numbers of eggs adhering to them. After hatching, the young are swept downstream into calmer water. Finding a similar situation in an English river, and unaware of the previous literature, Chris Mills recorded that 'It is perhaps surprising that the roach, typically a fish of still or slow flowing waters, will spawn on substrates adjacent to relatively fast flow.'

Predators

They are much hunted by Pike and when attacked the shoal of Roach will scatter instantly, diving into thick vegetation for cover. Young Roach are important food for Trout and Perch, as well as of diving birds such as grebes and Kingfishers - as

Richard Jefferies has recorded: 'Roach are here the kingfisher's most common prey. He chooses those about four inches long by preference, and "daps" on them the moment they come near to the surface.' David Carss and Keith Brockie found Roach to be one of the commonest items (21%) of the diet of Ospreys in Scotland.

Roach can be host to a large range of parasites, both internal and external. At least 53 species have so far been found on or in Roach in the British Isles. The cestode tapeworm *Ligula intestinalis* is a common parasite and can cause heavy mortality in Roach populations. Its life cycle involves two intermediate hosts, initially a planktonic crustacean, then a fish and finally an aquatic bird. The fluke *Diplostomum spathaceum* is also common where Herring Gulls are among predators of Roach. A common parasite of Roach is the Fish Louse, *Argulus foliaceus*.

Frank Buckland recalled that 'The roach is subject to a curious disease in which the scales turn jet black. ... The disease is often very fatal to whole shoals of roach.'

Growth

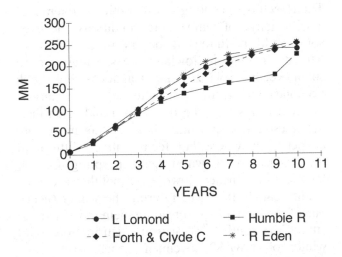

Growth of Roach in various Scottish waters (Mills 1969).

The fry are some 4.5-6.5 mm on hatching but they soon start to grow. Under favourable conditions they may reach 6-9 cm after one year, 9-12 cm after two years, 12-15 cm after three years (about the maximum length in stunted populations) and so on. The number of days when the water temperature is 14°C or above is a major factor in good growth. Males mature in their second or third years and females in their third or fourth years. Growth continues after maturity, though usually at a slower rate. Fish often live for 10-12 years, with females growing faster and living longer than males.

Derek Mills found that 'The growth rates of roach in some Scottish waters are much slower in the first two years of life than in some waters in England and Wales ... This is probably due to lower water temperatures resulting in a later spawning season and a shorter period of rapid growth. However, the growth rate of the older Roach from three Scottish waters compares favourably with that of roach from Llyn Tegid and Willow Brook and is better than that of roach in the River Birket and River Thames.'

Value

The Roach is a commercial species in some parts of Europe where it is caught in seine nets and traps. At one time it was important as a food fish in the Loch Lomond area; Mr Ure, a Killearn minister described in 1795 how 'the vast shoals of Braise which come up the River Endrick from Loch Lomond were at one time caught by nets in thousands.'

It is considered by some as the most important and popular of all sporting coarse fish and is much sought after by anglers both as a welcome addition to the bag during competitions and as a specimen fish requiring considerable skill to catch, though Juliana Berners recorded that 'the Roche is an easy fysshe to take.' and Isaac Walton called them 'Water Sheep' (in contrast to his term for Carp 'Water Fox') on account of their shoaling behaviour and apparent simplicity and foolishness. Possibly the wilyness of large fish in some populations is due to the fact that a great many individuals have been hooked a number of times and released. Small specimens are commonly used as livebait for Pike.

Isaac Walton reckoned that from the culinary aspect, the best feature of Roach is their spawn.

Houghton felt that 'The Roach is not much in request as an article of diet, the flesh being generally soft and woolly; nevertheless large specimens in September and October are not to be despised when nicely fried.' In fact, there is a great variation in flesh quality in this and some other coarse species. The author has eaten Roach from the clear waters of Loch Lomond, whereas those from the Forth and Clyde Canal are inedible.

It is an attractive species for the aquarium and pond, but, as various writers have noted it is not so easy to keep as Rudd, nor does it possess a golden variety. However, Bateman noted that 'The Roach ... if young, will live fairly well in an aquarium. He is rather a handsome fish, and by no means the "fresh-water sheep" which some people think him to be.'

Conservation

The Roach has no special protection in Scotland and is one of the few native species which appears to be under no threat at present.

References

There have been a few studies of the biology of this species in Scotland and many others area available from other parts of Europe.

Ali (1976, 1979), Bateman (1890), Berners (1496), Bolam (1919), Broughton & Jones (1978), Buckland (1880), Burrough (1978), Burrough & Kennedy (1979), Carss & Brockie (1994), Colquhoun (1880), Cragg-Hine & Jones (1969), Goldspink (1978, 1980), Hellawell (1971, 1972), Houghton (1879), Jefferies (1937), Linfield (1979, 1980), Maitland (1966), Mann (1973, 1975), Maxwell (1904), Mills (1969), Mills (1981), Regan (1911), Sweeting (1976), Ure (1795), Walton (1653), Wheeler & Easton (1978), Williams (1965), Wilson (1971), Wood & Jordan (1987).

'the Roche is an easy fysshe to take.'

Juliana Berners (1496) *A treatyse of fysshynge wyth an angle*

*****

Rudd *Scardinius erythrophthalmus* Bronn dhearg

'A Rudd of one or two pounds is a very beautiful fish,
with its bright red eyes and fins, and reddish gold body.'

William Houghton (1879) *British Fresh-water Fishes*

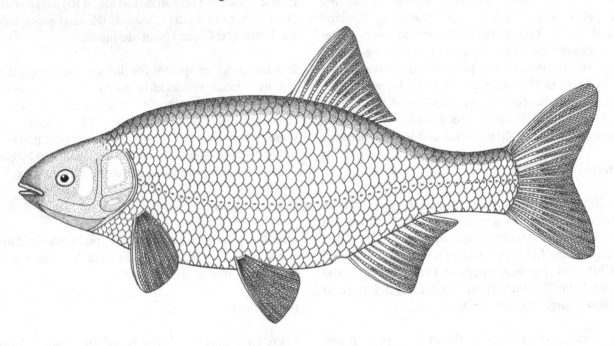

IZAAK Walton did not believe that the Rudd was a proper species and thought that it was a hybrid Roach x Bream: 'There is a kind of bastard small Roach that breeds in ponds, with a very forked tail, and a very small size, which some say is bred by the Bream and right Roach...' William Houghton disagreed '... it is very like the Roach in general appearance, and often mistaken for one. There is one character, however, which at once reveals the difference between the Rudd and the Roach; in the Roach, the origin of the dorsal fin is only slightly behind that of the ventral fin, while in the Rudd it is conspicuously behind it.' In his more loquacious way, Richard Franck agreed 'Whilst we paraphrase and discourse the roach, we but decipher and interpret the rudd; since Nature's laws are alike to both, for both have but one fate and period, though of different complexion of fin.' Though similar in appearance, they have quite different life styles.

There are several alternative names for the Rudd, including Finscale, Red Eye, Roach and Shallow. A literal translation of the scientific name is 'Red-eyed Scardine'.

Size

The size of adult Rudd is normally some 20-30 cm in length and 200-400 gm in weight. In particularly favourable habitats the species may grow up to 40 cm and one kg, and very occasionally fish larger than this are encountered. The present Scottish rod-caught record is vacant, with a qualifying weight of 0.9 kg (2 lb). The present British rod-caught record is for a huge fish of 2.041 kg (4 lb 8 oz) caught in 1933 by E.C. Alston in a mere near Thetford in Norfolk.

Appearance

The Rudd is a slim but fairly deep-bodied and laterally compressed fish. Large Rudd are decidedly hump-backed. The head is small and the mouth is terminal, but set obliquely and with no barbels. The body is covered in large cycloid scales of which there are 37-45 along the lateral line. Scales are used for age determination but care is required in determining the first annulus and in recognising subsequent checks or false annuli. There is a well developed keel between the pelvic fins and the anus. Characteristically, the origin of the dorsal is well behind that of the pelvic fins. The tips of the pectoral almost reach the bases of the pelvic fins. The caudal fin is well notched with pointed lobes.

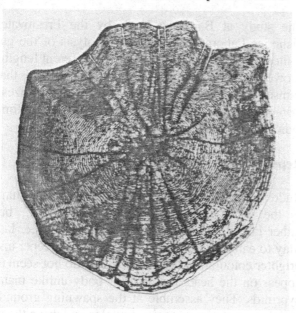

Scale from Rudd.

Young Rudd are rather silvery in colour, but the adults are a dark olive brown on the upper head and back grading to a yellowish silver on the sides and dull white on the belly. The eye has a beautiful golden-red iris with a red spot above. All the fins have a reddish hue, especially the ventral fins which may be quite strongly red. All colours become intensified at spawning time. Selective breeding has produced a golden form of this species which is a popular aquarium and pond fish. Houghton was impressed by its appearance: 'A Rudd of one or two pounds is a very beautiful fish, with its bright red eyes and fins, and reddish gold body.' and an anonymous poet obviously agrees:

> '... a kind of Roach all tinged with gold,
> Strong, broad and thick, most lovely to behold.'

Rudd are often confused with Roach with which they frequently occur. The Rudd can easily be distinguished by the upturned mouth, the position of the dorsal fin, the keel behind the ventral fins and the serrated pharyngeal teeth. Aflalo comments that 'Easiest of recollection, however, is the fact that the upper lip of this fish is horny and rigid, whereas that of the Roach can be pulled forward.'

Distribution

Recorded distribution of Rudd in Scotland.

The Rudd occurs throughout much of Europe and western Asia from Ireland in the west to the Aral Sea in the east. In the British Isles, it is native only to southeast England but has been redistributed by humans so that it is now common in much of Ireland, the midlands and south of England, the Channel Islands (Jersey) and a few places in Wales. Formerly, there were only two breeding populations known in Scotland - both of them in small lochs in the south-west, but in recent years there have been numerous introductions, again mostly in the southwest. However, it is not yet known in which of these sites they have managed to establish viable populations. The most northerly population known to the author is in a small lochan near Loch Rannoch, where several other coarse species occur.

Ecology

This species is a shoaling fish: in Ireland Niall Campbell has recorded an enormous static shoal of densely packed Rudd (mainly 10-13 cm in length) on two occasions during April, congregated in an artificially made inlet entering a small lake. Some confusion exists over its distribution in Ireland due to it being called 'Roach' in some areas. In some waters it can be very abundant - a shoal described by Niall Campbell was some 30 m long and 2 m wide - a conservative estimate put the number of fish at 100,000 plus. This event apparently took place every spring, but did not appear to be connected with feeding or spawning. It favours still water and is found mainly in rich lowland ponds and lakes, usually where there is an abundance of vegetation. It can stand relatively foul conditions better than some of the species with which it often co-exists. It occurs also in the lower reaches of some lowland rivers. In winter large shoals move to deeper water to await the spring.

Food

The newly-hatched fry feed on very small organisms such as diatoms and other unicellular algae, rotifers, protozoans and crustacean larvae. Rudd, on the whole, are mid-water feeders; as they grow, they continue to take some vegetable material in the form of diatoms and other algae, but most of the diet consists of planktonic crustaceans.

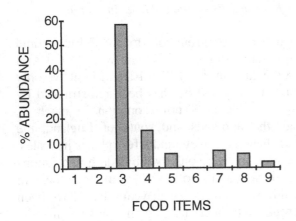

Food of Rudd in Barscobe Loch (1: snails; 2: crustaceans; 3: dragonfly larvae; 4: caddis larvae; 5: mayfly larvae; 6: midge larvae; 7: surface insects; 8: vertebrates; 9: others) (Freshwater Fisheries Laboratory 1994).

Later, as adult fish, they take benthic invertebrates such as crustaceans and insect larvae, especially those of caddis and midges, and often adult aquatic and terrestrial insects which have hatched or fallen on to the water surface. Filamentous algae and shoots of fine water plants may assume considerable importance in the diet. Very occasionally, large Rudd will eat small fish of other species, for instance in Ireland they eat their own

fry and Three-spined Sticklebacks. They are rather more active and less cautious in their feeding behaviour than Roach and, as well as being taken by traditional float and bait methods, they may be caught on wet or dry fly or by spinning.

The study at Barscobe Loch by the Freshwater Fisheries Laboratory noted that 'analysis of the gut contents of a sample of 22 Rudd of different lengths showed that invertebrates predominated and that damselfly nymphs were take by fish of all sizes.' Comparison with Trout taken at the same time indicated a considerable overlap in diet.

Reproduction

Rudd spawn from April to June in southern Britain as the water temperature rises above 15°C, but rather later further north and in Scotland (i.e. late May to early July). The males develop a deeper and brighter colouration though tubercles do not seem to appear on the head and anterior body unlike many cyprinids. They assemble at the spawning grounds and then start to pursue ripe females and drive them, often with much splashing into weed beds where spawning takes place near the surface. Individual males may eventually spawn with several females. Tate Regan noted that '... it spawns at about the same time as the Roach, repairing to weedy shallows for the purpose; at this season the shoals of Rudd make a characteristic noise by pouting, emitting air bubbles at the surface, which float on the water and then burst.' The sticky eggs, which are transparent and colourless or pale yellow in colour and 1.0-1.7 mm in diameter, adhere to the weed. The fecundity of the females varies with size and other factors but usually ranges from 90,00-230,000 eggs per female (i.e. some 108,000-211,000 eggs per kg). Hatching depends on the local temperature, taking from 5-10 days.

This species commonly hybridises with other cyprinid species to form intermediate but sterile hybrids. It has been known to do so with Bream, Silver Bream, Bleak and Roach.

Predators

Little has been published concerning the predators of Rudd but it can be assumed that, as well as coarse fishermen, many of the 'usual suspects' (e.g. Pike, Herons, Otters) are involved.

Rudd are well known as hosts for 'black spot disease' (*Posthodiplostomulum cuticola*) and heavy infestations can occur in some lakes. They are also common hosts to the Fish Louse, *Argulus*. Niall Campbell has recorded a Rudd population in Northern Ireland where every fish had 5-12 *Argulus* on its body. Trout placed in this lough soon died, covered with *Argulus*.

Some years ago, the ornamental pond at Culzean Castle, where a population of Rudd had thrived for many years, became (and still is) very overgrown with ornamental water lilies. Various options were considered to reduce their numbers and create more open water so that the fish could be seen. The suggestion offered by the Duke of Edinburgh, who was visiting at the time, was to use explosives to blast them out!

Growth

The young are only 4.5-6.0 mm on hatching and adhere by the head to the leaves of weeds for about a week until the swim bladder develops and the yolk sac is fully absorbed. They then start to swim and feed. Growth is very variable thereafter and dependent on local conditions. Some populations in small lochs consist of 'stunted' fish of only 10-15 cm which may be many years old, but in favourable waters growth is much faster and fish can be up to 9 cm in length after one year, 15 cm after two years and 20 cm after three years.

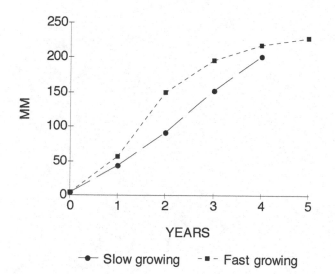

Growth of Rudd (Mann & Steinmetz 1985).

Male Rudd mature at three years of age and most females a year later. Thus they are all mature by their fourth year but continue to grow slowly thereafter. They may live for at least 17 years. Some studies show that, from maturity, females grow faster than males and live longer.

A study of Rudd in Barscobe Loch in Galloway by workers at the Freshwater Fisheries Laboratory at Pitlochry used a sample collected in June 1994 which were 11-23 cm in length and 16-173 gm in weight. 'Many of the fish were mature and ready to spawn among the sedges in the littoral zone of the loch. Scale reading suggested that the ages ranged from three to possibly nine years old. (The outermost age checks are difficult to determine in older fish of this size).'

Value

Although not fished commercially in Scotland, Rudd are important in some parts of eastern Europe where they are taken in seine nets and traps. Catches of up to 20 tonnes have been recorded in Russia in a single haul of the net.

They are not generally regarded as good eating however - though Houghton noted that 'In quality of flesh it is said to be preferable to the Roach.' Fred Buller and Hugh Falkus confirm this: 'Of the few who have tasted this fish, fewer still speak kindly of its flavour.'

The Rudd is a very popular sport fish in England and many other parts of Europe and they are also used as a bait species in fishing for Pike and Pikeperch. Sheringham believed that 'Were it as common as the chub it would be one of the most popular and sought after fishes, for it bites heartily and fights with power.'

Although less common than Roach, they are easier to keep in captivity; the author has successfully kept them in both ponds - where they have bred - and aquaria. The young, especially the golden variety, make very handsome aquarium fish. Bateman recorded that 'The Rudd ... when young will live very well in an aquarium. It is decidedly a handsome fish, handsomer than the Roach, with which it is often confounded. I do not know whether Rudd sleep more than any other fish, but whenever I took a light into my study at night they were the fish which seemed more certain to be asleep than any of the others.' It was a favourite with Jack Hems: 'The rudd is frequently at the surface. There it cruises about looking for food or enjoying the warmth of the sun. Its fins never fail to catch the eye: they are rich plushy red.'

Conservation

The Rudd has no special protection in Scotland nor is any needed for it is an alien species whose range is extending across the country from introductions by coarse anglers.

References

Little research has yet been carried out on Rudd in Scotland but several studies are available from other parts of Europe.

Aflalo (1897), Bateman (1890), Bean & Winfield (1995), Brassington & Ferguson (1975), Buller & Falkus (1988), Burrough (1978), Franck (1694), Freshwater Fisheries Laboratory (1994), Hartley (1947), Hems (1977), Houghton (1879), Kennedy & Fitzmaurice (1974), Mann & Steinmetz (1985), Regan (1911), Sheringham (1925), Svardson (1950), Walton (1653), Wheeler (1976).

'The rudd is frequently at the surface.
There it cruises about looking for food or enjoying the warmth of the sun.
Its fins never fail to catch the eye: they are rich plushy red.'

Jack Hems (1977) *A Garden Pond*

*****

Tench *Tinca tinca* Teins

'The Tench spawns in June
... when wheat is in flower.'

William Houghton (1879) *British Fresh-water Fishes*

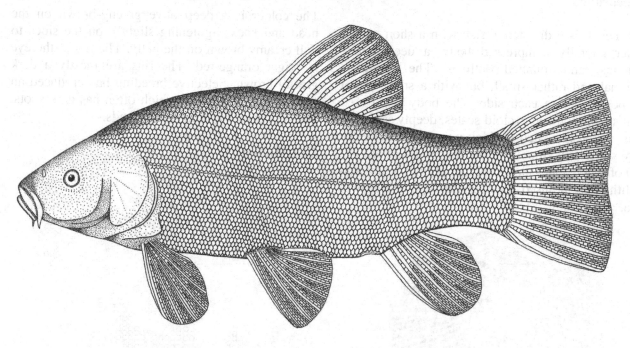

THE Tench is an intriguing, lovely, fish which has been successfully introduced to quite a number of lowland waters in Scotland. Herbert Maxwell noted that 'Folklore has gathered fondly round this fish, and many wonderful stories have been believed about its healing virtues, not only affecting other fish, but men also.' Many people will be surprised to learn that it is good eating; Lubbock describes how Tench may be caught rather like Trout, by a form of guddling, practised in Norfolk. Once the fisherman has marked where the fish is 'he insinuates one hand, which alone is used, under it, just behind the gills, and raises it gently, but yet rapidly, towards the surface of the water. ... and in the course of a favourable day one fisherman will easily secure five or six dozen.'

For centuries the reputed medicinal powers of the Tench in relation to other species of fish have earned it the surprising alternative common name of Doctor Fish. Juliana Berners noted in 1486 that 'A Tench is a good fyssh; and heelith all mannere of other fysshe that ben hurt yf they may come to hym.' Moses Brown put it thus in his *Piscatory Eclogues*:

'The Pike, fell tyrant of the liquid plain,
With ravenous waste devours his fellow train;
Yet howsoe'er by raging famine pined,
The tench he spares - a medicinal kind;
For when by wounds distrest and sore disease,
He courts the salutary fish for ease,
Close to his scales the kind physician glides,
And sweats a healing balsam from his sides.'

A literal translation of the scientific name is 'The Tench'.

Size

Most mature Tench in the British Isles are about 25-35 cm, though the species can grow as large as 70 cm in length and 8 kg in weight in favourable waters in Europe. The Scottish rod-caught record is 4.54 kg (10 lb) caught by P. Crookham in the Castle Loch, Lochmaben. The British record is 6.548 kg (14 lb 7 oz) for a Tench caught in a gravel pit in Hertfordshire by G. Beaven in 1993 - this is now presumably the world record, beating the previous Swedish Tench of just over 4.5 kg. Although heavier Tench have been caught in England, these were rejected as candidates for record status due to the fact that they were suffering from an abnormal condition termed 'dropsy', when the abdomen becomes much extended with accumulated fluid and the fish unnaturally heavy.

Appearance

The Tench is a distinctive fish with a short stocky rather laterally compressed body, a deep caudal peduncle and rounded outlines. The mouth is terminal and rather small, but with a small barbel set posteriorly at each side. The body is covered with small elongate cycloid scales, deeply set in the skin, which is coated in thick mucus and therefore very slimy and smooth to the touch. There are 87-120 of these scales along the lateral line. Scales and otoliths may be used for obtaining age and growth data.

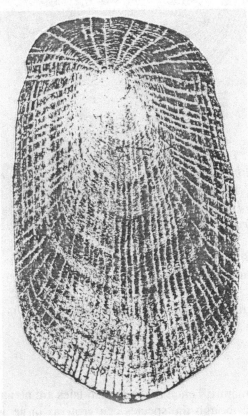

Scale from Tench.

All the fins are rather short and very rounded. The pelvic fins are of particular interest in this species, for in the males the second ray is very thickened and the whole fin is much longer - reaching back past the anus. The rays of this fin in the female fish are all normal and the fin does not reach the anus

when laid back. Thus it is possible to recognise the sex of most fish over about 15 cm by this feature - something which is not easy in most members of the cyprinid family outside spawning time. Males do not develop tubercles over the head and anterior body, unlike most other Cyprinidae.

The colour is a deep olive greeny-brown on the head and back, lightening slightly on the sides to dull creamy brown on the belly. The iris of the eye is bright orange-red. The fins are mostly a dark greeny brown. Selective breeding has produced an attractive golden form, which often has dark spots, and is popular for ornamental ponds.

Distribution

Recorded distribution of Tench in Scotland.

Tench are widely distributed throughout most of Europe except the north and also occurs across Asia into China. In the British Isles, like most cyprinids, it is native only to the south-east of England, but has been redistributed to many parts of the country so that it is now common in most parts of lowland England, Wales and Ireland. In Scotland it has been successfully introduced to a number of rich lochs in the central belt and a few ponds in other parts of the country, but, as in Ireland, may not spawn every year, owing to a high temperature requirement.

George Bolam noted that it 'Has long been a denizen of some of the ornamental ponds in the district, whence escapes into rivers have doubtless from time to time occurred; but nowhere truly indigenous. ... The earliest reference to Tench in the Borders is contained in Dr Johnstons's list, and is as follows:- "Hirsel Lough, - an artificial piece of water first filled in the month of December 1786." An occasional Tench has been netted in the Tweed as far back as the memory of the oldest fisherman carries. ... The last of which I have record was netted at Norham on 2nd August 1888. ... might have been an escape from his pond at Milfield, where for some years previously Tench and other fish had been kept without any serious obstacle to their casual descent to the Till.'

Leonora Rintoul & Evelyn Baxter record that Tench 'Were introduced into Duddingston Loch by the Earl of Abercorn before 1795 to stock his canals and ponds, but were eaten by pike or washed away by the flood. ... Dr Ritchie tells us that they were introduced into Pressmennan Loch.'

Ecology

This is a very shy fish, occuring mainly in quiet lowland lochs and very slow flowing rivers where there is plenty of weed cover and a muddy bottom. It is extremely tolerant of stagnant conditions and appears able to survive in ponds where low oxygen conditions in the heat of summer and under the ice in winter (oxygen levels may be as low as 0.7 mg/l) would eliminate other species. It can stand quite high temperatures of up to some 34°C. It is also very easy to transport around the country, surviving out of water in damp moss or a wet sack for many hours. Aflalo notes that 'The mud-loving Tench, in which the small scales are so embedded as to make it as slippery to the touch as an eel, thrives well in stagnant waters ...'

Ian Harman believed that 'Tench will live in any pond and the fish is extremely tenacious of life.'

Food

The fry eat protozoans, rotifers and small crustaceans from the plankton and among the weed beds. As they grow they depend more on bottom invertebrates such as worms, crustaceans, molluscs and insect larvae, especially those of midges, which they hunt mainly at night. They also eat fish eggs when available and regularly take in plant material.

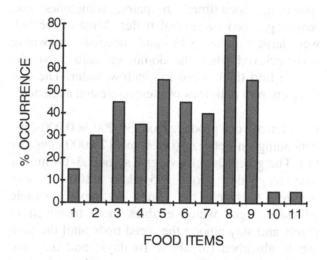

Food of Tench in Lake Tiberias, Australia (1: algae; 2: molluscs; 3: copepods; 4: cladocerans; 5: amphipods; 6: ostracods; 7: midge larvae; 8: dragonfly larvae; 9: mayfly larvae; 10: bugs; 11: others) (Weatherly 1959).

Tench are reputed to be able to penetrate some 7 cm into the bottom mud - somewhat less than reported for Carp. Bubbles rising to the surface are often a sign that Tench are rooting about on the bottom, which they do in a near vertical position. In captivity, 0+ Tench are actively feeding at 10°C. Little feeding, however, goes on during the winter in temperate areas when the fish stay near the bottom in a rather passive state.

Reproduction

Spawning takes place during the summer months, from late June until August, when the water temperature approaches 20°C. Houghton believed that 'The Tench spawns in June ... "when wheat is in flower.' They are probably the last of our species to spawn in the summer. Each female is driven by two or more males into thick weed beds and the fertilised adhesive eggs stick to the weed near the surface. The spawning act may be repeated a number of times until the female is spent. Edward Boulenger describes spawning in a more poetic fashion: 'The males are lazy fish but during courtship they become very active, assume the character of spirited lovers and have been described as dashing after their brides with such impetuosity as to drive them ashore.'

Walking along the shore of Lac Leman in June 2000, the author had an enjoyable afternoon watching Tench spawn there. Numerous fish were spawning, sometimes in pairs, sometimes one female pursued by several males. Most of the fish were large (at least 3 kg) and successful spawning was achieved when the dominant male drove the female into thick weed in shallow water. The fish were entirely oblivious of their interested audience!

Each female can produce from 30,000-900,000 eggs depending on her size (i.e. some 125,000 eggs per kg). The greenish-yellow eggs, some 1.3-1.4 mm in diameter, take about 3-5 days (100-120 day degrees) to hatch; the young fry have a simple adhesive organ which enables them to attach to plants and stay among the weed beds until the yolk sac is absorbed (in about 10 days) and they are feeding properly. Tench fry, even although their swim bladders are inflated within 24-48 hours of hatching, tend to remain inactive much longer than is usual for cyprinid fry, possibly up to a week after hatching. Tench and Crucian Carp are reputed to hybridise freely.

Predators

Little has been published concerning the predators of Tench but it can be assumed that, as well as coarse fishermen, many of the 'usual suspects' (e.g. Pike, Herons, Otters) may be involved. However, due to its cryptic colouring and bottom living habits it is probably less vulnerable that many other fish.

Growth

The fry are 4-5 mm at hatching but grow steadily if the water temperatures remain high, reaching some 4-8 cm (5-10 gm) at the end of their first summer, 10-15 cm (40-100 gm) at the end of their second summer and 20-30 cm (200-300 gm) at the end of their third summer - by which time most fish are starting to mature.

In some slow growing populations in Ireland, however, both mature males and females of 2+ years and only 12.5 cm in length were found. Spawning usually takes place during the third or fourth years and may be repeated every year

thereafter. Most fish live for 6-8 year and many much longer, 10-15 years of age being quite common. Individuals as old as 17 years have been recorded.

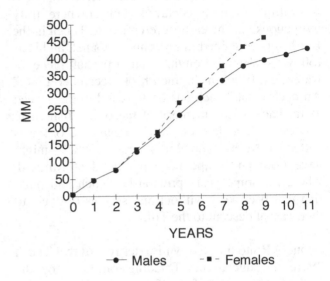

Growth of Tench in Coosan Lough, Ireland (Kennedy & Fitzmaurice 1970).

Value

There are few commercial fisheries for Tench, but the species is cultured in ponds in some parts of central Europe, yields of 20-80 kg per hectare being recorded. In most places, as in the British Isles, the Tench is more important as a sport fish and it has been introduced as such to other parts of the world e.g. New Zealand, Australia and North America.

The flesh is dark and has a strong flavour but is popular in some countries, e.g. France and Germany where the standard recipes for Carp are also used for Tench. However, it does not suit all tastes, for the Romans 'abandoned it to the common people, who alone feasted on it'. Others disagree. Richard Franck was enthusiastic: 'Let me tell you he's a delicious morsel.' - as was William Houghton: '... some people regard the Tench as excellent, others will not touch it; for my own part, I think a well fed Tench of about two pounds in weight one of the best freshwater fish that we possess, and that it does not in the least require rich and savoury sauces to make it palatable.' O'Gorman agreed: 'They are a firm and good fish to eat; some cooks stew them in wine, but they are by no means bad with parsley and butter.'

Herbert Maxwell also extolled its culinary virtues: 'Whole chapters might be devoted to deploring the heartless indifference our people show in preparing food for the table. Our neighbours across Dover straits understand thoroughly how much the pleasure of living is enhanced by skilful cookery. ... a certain *friture* of tench which I once had served up to me in a humble wayside tavern at Meung-sur-Loire - *voila des annees* - why, it lives as a landmark in my gastronomic experience.' Later the same author wrote: 'Meung-sur-Loire, a village made illustrious for ever as the scene chosen by Dumas to introduce the immortal d'Artagnan to his million readers. Arriving there from Orleans near midday on our bicycles, we were served with an excellent dejeuner, whereof the only features that remain bright in memory are the good red wine and a dish of small tenches delicately fried in oil.' In view of its muddy habitat, all this praise for the flavour of Tench is somewhat surprising.

Tench are still popular in parts of Europe. Alan Youngson recalls ' that "tinca" were on the menu in late May in Extramadura in western Spain, in the upper catchment of the River Tagus.'

The age-old legend about Tench being the 'doctor fish', whose slime possesses powers of healing for both fellow fish and for Humans has been passed from generation to generation and is very well known. It is apparent that both its slime and its powdered otoliths were used by early physicians.

It is also a fairly popular aquarium and pond fish, especially in its golden form, which Gregory Bateman was fond of: 'The Golden Tench ... is a very handsome fish and a great acquisition to an aquarium. ... It is of a most lovely yellow colour mixed with a gleam of gold.' Ian Harman: agrees: 'The Golden Tench is a highly attractive fish, its

colour being a fine deep yellow. The scales are of so fine a texture that the fish has the appearance of being made of metal, and has a sort of luminous glow caused by the mucous film which is always found on Tench.' The author has kept both golden and green forms successfully.

Derek Mills reports 'transferring Tench from Duns Loch to Duddingston Loch in 1967.' However, much thought should be given before any such translocations are carried out. The author remembers his disappointment when Tench were introduced by coarse anglers to Baldernock Pond near Milngavie in the late 1960s. An experimental population of Powan which he and John Flannagan had started there in 1964, and which was thriving, disappeared within a year.

Conservation

The Tench has no special protection in Scotland, nor does it need any as an introduced species which is still extending its range through introductions by coarse anglers.

References

Only a few studies have been carried out on this species in Scotland but published work is available from elsewhere in Europe, and even from as far off as Australia, by Alan Weatherley, where Tench have been successfully introduced.

Aflalo (1897), Bateman (1890), Berners (1486), Bolam (1919), Boulenger (1946), Burkel (1971), Franck (1694), Harman (1950), Houghton (1879), Kennedy & Fitzmaurice (1970), Lubbock (1845), Maitland (1977), Maxwell (1904), O'Gorman (1845), Rintoul & Baxter (1935), Rosa (1958), Weatherley (1959, 1962).

'... a certain friture of tench
which I once had served up to me in a humble wayside tavern ...
- why, it lives as a landmark in my gastronomic experience.'

Herbert Maxwell (1904) *British Fresh-water Fishes*

*****

Stone Loach *Barbatula barbatula* Breac beadaidh

'It is said to be good eating,
but I have never known it to be eaten by Scotch people.'

Peter Malloch (1910) *Life History of the Salmon ...*

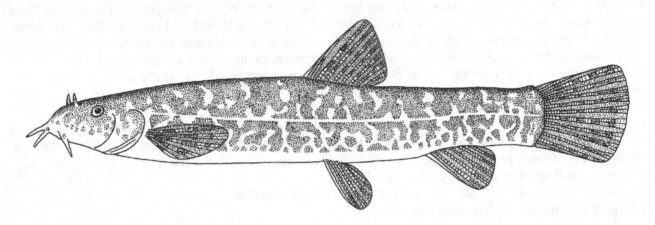

OF the two British loaches, the Stone Loach is much better known than its relative the Spined Loach as it is more widely distributed, usually occurs in fair numbers and is easily dislodged from its daytime hiding places under stones and thick aquatic vegetation. It is also less nocturnal in its habits. It is a fish of some personality that settles down in an aquarium and soon becomes tame, coming out of concealment in the daytime to be fed.

The Spined Loach (*Cobitis taenia*) occurs only in south-east England, and therefore George Bolam's record of this species in the Tweed catchment in 1919 is more than a little surprising: 'I twice took single examples in the Till at Weetwood upwards of thirty years ago, but could find no more, though often diligently looked for both there and in other places. It was therefore an agreeable surprise to find another solitary individual amongst a lot of Minnows and common Loaches brought home by my son and some other youthful friends in 1904. They had all been captured in the Low, near Bridge Mill, about six miles south of Berwick (where the common Loach is exceptionally numerous and attains a large size); but several subsequent visits failed to add anything to our knowledge of *C. taenia*.'

The Stone Loach has several alternative common names, including Beardie (a popular name in the

west of Scotland where the author grew up, and also in Fife) and Colley. George Bolam noted that in the Borders it is ' ... known amongst the younger generation under the names of Miller's Thumb, Beardie, Lotchie, Bessy Loach, etc., and on Tyneside as Hairygob.' John Carter recalled Katie Beardie in the Borders also. A literal translation of the scientific name is 'The Bearded Fish'.

Size

The Stone Loach is a small fish about 7-14 cm in length, only occasionally becoming larger than this. There are no rod-caught records.

Appearance

Behind the head, with its very small high-set eyes, the body is cylindrical, except for a short section near the tail which is flattened laterally. The mouth is set ventrally, as befits a bottom grubber, with six sensory barbels on the upper jaw - four in front (including the two shortest) and two at the corners - all of which are longer and much more conspicuous than those of the Spined Loach, with which it could be confused initially. The Stone Loach has no spines below the eyes. The body is distinctly slimy and bears small deep-set scales which do not overlap. The lateral line is most visible along the anterior part of the body.

The pectoral fins in the male are longer and more pointed than those of the female, but the difference is slight. During the breeding season the males develop minute tubercles on these fins and these are found permanently on all males over 7 cm. The patterning of the Stone Loach is very variable and often well-defined, forming ideal camouflage when lying among gravel and stones. Younger Stone Loach tend to have dark blotches or speckles on a pale background, while the older fish may be heavily blotched with speckled dorsal, anal and caudal fins. Large specimens may be dark olive to dirty yellow-grey.

Distribution

Stone Loach are found in clean rivers and the littoral zone of lochs throughout west, central and eastern Europe and across Asia to the Pacific Ocean. They also occur in some low salinity Baltic lagoons.

Recorded distribution of Stone Loach in Scotland.

In the British Isles they were originally indigenous to south east England but since the last Ice Age they have become widely dispersed throughout the rest of the country except northern Scotland.

Until recently, the most northerly catchment in which this species was known to occur was the North Esk. However, the author has recently had an authentic report from Chris Booth of an established population on Orkney, presumably having been introduced there by somebody local, certainly pre-1988. The first reports were from the Click Mill area but the species has now spread well beyond this. This is a major, and surprising, extension to the distribution of this species.

Ecology

Tate Regan noted that 'it generally occurred in small clear streams with sandy or gravelly bottoms' and research has shown that this species requires about the same level of dissolved oxygen (10-16 mg/l) as salmonids, being thus susceptible to quite low levels of pollution. Their visible discomfort or even death can serve as an 'early warning' of pollution in a stream. Changes in the nature of the river bed (e.g. the deposition of fine mud) can also lead to a decline in the loach population.

The European Pond Loach (or Weather Fish) *Misgurnus fossilis*, is well known for its ability to gulp down air and absorb oxygen through the wall of the alimentary canal which thus acts as an additional respiratory organ when the oxygen level in the water falls. Both British loaches appear to be able to do this too, but to a lesser extent. Ian Harman notes that 'Loaches are able to use the intestine as a supplementary breathing organism (sic) and this accomplishment comes in handy when, in droughts, their home becomes more or less dried up.' Edward Boulenger records that, in the wild, '... the loach is very sensitive to atmospheric conditions. Approaching thunder causes it to rush excitedly about in mid-water ...'

Food

Feeding takes place mostly at dusk or on dull days, when the animals root about on the bottom for food by forcing the mouth and head into the bottom sand and silt by a vigorous action of the posterior body and tail. In captivity, this characteristic behaviour can be observed as soon as the fish learn to feed in daylight and they can be seen to detect food at some distance. They are non-visual feeders, often sensing the presence of food by means of their sensitive barbels. Research has shown that, when feeding in small groups, Stone Loach detect and exploit a food source more rapidly and efficiently than when alone. However, when an individual Loach detects food and begins rooting about, its activity is sensed by other fish which then converge on the area. Gregory Bateman says that, in an aquarium, Stone Loach feel for their food with their barbels. It never starts 'in quest of food when it is first put into a tank but waits for an interval doubtless until the scent has been diffused through the water.'

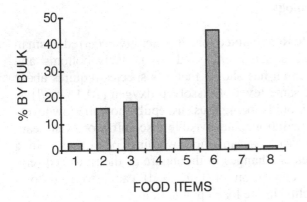

Food of Stone Loach in the Altquhur Burn (1: worms; 2: crustaceans; 3: mayfly larvae; 4: stonefly larvae; 5: caddis larvae; 6: midge larvae; 7: beetles; 8: others) (Maitland 1965).

Stone Loach of all sizes are often found together under stones and amid thick vegetation, often in the company of Minnows and sticklebacks. Francis Day noted that 'it browsed among long floating grass quite as commonly as under stones and that in rivers in flood or where very muddy, it sought refuge at the sides under banks.' The food consists mainly of bottom living invertebrates; midge larvae, the nymphs of mayflies and stoneflies form a major element in the diet, while worms, crustaceans, caddis larvae, leeches, molluscs and fish eggs are also taken. Feeding activity declines in winter, but in aquaria a small rise in water temperature at this time will induce them to take food.

Reproduction

Stone Loach spawn from spring to late summer when their yellow eggs, about 0.9-1.0 mm in diameter, are shed among gravel and vegetation to which the eggs adhere. The author has seen extensive batches of eggs scattered among the underside of submerged vegetation in the River Endrick. They hatch in about 14-16 days at temperatures of some 12-16°C. Egg-laying takes place at intervals during the spawning season. Studies of Stone Loach in the River Frome in Dorset showed that a female of 7.5 cm laid 10,000 eggs, at intervals, equivalent to 46% of her initial body weight, between late April and early August.

Knauthe observed the spawning of Stone Loach in Germany and in 1887 described how a dozen loaches circle round a deep hole, 'the smaller weaker males following the larger stronger females. Suddenly all the fish darted to the bank, where they squeezed themselves between the projecting roots of an over-hanging tree shedding ova and milt at the same time.'

After hatching, the fry very quickly adopt a benthic habit and are very difficult to see against the sandy bottoms on which they frequently occur.

Predators

Stone Loaches have many enemies, particularly Trout where they co-exist, but also Eels and large Bullheads as well as some water-side birds and mammals. George Bolam records a Trout in the Borders which was gorged with loach and has seen a Dipper killing one. P.H.T. Hartley found it in the stomachs of Pike and Eels. The stomach of a Greenshank killed on the River Irving in Cumberland contained 15 Stone Loach, whilst the Otter is also cited as a predator.

The preference shown by Brown Trout for Stone Loach used as bait during angling is one of the reasons for their present distribution as they became a popular live or dead bait for Trout and a spinning bait for Salmon (as recommended by Isaac Walton). The presence of Stone Loach on Orkney may have been due to introductions by anglers and they are now found in the stomachs of Brown Trout in the Loch of Hundland. Many of the older books on angling have accounts of Trout fishing with Stone Loach as bait in both Great Britain and Ireland. However, its occurrence in small streams in western Scotland not connected with any system which is important for angling remains unexplained. It seems that this species, like the Minnow, has quite good natural powers of dispersal.

Growth

The newly-hatched fry are about 3 mm in length and grow rapidly, reaching about 1.5 cm in length after some 5 weeks, when their barbels first become visible. 0+ Stone Loach may be seen in full daylight during the summer, in small dispersed shoals moving over the bottom. In the English Lake District they have been shown to reach over 11 cm after 5 years. Both sexes can mature at one year, and there appears to be a predominance of females in all age groups.

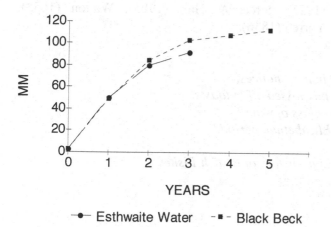

Growth of Stone Loach in two Cumbrian waters (Smyly 1955).

Value

Not only aquatic predators relish Stone Loach - in some parts of Europe its flesh is esteemed for its distinctive flavour. Attention is given to this aspect of the species by some early authors, including Isaac Walton who, influenced by a report by some 'learned physicians', claims the loach 'to be grateful to both the palate and stomach of sick persons'. Yarrell observed that 'In some parts of Europe, these little fishes are in such high estimation for their exquisite delicacy and flavour, that they are often transported with considerable trouble from the rivers they naturally inhabit to waters contiguous to the estates of the wealthy.' Even more surprising perhaps is that Latham, in his Large Dictionary observes: 'The Loach is said in most notices, to have been not unfrequently tossed off in toasts, or swallowed in a glass of wine, by the gallants of the Elizabethan period.' Mascall noted that 'They are fish holesome to be eaten of feeble persons having an ague, or other sickness.', whilst Frank Buckland

commented that 'From its slimy smoothness and activity it is very difficult to catch. The Winchester boys used to spear this fish with an ordinary fork tied to the end of a stick, and then pickle him with the minnows.'

R.D. Blackmore covered the same technique in his novel *Lorna Doone*: 'Being resolved to catch some loaches, whatever trouble it cost me, I set forth ... in the forenoon of St. Valentine's Day, 1675-6, ... Then I took a three-pronged fork firmly bound to a rod with cord, and a piece of canvas kerchief, ... and so went into the pebbly water, ... on the whole I had very comely sport of loaches, trout and minnows, forking some, and tickling some, and driving others to shallow nooks, whence I could bail them ashore.'

In Scotland, most writers have been less enthusiastic. Herbert Maxwell observed that 'People who speak from experience are enthusiastic about the delicacy of its flesh; but under ordinary conditions, the labour of collecting a dish of loaches must be out of proportion to its value, for the fish seldom reaches five inches in length, and is more commonly smaller even than that.' Peter Malloch commented that 'They invariably seek shelter under some flat stone, only to fall prey at the hands of some youth expert in the art of "guddling". It is said to be good eating, but I have never known it to be eaten by Scotch people.' It is certainly an easy fish to catch with a small handnet, as the author knows from his experience in the River Allander as a youth.

Gregory Bateman noted that 'Many people who keep aquaria declare that the Stone Loach ... is a delicate fish and will not live for any length of time in confinement; and they are right, for it will not continue very long in health in an ordinary deep tank. ... tanks of the least depth and the greatest area are most suitable for them; and ... I have found them to live in perfect health and apparently quite happily.' As a result of their habit of foretelling the advent of approaching depressions, they are sometimes kept in aquaria with a view to functioning as living barometers.' However Aflalo's method of capturing them is not recommended: 'During the day this little fish hides at the bottom, lurking beneath the stones, from which it may be dislodged in a half-stunned condition by a smart blow on the stone.' The author has kept this species satisfactorily in aquaria, the main requirements being, well-aerated water, a clean bottom of sand or fine gravel and an adequate supply of live food.

Conservation

The Stone Loach has no special protection in Scotland. It is regarded as a native species by the author, having probably reached here and largely redistributed itself by natural means - though there is no doubt also that it has also been moved about over the last century by anglers who have traditionally used it as both live and dead bait. However, it is reasonably widespread and abundant and has no need of any special conservation measures at the moment.

References

Only a few studies have been made of the biology of this species in Scotland and much of our knowledge depends on studies carried out elsewhere in Europe.

Aflalo (1897), Bateman (1890), Blackmore (1910), Bolam (1919), Boulenger (1946), Buckland (1873, 1880), Day (1880), Harman (1950), Hartley (1948), Hyslop (1982), Maitland (1965, 1966), Malloch 1910), Mascall (1590), Maxwell (1904), Mills *et al.* (1983), Regan (1911), Rumpus (1975), Smyly (1955), Street & Hart (1985), Walton (1653), Yarrell (1836).

'The Loach is said in most notices,
to have been not unfrequently tossed off in toasts,
or swallowed in a glass of wine,
by the gallants of the Elizabethan period.'

Frank Buckland (1873) *A Familiar History of British Fishes*

*****

Pike *Esox lucius* Geadas

'So dreadful a forest of teeth and tusks I think I never beheld.'

Thomas Thornton (1804) *A Sporting Tour* ...

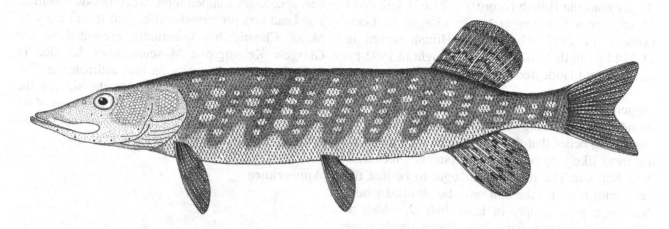

MENTION of the Pike conjures up quite different images for different people. Specimen pike anglers dream of eventually landing the giant which will break the Scottish record (itself of considerable controversy). Salmonid fishery managers generally view Pike as vermin and something to be got rid of at all costs, and thousands of Scottish Pike (regarded as inedible by the majority of the public, though most people have never tasted it) have ended their days netted or poisoned and buried in pits by the lochside. This was common practice at Loch Leven for many years. The fearsome reputation of Pike as 'freshwater sharks' has meant that some parents and also dog owners take care not to let their charges spend much time in known pike waters.

The name Pike derives aptly from the shape of this fish, with its long near cylindrical body and elongate duck-billed snout. Its mouth is large and wide, armed with rows of needle sharp teeth angled backward, making the escape of prey (and the removal of fish hooks - and fingers!) very difficult. There are a variety of common names in use in different places, including Gad, Ged, Gedd, Jack, Luce and Pickerel.

Such a formidable fish has attracted a variety of writers. George Orwell wrote that 'There were pike there too, and they must have been big ones. You never saw them, but sometimes one that was basking among the weeds would turn over and

plunge with a splash that was like a brick being bunged into the water.' This behaviour is very typical; on dozens of occasions the author has heard such splashes while walking along a river bank or loch shore - usually a fairly sure sign that Pike are present, even if they are not seen. Often there is a wake at the surface as the Pike swims off. F.G. Aflalo notes that 'Although a very active fish when on the feed, it is fond of basking at the surface.' A. & J. Lang agree 'Pike used to bask in the shallows here of a hot summer's day; perhaps even yet they do so. But I think these fish are more numerous now in the Loch of the Lowes than in St Marys.'

Poets too were fascinated by this predacious fish as indicated in *The Pike* by Edmund Blunden:

'And nigh this toppling reed, still as the dead
The great pike lies, the murderous patriarch
Watching the waterpit sheer-shelving dark,
Where through the plash his lithe bright vassals
thread.

The rose-finned roach & bluish bream
And staring ruffe steal up the stream
Hard by their glutted tyrant, now
Still as a sunken bough.

He on the sandbank lies,
Sunning himself long hours
With stony gorgon eyes;
Westward the hot sun lowers.'

Size

The Pike is one of our largest freshwater fishes, but there is a great deal of variation in adult size. In general, adult individuals range from 40-100 cm and 2-14 kg. The present Scottish rod-caught record (larger than the British record!) is 21.631 kg (47 lb 11 oz) for a fish caught by T. Morgan in Loch Lomond in 1947. The present British record is 21.234 kg (46 lb 13 oz) for a fish caught in 1992 by R. Lewis in Llandegfedd, in Wales.

Larger fish have been reported caught in nets or found dead. Among Pike anglers there is an interesting belief that the very largest specimen fish are most likely to be found in systems which also have Salmon. The reason is thought to be that the maximum size in Pike can only be attained where there is a good supply of large fish on which to prey. Adult Salmon have been found inside large Pike, and even where fresh run Salmon are too large for Pike to catch and swallow, there is always a plentiful supply of weakened kelts.

There has been much debate as to which is the largest Pike ever caught in Scotland. Colonel Thornton, writing in 1804, caught one which weighed 50 pounds after a struggle of an hour and a quarter by trolling in Loch Awe. It measured exactly 4 feet 4 inches from eyes to fork, and 'jaws and tail included could scarcely be less than five feet. So dreadful a forest of teeth and tusks I think I never beheld'. Herbert Maxwell mentions '... the measurements given of the great pike of Loch Alvie. This fish was 5 ft. 4 in. from the eye to the fork of the tail. The Colonel's scales only went to 29 lb., and he was obliged to calculate the weight of his prize, which he did at 47 lb. or 48 lb.' William Houghton records that 'One of the largest recorded British Pike was taken in Loch Ken, Kirkcudbrightshire ... It is said to have weighed over seventy-two pounds, and to have been about seven feet in length. ... It was caught about 1760 by John Murray, keeper at Kenmure, was taken by an artificial fly made of peacock's feathers;' The weight a Pike from Loch Lomond is mentioned by Harvie-Brown & Buckley as 79 lbs (see below). Fred Buller and Hugh Falkus record that 'Fishing in "Morgan's Bay", Loch Lomond, in 1967, Fred Buller hooked and lost a pike that may well have been bigger than Morgan's - the main line broke just above the leader swivel. Here in 1945, Tommy Morgan caught the biggest British pike of the twentieth century: 47 lb. 11 oz.'

The famous Endrick Pike, a female, was found dead in 1934 near the mouth of the River Endrick - a major spawning ground for Loch Lomond Pike. It was apparently trapped there when floods subsided. The head was preserved and kept at Ross Priory by Major Christie but eventually presented to the Glasgow Kelvingrove Museum when he died in 1969. From the scales, its age was estimated at 17+. Estimates of its weight, based on the size of the head ranged from 35-50 lb, with a best guess of 45 lb. Fred Buller commented that when it met its fate 'a magnificent reign of terror came to an end.'

Appearance

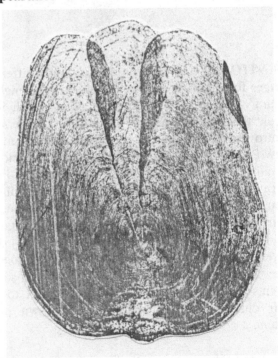

Scale from Pike.

Characteristically, Pike have large sickle-shaped teeth, hinged behind, on the head of the long vomer bone (palate) and pads of teeth on the tongue, while the lower jaws each bear 5 or 6 long strong canines, unequal in size and attached firmly to the jaw bone. All the teeth are set backwards, making the escape of prey difficult and also facilitating swallowing. The sharpness and set of these large teeth make Pike the only freshwater fish in the British Isles where anglers, to be sure of landing them, require to use a length of fine metal trace between the line and the lure. There are no teeth on the maxillary jaw bones.

The lower jaw, as in many piscivorous fish, is longer than the upper one, accentuating the Pike's predatory appearance and in accord with its sinister reputation. The eyes are set close together and are placed high and forward in the head thus giving good binocular vision, as befits a predator that hunts by sight.

The single dorsal fin is set well back on the body, above the anal fin. The strong scales are large and cycloid and set deep into the tough skin. The buried portions of the scales are lobed and there are 105-130 scales along the lateral line. An interesting anatomical feature of Pike is the presence of numerous fine Y-shaped intermuscular bones within the flesh which are not connected to the vertebral column. These reduce its culinary value.

The colour of Pike is very variable and is affected to some extent by size and habitat. Very young Pike are pale green and rather conspicuous, but they soon darken and display the typical banded effect when seen from above. In adults, the upper head and back are dark olive brown to green, grading on the flanks to pale green with oval yellow patches which usually form 14-15 broken or solid sub-vertical bands which curve forwards ventrally. This disruptive colour pattern is ideal for concealment among vegetation, as Herbert Maxwell remarks: 'Similarly the pike, gluttonous tyrant of the river and the loch, is so cunningly dappled with olive and grey as to match exactly the water weeds wherein it makes ambush. So effective is the camouflage that one standing beside still water will probably detect a pike from its shadow on the bottom before seeing the fish itself.' The belly is a pale creamy yellow, sometimes dotted with pale orange spots. The dorsal, anal and caudal fins are dark brownish with reddish yellow to orange mottling. The pectoral and pelvic fins are usually rather pale. The pattern of markings on Pike of 50 cm or more have been shown to be specific to individual fish and can be used for identifying at a later date without tagging or marking.

Distribution

The Pike is widely distributed throughout much of Europe, Asia and North America (where it is known

as the Northern Pike to distinguish it from other members of the family occurring there). It occurs in lakes and slow-flowing stretches of streams and rivers and is also found in parts of the Baltic Sea where the salinity is less than 10%. Its oxygen demand is comparatively low and so it can survive in habitats where there may be some degree of pollution, providing the oxygen level does not fall below 5.7 mg/l. It can stand very cold conditions, but water temperatures above 29°C are fatal. It is probably the most widely distributed piscivorous fish in the northern hemisphere.

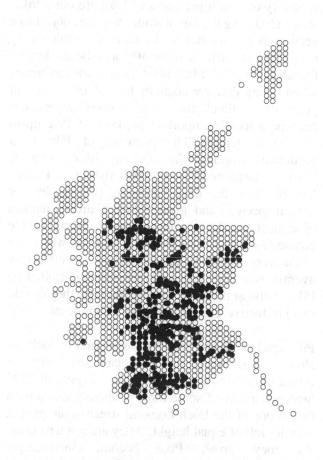

Recorded distribution of Pike in Scotland.

In the British Isles, the Pike was one of the indigenous species left in south east England after the last Ice Age. However, the very high price of Pike flesh in medieval times (due to scarcity) led many early writers to believe that the species had been recently introduced and was not yet widely established. Over the centuries it has been widely redistributed over much of Great Britain. In Scotland, Leonora Rintoul & Evelyn Baxter note that 'As long ago as 1521 John Major said Pike were common in the Forth and they are still to be found in almost all our lochs and a good many of our rivers.' It is less common, however, in the Highlands and very rare in the north west, the northern isles and the Hebrides, with the exception of two populations on Islay. In recent years, its distribution has been rashly extended by coarse anglers indifferent to its impact on native fish communities. F.G. Aflalo commented that 'There appear to be no Pike in Sutherland.'

Ecology

Probably the main predators of Pike are other Pike, although during the larval stage they are subject to a very high (sometimes 95%) mortality inflicted by invertebrates such as dragonfly and beetle larvae. Occasionally the 'Man bites Dog' situation arises when young Pike are taken by large Trout. In recent years in Scotland, the re-established Osprey has become a locally important predator of Pike up to 35 cm in length. The presence of Pike is a particularly important factor in any fish community. They eat large numbers of small fish of all kinds - though they do have apparent preferences for certain species - and affect the population dynamics of almost all the species involved. Many of the coarse fish species favoured by anglers, which are often very prolific, cannot attain a satisfactory average size unless their numbers are controlled by Pike, while at the same time the damage they can do to a productive Trout fishery can be immense.

Pike tend to inhabit the shallower parts of lochs or slow-flowing rivers but they can sometimes be found in quite fast-running water. Chrystal in 1927 'watched a pike of some three pounds or so, above a fall in one of the Loch Lomond streams, jumping at another fall of equal height.' They are not territorial. As they grow, Pike become increasingly piscivorous, reaching the top of the food chain in most communities in which they occur. They adopt a solitary and fairly sedentary way of life, concealing themselves until their prey (often another smaller Pike) is within striking distance, or stalking it stealthily before making a rapid rush forward to seize it. They will not pursue an intended victim very far if they miss, but will retire and wait for another opportunity. Unlike most freshwater species, except the salmonids, Pike will feed at very low winter water temperatures, often hardly above 0°C.

Pike populations are sometimes found in atypical habitats, e.g. high moorland lochs, having been introduced by fihermen to those with high population densities of small Trout, with the objective of reducing the Trout numbers so that the size of the remaining Trout would increase and become more attractive to anglers. Sometimes the two species have managed to co-exist, with the Trout much larger but very many fewer in numbers. Often, however, Pike have completely eliminated the Trout, and here, Pike have had to depend mainly on a diet of invertebrates and each other. Quite large Pike (1.4-2.3 kg) may be found with stomachs packed with invertebrates such as freshwater shrimps or water lice. Beyond a certain size or where invertebrates are scarce, Pike depend upon cannibalism for survival, and their population structure resembles that of a flat pyramid with a wide base, i.e. there are a few large Pike supported by a large number of small young Pike, the latter feeding on invertebrates.

A classic study of Pike was carried out by Bill Munro in Loch Choin - a 26.3 ha moorland loch in Scotland which contained only Pike, and which was poisoned in the autumn of 1955 to allow restocking with Trout. Virtually all the Pike were collected and examined. There were 2,078 altogether, weighing 140.9 kg, which represented a standing stock of 5.3 kg per ha. The largest Pike was a 12 year old female weighing 7 kg, 96 cm in length; the next two largest were both females of 9 years of age, each some 70 cm in length. The great majority of the Pike (1867) were between 10 and 15 cm (i.e. mostly 0+ fish). Age groups 3+ to 8+ consisted of only 98 fish in total. Females in all age groups had grown faster than the males. There were some mature fish of both sexes in the 2+ group, but none in the 1+ group. Invertebrates, mostly freshwater shrimps, were an important food of Pike up to 50 cm in length, whereas frogs and smaller Pike made up the bulk of the diet of larger fish. Many of the stomachs (33%) were empty - a common finding with Pike sampled in all kinds of habitat.

Interestingly, many years before Bill Munro's study and referring to a different loch, there is a piece in the The Field of 1889: '... somebody happened to say "Why don't you have a go at the pike on Loch Chon before you leave?" Why not indeed? ... About six o'clock I foregathered with Tommy. On reckoning up our joint bag we found that we had killed fifty-two fish, weighing in the aggregate 236 lb., a very fair day's work on such a piece of water ...'

Food

Richard Franck believed that 'He murders all he meets with.' However, even where apparently ample numbers of suitable prey fish are present, quite large Pike may still take considerable invertebrate food, for instance Niall Campbell found that, of 33 Pike (average length 59.3 cm) from Loch Tummel in Perthshire, 18% contained freshwater shrimps or lice, 28% midge larvae or pupae, 21% frogs or toads and 23% fish (Trout or Perch). In many other situations, however, Pike are almost entirely piscivorous on reaching about 50 cm. The conversion rate of food to body weight is about 5 or 6:1 but sometimes as high as 2:1. The annual intake of food for a Pike around 2 kg in weight has been calculated at some 21 kg.

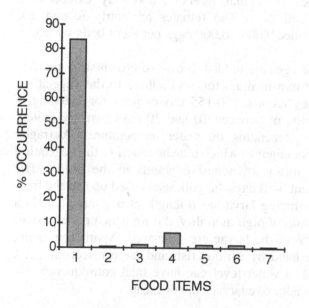

Food of young Pike in Loch Choin (1: Gammarus; 2: pea shells; 3: caddis larvae; 4: mayfly larvae; 5: alderfly larvae; 6: beetles; 7: Eurycercus) (Munro 1957).

Adult Pike will try to eat any animal, alive or dead, that is not too big to swallow - indeed Pike are sometimes found dead, apparently choked by an extra large food item. A Pike of 6.3 kg has been known to swallow a Pike of 3.2 kg. W.J. Gordon's opinion was that ' It is quick of sight and hearing, ferocious, audacious, and persistent, and mostly carnivorous, but will eat anything, even model yachts.' As well as fish, which are swallowed head-

first and are usually the main food, they take small mammals (including Rats, Water Voles and Water Shrews), amphibians, crayfish and wildfowl. In fact, it is believed that they can make serious inroads into wildfowl populations by taking their young, or in some habitats even the adults, such as Moorhens and Dabchicks. If a prey fish is too long to be swallowed completely, the Pike will swim around with the tail of the former protruding from its mouth while the front end is being digested. There authentic records of a Pike grabbing a hand trailing over the side of a boat and of another occasion when a dog was attacked while it was wading. Pennell, in his *Book of the Pike*, gives an authenticated account of a boy bathing in a pond near Ascot being bitten by a Pike, both hands being seized in turn and severely bitten. The fish, found to measure 41 inches in length, was very lean, and it was concluded that it was starving, having devoured all the other fish in the pond.

There are many records of the Pike's voraciousness. Derek Mills 'remembers Pike caught in Loch Odhar (Strathbran) with its stomach full of fully fledged Black-headed Gulls and in August 1960, a Pike caught in Loch Ussie with six ducklings in its stomach.' Keith Ritchie recalls that Nigel Forteith was fishing in Loch Freuchie, by Amulree, in July 1992, when he saw a fully-grown Moorhen disappear from the surface. Continuing to fish, he eventually landed a large Pike which, when opened up in the nearby college laboratory contained a Moorhen. The latter was alive and returned to the loch apparently unharmed. Ross Doughty's tale is that 'On June 8 1978 I examined a large (ca 5 kg) Pike which had been found two days earlier in a moribund state in the Clyde Estuary at Yoker by a member of the public. This was at a time when the discovery of any fish in this part of the Clyde was a notable event! Curious to find out what it had been eating, I opened up the stomach to find it contained only a ladies' handkerchief. There was no sign of the lady!' David Carss records that 'in the summer 1993, electrofishing for eels in the Dinnet lochs, I caught a fair-sized Pike. On examination, this fish was a dye-marked one I'd encountered before. It had been caught on a spoon and marked by Martyn Lucas (Aberdeen University) a few years previously. Before 1993, I'd also caught it by spinning. Later I caught the fish in a fyke net and subsequently in a perch trap too.' Clearly, this fish had not learned from its experiences!

Pike are often blamed as feeding largely on salmonids. However, from the Tweed, George Bolam recorded that '... out of eighty-nine Pike examined only eight were found to contain Trout, and in only one could the remains of Smolts (three) be identified.' However, Frank Buckland believed that 'Pike are terribly voracious rascals, and do much injury to salmon and trout fishings. ... I have also received from Mr Speedie, salmon tacksman, of Perth, a jack containing five salmon-smolts and three trout.' Some years before, 'Mr. Francis Crump, of Killin, sent me, in April, 1870, a box containing two pike, the two fish together weighing 19lbs., exactly in the same position as when gaffed by the boatman on Loch Tay. He says 'We saw a considerable commotion in the water, and on approaching to discover the cause, the fish appeared to be fighting, and merely sank a short distance below the surface. The gaff penetrated both their heads. They both lived for some hours after they were in the boat ...' I have made a cast of these fish as they were - never having been separated - which is in my museum.'

Herbert Maxwell summed up the diet succinctly: 'Pike are habitually cannibals; quantity, not quality, is what they demand, and the only line they draw in diet is at what cannot be swallowed.' He further recorded that 'In April, 1870, two boatmen on Loch Tay, noticing a disturbance in the water, rowed to the spot and saw two fish, which they supposed were fighting. With a single stroke of the gaff, both fish were drawn into the boat, when it was found that they were a brace of pike, weighing together 19 lb. and nearly equal in size. The head of one was firmly fixed as far as the pectoral fins within the jaws and gullet of the other.' As a sequel to this story, William Houghton records that 'There is also a cast of one Pike being swallowed by another; the pair weighed nineteen pounds; both were found dead in Loch Tay in 1870.' Curiously, in the *Fishing Gazette* of 17 April 1920 there is a photograph of two pike (one eating the other) found dead on the shore of Loch Tay, near Killin.

Reproduction

Herbert Maxwell observed that 'The pike is a solitary fish, except during the brief season when he associates with a mate for reproductive purposes. Indeed, he adopts very effective measures to secure the solitude he desires by swallowing all the smaller individuals of his kind into which he can set his teeth.' Pike come together to spawn from February to May, depending on which part of the country, when the water temperature is rising and between 4 and 11°C. This is a period when Pike may make long migrations from feeding areas to spawning grounds, behaviour which is exploited by fishery managers wishing to control Pike numbers in salmonid waters, as they are then particularly vulnerable to gill netting. In Loch Lomond, many of the Pike gather in a weedy area beside the mouth of the River Endrick - a favourite area for netting in times past. Males arrive at the spawning grounds first, selecting shallow areas of water with thick vegetation, or, if the water level is high, flooded terrestrial vegetation is also suitable to spawn over. The young fish spawn first and the spawning act itself is usually very vigorous, with one or more males and a female rolling over in excitement. Eggs are scattered at random, aided by the activity of the spawning pair stirring the water. Spawning takes place at regular intervals and may extend over several days. The females are fairly fecund and produce 9,000-30,000 eggs per kg of body weight.

The eggs are reddish brown to brownish, some 2.3-3.0 mm in diameter, and adhere to the vegetation. They require 150-155 day-degrees for hatching, a period of between 10 and 30 days (usually around 15), depending on water temperatures. A fragile larva emerges which attaches itself to the vegetation by means of adhesive glands in the head. After about 6-10 days the yolk sac is used up and the free-swimming larva has a length of 6-8 mm. This is a period of high mortality, during which perhaps over 90% of the larvae are destroyed. Apart from being predated by various fish and large insect larvae, a fall in water level can have fatal consequences, as the adhesive larvae cannot escape.

Predators

Undoubtedly one of the main predators of Pike are other Pike! However, as noted above, this very cannibalism enables the species to maintain populations when no other fish are present. Another increasingly important predator is the Osprey, which is able to take advantage of the Pike's habit of basking at the surface, where it is very vulnerable to Ospreys plunging from the sky. David Carss and Keith Brockie found Pike to be a significant item (11%) of the diet of Ospreys in Scotland.

Of course there are other predators too: C. Kearton records that a Kingfisher 'was on one occasion discovered unable to fly but for a short distance, ... and upon being caught and examined was found to have a young pike protruding from its gullet. Directly the fish - which measured no less than four and three-quarter inches in length - was removed the bird flew away none the worse for its experience.'

In North America, the tapeworm *Diphyllobothrium latum* infects Pike and if the flesh is not properly cooked this parasite can be transferred to humans.

Growth

Once free-swimming, the fry start to grow very rapidly, feeding first on zooplankton, but are soon large enough to take insect larvae. Within one month a pikelet can be 4 cm in length. By the time they are a year old they are 8-10 cm in length and fish have become an important item in their diet. A rapid growth rate is maintained for the first two years of life, during which an estimated daily intake of about 3-5% of body weight is required.

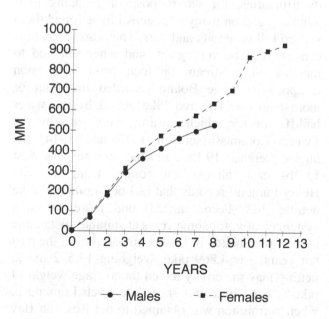

Growth of Pike in Loch Choin (Munro 1957).

Female Pike may live longer and grow larger than males, and few of the latter ever exceed 4.5 kg. As a result, in some habitats, male Pike may be an important item in the diet of large females. Some male Pike are mature at 2+ years of age and all are mature at 4+ years. Females mature at 3+ or 4+

years. Age is usually determined from the annual growth zones found on the opercular bones, though the scales are also useful sometimes, as with most other fish.

Value

Pike were once more highly esteemed in Britain than they are now. Yarrell recorded that 'Edward the First ..fixed the value of Pike higher than that of fresh salmon, and more than ten times greater than that of the best Turbot or Cod.' Herbert Maxwell pointed out that, in his day 'There is a limited market for pike and the price to the netter varies considerably. In 1918 the fish fetched from tenpence to one shilling per pound. ... The best market for pike is found in such Midland towns such as Manchester, where there is a large Jewish community.' However, though eaten rarely in Scotland, large amounts of Pike are consumed in North America and Europe where they are of considerable commercial value. There are many, often complicated, recipes for its culinary use. The farming of Pike takes place in parts of Europe.

Edward Boulenger records that 'The pike once shared the carp's popularity as an item of our pharmacopoeia. Its jaws and heart were used to allay haemorrhage, to abate fevers, cure agues, keep off the plague, etc. - but the fish's bite was regarded as being invariably venomous.'

To many coarse fishermen, Pike are the ultimate quarry, and good pike waters are an important tourist attraction to anglers from many parts of Britain and Europe. They are regarded as sporting fish and fight well on a light tackle, often jumping a number of times. The initial runs of a large Pike on being hooked may be much faster and longer than those of a Salmon. In many waters they are returned after weighing and thus may become increasingly wily, consequently making their recapture a great angling achievement. In salmonid waters, in contrast, they are regarded as evil vermin and are controlled if at all possible by angling, netting, long lines, poisoning and electric fishing. However, even salmonid fisherman consider them sport, as Thomas Stoddart indicates in his *Angler's Companion*: ' With regard to fly-fishing for pike, I used to practise it many years ago, with tolerable success, on dull and windy days, in a shallow loch in Fife.' Herbert Maxwell recalls another way of catching Pike: 'May

... This is a favourite time for noosing basking pike out of the backwaters of our trout-stream, ... It really is a very delicate art, especially when the fish are small...'

The flesh of Pike is off-white in colour and has a distinctive flavour. It is considered unpalatable by most Scots, but elsewhere is treated as a great delicacy and can be a very expensive fish to buy. Peter Malloch considered that 'Those found in stagnant lochs are not at all palatable, whereas those caught in clear lochs and rivers where food is abundant are usually quite good.' They are also favoured in some countries, where they are cropped commercially, because they are converters of small useless fish species into well-flavoured Pike flesh. The intermuscular bones are a nuisance when eating large fish but can be ignored in small fish by determined diners. Thomas Stoddart, who should have been a connoisseur in salmon, goes so far as to say that he considered the Teviot Pike 'preferable to the general run of salmon captured in that river.' However, Herbert Maxwell disagrees: ' ... but, all assertions to the contrary notwithstanding, it affords unpalatable, or at best, insipid food. Colonel Thornton ... wrote enthusiastically of the excellence of Highland pike on the table; but the utmost that ordinary culinary cunning can do is to make this fish a neutral vehicle for savoury stuffing and toothsome sauce.' The low opinion which Ausonius has of this fish is clear:

> 'The wary luce, midst wrack and rushes hid,
> The scourge and terror of the scaly brood,
> Unknown at friendship's hospitable board,
> Smokes 'midst the smoky tavern's coarsest food.'

Pike make interesting fish in aquaria and can grow quite rapidly if given enough food. If inadequate food is given they rarely stunt like other fish but become emaciated and die. Captive Pike become quite tame but they must only be kept with fish of their own size - a point noted by Gregory Bateman: 'Small Pike or Pickerel can easily be kept in an aquarium, but of course it is almost unnecessary to say that they should have no companions. ... The Pike soon becomes very tame, and will learn to know the tank in which his food is kept...' The author recalls one Pike which was kept in an aquarium at the University of Glasgow in the 1960s and looked after by John Flannagan. It was fed regularly and lived for many years - appearing at several of the Scottish Aquarium Society's shows, as part of a demonstration organised by John and the author for the University.

Conservation

Pike have no particular protection in Scotland and the species is not under any particular threat. There is a strong divergence of opinion among anglers regarding this species and many attempts have been made to eradicate it from particular waters. For example, Robert Burns-Begg notes that, in 1864, 1571 Pike were killed in Loch Leven: 'Their spawning season occurs in the months of March and April ... This period forms the golden opportunity for destroying this insatiable foe of the trouts, and it is always turned to the best possible account.' Harvie-Brown and Buckley record that 'In the five years to 1890 the Improvement Association of Loch Awe destroyed 1600 Pike, with the result that not only the number but the size of these fishes was greatly reduced.'

Peter Malloch also agrees with their destruction: 'I have had occasion to net pike on the River Tay and its tributaries, for the purpose of reducing their numbers, and on many occasions I have found them stuffed full of smolts and parr. The destruction thus caused must be very great. and when allowed to increase in a stream or loch trout very soon disappear.' George Bolam recorded that 'Out of more than one hundred Pike netted by the water bailiffs, or for which rewards were paid by the Tweed Commissioners in 1913 and 1914, the largest weighed 19 lbs., and the four next heaviest 15 lbs. each, all of these coming from the Till.' Henry Lamond records that in Loch Lomond: 'Pike netting has been carried on, more or less systematically, for some years at suitable spots ...the total of pike netted in Loch Lomond during the past ten years.... is 4,869 pike, weighing 13,572 lb., ... netting may materially lessen the average weight of pike ... even in a vast sheet like Loch Lomond ... when permission was obtained to net Rossdhu Bay the average weight of pike taken there was found to be 7 lb. The following season this fell to 3 lb.'

In contrast, Charles St John considers that although 'Like a shark, when hungry, the pike swallows anything and everything which comes with reach of his murderous jaws.' ... 'I am strongly of the opinion that pike deserve encouragement in all Highland lakes where the trout are numerous and small.' Elsewhere he records that 'It is a fallacy to suppose that pike are detrimental to the sport of the fly fisher, that is, in a Highland lake where there is depth and space enough for both kinds of fish to live and flourish. Of course, pike kill thousands and tens of thousands of small trout, but the fault of most highland lakes is that there are too many trout in them ... Pike keep down this overstock. There are still plenty and more than plenty of trout remaining in the water, and of a better size and quality than when they were not thinned.' Although written in 1891, this attitude was by no means new, for more than 400 years earlier, Juliana Berners had commented: ' The pyke is a good fysshe: but for he deuouryth so many as weel of his owne kynde as of other: I loue hym the lesse.'

A final word of warning: Pike can injure people! Niall Campbell recalled that 'my colleague Gordon Struthers was once so badly bitten by a large Pike (18 lbs) that I had to take him to the District Nurse for first aid.' The author also has been bitten badly, and more than once, by large Pike from Loch Lomond.

References

There have been several useful studies of Pike in Scotland, as well as in other parts of Europe and North America. Its angling and culinary importance as well as its reputation as a predator have lead to considerable interest in this species.

Aflalo (1897), Bateman (1890), Berners (1496), Beukema (1970), Bolam (1919), Boulenger (1946), Bregazzi & Kennedy (1980), Buckland (1880), Buller & Falkus (1988), Burns-Begg (1874), Campbell (1955), Carss & Brockie (1994), Cholmondeley-Pennell (1866), Chrystal (1927), Fabricius & Gustavson (1958), Fickling (1982), Fitzmaurice (1983), Franck (1694), Frost & Kipling (1968), Gordon (1902), Harvie-Brown & Buckley (1892), Healy & Mulcahy (1980), Houghton (1879), Kearton (1922), Kipling & Frost (1969), Lamond (1931), Lang & Lang (1923), Maitland (1990), Malloch 1910), Mann (1976, 1982), Mann & Beaumont (1980), Maxwell (1904), Munro (1957), Orwell (1929), Rintoul & Baxter (1935), Shafi & Maitland (1971), St John (1891), Stoddart (1847), Svardson (1950, Thornton (1804), Treasurer (1980), Yarrell (1836).

*'I saw, dimly
Once a big pike rush,
And small fish fly like splinters.'*

D H Lawrence (1923) *Fish*

*****

Sparling *Osmerus eperlanus* Dubh-bhreac

'... a fragrance as of cucumbers and violets diffuses itself...'

Herbert Maxwell (1904) *Memories of the Months: First Series*

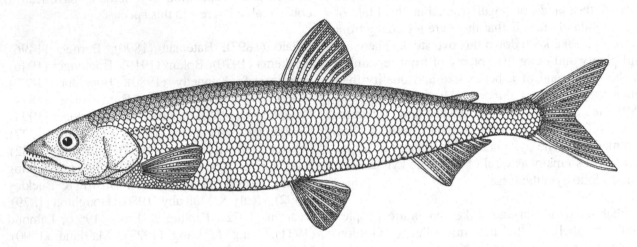

NOT many people would associate Herbert Maxwell's fragrance of 'cucumbers and violets' as something concerning fish, yet those who know and admire the Sparling would agree that this fish is something special - both to observe, to smell and to eat. He continues: 'One mild November morning I enjoyed the best dish that I can recollect ever to have eaten at breakfast. It was in an excellent little hotel in Creetown. We had been down to the tidal estuary before sunrise to watch the nets drawn for smelts (or sparlings as we call that pretty fish in Scotland), and we carried some of them up to the hotel and had them piping hot on the table before they had been half an hour out of the water. ... Smelts are more perishable than any other British fish, and a few hours suffice to dissipate their peculiar aroma ... No one, therefore, can have a notion of the subtle toothsomeness of the smelt who has not treated it as we did ours on that far-off morning at Creetown.'

On the other side of Scotland, beside the estuary of the River Tay, Clement Gunn, a doctor in Newburgh agreed: 'Lastly, to complete this gastronomic paradise, there were the sparlings, which followed the salmon, and were in excellent condition just when the salmon fishing closed. These were elongated silver fish, semitransparent, and a great delicacy. They were packed into flat boxes as soon as caught, and like the salmon, hurried off by special train to the London market.'

Other names used for this species are Spirling, Sperling, Smelt and Cucumber Smelt. A literal translation of the scientific name is 'Scented Smelt'.

Size

The Sparling is a small to medium sized fish whose adult size varies greatly according to habitat. The normal range in length is 10-20 cm but some fish reach 30 cm. A fish about 15 cm long weighs some 30 gm. In general, fish from non-migratory freshwater populations are much smaller than those which have lived in the sea. The present British rod-caught record stands at 191 gm (6 oz 12 dm) for a Sparling caught by G. Idiens near Fleetwood in 1981. The present Scottish shore record is for a fish which was caught by M. Ettle at Riverside, Dundee, which weighed 149 gm (5 oz 4 dm). The largest female recorded by the author from the River Cree was 189 gm and 290 mm, and the largest male was 176 gm and 272 mm, both caught in 1994.

Leonora Rintoul & Evelyn Baxter, writing in 1935, noted that in the Forth, 'Up to January those taken usually measure from four to six inches [100-150 mm] in length, while from that time to the end of March the larger ones make their appearance and the young ones disappear.'

More recently, the lengths of adult Sparling taken around spawning time in Scotland by the author show clear differences among fish from the three remaining populations. Cree Sparling are typically from 175-250 mm, Tay Sparling from 150-200 mm and Forth Sparling from 125-175 mm. However, there is recent evidence that the length of adult Sparling in the Forth is increasing since they reappeared in 1989; occasional specimens over 200 mm are now taken there.

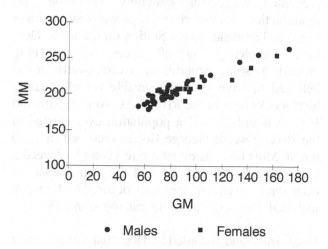

Length / weight relationship in adult Sparling from the River Cree in March 1992.

The mean lengths and weights of adult females were consistently greater in the River Cree in 1990 than those of males. Peter Hutchinson found the same with the adult Cree fish in the early 1980s but noted that for the 1+ age groups the reverse was the case.

Appearance

The Sparling is a small rather slender fish with large scales and eyes, a large mouth with a projecting lower jaw and obvious teeth on both jaws. Aflalo notes that 'This fish has a large mouth armed with sharp teeth ... It spawns in spring and early summer, having a preference for shedding its spawn in stormy weather.' It can sometimes be confused with Powan and Vendace - which have small mouths and no teeth - and with salmonid smolts which have small scales and at least a few spots of some kind. In addition to teeth on both jaws, there are teeth on the tongue and on the vomer bone on the roof of the mouth. Like the salmonids and coregonids it has a fleshy adipose fin between the dorsal fin and the tail fin. The paired fins are moderately developed as is the anal fin. The body is well covered with shiny

scales and the lateral line itself is incomplete, occupying only about 10 scales.

Scale from Sparling (courtesy Peter Hutchinson).

Live fish are somewhat translucent in appearance. In colour, the top of the head and the back are a rather variable dark grey-green which shades into a silvery green stripe along each side and a silvery white colour below. Herbert Maxwell's description is good. 'The creature is of a fairy-like beauty when freshly landed, the colour on the back varying from sea-green to palest brown; the sides are faintly tinged with yellow, violet, or rose, shot with silvery gleams; but as the scales are devoid of pigment, the body of the fish is translucent, the bones and internal organs being discernible through the skin.'

They have a very characteristic cucumber-like smell (and are sometimes called Cucumber Smelt). This odour accounts for their popularity as a food fish: William Houghton notes that 'The Smelt or Sparling is, perhaps, one of the most delicate in flavour of all fish that swim, whether in salt or fresh water.' whilst Willughby's opinion is expressed in these terms: "'Carne est molle et friabile, sapore delicato, gratissimum violae odorem spirante.'" Herbert Maxwell concludes: 'About that fragrance there can be no doubt in the mind of anyone who has seen the fish landed. It is perceptible under a gentle breeze at a distance of nearly one hundred yards. Some have compared it to the scent of violets, others to that of cucumbers; but my own olfactories detect in it nothing more refined than the smell of fresh rushes.'

Distribution

The Sparling occurs from southern Norway around the western coast of Europe (including the Baltic Sea) to north west Spain. It is found in coastal waters and estuaries, migrating into large clean rivers at spawning time. The species is tolerant of wide salinity changes and there are several non-migratory, purely freshwater, populations in large freshwater lakes in Finland, Sweden and Norway. The Romans were supposed to have cultivated Sparling in freshwater ponds. The sole freshwater population in the British Isles, in Rostherne Mere in Cheshire, became extinct in the 1920s, probably as a result of eutrophication.

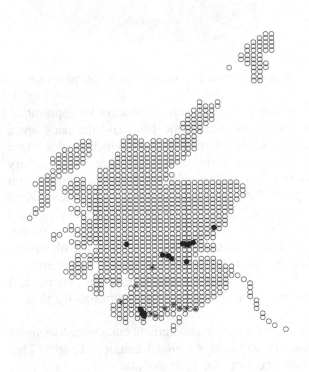

Recorded distribution of Sparling in Scotland. Asterisks represent extinct stocks.

Elsewhere in Great Britain, the Sparling was once a common estuarine species, occurring in most larger rivers from the Clyde and Tay southwards. It supported small commercial fisheries in the estuaries of some of these rivers. Over the last century, the species has gone into decline and has disappeared from many rivers. In spite of its return to a few rivers (e.g. the Forth and the Thames) it has been regarded by the author as significantly threatened for over three decades but little conservation action has been taken to date.

Sparling populations have been recorded from at least 15 rivers in Scotland (Almond, Annan, Bladnoch, Clyde, Cree, Dee, Esk, Fleet, Forth, Girvan, Lochar, Nith, Stinchar, Tay and Urr), but

over the 20th century the species has suffered a severe decline and has disappeared from all its former sites except the Rivers Cree, Forth and Tay. These populations must now be regarded as having high conservation importance.

There are probably several different reasons for this decline. In some rivers (e.g. the Clyde and the Stinchar), pollution in the lower reaches has prevented successful spawning migration and reproduction. In other rivers, high weirs and barriers (e.g. the Tongland Power Station on the River Dee) have completely cut off access to spawning grounds. In some estuaries and rivers, overfishing is believed to have been responsible for eliminating local stocks (e.g. in the Rivers Esk, Annan, Nith and Urr). It is not clear if a population ever existed in the River Tweed; George Bolam recorded in 1919 that it 'Must be considered a rare visitor to Tweed .. I have been assured by old fishermen that an occasional specimen does still occur. Dr. Johnston included it as very rare in his catalogue in 1838 ...'

Frank Buckland recorded in 1873 that 'In my report on the salmon fisheries of Scotland, 1870-1, when writing specially on the fisheries of the Solway, I made the following remarks. "There exist in the estuaries of many of the Scotch rivers, especially at Wigtown Bay and the mouth of the Nith, a fish called the smelt or sparling (*Salmo Eperlanus*). ... They spawn exactly at the head of the tideway among the small shingle-stones; they never go further up the river than the brackish water extends.'

A fishery for Sparling used to take place at Erskine on the Clyde Estuary but the population was last reported there in 1845. Although a single specimen was recorded by Anne Henderson of the Clyde River Purification Board (now SEPA) in one of the Glasgow docks in the 1980s, none has been found since then.

River Cree

The population of Sparling currently found in the River Cree and its estuary is the sole survivor of several populations which formerly occurred along the south and west coasts of Scotland. Although the population is still surviving, and is the only one in Scotland to have been studied in some detail, it faces an uncertain future with a number of serious threats to its existence. At one time numbers were very high: in the 1980s several of the older people in Newton Stewart who were interviewed by the author said they were once so abundant 'you could almost walk across the river on their backs'.

Catches in the local fishery for Sparling have declined considerably in recent years although netting still continues. The present status of the Cree population may in part be due to overfishing in the past but may also be related to habitat changes in the spawning area caused by pollution from local sewage discharges, acidification from the upper parts of the catchment and physical barriers to its migration such as the weir in Newton Stewart and recent obstructions placed in the river by anglers.

Although no estimates of the size of the Cree population have ever been attempted, some data are available which indicate the minimum numbers in the adult stock. Catches of up to six tonnes have been made by the fishery in the recent past. The average weight of fish collected in the author's study from 1991-94 was 116.35 g (= 8595 fish per tonne) and so the minimum size of the adult population in the River Cree in some years at least (e.g. 1984) must have been well in excess of 50,000.

River Forth

The River Forth at one time supported an enormous population which was the basis of an important fishery, both for local needs and as a luxury item for the Edinburgh market. The spawning run during the 1800s was apparently spectacular. Leonora Rintoul & Evelyn Baxter recorded that 'Near Alloa Sperlings are taken in great numbers in the Firth. Parnell says "In the month of March these fish ascend the Forth in large shoals ... The young can be taken at Alloa throughout the summer months, but the larger specimens are only met with during the season of spawning."'

A Scottish Office review of the fisheries of the Firth of Forth, noted that: 'Data on the catches from this fishery are available from 1891 to 1974 .... This fishery yielded an average of 15 tonnes of fish each year up to the mid 1910s, from then until the mid 1940s the annual catch was around 7.5 tonnes. Following a short period of higher catches in the late 1940s the catches declined to zero by the mid 1950s, during the 1960s there was a slight recovery

in the fishery but it again declined to zero in the mid 1970s. Although the catches of Sparling in the area have never been high, this species, particularly in the late 19th century, was highly prized and commanded a high price.'

The species then disappeared from the Forth until the author, in 1983, introduced a few hundred fry to a pond in East Lothian, connected to the sea by a small burn. These fish were hatched from eggs stripped from fish taken during the spawning run in the River Cree. In 1989, a single specimen was recorded by the Forth River Purification Board in the Forth Estuary and three further specimens were taken in 1990. The exact origin of the new stock is unclear (i.e. Cree or Tay) but the Sparling is now common again in the Forth Estuary, being taken regularly in monitoring trawls by SEPA and in commercial boom netting for Sprats near Kincardine-on-Forth; it is entrained in considerable numbers on screens at Longannet Power Station. Here, entrainment rates of up to 3,312 Sparling per day were recorded by the author in 1994, so the population must now be sufficiently large to be able to stand such mortality.

River Tay

Sparling are known to have occurred in the Tay Estuary for many years (they were common in 1838) and have been the basis of an erratic fishery there. This diminished in size until the 1970s when only two boats were fishing part time, but catches have increased in recent years and, until recently, Sparling were regularly sold by fishmongers in Perth and in the villages around the estuary. In the early 1990s the annual catch obtained by the late Charlie Johnston (one of the most knowledgeable of the sparling fishermen) was about 10-15 tonnes. Unfortunately, the fishery seems to have gone into decline in recent years.

Although few statistics are available from the Tay, the catch in 1991, when fish were obtained for study by the author, was at least 15 tonnes. This, with an average weight of 47.5 g (much smaller than the Cree Sparling), represents some 315,889 adult fish.

Ecology

This fish clearly favours clean estuaries except at spawning time, and though it has the salinity tolerance noted above it, is apparently very susceptible to pollution and perhaps other stresses created by humans. Many estuaries which formerly had large populations lost them as pollution increased and the species is much less common than formerly. However, its return to waters such as the River Thames and Forth as they become cleaner is a welcome sign and one which it is hoped will be repeated on other rivers.

The ecology of Sparling in the River Cree has been studied by Peter Hutchinson, Alex Lyle and the author; these studies are the basis of much of our present knowledge of this species in Scotland. Elsewhere, Scot Mathieson recalls catching large Sparling in a creek in the Kincardine salt marsh in February 1994 on a night-time high tide. The largest fish was 219 mm and its gut contained Common Shrimps. Although this was the only occasion that adult fish were taken, juvenile Sparling were found to be entering creeks in spring and summer 1995 (on 10% of 70 tides sampled over a 15-month period), particularly on dawn tides.

Food

Sparling fry are very small at first and feed on minute zooplankton, probably mainly protozoans and rotifers. As they grow they take larger planktonic crustaceans and some bottom animals, eventually becoming quite voracious predators taking larger crustaceans (shrimps and mysids) and young fish such as Sprat, Herring, Whiting and gobies.

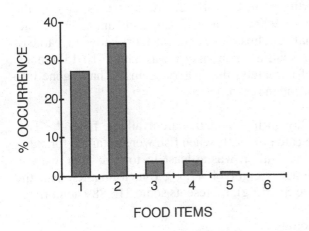

Food of Sparling in Lake Vanern, Sweden (1: cladocerans; 2: copepods; 3: mysids; 4: insects; 5: fish; 6: others) (Nilsson 1979).

Reproduction

Parnell recorded that they '... deposit their spawn in fresh water; this they shed in immense quantity about two miles below Stirling Bridge, when at that time every stone, plank and post appear to be covered with their yellowish coloured ova.'

Coastal and estuarine Sparling are generally regarded as migrating up rivers into fresh water to spawn during spring. This is certainly the case in the River Cree and was also the case in the River Forth in earlier times when the population there was large. However, since its recovery in recent years the present spawning grounds in the Forth have not been located, despite searches there from 1991-94 as part of the Operation Brightwater project. Similarly, it has not yet proved possible to identify the spawning grounds of Sparling in the River Tay. It is suspected that, in both these rivers, the Sparling spawn, unseen, over coarse gravel beds in the relatively inaccessible freshwater parts of the upper estuaries.

On reaching maturity, adults migrate up the estuaries and into the lower reaches of rivers in March and April. Usually the run in each river occupies only a few days, but during that time the spawning activity becomes furious and the sticky eggs attach themselves to everything on the river bed - gravel, stones, weed and sticks. Alex Lyle recalls 'The amazing experience of being in among a shoal of spawning Sparling in the River Cree at Newton Stewart one night in March during the 1990s. The fish were in such a spawning frenzy that they were throwing themselves up on the bank.' Sometimes the river level drops subsequently and leaves many eggs stranded to dry out, occasionally seen as a white band along the river's edge.

Peter Hutchinson found that 'The spawning runs which started in early March when water temperatures were greater than 5°C were characterised by marked changes in, age and sex ratio. The fish enter the spawning stock at 1+ years of age and, in common with other estuarine smelt populations, are highly fecund. The spawning stock was composed mainly of two age-classes (1+ and 2+ years) ... Spawning, which was interrupted during periods of high flow, lasted about 1 week. Male fish tended to remain on the spawning grounds after females had vacated the area.'

On several occasions the author has watched shoals of spawning Sparling in the shallow clear waters of the River Cree at Newton Stewart. Sometimes the fish just maintain themselves in the current, heads facing upstream, perhaps waiting to spawn after nightfall. At other times, usually in faster flowing water, they seem to be more active; periodically individual fish are seen to 'flash' as they turn on their sides. The precise meaning of this 'flashing' is not clear.

Sparling are highly fecund - 16 Cree fish examined by Peter Hutchinson in the 1980s had a mean fecundity of 56,603 eggs per female. In a later study by Alex Lyle and the author, the fecundity of 32 smaller female Sparling caught by commercial fishermen in the Tay Estuary was estimated at 30,360 eggs per female. The pale yellow eggs (0.9 mm in diameter) hatch in about 20-35 days according to the local temperature. The eggs are often laid in quite fast flowing water, but as Alex Lyle recalls 'they have an astonishing adhesive quality - when fish from the River Cree were being stripped during the 1990s, as soon as the eggs struck anything after stripping - egg trays, tiles, buckets, boots, etc. - they stuck there firmly.' After hatching, the young are swept quickly down into the upper estuary where they start to feed. Though there is considerable mortality among the adults during spawning, many do manage to return to the sea where they recover to grow further and spawn again in subsequent years.

Predators

The adults are very vulnerable to all kinds of predators (including humans) at this time, especially when they strand themselves on dry land. A fairly sure sign that a spawning run is under way is an increase in the numbers of piscivorous birds in the area, numerous Cormorants appearing in the upper Tay Estuary at this time. One memorable March morning on the River Cree in the 1990s, after there had been an intensive spawning the previous night, the author and Alex Lyle recorded 'Finding numerous Sparling in the River Cree some way below Newton Stewart in the 1990s (March) with their heads under stones at the edge, but bodies exposed, as the water receded with the tide. Lifting any of these fish out and into the river had little effect for it immediately swam to the nearest stone and stuck its head under it again - ostrich style.'

Growth

Growth is a very variable process in this species according to local conditions. In suitable estuaries the young may reach 10 cm by the end of the first year and some 15 cm by the third year, by which time they have all started to breed. They may live for several years beyond this, reaching lengths of about 20 cm at 6 years of age. The males are usually smaller than the females and average weights are about 28 and 36 gm respectively.

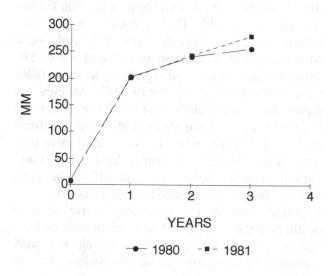

Growth of Sparling in the River Cree (Hutchinson 1983).

Of the existing Scottish stocks, the Cree population is the most vulnerable at spawning time, when it is very exposed and is readily accessible to netting. Moreover, this population (like most others) has few age classes: Peter Hutchinson found the age structure in spawning fish to be 61.3% 1+, 37.6% 2+ and 1.1% 3+, with 3+ fish present in only half the samples taken during the 1980 and 1981 spawning runs.

Value

In the British Isles, there are still several commercial fisheries for Sparling which rely mainly on their vulnerability during the short spawning run to catch them (sometimes in enormous numbers) in traps and nets. In some parts of Europe they are caught in the estuaries in drift nets and trawls and they are sold either fresh or smoked. The author can verify that they are delicious to eat, as is the roe when lightly fried. Although only three populations remain in Scotland, all are the subject of fisheries. On the River Cree in some years up to six tonnes of Sparling (about 55,000 fish) are taken from the spawning run - probably a high percentage of the population there and undoubtedly a threat to its existence.

The value and history of the Sparling is better known for the Solway than for any of the other estuaries, thanks to Herbert Maxwell who wrote in 1897: 'Londoners have no idea of the real excellence of smelts. To be eaten in perfection, the fish should be carried from the nets into the kitchen and served forthwith. There comes to mind a quiet fishing village on Solway shore. It is early on a winter morning, but the air is still and warm. The small-meshed seine is coming slowly to shore, filled with a shoal of pearly "sparlings". As they are drawn from the muddy water of the estuary, a fragrance as of cucumbers and violets diffuses itself, plainly perceptible at a distance of more than fifty yards. The fish, as soon as landed, are neatly packed in small boxes and sent off to the great towns, where they command a high price - 3s or 4s. a pound. But by the time they arrive at their destination they will have lost much of their delicate texture and exquisite flavour, which we shall presently be savouring in the inn parlour close at hand.'

He held the same high opinion in 1904: 'The flesh is exceedingly delicate, with a flavour unlike that of any other fish. ... Little do townspeople understand of its excellence, for smelts are more perishable than most other fish. To eat them in perfection one should rise betimes on a fair morning in October, and watch the nets being drawn ashore. To the distance of nearly one hundred yards the gentle breeze wafts the fragrance of the catch; fill your basket with the spoil, carry it straight to the kitchen, have them fried without cleaning, and sit down and enjoy such a breakfast as all the resources of the Mansion House could not furnish.'

Years before, J.M., writing in *Land and Water* in 1866 observed that 'So delicious and delicate a dainty is the smelt, that the only wonder is, how it is he has not become more esteemed and more sought after by the *gourmand* and the invalid.'

The price appears to have held between Maxwell's writings: 'I have spoken above of the sparling as a

valuable food fish. In support of that may be quoted a note which I have of the returns from the sparling-fishing in the Cree ... In the seven months from 1st September to 31st March 1901, the four fishermen ... accounted for sparlings to the amount of £300.' Note: at ca 3/6d per pound, this would be about 857 pounds (< 1 tonne).

The traditional method of fishing for Sparling in the Cree was described by the Fishery Board for Scotland in 1883. 'The sparling net commonly used in this neighbourhood is about 140 to 150 yards long. The mesh is so small as to catch shrimps, being 42 to 45 rows to the yard in the bosom, and 45 to 55 rows in the wings. Actual measurement of the mesh of a new net showed the mesh to be 5/8-inch. In the estuary of the Cree the practice prevails for those who have right of salmon fishing to let the right of sparling fishing. The result is that certain authorised men, who are at the same time the lessees of the salmon fishing, fish in the tidal waters of the Cree and in the upper part of Wigtown Bay, while other men fish as they choose further down the estuary beyond the limits of the salmon fishing.'

The situation in the Annan was described differently. 'A number of men and boats fish for sparlings every year, commencing in the month of August and continuing through the winter, unless stopped by hard frost. The sparling net is so small in the mesh that it takes the smallest of fry, and it is used only in the low water channel and mostly during the night time. Great numbers of salmon and herling are taken in these sparling nets, and although several convictions are got every year against the parties who use them, there is great slaughter of salmon during the close season which is most difficult to get at.'

In the Tay, arising out of a case for interdict against sparling fishers tried before Lord Trayner, a local close time for this kind of fishing was established from 25th February to 25th September in each year. The method of fishing is somewhat different in the Tay, boom nets being employed from boats anchored in the tideway. However, it is apparent that at one time Sparling were taken by fixed nets in the Tay. John Buchan includes this account in his description of a tussle with a large Salmon: 'He turned out of the Earn and went down the channel in a succession of long steady runs ... In an hour and a half the boat had drifted down into a most dangerous neighbourhood - the poles and ropes of a row of sperling nets, standing on a sandbank in five or six feet of water, with each rope surrounded by a drift of surface rubbish.'

The late Charlie Johnson was one of the last of the sparling fishermen in the Tay and was most helpful to the author in obtaining samples: 'At the turn of the century there used to be 40 boats here at Newburgh. My boat, the *Fait*, is one of the last pair. ... and there used to be a demand in the Netherlands. Dutch boats came over to fish for them here after the war. ...You can smell them in the water. It's just like cucumbers. If you're fishing the Tay, and you catch them with other fish, then the whole netful will smell of cucumber.'

Further south, yet another method of fishing for sparling was at one time employed, as described by W.A. Dutt: '... near the junction of the Yare and Wensum, two or three cast-netters may still be found in spring, industriously netting from dusk until late in the evening or early in the morning. Each boat contains two men, one of whom, by "back-watering" with his oars, keeps the boat from being carried downstream ... while the netter ... stands at the stern on a little deck made of half a dozen boards. Over one arm is slung the lead-weighted net, which from time to time, with a swaying turn and a sudden jerk, he casts upon the water in such a way that it expands to its full circumference as it falls, ... For a few moments the net is allowed to sink in the deep water ... and as it sinks its leads cause it to assume the shape of an inverted funnel; then it is carefully drawn towards the side of the boat, the funnel of net meanwhile closing as the leads come together, and by a quick final movement it is lifted on to the little deck at the stern of the boat. ... there is always a chance of a fair catch being made, or even a good one, which will reward the patient fisherman for many fruitless casts.'

The culture of Sparling is mentioned by W. Yarrell, who noted that 'Colonel Meynell ... kept Smelts for four years in a fresh-water pond, having no communication with the sea: they continued to thrive, and propagated abundantly. ... When the pond was drawn, the fishermen of the Tees considered that they had never seen a finer lot of Smelts. There was no loss of flavour or quality,' William Houghton expands on this 'Experiments

have been made occasionally in this country to retain it in ponds ... I have often thought that modern pisciculturalists might profitably turn their attention to the cultivation of a fish of such a peculiar and delicate a flavour as the Smelt or Sparling.' So far as the author is aware, no-one has attempted this in recent times.

Conservation

Threats

It is clear that the status of Sparling in Scotland has declined markedly over the last century and that the current number of populations (three) is too low to be confident of the future survival of this species here. The main threats to the Sparling, which account for its demise in previous locations and present dangers for existing stocks, are overfishing and pollution, obstruction of spawning migrations by dams and weirs, and physical destruction of spawning and nursery areas by engineering works. High levels of oxygen are known to be of importance during the period of spawning and larval development.

In 1882, following the collapse of the smelt fisheries in the inner Solway, the Solway White Fishery Commission recommended a close season for smelt extending from 1 February to 1 August in order to protect migrating smelt (and salmon smolts). However, A.T. Masterman considered that a total ban on the exploitation of migrating smelt was out of line with the protection afforded to other anadromous salmonoids, particularly as the most valuable smelt are those containing roe.

Most of the historical data on Sparling in Scotland are from the Solway area, where the extinction of stocks, with the exception of the Cree, was almost certainly due to overfishing, although pollution and river barriers were also likely contributory factors. A century ago (1897), Herbert Maxwell - a real visionary of his time - wrote of his concern: 'There is no close time provided for these fish by law; hence in some rivers - the Annan and the Nith, for example - where they were plentiful in former times, they have been netted to the verge of extinction, and the industry, once profitable, no longer pays. All fishermen to whom I have spoken on the subject agree that a close time is desirable; but so long as it is not made binding upon all alike, none is willing to hold his hand while others may be fishing at other parts of the same stream.'

And in 1904: '... in consequence of no close-time being provided for them, they are netted during the spawning season, which has resulted in the ruin of a remunerative industry in some rivers such as the Annan and the Nith. It is gratifying to observe that the Royal Commission on Salmon-Fisheries in their report just issued (August 1902) recommend the establishment of a yearly close-time. In the Cree, one of the Solway rivers, the salmon fishings passed in 1900 under the management of an angling association, ... A close time, accordingly was instituted from April 1st to September 1st, with the entire concurrence of the fishermen, who had difficulty in conveying these delicate fish to the market in warm weather. The result of this regulation has proved very satisfactory.'

Twenty five years later (1922), Herbert Maxwell was less optimistic: 'we reported strongly in favour of a statutory close-time for the protection of this valuable food fish during the spawning season. This recommendation was subsequently approved and reiterated in the Report of the Royal Commission in Salmon Fisheries, 1902, and it is much to be regretted that legislative effect has not yet been given to these recommendations. Every witness before my committee spoke of the urgent need for such a close-time. Sparling-fishing, once a remunerative industry in the Annan, had been brought to an end through indiscriminate netting during the spawning season.'

Since then, with the exception of the Cree stock, Sparling have disappeared from the Solway and over the last century, threats additional to overfishing have increased. Changes in land use practice in agriculture and forestry have led to the enrichment of rivers and estuaries and to altered hydrology. Pollution from domestic and industrial sources and acidification from the atmosphere has also contributed to habitat damage. In addition to threats from overfishing, the Cree stock has to face problems of restricted access upstream because of a weir at Newton Stewart, pollution of the spawning grounds from a sewage discharge and more recently, further restriction to migration and disturbance of the spawning area because of several groynes placed in the river channel by anglers, supposedly to improve fishing.

Social aspects

There is no doubt that, as Herbert Maxwell indicated, the sparling fishery of the Solway was an important one until the last few decades of the 20th century and, as with the Vendace at Lochmaben, many rituals and local traditions were linked to it. The Minnigaff Kirk Session Minutes for the 1730s record that one man was charged with sleeping with one of the female servants at Machermore. The man's defence was that it was traditional that the men coming for the 'spirling harvest' should sleep with the female farm servants 'time immemorial, past memory of man'.

The Penninghame Parish records for March 21st 1731 note that the Kirk Session had been informed that 'some people in Newton Stewart and Barbuchanie had gone out to the spirling fishing on the Sabbath night, which practice they judge sinful and offensive and therfore to appoint their officer to cite Patrick McDougall, shoemaker, William McGill, weaver and John Bell, labourer to the session to meet this day fourtnight.' Two of the men were found guilty and 'They were call'd in and rebuked accordingly.' In addition 'this to be intimate from the pulpit and also that warning be given against the like as to any sort of fishing or any worldly business.' (on Sunday).

Conservation action

A Species Action Plan for the Sparling has been prepared by the author and it is hoped that this will be the basis of a future conservation strategy for this species in Scotland. Sufficient data are already available for specific conservation proposals to be acted upon with the objective of restoring the former status of this valuable fish, which could be of increasing economic importance if stocks are managed wisely. Prospects now look better for the population in the Forth Estuary and the Forth River Purification Board (now SEPA) is to be congratulated on the visible (and economically important) conservation reward for its efforts in former years.

One aspect of the proposals is that there should be proactive conservation management of the population in the River Cree. This will involve habitat restoration and management of the spawning grounds, making sure that spawning Sparling have access to these grounds, ensuring that any future commercial fishing is sustainable, and implementing an adequate scheme to monitor the status of the population.

Other measures proposed relate to the recovery of stocks through a suitable translocation programme to rivers where the Sparling previously occurred. The recent research has shown that large numbers of Sparling eggs can readily be obtained during the spawning run and that these can be transferred and hatched elsewhere. Assuming that the factors responsible for the extinction of the species have been, or could be, eliminated, then a careful programme of translocation should be attempted. Obvious rivers to which stock from the River Cree could be transferred are the Annan, Fleet, Nith and Urr. This is not a new idea, for in 1882 the Solway White Fishery Committee also expressed the view that 'the attention of the Fishery Board of Scotland might be directed to the introduction of this excellent fish (smelt) in estuaries where at present they do not exist'.

Given the success of such conservation measures, new sparling fisheries may eventually be created for the sensible sustainable use of this once valuable resource. The opportunity exists to create a programme of action where conservation enhancement of Sparling and legislative controls on its exploitation can proceed in a structured and sustainable way for the benefit of all. The collapse of the Sparling populations in the Forth and in the Solway, preceded as the latter were by the warnings of Herbert Maxwell as early as 1897, should not be forgotten.

Commercial fishing has not been the only problem on the River Cree in recent years. Peter Hutchinson recalls 'seeing Sparling being netted by local folk -

to be used as a fertiliser for rose bushes' - even after he had explained the background to their rarity and conservation status.

In summary, conservation action urgently needed for the Sparling in Scotland is as follows:

(a) Legislation to give protection to this species which is endangered in Scotland.
(b) Monitoring of all three extant populations in the Rivers Cree, Forth and Tay.
(c) Habitat survey, restoration and management of traditional spawning grounds.
(d) Management of the fisheries to ensure that they are sustainable.
(e) Translocation of stock to restore populations in formerly occupied rivers.

References

Most of the research carried out in Scotland by the author and others has been published. In addition, valuable information is available from studies elsewhere in Europe.

Aflalo (1897), Banks (1970), Bolam (1919), Borchardt (1988), Buchan (1917), Buckland (1873), Dutt (1906), Ellison & Chubb (1968), Gibson (1984), Gladstone (1912), Gordon (1921), Gunn (1880), Henderson & Hamilton (1986), Houghton (1879), Howard *et al.* (1987), Hutchinson (1983), Hutchinson & Mills (1987), Jilek *et al.* (1979), Lyle *et al.* (1996), Maitland (1966, 1974, 1979, 1985, 2003), Maitland *et al.* (1980), Maitland & Lyle (1990, 1991, 1992, 1996), Masterman (1913), Maxwell (1897, 1904, 1922), McLusky (1978), Mitchell (1990), Naesje *et al.* (1987), Nilsson 1979), Parnell (1838), Rintoul & Baxter (1935), Scott & Brown (1901), Service (1902), Sinclair (1779), Wood (1791), Yarrell (1836).

'Sparling-fishing, once a remunerative industry in the Annan, had been brought to an end through indiscriminate netting during the spawning season.'

Herbert Maxwell (1922) *Memories of the Months: Seventh Series*

*****

Vendace *Coregonus albula* Pollan

*'The inhabitants of Lochmaben look upon the Vendace as a mysterious fish
peculiar to their lakes, a fish in which they take a pride,
and one concerning which there are many curious traditions.'*

Tate Regan (1911) *British Freshwater Fishes*

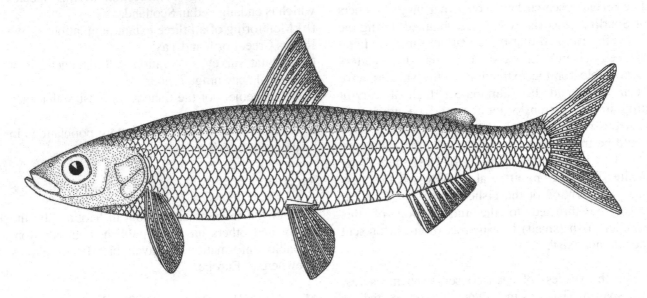

IN 1854, R Knox wrote that in ' ... the Castle Loch of Lochmaben; ... Pike abound, but chief and foremost is the vengis or vendiss, thought to have been transplanted to this lake by the early monks. ... The vendiss live on minute microscopic entomostraca, a kind of shellfish, declining all other food. ... The vendiss, shorn of its mystery, is still fished at a particular time of the year, and ate in season - a kind of whitebait dinner for some old-fashioned people. It brings them together - suspends for an hour or two the little animosities which all farmers and small lairds have to their next neighbours - and benefits the inns. ... Abhorring clubs and whitebait dinners, I always avoided these vendiss clubs, but I have often eaten the vendiss when just from the lake. It is a moderately good fish to eat, but not to be compared to the Loch Leven trout.'

The Lochmaben Vendace, although one of the rarest fish in Scotland, has more cultural associations and local traditions than almost any other Scottish species. The first written account of the Vendace is by Robert Sibbald in his 'Scotia Illustrata', published in 1684, who gave it the name *vandesius* - 'Pifcis in Lacu mabano, Vandefius. In eodem Lacu

Gevandefius'. In 1831, William Jardine pointed out that the Vendace was a separate species from other *Coregonus*. As far as the common name is concerned, Herbert Maxwell believed that 'Pennant was no doubt correct when he interpreted the name 'vendace' as representing the French vaudoise, or vendois, a dace, ... The French origin of the name has tended to strengthen the tradition, assuredly groundless, that the vendace was introduced to Scotland from the Continent by Mary Queen of Scots.'

The association with royalty continued, for William McDowall in his *History of Dumfries* records that 'King James VI was offered vendace at a banquet in his honour in Dumfries in 1617. Seemingly, they had been chosen by the chef with an assurance that they were a local delicacy which would prove favourable to the royal palate. James, thinking they emitted a peculiar smell and had an inauspicious appearance, viewed them with almost as much horror as was felt by his ancestor Macbeth when the ghost of Banquo glided in to disturb the feast at Glamis. Starting to his feet, he shouted "Treason", and it was not until the offending dish was removed that he resumed his seat and his equanimity.'

Thereafter, the Vendace was known in some quarters as the 'Treason Fish'.

The Vendace was originally regarded by local people as unique to that area; in 1907 R Service recorded that 'At a recent meeting of this Society he had the privilege of reading a paper upon "The Vendace", in the course of which he had occasion to allude to the fact that the Lochmaben folks had not any monopoly of the vendace as was often supposed, because it was also found in Bassenthwaite and Ullswater. A post or two afterwards brought a copy of a paper ... by his friend, Mr C Tate Regan ... who had ... found sufficient structural differences between the vendaces of these separate localities to justify the naming of the English fish as a distinct species ... On the present occasion he did not propose to discuss the structural characteristics on which the new species was founded, but wished to congratulate Provost Halliday and the town clerk, Mr Rae, whom he saw present, on the fact that for the first time they might safely assure the good folks of Lochmaben that no one would now dispute with them the sole possession of at least one species of vendace.'

Sadly for Lochmaben folk, many decades later, the original scientific name of *Coregonus vandesius* was superceded by *Coregonus albula*, the common European Vendace, when Andrew Ferguson showed in 1974, using modern biochemical techniques, that these two fish were really the same species. Instead of being referred to, as formerly, the Lochmaben Vendace (or south of the Border as the Cumberland Vendace), it is now commonly known simply as the Vendace (or in some circles as the Cisco a name derived from North America, where similar species are found). A literal translation of the scientific name is 'Angle-eyed Whitefish'.

Size

The Vendace is normally regarded as a small to medium-sized fish, but its size can be very variable between populations, some of which can be quite stunted, with fish maturing in their first year at 10-12 cm. The normal adult size is some 20-25 cm (100-150 gm), but in some European lakes it may reach a length of 35 cm (500 gm). Exceptionally large specimens of 45 cm (1 kg) have been reported from Lake Ladoga in Russia. This fish does not

appear on the British rod-caught record list, originally because of its rarity and the fact that it practically never takes bait (see below), and more recently because it is now fully protected in Great Britain.

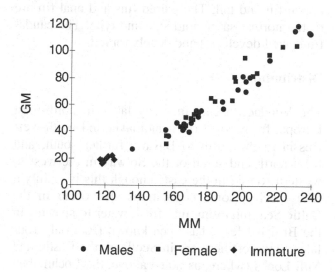

Length / weight relationship in Vendace from Loch Skene in July 2003.

Appearance

The head and back of the Vendace are a dark greenish blue grading to a silvery green along the sides, and eventually to a whitish silver on the belly. There are no spots. The body is streamlined and laterally compressed. The head is rather small and pointed, with the mouth opening slightly upwards. There are no teeth. The leading lower jaw is very characteristic: this feature alone distinguishes the Vendace from all other members of this family in Europe.

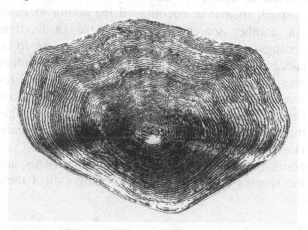

Scale from Vendace from the Mill Loch.

The body is covered by relatively large simple scales, of which there are some 70-90 along the lateral line. All the fins are rather transparent but darkened somewhat near the tips. The dorsal fin, which is short, is set midway along the back. The adipose fin is well-developed and midway between dorsal fin and tail. The paired fins and anal fin are of the normal salmonoid size and type. The caudal fin is well developed and deeply forked.

Distribution

The Vendace occurs in many lakes in north-west Europe, from northern Scandinavia and north-west Russia in the north to Bavaria further south, and from north and south of the Solway in the west to western Russia in the east. Though this is mainly a lake species, some populations also occur in the Baltic Sea, migrating into fresh water to spawn. In the British Isles it has been known from only four lakes. Two of these are in Scotland, the Castle and Mill Lochs (where, as noted above, the Lochmaben Vendace was originally described as a distinct species and subspecies - *Coregonus vandesius vandesius*), and two are in England, Bassenthwaite Lake and Derwentwater (the Cumberland Vendace - *Coregonus vandesius gracilior*).

The population in the Castle Loch, Lochmaben became extinct at the beginning of the 20th century, none having been recorded since shortly after a new sewage works was opened there in 1911. However, though it was also presumed extinct in the Mill Loch by Emile Dottrens and others in the 1950s ('Le Dr Slack a vainement cherche pour moi le Vendace du Loch Maben. Cette forme doit etre consideree comme eteinte.'), the discovery of Vendace in the stomach of a Pike encouraged the author to carry out another search. After a couple of fruitless nettings, using the same nets as were known to be successful for Powan in Loch Lomond, the author was given a piece of the original netting used for Vendace. After ordering equivalent meshes from a firm in Norway, specimens were finally caught there by the author in 1966 and in subsequent years. However, in spite of attempts to save this population, none has been seen since the 1970s, and the species is now regarded as being extinct there too.

The Vendace, over its whole range, seems to occur in lakes of any size from a hectare or so upwards.

These lakes are often quite rich (mesotrophic) and are rarely exceedingly oligotrophic. Depth seems to be not too important, though probably several metres is needed giving freedom from summer heat stress and oxygen lack as well as winter kill. Strong competition and/or predation from Pike, Perch, Roach and other species is probably very harmful unless a good deep/open water niche is available. The size of individual populations of Vendace seems to fluctuate greatly from time to time for reasons which are unknown but may well be climatic.

Recorded distribution of Vendace in Scotland. The solid circles are extinct sites; the asterisks recent introductions.

The original habitat of Vendace in the Castle Loch was described by William Graham in 1865: 'The loch is fed by two small rills. ... It abounds in a variety of fish (not all "uncouth fische" as they are called by the fable-loving Boece), including the vendace, the rarest and most elegantly shaped of all the finny tribe, and only found in Britain in this loch, and in the Mill Loch, on the north-west side of the town. Besides the vendace there are pike, perch, two species of loch trout, bream, roach, skelly or chub, greenback, eels, beardie, minnow and banstickle.'

More poetically, however, Graham's description was preceded by the second verse of a poem by George Washington Anderson read at the picnic at Lochmaben Castle, 12 July 1824, in celebration of King Robert the Bruce's birthday:

'Sweet loch, the stalking heron feeds,
By margin of thy stream,
There the heart-crested Vendace breeds,
Of epicure the theme.
Perch, Roach and Bream in myriads swim,
and sport among the reeds.
The hungry ged has made his bed,
Basking on thy broad meads.'

Ecology

The Vendace is a delicate pelagic fish which lives mainly in shoals in the open water offshore areas among the plankton. Echosounding studies by the author at the Mill Loch in 1966 showed that it remains in deeper (dimly lit) water during the day rising to the surface at dusk to feed, then descending at dawn. Although William Graham referred to 'the Vendace burn', running into the Castle Loch, there is no evidence that the Vendace ever entered it.

Food

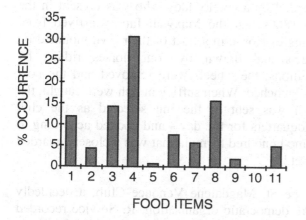

Food of Vendace in Lake Vanern, Sweden in 1973 (1: Bosmina; 2: Daphnia; 3: Limnosida; 4: Bythotrephes; 5: Leptodora; 6: Eurytemora; 7: Heterocope; 8: Diaptomus & Cyclops; 9: midge larvae; 10: others; 11: aerial) (Nilsson 1979).

After absorbing their yolk sac over the first few days of life, the Vendace fry start to feed on small plankton (mainly crustaceans), and zooplankton forms their main diet throughout life. They may supplement it from time to time with benthic animals which appear in the water column (e.g. midge larvae and pupae) or terrestrial insects which are trapped on the surface, and some eggs may be eaten at spawning time.

Peter Malloch noted in 1910 that 'No angler has yet been able to catch one with a hook, and they are usually caught with a sweep net.' However, though they were almost always caught by net, John Gardiner records that on one occasion two were caught using bait.

Reproduction

Spawning takes place over gravelly/stony shores during winter, probably late November into December, though the actual timing is likely to vary from site to site (and perhaps from year to year) by a week or so. The mature males gather early on the spawning areas and as females ripen they join the males and spawn together in midwater, the fertilised eggs dropping down into crevices among the stones and gravel.

The fecundity of females varies with size, and ranges from 1,500 to 5,000 eggs per adult female. Exceptionally large females from some Russian lakes have contained over 20,000 eggs. The eggs are some 1.5-1.8 mm in diameter and yellowish in colour. The incubation period varies with temperature, but experiments by the author and Alex Lyle have shown that it is normally 90-120 days at 1-4°C.

Predators

Possibly the most notable predator of Vendace is Pike, for, as is the case with Powan, most terrestrial predators (e.g. Herons and Otters) are unable to reach it because of its habit of remaining in deep water during the day. However, when it comes into shallow water to spawn it is then vulnerable to Pike which stay around the spawning beds during the reproductive period. Its eggs too are undoubtedly on the menu of several fish species and even some invertebrates.

Growth

The young possess a small yolk sac on hatching, but, like other whitefish, they are free and able swimmers immediately. They are normally some 7-9 mm in length but soon start to grow rapidly as the surface waters of their loch warm up and zooplankton become available.

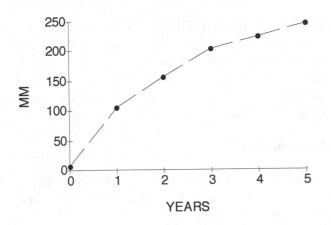

Growth of Vendace before extinction in the Mill Loch, Lochmaben.

At the end of the first year they reach a length of 8-10 cm. After two years they may be 14-18 cm, at which time many are mature and growth slows, so that most adults only reach some 20-25 cm, even after 5 or 6 years, which is the normal length of life.

Value

The Vendace is an important commercial species in some parts of Europe where it is caught mainly in gill nets, but sometimes in traps and seine nets where there are migrations at spawning time. Its flesh is white and rich in oils, thus making it good eating. It is sometimes smoked and small specimens may even be eaten raw. The author has vivid recollections of sitting in a farmhouse kitchen in Finland, after saying how much he was looking forward to his first taste of Vendace (assumed cooked), then crunching his way through an entire raw fish and attempting to look as though he was enjoying it as much as his host apparently genuinely did!

The netting at Lochmaben provided a gainful occupation to local fishermen, proof of which is provided by a receipt for Vendace dated early in the 20th Century by an Edinburgh fishmonger. This receipt can be seen in the Dumfries Museum. Most commercial fishermen gave up fishing at the beginning of the 20th Century. The last fishery was operated by a Mr Smith of Lochmaben, whose son, John Smith, gave the author a piece of one of the last vendace nets. It is a gill net made of fine cotton, with a mesh of 22 mm knot to knot. Such nets were always fished on the bottom of the loch.

Lochmaben traditions

Tate Regan recorded in 1911 that, although formerly held entirely by the Crown, 'The right of fishing in all the lochs was granted to the Burgh of Lochmaben by a charter of James VI, and this privilege is still enjoyed by the townspeople.' However, with the extinction of the Vendace in the 1970s the local community has lost an important part of its traditions and culture. In times past it was an important fish at Lochmaben, where an annual festival was arranged each summer by two local 'Vendace Clubs', during which Vendace were caught in fine gill nets and then cooked and eaten outdoors while athletic contests took place.

Service noted in 1907 that 'The vendace was of old held in high repute as a delicacy, and two local clubs were wont to hold an annual feast at Lochmaben for the discussion of the same. ... The small Vendace Club, mainly made up of local aristocracy, was still in existence in 1869, but was wound up in 1870 or 1871.' Robert Fraser, a former Provost of Lochmaben recalled one incident: 'It is recorded by a county lady who was present in the late 1860s that the Maxwell family arrived from Terregles House in a boat of their own mounted on wheels and drawn by four horses ridden by postilions; the wheels were removed and the boat was launched. When sufficient fish were caught, the catch was sent to the inn selected as the club headquarters for the day - and cooked according to a time-honoured formula that was a closely guarded secret.'

Of the St. Magdalene Vendace Club, a decidedly more democratic organisation, R. Service recorded that it ceased shortly before the more aristocratic society. 'Members of this Club enjoyed their sport in a different fashion, for after fishing the lochs for vendace in the usual way, they held a meeting for Border games, etc., and some thirty-five to forty years ago this was rather a big event.'

On some occasions, the two clubs must have participated in fishing at the same time for *The Scotsman* of Wednesday, July 25, 1855, reported: 'VENDACE AT LOCHMABEN.- On Thursday last, at Lochmaben was held the annual gathering of the St Magdalene Club, whose objects are at set seasons to fish the lochs for vendace and other finny tribes, and socially to masticate the same when cooked. When the fishings commenced in the forenoon, a considerable crowd of members and others had collected on the shore of the Mill Loch. The first haul of the net brought up about 28 dozen of vendace - some of them small, but mostly seven or eight inches in length, which is about their full size. Another cast or two of the net added only a few dozen to the previous number, and in the Castle Loch, which was next tried, not a solitary vendace came within range. Meanwhile on the Castlehill were being carried on such pastimes as foot-racing, wrestling and putting the stone, and races between grown-up girls and even stout dames for prizes of bonnets, caps, &c. About five o'clock the outdoor recreations terminated, and the members of the St Magdalene's adjourned to a dinner at the Commercial Inn, in which the vendace formed, as a matter of course, a principal dish. About half-past eight the party broke up, and thus ended the anniversary gathering of the St Magdalene's Fishing Club of Lochmaben.'

Though both clubs ceased to function during the 1870s - probably because of the decrease in numbers of the Vendace - Tate Regan recorded in 1911 that 'about four years ago one was re-established, but so far has not had a very successful career, the annual netting procuring only about half a dozen Vendace on each occasion. When I was at Lochmaben, it was explained to me that this probably did not indicate that the fish were getting scarce, but was due to the fact that the fishermen had not had the good fortune to strike a shoal.'

William Jardine was one of the local aristocracy and recorded that 'The vendace is well known to almost every person in the neighbourhood; and if among the lower classes fish should at any time form the subject of conversation, the Vendace is immediately mentioned, and the loch regarded with pride, as possessing something of great curiosity to visitors, and which is thought not to exist elsewhere. They are considered a great delicacy, resembling the Smelt a good deal in flavour; and though certainly very palatable, the relish may be somewhat heightened by the difficulty of always procuring a supply. During the summer fishing parties are frequent, ... and a club, consisting of between twenty and thirty of the neighbouring gentry, possessing a private net, etc., meet annually in July to enjoy the sport of fishing and feasting on this luxury.' Yarrell tells us that in the autumn of 1840, when he was staying with Sir William Jardine, he saw some Vendace caught, and partook some at dinner. He considered " the fish quite entitled to all their character for excellence.' An anonymous writer about the same time agreed: 'A dinner or supper of vendace. with other combinations of fish or fowl, is a feast for gods and men. They melt in the mouth as a sweet, and a wee drappie from the town cellars to wash it down is the height of gastronomic pleasure.'

A report by 'Mabie Moss' in *The Dumfries Courier* of December 18, 1895, noted that 'Although no public information has appeared on a subject which, when the Vendace Club and the St. Magdalene Vendace Club were both in their palmy days, was of much interest to a large number of people, it is still very satisfactory to know that the peculiar fish, which the Lochmaben folks faithfully believe is to be found nowhere else than in their lochs, is still fully holding its own against adverse conditions. For some years past the Vendace were understood to be diminishing, probably on account of the superabundance of pike. A few weeks ago a more than ordinarily successful fishing was had with the nets. It is doubtful, however, if the policy of allowing netting for this fish, so near to its spawning period, is a very wise one.'

Conservation

In the late 1960s the author, having verified the continued presence of Vendace in the Mill Loch in 1966, could see that its days were almost certainly numbered, reported that 'it could quite easily face extinction in the relatively near future'. Threats to the population included: over-enrichment of the loch, the introduction of alien aquatic plants and coarse fish, a high incidence of fish lice - probably brought in with coarse fish - the construction of a new housing estate in a field adjacent to the loch, and the lowering of the loch level to allow a path around the loch. In order to save the species from extinction in Scotland it was decided to try to create 'safeguard' population in suitable lochs in the area. Unfortunately, the Nature Conservancy (for which the author worked at the time), though prepared to designate the Mill Loch a Site of Special Scientific Interest, did not feel able to sanction such a project so it had to be carried out at weekends and with minimal resources. Working with Ken East, the plan was to collect eggs from the Vendace during its spawning period in December and introduce these to hill lochs in Galloway which were assumed to offer a safe haven for the Vendace. Unfortunately, these very lochs were at that time, unknown to the conservationists, suffering from the effects of acid rain, and in the next few years even their native Brown Trout were to disappear - along with any Vendace which may have hatched from the introduced eggs.

The designation of SSSI by the Nature Conservancy, which was really a paper exercise, did little to stop the Mill Loch deteriorating and by the mid-1970s it was apparent from several nettings that the Vendace had become extinct. Naturally this was a great disappointment to the author and contrasted sharply with the earlier thrill of rediscovering the Vendace some 10 years earlier. Some years later, the Vendace came under the protection of Schedule 5 of the Wildlife & Countryside Act 1981, and later from Annex 5a of the Habitats Directive 1992, and Schedule 3 of the Conservation Regulations 1994. All of these were too late to help the Vendace in Scotland, but did give some impetus to the conservation of the only two populations in England - in Bassenthwaite Lake and Derwentwater.

There have been occasional reports of Vendace from the Castle Loch but all those investigated by the author have proved false. On two occasions specimens or photographs were available for examination - both fish had adipose fins, but the first proved to be Sparling, the second Grayling. It was assumed in both cases that these fish had been used as bait by pike fishermen and left behind at the loch. However, the report of Vendace which appeared in the newspapers enabled the owner, when he was selling the loch some years later to advertise, controversially, that it still contained the 'rare Vendace'.

In the mid-1990s, with the impetus of this new conservation legislation and the increasing interest

in restoring biodiversity, Scottish Natural Heritage funded the author and Alex Lyle to carry out a project which would restore the Vendace to suitable waters in Scotland. After a careful selection process, two suitable waters in south-west Scotland were chosen and in the winters of 1996 to 1998 eggs, obtained by stripping ripe fish from the two populations in the English Lake District, were incubated and hatched indoors in Scotland. The young from Bassenthwaite Lake stock were then introduced to Loch Skene, above the Grey Mare's Tail near Moffat, and from Derwentwater stock to Daer Reservoir in the upper Clyde valley.

Alex Lyle, on watching the orientation of Vendace fry being released into Loch Skene in March and April of 1997 as part of the introduction scheme there, commented that 'No matter in which direction the fry were released from their container, they always orientated immediately toward the dark deeper water and swam away in that direction.' This casual contemporary observation was interesting in relation to the much earlier comments of Jardine: 'They are most successfully taken during a dull day and sharp breeze, approaching near to the edges of the loch, and swimming in a direction contrary to the wind.'

In 2004, Loch Skene was netted by the author and Alex Lyle and a healthy population of Vendace of mixed ages was identified – an exciting result! Similar netting in Daer Reservoir was unfortunately unsuccessful, but further stocking with eggs and 25 adult Vendace from Derwentwater was carried out here in 2005. Time alone will tell whether this conservation exercise has been successful, but it has served to disprove one myth. In 1912, Peter Malloch noted of the Vendace that 'This fish ... is said to be so delicate that it dies as soon as taken from the water, and cannot be transferred to any other loch.' Not only were Vendace fry transported up the steep and difficult path from the road to Loch Skene, but on one occasion a batch of juvenile fish, reared in the laboratory, were also successfully transported from Edinburgh and released there. Further conservation efforts are needed if the Vendace is to be saved for the future as a native species in Scotland, and these may be summarised as follows:

(a) Regular monitoring of the introduced stocks should be initiated.

(b) Further stock should be obtained from Cumbria to diversify the Scottish gene pools.

(c) An additional 2-3 waters in southwest Scotland should be chosen for additional populations, perhaps using the Scottish stock as a source.

(d) The possibility of restoring the Vendace to one or more of the Lochmaben lochs should be subjected to a full cost-benefit analysis. If successful, this could lead to a restoration also of the local culture and traditions surrounding this fascinating species.

References

Little research has been carried out on the Vendace in Scotland, but many studies are available from other parts of Europe, especially Finland where the Vendace is an important commercial species, and recently in England where Ian Winfield is monitoring and researching it.

Aass (1972), Dembinski (1971), Dottrens (1958) Ferguson (1974), Fraser (1833), Graham (1865), Haram & Pearson (1967), Jurvelius *et al.* (1988), Knox (1834, 1855), Malloch (1910), Maxwell (1904), McDowall (1867), Moss (1895), Maitland (1966, 1967, 1970, 1982, 1989), Maitland & Campbell (1992), Maitland *et al.* (2003), Nilsson 1979), Pennant (1761), Regan (1911), Service (1907), Sibbald (1684), Winfield *et al.* (1996), Yarrell (1836).

'The Vendace is a great delicacy, resembling the Smelt in flavour, but its reputation as a food-fish is enhanced by its restricted distribution and the difficulty of procuring it.'

Tate Regan (1911) *British Freshwater Fishes*

*****

Powan *Coregonus lavaretus* Pollan

'... it was an interesting sight to behold this brilliantly shining specimen,
iridescent with the most delicate colours,
as it lay on the grass just after it was taken out of the water.'

William Houghton (1879) *British Fresh-water Fishes*

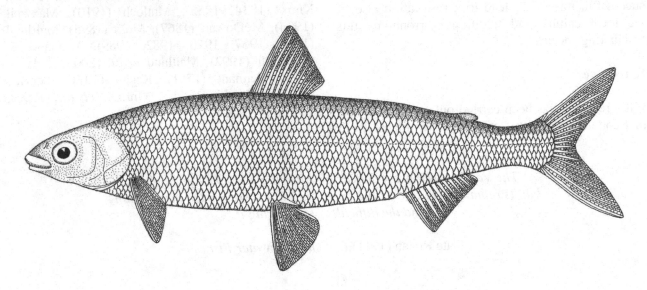

THE Powan is one of three rather similar species of fish which occur in the British Isles (the other two are Vendace and Pollan); there is some confusion among them because of the various common names which these fish have acquired. Sometimes called the Freshwater Herring, the Powan is also known as Guiniad or Gwyniad in Wales (where it occurs in just one lake) and as Schelly or Skelly in England (where it occurs in four lakes). Unfortunately, the name Skelly is also applied to Chub in northern England. Even more confusing perhaps is the Gaelic name for Powan - Pollan - which is also the common (and part of the scientific) name for the Pollan of Ireland, *Coregonus autumnalis pollan*, a species found nowhere else in the British Isles or indeed in western Europe. It is when such difficulties arise that scientific names are of such value. Most of the common names are derived from Gaelic terms meaning 'white' or 'shining'. A literal translation of the scientific name is 'Angle-eyed Lavarete'.

The author has recently written a short paper, based on local information and first-hand accounts, concerning the incident when bombs were dropped into Loch Lomond during the Second World War.

The heather in the Ben Lomond area had been set alight to confuse the Luftwaffe into thinking this was a town. Several bombs were delivered there and at least one of them fell into the loch and exploded. Next morning there were hundreds of Powan (as well as Atlantic Salmon, Brown Trout and other species) floating in the water - a welcome addition to local rations at that time.

Size

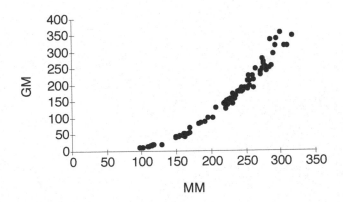

Length / weight relationship in Powan from Loch Eck (Maitland 1994).

The adult size of Powan is normally some 30-35 cm (300-400 gm), but in some European lakes fish up to 70 cm (8 kg) have been recorded.

The present British rod-caught record is for a fish of 0.95 kg (2 lb 1 oz) caught in 1986 by S.M. Barrie in Haweswater (Cumbria), but recording has been discontinued because it is now illegal to catch this species in Great Britain (Wildlife and Countryside Act 1981) without a permit from the appropriate country conservation agency - Scottish Natural Heritage in the case of Scotland.

Appearance

The body is well-built, elongate and laterally compressed. The head is small, as is the mouth, which is inferior in position and has no teeth. The body is covered by large shiny scales of which there are 84-100 along the lateral line. The adipose fin is large and fleshy whilst the paired fins are well-developed. The tail fin is large with a well-marked fork. In Loch Lomond (but not apparently in Loch Eck) a small percentage have an extra (supernumerary) pelvic fin between the two normal pelvics. This unusual feature was first discovered in Powan by F. Gervers in 1951 and still occurs in about one percent of the Lomond population.

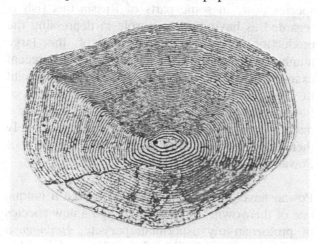

Scale from Powan.

The head and back are dark bluish grey which grades to greenish grey along the sides and eventually to a silvery whitish yellow on the belly. The iris of the eye is white. There are no significant spots, though sometimes fine black specks are evident. The single fins (dorsal, adipose, anal and tail) are mainly dark, whereas the paired fins are

darkened only towards their tips. The flesh is pure white.

Like some other species (for example the Sparling and the Grayling), the Powan is remarkable for its smell. Alfred Brown describes the Powan as having a strong odour like the Sparling, 'so that in the summer and autumn when they float in large shoals at the surface the surrounding air for a distance is tainted with their scent.' Writing some years later, in 1931, Henry Lamond seems to agree: 'While alive, and even when cooked, the fish seems to have a scent and flavour faintly reminiscent of cucumber.' Over many years, the author has handled thousands of these fish and they certainly have a characteristic odour, which always seems fresh and pleasant.

Distribution

Recorded distribution of Powan in Scotland. The asterisks are the recent introductions.

The Powan is widespread across much of north west Russia, Finland and Sweden. It occurs also in several other countries (Norway, Switzerland, Germany, Poland and France) but only in certain areas - usually in mountainous alpine lakes. In the British Isles it is a relatively rare fish with a scattered distribution and only seven natural populations. In Scotland it occurs naturally only in Loch Lomond and in Loch Eck, where it can, however, be abundant at times - Henry Lamond recording in 1931 that 'The powan net on one occasion could not be drawn to the boat owing to the mass of fish it contained, and netting had to cease lest the boat be swamped by the quantity of fish already in her.'

As discussed below, two 'safeguard' populations have recently been created by the author and Alex Lyle in Loch Sloy and in Carron Valley Reservoir, both within the catchment of Loch Lomond.

Ecology

Typically, Powan - as befits their 'ice age relict' image - occur in relatively large deep lakes with clear well-oxygenated water. During daylight, outwith the breeding season, adult fish stay in relatively deep water - on the bottom if they are near the shore or at depths of 20-30 m if in deeper water. At dusk they rise into shallower water, often coming right up to the surface and in to the shore. At dawn a reverse migration takes place.

Food

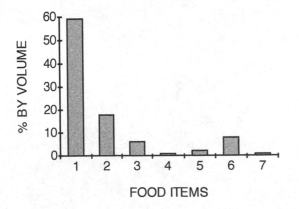

Food of Powan in Loch Lomond (1: cladocerans; 2: copepods; 3: isopods; 4: mayfly larvae; 5: water bugs ; 6: midge larvae; 7: surface food) (Slack et al. 1957).

Powan feed mainly on zooplankton (especially crustaceans) when they are young. As they grow they start to feed more on bottom invertebrates (e.g. midge larvae, molluscs), though still relying a great

deal on plankton - especially during the summer months. At spawning time, many of them eat considerable numbers of their own eggs. In Llyn Tegid, Gwyniad concentrate on bottom fauna from December to July, changing to midwater and surface feeding during the rest of the year.

Predators

In Loch Lomond, Powan are known to have a number of predators, one of the most important of these being Pike, which prey on them especially during their spawning period when they are very vulnerable in shallow water. Fewer Powan are eaten by Pike during the summer months, but they are then attacked in large numbers by River Lampreys which leave characteristic oval flesh wounds along the backs. Usually about 30 per cent of adult Powan in the loch carry the scars of such wounds. The Ferox Trout of Loch Lomond were once referred to as 'powan eaters'.

Powan eggs, too, have many predators, including adult Powan themselves. In some areas large caddis larvae have been shown to be significant predators. A new threat may be that of Ruffe, which have recently been introduced to Loch Lomond and are already abundant there, being one of the commonest species now. In some parts of Russia this fish is regarded as having a major role in depressing the production of whitefish because of the large numbers of eggs which it eats. A recent examination of the stomachs of many Ruffe caught in Loch Lomond during the spawning season of Powan (January) confirmed that Powan eggs are a major item of Ruffe diet at this time. It seems likely therefore that they will depress the numbers of Powan in the loch.

Powan have a variety of parasites - even a unique one of their own, for some years ago a new species of protozoan myxosporidian parasite, *Henneguya tegidensis*, was described from Gwyniad in Llyn Tegid. It occurred there in about 5 per cent of the fish examined, producing intramuscular cysts causing swellings on the body. Several other parasites, including tapeworms and trematodes are known to occur commonly in Powan in Loch Lomond.

The Powan population seems to be able to withstand the depredations of its parasites: it is rare to find dead fish in any numbers. An exception to this occurred during the late 1960s, when Powan started dying in large numbers and could be found dead all along the shoreline. The cause was found by Ron Roberts (1970) to be a disease called 'bald spot', which caused lesions on the skin, particularly on the head, and was obviously fatal to many fish. In June, 1968, during this major mortality of Powan, thousands were washed on to the shore. John Mitchell observed that 'during the mass mortality of Powan at Loch Lomond in June 1968, when thousands were washed onto the shore, Buzzards were coming down regularly to the edge of the loch and feeding on dead Powan there. Several Powan were found at a nest on one of the islands.'

Coregonus lavaretus has suffered catastrophes in other waters. Norman Ellison records '... a remarkable stranding of Schellies along the western shores of Ullswater ... during the fortnight ending 29 January 1966 ... The wind blew on to the west shore of the lake causing the spawning areas to be in a state of turmoil for most of the month. At least 400 dead Schellies were washed ashore and counted, with many others seen floating or carried into nearby fields partly eaten by crows and gulls.' This was 'by far the greatest catastrophe suffered by this species within living memory.'

Reproduction

Near spawning time the fish move much nearer to the shores, shoaling over the sublittoral areas there. At these times, as just noted, fish may get washed ashore and stranded, sometimes in considerable numbers. Thus in the January 1966 catastrophe, large numbers of Schelly were washed ashore in Gowbarrow Bay in Ullswater after a strong easterly wind. A similar event occurred in January 1967, on another shore, after a strong southerly gale.

Spawning takes place during winter, usually starting in late December and finishing in early February. At dusk the adult males come on to the spawning grounds, which are characteristically gravelly shallows off headlands or offshore reefs. They shoal there in large numbers, spawning with those females which move in each night as they become ripe. The fertilised eggs fall to the bottom and lodge between the crevices among the stones and gravel there, being slightly adhesive. The eggs are some 2-

3 mm in diameter and pale yellow in colour. The fecundity of the females varies with size, but is usually 2,000-11,000; very large females may have up to 24,000 eggs. Incubation takes about 90-100 days (340-420 day-degrees) at winter temperatures in the lake, the optimum temperature being 6°C.

Hermaphrodite specimens have been caught in Loch Lomond by David Scott, one in 1973 (295 mm) and another in 1986 (197 mm). The first was caught at spawning time and was apparently functional as both male and female. David points out that this is very rare in salmonoid fish.

Growth

The young are some 9-11 mm at hatching and have a small yolk sac. However, they are strong swimmers and immediately move off into the pelagic areas where they spend much of the rest of their lives. In captivity they become quite tame and do not react to movements outside their tanks, readily taking food from the bottom.

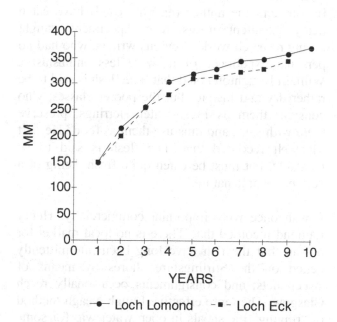

Growth of Powan in Loch Lomond and in Loch Eck (Brown & Scott 1994).

Growth is variable but in favourable lochs the young can reach lengths of 10-12 cm after one year and 15-20 cm after two. After two years they may start to mature and usually have an adult length of 25-35 cm at 3-5 years of age. Most fish die after 6 or 7 years, but some live to 9 or even 10 years.

Value

The Powan is an important commercial species in most of the countries in which it occurs - Scotland being an exception. It is caught in traps, seine nets and gill nets and is especially vulnerable at spawning time, when it migrates and masses on spawning grounds in great numbers. In recent years in Scotland it has had no commercial or sporting significance, other than, prior to protection, to a few specialist anglers who used ledgered maggot fished on the bottom in deep water. Occasionally they are taken on fly by trout anglers, as recorded by Henry Lamond in 1931: 'it is the rarest possible occurrence to catch one fairly with the artificial fly. I had only once seen one being actually hooked and landed. ... But on 23rd August, 1930, I had the good fortune to catch one while fishing for sea-trout. It rose and took the fly as it alighted and, when netted, the fish was seen to have taken the fly fairly into its mouth.' In fact, Flora Slack used to catch them by fishing a small white dry fly in the evening.

The flesh of Powan is white and oily with a good flavour and the author (and his cats!) have eaten many excellent meals from specimens caught during research work. Previous writers, who had no personal experience of it, were less enthusiastic. William Houghton noted that 'The flesh is said to be rather dry and insipid; but the poorer classes, who consider them as Fresh-water Herrings, preserve them with salt , and thus use them as food.' Herbert Maxwell recorded that 'The flesh is said to be palatable, but must be eaten quite fresh, being of a very perishable nature.'

Powan once were important commercially. Henry Lamond recorded that 'There is no local market for powan, but the fish have long been intermittently netted on the Stirlingshire shores by means of sweep nets, and consignments occasionally reach Glasgow. The more effective Lough Neagh method of "ringing" the shoals in open water was for some seasons operated commercially in the loch ...The fish were consigned to England where they were sold, chiefly in the Midland towns, as "pollen" and sometimes even as "grayling". ... For the seasons 1918-21 the price reached eightpence and sometimes ninepence per pound.'

During both World Wars, however, it was fished commercially in Lochs Lomond and Eck and several hundred thousand fish were taken. Henry Lamond noted that 'During the closing months of the War, when there was a shortage in the food supplies of the country, powan fishing was intensively prosecuted by request of the Government.' The minimum annual catch between 1916 and 1926 was 10,000 fish (1916), the maximum catch 51,500 (1918). 'In season 1926 prices fell to fourpence and even threepence per pound, and fishing was abandoned.'

Powan were once cultivated in ponds at Rossdhu on the west shores of Loch Lomond by the Colquhoun family. During the 5th annual excursion of the Scottish Arboricultural Society on 8 August, 1882, it was recorded that 'A short visit was then made to the fish ponds, very neatly laid out and surrounded with flowered terraces and lawns. In these ponds are preserved and bred the Powan or fresh-water herring, a beautiful silvery fish something like the herring in shape, and only found elsewhere in Great Britain in Loch Maben in Dumfriesshire. The antiquarian, in re the Powans, said that these fish are the "legitimate but somewhat transmogrified descendants of the herring tribe, which happened to be 'cribbed, cabined and confined' when the loch got the rise in the world which altered its character," but the majority who heard the statement took it as advised, *cum grano salis*.'

Conservation

Although the Powan has always appeared to be abundant in the lochs in which it occurs, it has always been vulnerable because there are just two populations. A major epidemic, catastrophic pollution (for example from a tanker plunging off the A82, as has happened several times in the past - fortunately never with a toxic load), or some other unforeseen event, could wipe out either, possibly both, populations quite rapidly. For this reason the species has always been very vulnerable in Scotland. Such concern is not unrealistic, as the major mortalities from bombing, disease and commercial fisheries, described above, indicate.

In 1964, the author started experiments with the introduction of Powan to other sites with the release in a small pond near Milngavie of a few hundred fry. The intention was not to start a permanent population, for there was no suitable spawning habitat, but just to gain some experience of the rates of growth and the way in which these fish might adapt to this new habitat - a small lily-fringed pond with a soft muddy bottom. In fact the fish did extremely well and Adrian Hopkins, a parasitologist colleague, was astonished when he netted the pond a few years later (not knowing of the introduction) to find a healthy population of Powan there! The fish had grown well and were actually in much better condition than fish from Loch Lomond, for, introduced as unfed fry, they had no internal parasites, nor had they any scars from attacks by lampreys. Additionally, and much to the author's surprise, young fish appeared each year, for it appeared that a few eggs managed to survive by being caught up in submerged plant growth. Unfortunately, this experiment was brought to an end a few years later when anglers introduced Tench to the pond, after which the Powan died out.

Many years later, the Powan and a few other species, were given legal protection by their inclusion in Schedule 5 of the Wildlife & Countryside Act 1981 and eventually in Annex 5a of the EC Habitats Directive 1992. However, though this was welcome recognition of the vulnerability of the species, it did little to help, for the problem was not people killing Powan, but that it was vulnerable to pollution and other threats - especially since it occurred in only two lochs. For this reason, the author and Alex Lyle started a translocation exercise in the 1980s to safeguard the future of this fish in Scotland by creating two new populations. Two apparently suitable sites were selected - Loch Sloy and Carron Valley Reservoir - and permission was obtained from the then owners, North of Scotland Hydro-Electric Board and Central Regional Council respectively. Eggs were obtained from Powan at spawning time in Loch Lomond and hatched in the laboratory. Several thousand unfed fry were introduced to each site between 1988 and 1990, and it was soon clear that these had done well, for adults were netted in Loch Sloy in 1991, and were being caught by anglers in Carron Valley

Reservoir. It is now known from netting carried out by Liz Etheridge and colleagues in 2006 that Powan have established very successful populations in both Loch Sloy and in Carron Valley Reservoir.

Thus instead of just one population of the Lomond stock there are now three, all within the Lomond catchment. If a similar project were to be carried out for the Loch Eck stock (there is at least one suitable water in the area - Loch Tarsan) the future of the Powan in Scotland would be reasonably secure.

In summary, the conservation actions required for conservation of the Powan in Scotland are:

(a) Regular monitoring of populations in Lochs Lomond, Eck, Sloy and Carron Valley Reservoir.
(b) Translocation of stock from Loch Eck to start two new populations of Powan in Argyll.
(c) A study of the feasibility of a small scale sustainable fishery for Powan in one site.

References

There have been several studies of the ecology of Powan in Scotland and elsewhere in Europe.

Anonymous (1883), Ausen (1976), Bagenal (1966, 1970), Bodaly *et al.* (1991), Brown (1989), Brown & Scott (1987, 1988, 1990, 1994), Brown *et al.* (1992), Copland (1956), Dabrowski *et al.* (1984), Dottrens (1958), Ellison (1966), Ellison & Cooper (1967), Ferguson (1974), Ferguson *et al.* (1978), Fuller *et al.* (1976), Gasowska (1965), Gervers (1954), Haram (1965), Haram & Jones (1971), Haram & Pearson (1967), Hartley (1995), Houghton (1879), Lamond (1922, 1931), Maitland (1967, 1969, 1970, 1980b, 1982), Maxwell (1904), Mitchell (1984), Nicholas & Jones (1959), Parnell (1838), Pokrovskii (1961), Pomeroy (1991), Regan (1908), Roberts *et al.* (1970), Scott (1975), Slack (1955), Slack *et al.* (1957), Svardson (1956).

*'During the closing months of the War,
when there was a shortage in the food supplies of the country,
powan fishing was intensively prosecuted by request of the Government.'*

Henry Lamond (1931) *Loch Lomond*

*****

Rainbow Trout *Oncorhynchus mykiss* Breac-dathte

*'A steelhead is a rainbow that lives in the salt water and runs up fresh-water streams to spawn.
... But one thing seems assured ... the steelhead is the most wonderful of all fish.'*

Zane Grey (1928) *Tales of Freshwater Fishing*

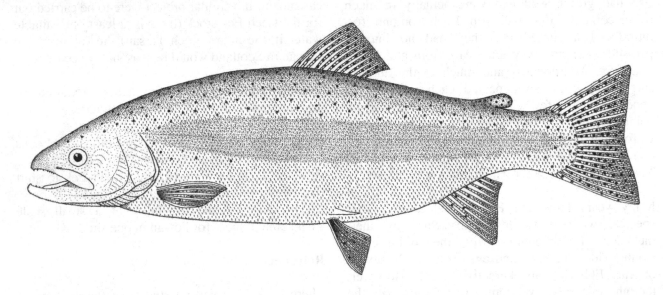

STRICTLY speaking, the Rainbow Trout is not actually a 'trout' but a Pacific salmon, since it is now placed in the genus *Oncorhynchus*, all members of which are native to the drainages of the Pacific Ocean. Thus, although it is one of the most popular species with anglers in Scotland, the Rainbow Trout is not a native, nor has it been even moderately successful in establishing itself here - even though it is well over a century since it was first introduced, and has since been stocked in hundreds of waters. During the 1960s, however, it was uncommon to find Rainbow Trout in the wild; the author remembers his surprise at being sent a very fine specimen for identification, caught by an angler in the River Blane. Nowadays, escaped fish seem to be everywhere.

The name of Rainbow Trout is likely to remain, as are the other two common names by which this species is known - Steelhead Trout and Kamloops Trout. A literal translation of the scientific name is 'Mykiss's Hookbeak'.

Size

There is no 'average' size for Rainbow Trout in the British Isles, due to the fact that nearly all populations are artificially maintained, and therefore size is dependent on the size of fish when introduced, and on the stocking density. On the whole, fishery managers aim to produce a Rainbow Trout for the angler of around 340-680 gm when stocking 'natural' fisheries (i.e. those into which small Rainbow Trout are introduced and allowed to grow) with hatchery produced fingerlings. However, due to the tremendous increase in the scale of the Rainbow Trout farming industry, trout of all sizes up to some 9 kg can be purchased for stocking, making the catching of 'trophy' Rainbow Trout rather a pointless achievement. A high proportion of the large Rainbow Trout used for stocking angling waters are elderly, retired broodstock and the objective seems to be to put them into a fishery and then angle for them before they die of old age or starvation.

In large lakes outwith the British Isles, a proportion of the population of Rainbow Trout may become 'ferox' like some Brown Trout, and reach a large size on a piscivorous diet. The world record Rainbow Trout is claimed to be a 19.1 kg fish from Alaska, while in New Zealand (where they are not of course native) fish up to 9.5 kg have been taken. Anadromous Steelhead Trout can reach over 120 cm in length and 16 kg in weight. The Scottish rod-caught record seems to be for a fish of 9.645 kg caught in Loch Awe in 1986: this was probably an escapee from a local cage farm in the loch. The British rod-caught record at the moment is (a) for a cultivated fish: 16.74 kg (36 lb 14 oz 8 dm), caught in 1995 by C. White at Dever Springs Trout Fishery in Hampshire, (b) for a resident fish: 10.921 kg (24 lb 1 oz 4 dm), caught by J. Hammond in 1998 in Hanningfield Reservoir, Essex, and (c) for a wild fish: vacant. The last-mentioned is indicative of the status of this fish in the wild.

Some anglers, including Zane Grey, rate them as the finest of sport fish: 'The steelhead in question was certainly a magnificent fish. He was thirty-two and one-half inches long and weighed over eleven pounds, being the largest I ever saw.' The more normal size of freshwater fish in the wild in North America is described by S.E. White: 'The average mountain rainbow, the fellow you expect generally when you cast into a likely-looking pool, runs from eight to fifteen inches, and is game for his weight.'

Appearance

Rainbow Trout are easily distinguished by the iridescent pink to red band that runs along each side from the head to the tail - i.e. the rainbow. Also they have no red spots, but instead many black ones on both body and fins, including the adipose and tail fins. Usually they are more silvery than the average Brown Trout, but individuals, especially the older ones, can be as dark as Brown Trout.

The scales are small, with some 100-150 scales along the length of the lateral line. Naturally hatched young Rainbow Trout, which may occasionally turn up in the wild, are silvery little fish, bluish green dorsally and bearing 5-10 widely spaced oval parr marks along their sides as well as a smattering of small black spots. Their dorsal and anal fins have whitish yellow to orange tips and the adipose fin is edged with black. They are easily distinguished from Brown Trout of the same size.

The teeth, as in Brown Trout, are well-developed on both jaws and are present on both the head and the shaft of the vomer bone on the roof of the mouth. When ripening sexually, the head of the male elongates and the background colour of the body becomes darker, contrasting strongly with the

increased brightness of the pink-red bands along the sides.

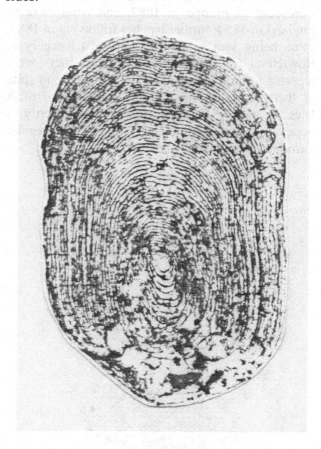

Scale from Rainbow Trout.

In recent years, in order to produce something different for the supermarket and for the angler, a golden form with no black pigment has been developed.

Distribution

The Rainbow Trout is native to the Pacific coast of North America, from Alaska south to north-west Mexico. It is found mainly in the river systems that drain the coastal ranges and the Rocky Mountains west to the Pacific Ocean. It also occurs in many rivers on the Russian side of the Pacific. The many races of the Rainbow Trout include well known ones such as Steelhead Trout (which parallels the behaviour of our Sea Trout) and Kamloops Trout.

Because of their fine sporting qualities and food value, Rainbow Trout have been widely redistributed throughout North America and much of the rest of the world, including the southern hemisphere. They were successfully introduced to the British Isles and to Europe towards the end of the 19th Century. The first consignments of eggs were sent to Europe in 1882, and from there to England in 1884, further batches following in 1885, some being sent to James Maitland's hatchery at Howietoun in Stirlingshire, where they were reported to be breeding by 1887. However, in spite of their very wide distribution within the British Isles for more than a century, the vast majority of populations are maintained only by the regular introduction of hatchery-reared fish.

Recorded distribution of Rainbow Trout in Scotland.

For 'put-and-take' fisheries, Rainbow Trout are a better investment financially than Brown Trout, for the former are more easily captured. Also, in recent years, hatchery-produced Brown Trout have been difficult to obtain, whereas almost anywhere in the country there is a convenient Rainbow Trout farm with stock for sale. An increased distribution of Rainbow Trout has also occurred as a result of escapes from both inland freshwater farms and marine pens. The author recorded one such escape in the 1960s - a rare event in those days. However, he recently spoke to one Rainbow Trout farmer in Galloway who admitted that about 2,000 fish

escaped each year from his floating cages into the loch and river in which the cages were situated. When Rainbow Trout escape from sea pens (or are dumped in the sea for some reason) many seem to move quickly to the nearest estuary and ascend into fresh water. Although there are regular reports of Rainbow Trout being caught in coastal waters or in estuaries, there is no evidence so far of runs of Steelhead Trout becoming established anywhere - nor is this likely until breeding populations establish in fresh water somewhere.

Out of the many hundreds of waters in Scotland in which Rainbow Trout have been stocked over the last century, no self-sustaining populations have arisen, although there are a number of waters where they have spawned once or twice after introduction, but then died out. The Loch of Menteith is a good example, for when Rainbow Trout were stocked there during the 1960s, they were found by Tom Stuart to be spawning in one of the tributary burns a year or so later. Little recruitment seems to have come from this, however, and the population can only be maintained by stocking. Sporadic recruitment does appear to happen occasionally and very young fish have occasionally turned up in Loch Fitty. Since Rainbow Trout are notorious for escaping from waters into which they have been introduced, elaborate devices are sometimes installed across loch outflows to prevent this.

About 1940 Barton Worthington sent out a questionnaire to survey the distribution of Rainbow Trout in Great Britain. The survey was repeated by Winifred Frost in 1971. Of the 55 or so waters identified by Worthington as having Rainbow Trout, only one was in Scotland - at Howietoun. The 1971 survey, however, which incorporated distribution work carried out by the author, indicated that there were 'nearly 100 waters in Scotland into which Rainbow Trout had been introduced.' The more recent study found that while at some 40 sites spawning had taken place, at only five were the populations self-sustaining. None of these populations is in Scotland. Writing in 1910 about Rainbow Trout, the situation was well summed up by Peter Malloch: 'I have had ten years' experience of them in Scotland, and have introduced them to many lochs. They did well for three or four years and weighed about 1 1/2 lbs; but after that we saw few of them. Some of them spawned, but never in sufficient numbers to establish themselves.' A further survey of the distribution of this species in Scotland was carried out recntly by Andy Walker, who showed that it is now very widespread, thanks to regular stocking.

Ecology

The reasons why Rainbow Trout do not establish populations in the British Isles are not clear. It has been suggested that they establish more readily in mineral-rich eutrophic waters, such as those in England where self-sustaining populations occur. However, one of the permanent populations in Ireland, at Lough Shure, Aranmore, and the only one in Scotland where they may be establishing, are both in acid, mineral-poor waters. It is believed essential for them to have running water for spawning: this would preclude them from establishing in lochs without an adequate inflow or outflow. Also, it is thought that the newly-hatched fry are at a disadvantage in spawning and nursery streams shared with Brown Trout where the latter hatch earlier and are therefore established territorially, providing too strong competition for the later-emerged Rainbow Trout. In their native waters and in most countries where they have been introduced, Rainbow Trout populations are established in a very wide range of habitats, some very cold in winter, others at the upper limit of their temperature tolerance (ca 24°C) in summer, in both mineral-rich and mineral-poor waters.

Rainbow Trout seem to be most successful in their native habitats at summer temperatures around 21°C, but lake-living fish can stand higher temperatures than this if they are able to retreat at times to cooler, well-oxygenated waters. In the British Isles, Rainbow Trout have been found to be most active at 16°C. They grow more rapidly than Brown Trout, but overstocking leads to populations of stunted individuals. In lochs with mixed populations of Brown and Rainbow Trout, the fact that the latter show better catch returns is probably due to the fact that they are easier to catch, perhaps because they are less cautious than Brown Trout. This seems to be true for these species in all countries where they are both angled.

Food

Most Rainbow Trout in this country have spent some time in a hatchery and are thus likely to be tame. One interesting observation about their diet is that their stomachs often contain green vegetable matter and indigestible objects such as pieces of waterlogged wood and not infrequently cigarette filter tips. Neville Morgan remembers fishing for Sea Trout at the mouth of a river on the West Coast of Sutherland, where he caught two Rainbow Trout which had apparently escaped from a nearby fish farm. Opening them up to find what they had been eating, he was astonished to find the stomachs of both fish full of cigarette filter tips. Apparently these were a good imitation of the food pellets which they were used to eating and are often tipped off bridges into the river below by tidy-minded anglers' spouses who spend the long hours of waiting tidying up the family car and emptying ash trays.

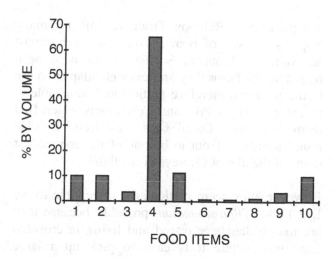

Food of Rainbow Trout in Llyn Alaw, Wales (1: algae; 2: plants; 3: Asellus; 4: Gammarus; 5: bugs; 6: beetles; 7: midge larvae; 8: others; 9: fish eggs; 10: detritus) (Hunt & O'Hara 1973).

The food and feeding behaviour of Rainbow Trout is relatively similar to that of Brown Trout. However, in a study of the food and feeding behaviour of Rainbow Trout living alongside Brown Trout in Loch Awe, Willie Duncan found that the former have a high dependency on uneaten pellets coming from fish cages in the loch. 'When they move away from the cages they consume large quantities of non-conventional prey items, this is due to the low repertoire of search prey images that they encounter in the fish farms. The native species (Brown Trout) largely consume conventional prey at and away from the fish farms depending upon its seasonal availability.'

Reproduction

In their native waters in North America, Rainbow Trout spawn from March to August, the main period being from mid April to late June. However, there are also winter spawning populations which breed from late December until early spring. In the British Isles, Rainbow Trout seem to be mainly early spring spawners, from February to March; at the Loch of Menteith, the species spawned in an inflowing stream on 26 March. These eggs hatched on 30 April. The orange yellow eggs (some 3-5 mm in diameter) are laid in a redd in the usual salmonid manner. Females produce about 1760-2640 eggs per kilogram of body weight. Their reproductive behaviour seems similar to that of Brown Trout.

Predators

The parasites of Rainbow Trout are similar in many respects to those of Brown Trout but since almost all Rainbow Trout in Scottish lochs have been reared in fish farms they are less well adapted to life in the wild and therefore particularly vulnerable to predators such as Pike and Cormorants which find them easy prey. David Carss and Keith Brockie found Rainbow Trout to be one of the commonest items of the diet of Ospreys in Scotland.

However, they seem particularly susceptible to the Eye Fluke, *Diplostomulum*, probably because they are mainly hatchery reared and living in crowded conditions where it is easy to pick up a large parasite burden before being released into the wild. Badly-infected individuals assume an overall dark colour and swim blindly, close to the surface of the water. This makes them readily available to fish eating birds (which are necessary secondary hosts in the parasite's life cycle), thus improving the chances of breeding success for both bird and fluke.

Growth

The age of Rainbow Trout can usually be determined fairly readily from scale reading. In Scotland, the species lives for only 3 or 4 years on average and in their native waters a similar life expectancy seems normal. The anadromous form, however, often lives for 6 to 8 years of age. It is often stated that genuine wild Rainbow Trout are much stronger and more successful in the wild than hatchery produced ones; Niall Campbell, who had

angled for indigenous specimens in British Columbia and naturalised fish in New Zealand, wholeheartedly agreed.

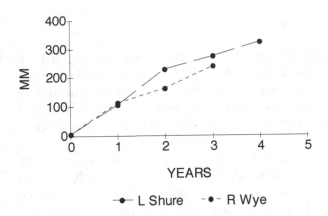

Growth of Rainbow Trout in two different waters (Frost 1974).

Value

In spite of their lack of breeding success in the wild, there is no doubt that Rainbow Trout will continue to be a common species in Scotland. Their ease of domestication, high conversion rate of food to flesh, sporting qualities and palatability will ensure their survival, both for food and for angling. In January 2001, *Trout News* reported that, in Scotland, 'Rainbow Trout were produced from 68 sites involving 54 companies with an overall production of 5,834 tonnes in 1999' This was an increase if 921 tonnes on the previous year. Production of table fish in 1999 'amounted to 4,857 tonnes representing an increase of 788 tonnes on the previous year and accounting for 83% of total production. Fish weighing up to 450 gm and over 900 gm made up the bulk of table production representing 65% and 32% of the total respectively. ... Production for restocking increased by 131 tonnes to 975 tonnes representing 17% of the total production.'

Many anglers who have caught Steelhead in North America or read accounts such as those of Zane Grey, would understand his desperation for a spate which would induce the fish to run: 'An Oregon storm in October was not to be desired by campers. Nevertheless, so keen was my eagerness for a run of steelhead, such as I had heard about, that I went on praying for rain.' However, just as there is no sign of the freshwater form establishing itself in Scottish lochs, there is equally little chance of there ever being runs of Steelhead in Scottish rivers.

Conservation

Since the species is not native to Scotland and is not threatened within its natural area of distribution, no special conservation measures are required for this fish here.

References

There have been few studies of Rainbow Trout ecology in Scotland, but there is a wealth of literature, both from countries where it has been introduced and from its native North America.

Carss & Brockie (1994), Frost (1940, 1974), Grey (1928), Hunt (1972), Hunt & O'Hara (1973), Kennedy & Strange (1978), Lever (1977), MacCrimmon (1971), Maitland (1887), Maitland (1966), Maitland & Campbell (1992), Malloch (1910), Narver (1969), Scott & Crossman (1973), Stuart (1967), Walker (2003), Worthington (1941).

'It was a rainbow of about three pounds weight,
of a singular bluish-bronze color on the back,
pale silver underneath, with dark flecks,
and a bar of rainbow tints down the sides.'

Zane Grey (1928) *Tales of Freshwater Fishing*

*****

Atlantic Salmon *Salmo salar* Bradan

'Mull ...In this ile there is twa guid fresche waters, ane of them are callit Ananva,
and the water of Glenforsay, full of salmond, with uther waters that has salmond in them,
but not in sic aboundance as the twa foresaid waters.'

Donald Munro (1549) *Description of the Western Isles ...*

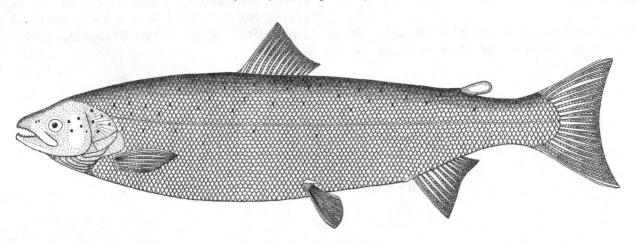

SOMETIMES described as 'The King of Fish' the Atlantic Salmon has always been one of the most important fish species in Scotland. This is exemplified by the fact that there are many different pieces of legislation dealing with Salmon in Scotland and only a few which are relevant to the other species. Yet in spite of this mound of paper, the Salmon appears to be more threatened now than ever it has been, clearly suggesting that it is better management and not more bureaucracy which is required. As befits its importance, there is a wealth of literature of all kinds which deals with this amazing fish. Its value to anglers is summed up by Seton Gordon: 'A January clean-run salmon when he is landed lies on the bank like a bar of burnished silver; he more than repays hours of cold and discomfort beside the river and (in pre-war days) his capture was celebrated by a good dram of neat *uisge beatha*, shared between the angler and his gillie.'

To many enthusiasts in Scotland, the Atlantic Salmon is the only fish worth talking about, as Scrope recalls: 'Having just caught a two pound trout he laid it in front of the local angler. "There's a fine fish now - a perfect beauty!" "Hoot toot! That's no a fish ava." "No fish, man! What the deuce is it, then? Is it a rabbit, or a wild duck, or a water rat?" "Ye are joost gin daft. Do ye no ken a troot when ye see it?" I could make nothing of this answer, for I

thought a trout was a fish; but it seems I was mistaken.' More recently, Diana Rigg caught an excellent Sea Trout in the River Conon, aided by her gillie, Jock. 'Oh what a beautiful fish. I've never seen such a beautiful fish' she enthused. 'That's no' a fish' came the curt response. 'No?' 'No. A sea trout is a Sea Trout. Only a Salmon is a fish'.

Salmon have been responsible for town planning! Frank Buckland noted: 'When monasteries were first established, previous to cathedrals themselves being built, the founders selected sheltered spots where, for the most part, they could get a plentiful supply of fresh-water fish, especially Salmon, for the use of the table on fast-days.'

Unlike the Brown Trout, the Atlantic Salmon has relatively few common names, apart from those used to describe different stages in its life history. Braddan and Springer are occasionally used alternatives and, in the Borders, John Carter wrote that Kippers and Rowaners (male Salmon) and Gilch (Grilse) are used. A literal translation of the scientific name is 'A Kind of Leaping Trout'.

Size

The Atlantic Salmon is the largest of the Scottish salmonids, but its size is very variable. Adults returning to fresh water can be as small as 40 cm in length, weighing only about 0.9 kg, or as large as 125 cm and 22.7 kg. However, most fish are around 60-75 cm and 2.7-4.5 kg. The British rod-caught record is for the famous fish caught in 1922 by Miss G W Ballantine in the River Tay. This magnificent fish weighed 29.029 kg (64 lb). The world record, for an Atlantic Salmon caught in Norway in 1925, is 35.89 kg.

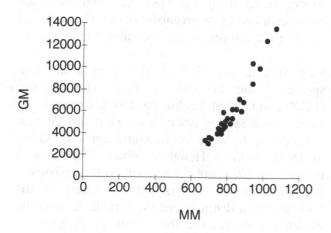

Length / weight relationship of adult Atlantic Salmon from Loch Lomond.

Larger fish, however, than the rod-caught record have been taken in Scotland. In 1910, Peter Malloch noted that 'The largest salmon caught in British waters during the last thirty-six years was caught in the nets on the Tay at Newburgh in 1872. Its length was 4 feet 5 inches, its girth 2 feet 7 inches and its weight 71 lbs. ... Fish between 50 and 60 lbs in weight are often caught in the nets on the Tay, while a few between 60 and 65 lbs are sometimes captured; but beyond this weight fish are rare. I have noticed in our fish house as many as forty fish over 40 lbs in weight, all caught in one day with the nets.' Later, William Calderwood stated that '... the record weight for a Scottish salmon is 84 lb., but the fish was taken in the estuary of the Tay, though both in the Dee and in the Tay fish of over 70 lb. have been taken in fresh water.'

John Buchan gives additional information on one of Scotland's largest Atlantic Salmon: 'The following year a large fish was taken in the nets at Newburgh ... The Newburgh salmon, of which there is a cast in the South Kensington Museum, weighed seventy-four pounds, and was the largest salmon at that time ever known to be taken.'

Appearance

Sometimes there is confusion between smaller specimens of this species and Sea Trout, but in Atlantic Salmon, unlike Brown Trout, the maxillary bone usually reaches to about the middle of the eye, occasionally to the hind edge in large specimens; in Trout it reaches well beyond. Also, unlike in Trout, the head of the vomer bone bears no teeth and the shaft only a few deciduous ones in a single row. The Salmon has well-developed teeth on both upper and lower jaws. The body is well-covered with cycloid scales which number 109-130 along the lateral line. Occasionally, large silver fish, taken along with adult Salmon, have characteristics between Salmon and Trout. These are probably natural hybrids, a small proportion of which occur in many wild populations.

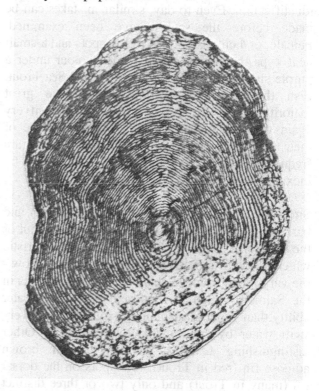

Scale from Atlantic Salmon.

As male, or 'cock' Salmon mature, a pronounced hook, or 'kype', develops on the lower jaw, the tip of which becomes upturned and eventually fits into a corresponding hollow in the upper jaw when the mouth is shut, while the whole skull elongates considerably. With the coincident colour changes, males look completely different from when they left the sea. At this time too, the scales erode, become embedded in the skin and are difficult to remove.

When fresh-run from the sea, the sexes are similar in colour with bright silvery sides and a silvery white belly, though the back may range from brown, through olive green to dark blue. Previously spawned fish may be quite heavily spotted, especially on the gill covers. As they ripen for spawning in fresh water the pristine 'bar of silver' appearance fades to a bronze-pink. Cock fish colour up considerably, red trout-like spots appearing on the sides over a mottled background of browns, reds and purple - earning it such names as 'Tartan Fish' or 'Kipper'. Their appearance at this stage, including the kype and elongated head, is similar to that of a large old cock Trout, either Brown or Sea, and before advances in scale reading made it possible (i.e. up to the early years of the 20th Century), the scales of Salmon and Trout could not be distinguished, and frequent mistakes were made in identification. Even to-day, similar mistakes can be made before the scales have been examined. Female, or 'hen', fish also develop spots and a small neat kype, becoming a dull leaden colour under a purple sheen, again not unlike large hen Sea Trout. Fish that survive spawning, 'kelts', the great majority of which are hen fish, become silvery again with black or dark blue backs. Because of their silvery appearance, well-mended kelts are frequently mistaken for fresh-run fish by inexperienced anglers.

Small salmon fry, between the larval (alevin) and parr stages are difficult to distinguish from Trout of the same size. There is, however, one characteristic which becomes more pronounced as the fish grow - the very broad rounded shape of the pectoral fins in the Salmon, which is supposed to give a greater ability than Trout in maintaining themselves in very swift water by 'holding on' to the bottom. Other distinguishing features include: a dull brown adipose fin (red in Trout), few spots on the dorsal fin (many in Trout) and only two or three distinct spots on the gill cover (many in Trout), tail deeply forked (less forked in Trout). Parr marks along the side are normally more distinct in Salmon than Trout.

Each life stage in the sequence from egg to adult is quite distinct. The eggs hatch into alevins, which, when the egg sac has been absorbed, start to swim and are called fry. The next stage, parr, are still quite different from the adults and characteristically they have a row of 8-12 dark-grey to blue thumb-prints or parr marks along the body. Between each of these is a red spot. There may also be a sprinkling of small black or brown spots over the back and sides and a few red spots below the lateral line. The leading edge of the pectoral and pelvic fins may be white and the adipose fin a leaden colour, rarely appearing brownish or red as in Trout. As the time for their seaward migration approaches, the parr undergo a process of 'smoltification' when the parr marks become masked by a coating of silvery material in the epidermis and the back becomes a brown or greenish-brown. The scales of these smolts are very easily shed at this time.

Parr were at one time described as a different species; Walter Scott in *The Fair Maid of Perth* (1828) commented 'Eachin resembles Conachar no more than a salmon resembles a parr, though men say they are the same fish in a different state.' Later, in 1879, William Houghton noted 'There are, I believe, some people who will still persistently maintain that the parr is not the young of the Salmon, but a distinct species, notwithstanding the evidence derived from the careful experiments of Mr. John Shaw, of Drumlanrig, in the years 1833-6, who proved beyond a shadow of doubt that the parr was a young Salmon.' In spite of the classic experiments by John Shaw and others, there were still disbelievers such as James O'Gorman '... I have ... seen extracts from a treatise, by a Scotchman, on this subject, and in this treatise it is gravely asserted, first, that salmon breed with a fish called par; next that salmon do not become what we call fry ... but have a kind of tadpole existence ... Now, to my mind, these are monstrous doctrines and, I think, incapable of being proved ... '

Distribution

The Atlantic Salmon, as its common name implies, is an anadromous species of the north Atlantic and is widely distributed along the Atlantic seaboards of northern Europe and North America, running into most of the suitable clean rivers. It occurs in southern Greenland and in Iceland as well as in rivers flowing into the White and Baltic Seas. It ranges as far south as the north of Portugal in Europe and the northern Atlantic coast of the USA in North America. Some non-migratory ('land-locked') populations occur naturally in North America and Scandinavia. The species has been introduced to New Zealand and South America.

In Scotland, Salmon are found in nearly all suitable clean running (and some standing) waters where they are not excluded by physical barriers or pollution, and where there are adequate spawning and nursery areas to provide sufficient recruitment of young fish to maintain the population. The excellent map produced by Ross Gardiner and Harry Egglishaw shows this clearly.

In suitable habitats, both the young freshwater stages and the returning adults can be abundant: Herbert Maxwell noted that 'Into some rivers salmon enter in greater or lesser numbers in every month of the year, nor is this dependent upon the size of the river, because it is the case both in the mighty Tay and the puny Thurso.' However, due to commercial exploitation (both legal and illegal) in the open sea, off the coasts and in the estuaries and rivers, possibly coupled with an outbreak of disease and parasites (some of them derived from commercial fish farms) or intense predator pressure, the number of adults surviving to reach their natal spawning stream may be drastically reduced at times.

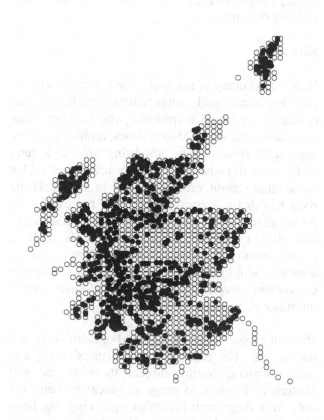

Recorded distribution of Atlantic Salmon in Scotland.

J. Lapslie, writing about the Parish of Campsie in the Old Statistical Account of Scotland in 1793 remarked that 'I have not heard of a single salmon being seen in our river [Clyde] for 18 years; whereas, in former days, they were so plenty in spawning time, that it was customary, though unlawful, for the country lads to go out with torches made of the dressings of lint, and with long spears to kill considerable quantities of these foul fish.' Another account, also in 1793, for Glasgow, describes the importance of Salmon in previous times: 'The first branch of trade, in which the citizens of Glasgow engaged seems to have been the curing and exportation of salmon, caught in the River Clyde. About the middle of the 16th century, they appear to have had vessels at sea, which made attacks on the shipping of the English. In the end of the 16th, and beginning of the 17th century, they, together with the inhabitants of Renfrew and Dumbarton, prosecuted the fishery of salmon and herring with great spirit, and to a considerable extent. In the reign of Charles the II we find a privateer was fitted out in the Clyde, to cruise against the Dutch. In the latter end of the last century, the merchants of Glasgow continued to export their salmon and herrings, the principal market for which was France.'

Atlantic Salmon expend substantial amounts of energy in their efforts to reach their spawning grounds. John Colquhoun confirms this: '... the struggle they hazard to gain their object is almost incredible. I have watched a shoal boring up the Knock for Loch Baa when they had to turn on their sides to force themselves through the shallows, and the noise caused by these exertions could be heard at a considerable distance. At first I fancied some creature was fording the stream, and, on walking up saw monsters of from 10 to 25 lb. fighting against water only a few inches deep.' In relation to waterfalls as barriers, Tate Regan noted that 'Opinions differ as to the height which a Salmon can clear at a single spring, but most authorities think 10 feet about the limit, and direct falls of more than this height are impassable.' The leaping behaviour of Atlantic Salmon is legendary. Derek Mills recalls that 'On many occasions at the Orrin Falls, Salmon were seen to leap successfully and one fish was photographed in 1958. Before modifications were made downstream to lessen their height, Salmon had to make a vertical leap of 12 feet in order to clear them.'

Less well-known are the jumping antics of parr. David Harley recalled that on 25 September 1999, SEPA received a report of a suspected fish kill in the Water of Feugh at Bridge of Feugh, Banchory. It transpired that 37 dead juvenile salmonids (parr) had been collected from 2 small rock pools, and parr had been observed jumping out of the large pool at the base of the falls, and on to the rocky banks where they were collecting in the 2 pools. Adult salmon were leaping the falls at the time and there were no visible signs of pollution. A sample of the dead parr revealed that approximately half of the fish were sexually mature 'precocious males', which are known to migrate upstream with the adults. It was therefore likely that the 'kill' was due to behaviour and that the parr had died as a result of asphyxiation due to overcrowding, and not from a pollutant.

Adults too can make mistakes as Ross Gardiner recalls: 'About May 1993, a fresh-run Salmon in Loch Faskally jumped right out of the water and beached itself on the bank just opposite the Freshwater Fisheries Laboratory. After flapping around for quite a while, it managed to wriggle back into the water.'

Willie Shearer recalls the excitement of tagging and recapturing Salmon in the early days of research at the Freshwater Fisheries Laboratory at Pitlochry (then called the Brown Trout Research Laboratory) and just at the time the Greenland feeding grounds were being discovered. One such recapture was in October 1956 south of Maniitsoq, West Greenland of a female Salmon tagged in November 1955 after stripping at Loch na Croic on the River Conon system in Ross-shire. In the late 1960s, a native fisherman jigging for Cod through a hole in the ice off the coast of East Greenland near Ammassalik caught a Salmon which had been tagged as a smolt on the North Esk some 18 months earlier. On 27 February, 1969, a sweep-net, operating a short distance above the head of tide on the River North Esk, caught a female grilse kelt weighing approximately five pounds and bearing a Swedish tag. An enquiry to Sweden elicited the information that this fish was tagged as a smolt and was released on May 5 1967 in the River Lagan, situated in northwest Sweden and flowing into the Kattegat. All the available evidence suggested that this fish had spawned in the River North Esk in the autumn or early winter of 1968 and that when caught it was returning to the sea.

Although Atlantic Salmon disappeared from several Scottish rivers due mainly to pollution, river barriers and general habitat destruction, towards the end of the 20th century a number of rivers in which Salmon had not been seen for many decades started to show signs of recovery. Most notable among these was the River Clyde where adult Salmon were first seen in 1983 (after an absence of 80 years) negotiating the weir at Blantyre. These may have resulted from a stocking of the King's Burn at Rutherglen in 1980. At this same place, Desmond Hammerton recalls that, in 1983, numbers of Salmon became trapped in the pool below the weir and numerous local folk were seen trying to catch fish with anything available – from their bare hands to shotguns and even a crossbow! In the River Carron (Forth catchment) too, stocking in the upper reaches started in 1985, and may have been reponsible for the small runs of adults which started there in 1989. However, in several other rivers where no stocking had been carried out, small runs of fish have appeared. These include the Rivers Gryfe, Leven (Fife), Avon (Forth), Almond (Forth) and Esk (Lothian).

Ecology

Homing accuracy is not quite one hundred per cent and very occasionally adult Salmon, marked as parr or smolts, are found at spawning time in rivers other than their natal ones. Also at times, returning adults, apparently frustrated by not being able to ascend their home river because of very low water or for some other reason, cruise patiently in shoals off the river mouth for a considerable time but eventually move off to find another river which it is possible to ascend. On maturing, Salmon which have been experimentally deprived of their olfactory senses somehow still locate fresh water and will run up any convenient river. Possibly they follow other, unimpaired, Salmon.

Howlett's explanation of the need of salmon for salt waters was 'For Salmon being Fish of Prey, and great Feeders, Nature directs them to the salt Waters, as Physick to purge and cleanse them, not only from their impurities after Spawning, but from all their muddy terrene Particles and gross

Humours, acquired by their extraordinary, excessive Feeding all the summer in fresh Rivers.'

After spawning, most fish, then known as kelts, die in the river on their way back to the sea. Many probably end up as carcasses on the river bank, as exemplified by this account from Willie Shearer: 'During the second week of November 1957 a salmon tag was returned after having been found in the garden of a house in Milltimber, Aberdeenshire. This tag had been attached to a salmon kelt released in the River Feugh, a tributary of the Aberdeenshire Dee, four years previously. Further enquiries revealed that a clump of primroses removed from the banks of the Feugh had been planted in the garden and it was in the centre of one of these plants that this tag had been found.'

Food

In fresh water, the food of salmon fry and parr is mainly invertebrates, especially insect larvae and molluscs, but during the summer months, especially in windy weather, aerial and terrestrial insects falling on to the water surface become important. If there are lochs in the river system, some parr may well make their way into these and occupy the littoral zone along the shore, provided that there are not too many predatory fish or competitors.

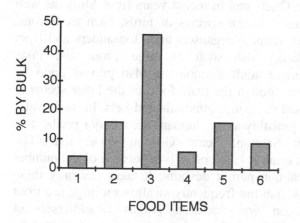

Food of juvenile Atlantic Salmon in the Altquhur Burn (1: crustaceans; 2: stonefly larvae; 3:mayfly larvae; 4: caddis larvae; 5: midge larvae; 6:others) (Maitland 1965).

The food of Salmon at sea consists of pelagic crustaceans such as euphausid shrimps, prawns, small squid and a wide variety of other fish such as Sand Eels, Sprats, Herring, Lantern Fish and in the

far north, Capelin. The presence of some of these food organisms indicates that the feeding range of Salmon extends down to 300 m, although most feeding seems to take place in the surface layers. Frank Buckland notes that 'Mr W. Campbell records that he once had a wonderful haul of salmon at Islay, in the estuary. He landed 716. As the net approached the shore, he saw the fish discharging the contents of their stomachs, which consisted of small eels.'

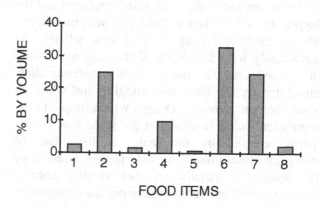

Food of Atlantic Salmon in the sea - off the Faroes (1: euphausids; 2: amphipods; 3: molluscs; 4: capelin; 5: sand eels; 6: sand lance; 7: other fish; 8: others) (Freshwater Fisheries Laboratory 1992).

Though not supposed to feed as adults in fresh water, there are exceptions. For example, William Calderwood recorded that 'In 1885 a small trout was taken from the stomach of a grilse caught in the river Thurso.' Interestingly, there are two instances of fresh run Salmon caught in Loch Tay with Arctic Charr in their stomachs. In the most recent case, a 12.7 kg spring fish was found to contain a Charr of 0.34 kg.

Reproduction

In Scotland, Salmon normally spawn between late October and early January - although earlier and later instances of spawning are regularly reported. Peak spawning time is usually in November. After reaching the spawning grounds, the hen fish selects an area of suitably sized gravel or smallish stones in swiftly flowing, usually shallow, water in which to cut her nest or redd. Because of these special requirements for the siting of redds, most spawning takes place in the upper reaches of rivers and their tributaries, but in pollution free, swift flowing rivers, spawning may take place from just above the upper tidal reaches (sometimes even below) all the way upstream, wherever the bed is suitable.

The spawning behaviour of Salmon is well documented. The hen fish, accompanied by the local, dominant, cock fish, cuts out a redd by turning on her side and flexing her body and tail up and down in short bursts of powerful activity. The upward stroke creates a sucking action lifting the gravel and stones off the bottom to be carried downstream by the current. The larger the hen fish the larger are the stones which are displaced and the deeper the redd. Even a small fish will excavate a hollow 15-20 cm deep. During this activity she periodically tests the depth of the depression with an erected anal fin until she is satisfied. Then, stimulated by the cock fish nudging and vibrating beside her she releases her eggs which drop into the depression, being fertilised at the same time by the sperm released by the male. A high rate of fertilisation is normal. The redd is then filled in by the female - usually by her cutting another depression just upstream, where the spawning act is repeated. Several redds may be excavated, then covered, and a large and conspicuous area of clean looking gravel may be turned over before spawning has been completed.

During all this activity the cock fish regularly chases away all intruders, including other cock fish as well as small Salmon and Trout, which gather to feed on any eggs which may be washed downstream. In addition there are often present precocious mature male Salmon parr in attendance, dashing in and releasing their sperm at the same time as the adult cock fish and fertilising some of the eggs.

After spawning, the hen fish drop downstream into quieter deeper waters and a small proportion of them (the number varying from river to river) may eventually reach the sea to begin feeding again after their long fast and eventually become 'mended kelts'. Previous spawners make little contribution to the stock in a river and may represent as little as 0.5-3.0 per cent of the season's run in eastern Scottish rivers. The cock fish usually remain in the vicinity of the spawning grounds for some time and try to mate with other hen fish, but eventually they become very weak and emaciated, and often attacked by fungus. Very few cock kelts ever reach the sea to recover sufficiently to spawn again.

The fecundity of female Salmon can be rather variable, but most fish produce around 1,100 eggs per kg of body weight. Eggs of individual fish also vary in size and can range from 5-7 mm in diameter. As well as varying from fish to fish, the egg size can also vary significantly among different populations. Previously spawned fish do not produce as many eggs as maiden fish, also their eggs may be pale yellow in colour and not the rich orange-red of first time spawners. Derek Mills and his colleagues showed that '... there are significant differences in fecundity among the various rivers. The average egg counts corresponding to a fish of length 70 cm, this being the average length of all salmon handled, on each of the six rivers are:'

R Lyon	4943
R Blackwater	5117
R Garry	5370
R Dee	5495
R Conon	5572
R Meig	6067

Predators

Mortality at sea due to predation is high. Salmon fall prey to seals, whales and large fish such as sharks and tuna, and, especially during their early stages at sea, Saithe and Pollack. In fresh water, young Salmon are also subject to predation pressure from Otters and in recent years from Mink, as well as from several species of birds, such as Herons, Cormorants, Mergansers and Goosanders and from predatory fish such as large Trout and Pike. Returned adult Salmon are also preyed upon by Otters, though the main food of the latter seems to consist of young salmonids and Eels. In fresh water and probably at sea, humans are a major predator of adult Salmon. Seton Gordon wrote that 'The cormorant is hated on a river because of the number of salmon parr it devours. It takes not only these small fish but frequently swallows a large sea trout and on one occasion Angus MacPherson of Invershin took from the gullet of a cormorant he had shot a grilse of between five and six pounds.' Even cannibalism occurs: Derek Mills reports that 'a cock fish, caught illegally in autumn on the Manor Water in the 1970s, was found to have a salmon parr in its stomach'.

Salmon are hosts to many parasites, both external and internal. The most familiar are the copepods known as 'sea lice' and 'gill maggots'. Sea lice become attached to their host at sea and drop off in fresh water within about 3-5 days - thus their presence is an indication that their host is fresh run. Gill maggots on the other hand become attached to their host while in fresh water, but can survive in sea water and may be present on a Salmon throughout its adult life. Salmon are also subject to a number of, often fatal, bacterial and viral infections. Furunculosis is a familiar example of the former which quickly kills fresh-run adults, in certain conditions reaching epidemic proportions - as can the viral infection UDN (Ulcerated Dermal Necrosis) which has been endemic in many rivers in the British Isles since 1967. Mass mortalities of Salmon and Sea Trout resulted from the early outbreaks of this disease.

Growth

In the redd, the embryos develop slowly and hatch in March or April, some 110 days after spawning at temperatures around 4°C. However, the alevins remain in the redd for several weeks more until their substantial yolk sacs have been absorbed. They then emerge from the gravel as fry, usually in early May. The times of hatching and emergence are wholly dependent on the ambient temperatures and can vary from year to year. The newly emerged fry remain in the shallow flowing water over their redds before dispersing gradually.

The growth of fry to the parr stage is slow in most rivers; normally 2-3 years elapse before the latter are ready to become smolts and make their way to the sea - usually at lengths of 12-16 cm, though smolt size can actually vary from 10-20 cm. During their growth phase, the parr have become well dispersed throughout the river, tending to occupy the riffles and runs, rather than the pools and quieter stretches favoured by Trout. In some situations, parr will move upstream to find good feeding stations. In the richer sections of a river, or where individuals have found a particularly good feeding station, parr may reach the smolt stage within one year. At the other extreme, in high cold northern tributaries, parr may still be only 4 cm or less a year after they have emerged from the gravel. In such habitats, parr, if they do not drop downstream, may be 4-6 years old before smolting.

Smolting normally takes place between April and June, as exemplified by the traditional couplet:

'The first spate of May
Takes the smolts away'

However, a small number may reach this stage in the autumn and drop downstream. David Dunkley recalls his surprise at finding a perfect salmon smolt in the Kinnaber Mill Trap on the North Esk in the autumn - 'bright silver; long and slender; long, pale yellow pectoral fins with darkened tips; deeply forked tail with a dark band at the posterior edge'. However, it was the 9 October during the early 1970s - a time of year when one would not expect to see a smolt.

A minor downstream migration of parr from the upper tributaries may also take place at this time, as the first stage of a general smolt migration the following spring. It is not imperative for the survival of a smolt that it reaches salt water. Where juvenile Salmon have been prevented from migrating downstream by a dam or even by being placed as fry in a loch with no exit, if the feeding is good, both sexes will grow and eventually develop mature gonads and the author has seen several of these 'landlocked' fish: for example from Loch Glashan in Argyll and Loch Ettrick in Galloway.

Once in the sea and feeding, growth is very fast and Salmon returning to fresh water after one winter in the sea (now known as grilse) average around 3.5 kg. This average may vary considerably during the season from small fish early on to substantial fish of well over 4.5 kg later on, depending on the length of time spent feeding in the sea. Willie Shearer recalls that 'The largest fish are not necessarily the fastest growers. Two grilse caught in Montrose Bay on 25 July and 28 August, 1972 weighed 14.3 and 14.4 pounds, respectively. These fish had condition factors of 1.27 and 1.23 which in terms of weight were 11 and 13% heavier than accepted averages of fish of these lengths. Scale samples showed that both fish had spent two years in fresh water before going to sea as smolts in the spring of 1971. The larger fish was 13.5 cm and the smaller 13.0 cm at this time; final lengths - 14-15 months later - were 81 and 80 cm respectively. Both fish had increased their weight almost 100 times in this period.'

Salmon returning to fresh water after two or more winters in the sea make up the bulk of the so-called 'spring fish', much sought after by anglers to whom they are particularly vulnerable (because they spend so long in the river) and valuable (because they are normally large). The first of the spring fish may enter their river in the late autumn and in some years a substantial stock may have built up in fresh water by the time the official angling season opens early in the following year. However, their main runs take place normally in the late winter and early spring. Later running, two sea-winter, fish are usually referred to as 'summer fish'; again a small proportion of these may have spent 3 or even 4 winters in the sea, providing the largest individual Salmon of all - up to 14 kg or even more in weight. This run of summer fish into the river may continue until late autumn, the fresh run 'autumn fish' - a dullish silver in colour - arriving almost ripe to spawn, Thus Salmon due to spawn the coming winter and those not due to spawn until the following one may be in the river at the same time. In fact, in our larger rivers, fresh run fish may be present in variable numbers all the year round.

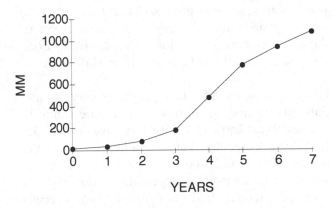

Growth of Atlantic Salmon in fresh water and the sea (Nall 1930).

The average age of most adult Salmon in the British Isles is some 3-5 years, i.e. 2-3 juvenile years in fresh water plus 1-2 years in the sea. In exceptional cases, 8 year old fish can occur, while some individuals at the extreme northern limit of their range could be even older. While in the sea, Salmon travel great distances, many migrating to feeding grounds at the edge of the Arctic Sea and along the east and west coasts of Greenland where Salmon from North America, the British Isles and Scandinavia mingle. They appear to concentrate in sea areas where the surface temperature is between

4 and 8°C. Their methods of navigation to these feeding grounds and then back to their natal rivers are not yet clearly understood, but it is thought that at least three elements are involved - the use of currents in the open sea, odours (known as pheromones) and other chemicals in the water on approaching the home coast, and finally, from the river mouth upstream, both visual and pheromonal/chemical cues. It has been shown that Salmon do not have a genetic link with the river of their birth and that their ability to return to it at maturity depends entirely on the sequential imprinting of various cues that they received on their way downstream, through the estuary and out to sea as juveniles.

Value

Mediaeval laws regulating Salmon exploitation demonstrate the value of this resource, and the early respect that humans have held for this fish from early times. Its size, food value and anadromous habit made it ideal as a source of food for primitive humans, using simple devices to harvest this dependable crop - which was also abundant in many rivers, for example Leonora Rintoul & Evelyn Baxter record that '... in 1521 John Major said the River Forth abounded with Salmon.' Harvie-Brown & Buckley recorded that 'One day, this season - August 1888 - one angler took 54 salmon to his own rod in the Grimersta.' Ancient trapping devices can still be seen in some rivers.

However, the life cycle of anadromous fishes is always at risk when rivers are used for other purposes, such as power, water supply and the disposal of sewage. Thus, since the Industrial Revolution in Europe and its counterpart in North America, very many populations of Salmon have been exterminated as the vital link between the sea and the spawning and nursery areas became severed. In latter times, however, efforts have been made in some countries to ensure that developments do not totally disrupt the movement of Salmon in rivers; fish passes or 'salmon ladders' have been installed in many recent barriers to allow fish to move up and down stream. In addition, the laws relating to pollution control have been strengthened and many rivers are becoming cleaner as a result. The recent natural return of the Salmon to the River Clyde after an absence of more than 100 years (because of pollution) is a very welcome sign.

John Buchan noted that '... salmon spearing, which Sir Walter [Scott] loved, brings its votaries within the danger of the law.' '... he preferred spearing salmon with a leister, which is a nefarious but highly exciting business ...' Problems with the management of salmon stocks are not new. St John noted that 'It is a matter quite beyond doubt that salmon are decreasing every year in most of the Scottish rivers. With short-sighted cupidity, these valuable fish are hunted down, trapped and caught in every possible manner; and in consequence of this reckless destruction the proprietors of some salmon rivers will, before many years have elapsed, lose the high rents which they now obtain from sportsmen and speculators.'

There have been many tributes in the past to the Atlantic Salmon and there is no doubt that this one species has been, and hopefully may continue to be, of importance to the Scottish economy. In 1867, Douglas Maclagan wrote a 12 verse poem, called *Saumon*, which he recited at a dinner of the Aesculapian Club in Edinburgh. Here is just one verse:

> *'O! mark him rinnin' frae the tide,*
> *In blue and silver braw, man;*
> *The ticks upon his gawsy side*
> *Shew him a new-rin saumon.*
> *And though he 'scape the Berwick net,*
> *The Duke at Floors an' a', man,*
> *There's mony a chance remainin' yet*
> *To catch that bonnie saumon.'*

Kelt Salmon, in contrast, are less attractive - certainly they are poor eating! David Carss recalls that 'in the winter 1995, colleagues and I had spent countless freezing nights radio-tracking Colin - a male otter on the River Dee. Like animals we'd tracked in the past, Colin was taking spent kelts and leaving much of the carcasses uneaten on the shingle. One of the these, a particularly fine specimen, became the centrepiece of a dinner party - it looked magnificent in the candlelight but tasted of nothing.'

Dame Juliana Berners remarked that 'The samon is a gentyll fysshe: but he is comborous for to take. For comoynly he is best in depe places of grete ryuers.' Atlantic Salmon provide anglers with the 'King of Sport' and throughout its range, except where owned by the State, salmon fishing rights have high values and may be leased at high rents or change hands for immense sums. However, the formidable pressure groups representing angler interests are of benefit to the Atlantic Salmon in a society which still otherwise puts quick profit before the future of a long-term natural resource. The future of commercial salmon fishing in estuaries and around the coasts of the British Isles, once a highly profitable industry, is dying out due to the continuing rapid expansion of salmon farming and to buy-outs by salmon 'conservation' organisations (i.e. anglers!).

Frank Buckland recorded that in 1871 'The Scotch Fisheries have been valued at about £300,000 (per year), and this is, in my opinion, within the mark.' More recently (2004), it was estimated by Alan Radford and his colleagues that anglers spent £73 million fishing for Atlantic Salmon and Sea Trout in Scotland.

Conservation

The Atlantic Salmon is one of the few fish to have an international organisation concerned with its conservation. The North Atlantic Salmon Conservation Organisation (NASCO) is an inter-governmental grouping of all countries with native stocks of Atlantic Salmon. Its headquarters are based in Edinburgh.

In Scotland, the Atlantic Salmon is protected by a wide variety of legislation, some old, some relatively new. It is not listed in Schedule 5 of the Wildlife and Coutryside Act, but recently it has been added to Annexes IIa and Va of the EC Habitats Directive 1992 and Schedule 3 of the Conservation Regulations 1994 (in fresh water). This new conservation obligation requires member states to (a) designate sites to form part of the 'Natura 2000' network comprising Special Areas of Conservation (SACs), (b) protect such sites from deterioration or disturbance with a significant effect on the nature conservation interest (and take steps to conserve that interest), and (c) protect the species of Community interest listed in the Annexes to the Directive. There are six SACs in Scotland where the Atlantic Salmon is the 'qualifying' feature and eleven SACs where this species is a 'primary' feature. The Rivers Bladnoch, Dee (Aberdeen), Endrick, Moriston, Oykel, South Esk, Spey, Teith and Tweed are examples of these SACs.

A major conservation project funded by the European Community is at present (2006) under way in some of these catchments - 'The Conservation of Atlantic Salmon in Scotland Life Project'.

A future programme of research on the Salmon in Scotland should include:
(a) A national framework for the management of Salmon in Scotland.
(a) Proper monitoring of angler catches throughout Scotland.
(b) Monitoring of juvenile stages (fry and parr) in selected rivers.
(c) Increased research on Salmon in the sea.
(d) Increased research on relationships between salmon farming and wild fish.

References

There has been extensive research on many aspects of the ecology of Atlantic Salmon in the past and only a selection of relevant references are included here.

Banks (1969), Banks (1989), Berners (1496), Bolam (1919), Buchan (1917), Buck & Hay (1984), Buck & Youngson (1982), Buckland (1873), Calderwood (1930, 1945), Child et al. (1976), Colquhoun (1878), Egglishaw (1967, 1970), Egglishaw & Shackley (1977), Elder (1965), Freshwater Fisheries Laboratory (1992), Gardiner (1974), Gardiner & Egglishaw (1986), Gardner (1971, 1976), Gee et al. (1978), Gordon (1944), Hansel et al. (1987), Hansen & Pethon (1985), Harvie-Brown & Buckley (1889), Hawkins & Johnstone (1978), HIDB (1985), Houghton (1879), Johnston (1904), Jones (1959), Laughton (1989), MacCrimmon & Gots (1979), Maitland (1965, 1985, 1990), Malloch (1910), Maxwell (1904), Metcalfe et al. (1989), Mills (1964, 1971), Morrison (1983), Munro (1969), Myers & Hutchings (1987), Nall (1930), Netboy (1968), Payne et al. (1971), Pope et al. (1961), Pyefinch (1955, 1969), Radford et al. (2004), Reddin & Shearer (1987), Regan (1911), Rintoul & Baxter (1935), Scrope (1843), Shaw (1836, 1840), Shearer (1972, 1984), Smith (1962, 1964), Solomon & Child (1978), Stabell (1984), St John (1891), Thorpe (1977, 1987), Thorpe et al. (1981, 1992), Thorpe & Mitchell (1981), Thorpe & Morgan (1978), Tytler et al. (1978), Wankowski (1979), Went (1976), Wilkins (1972).

'Wi' flesh as pink as rose in June,
and curd as white as snaw, man,
And sappy bree they boiled him in -
Oh, that's what I ca' saumon.
To best of friends I canna wish
That better should befa', man,
Than juist to hae as guid a dish
As we hae wi' our saumon.'

Douglas Maclagan (1867) *Saumon*

*****

Brown Trout *Salmo trutta* Breac

*'It is thus most confusing that in almost every locality
the same fish should be called by different names...'*

Peter Malloch (1910) *Life History of the Salmon, Sea Trout ...*

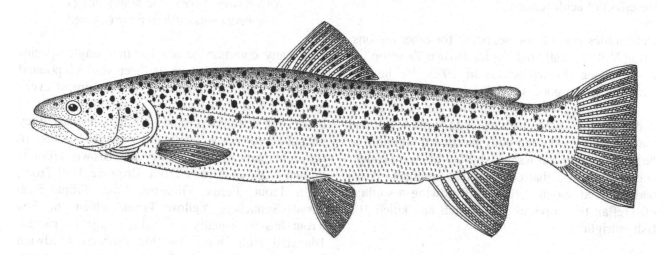

THE enormous variety of form of this species is exemplified by the large number of local names used. Trout which spend some of their lives in the sea (Sea Trout) and those which are purely freshwater (Brown Trout) are now known to be one and the same the same species, although they are awarded sub-specific status by some authorities as *Salmo trutta trutta* and *Salmo trutta fario* respectively. The situation was summed up by Edward Boulenger in 1946: 'Some of the many local varieties of trout have been given specific distinction by certain naturalists, but the best authorities assign the many forms, including the migratory sea-trout or salmon trout, to a single species.' Thus, 'The Loch Stennis Trout (*Salmo orcadensis*). Probably a non-migratory species, from Loch Stennis, in the Orkney Islands.' described by William Houghton in 1879, is, along with all the other 'species' (e.g. *Salmo levenensis*) no longer regarded as a separate species.

However, as Charles Holmes notes, in *The Tarn and the Lake*: '*Salmo fario* may, for the ichthyologist, be no more than a single species, with perhaps a few local variants. To the trout fisherman, however, he is no mere trout, one and indivisible, ... the trout is no single, common, identical, definite, determined and measurable fish, but rather ten thousand tantalizing, distinct and different devils.'

Many other observers have noted curiosities in the variety of form in this species. In 1911, Tate Regan noted: 'The River Smoo in Sutherlandshire plunges through a hole in the roof of a cave and falls about 40 feet into a dark pool below; from the cave pool to the sea is a distance of only about 30 yards, but in 1876 Mr. Neil Campbell caught some Trout in this stretch and transferred them to the water above the falls, where none had previously existed. In 1882 Mr Harvie Brown caught quite a number of these fish, which were remarkable for a line of large irregular blotches of bright crimson along the middle of the side, sometimes confluent to form an undulating band, whereas the ancestral form was said to be very ordinary looking.'

Peach recorded another aberrant form. The so-called tailless trout of Islay 'were sent ... by Mr. Colin Hay, distiller, of Ardbeg Islay, taken in Loch Namaorachin, about 1000 feet above the level of the sea; ... It is ... so shallow that a man can wade through it; the bottom is quartz rock, ... Several other lochs are near it in which trout are plentiful, but none "tailless".' Harvie-Brown describes them from another Islay loch: 'Perhaps the most remarkable variety in the region is the so-called Tailless Trout of Islay, which has rendered permanent a strange malformation. Loch Finlagan, in which it is found, is of no great extent, and yet this interesting deformity has managed to perpetuate itself within its precincts.' This unusual fish still occurs on Islay to this day. During work on the impact of acid rain on fish in Scotland, the author and Alex Lyle electrofished peat pools on the Silver Flowe NNR, Galloway, in July 1987 and found Brown Trout with similar deformities attributed to the effect of acidification.

Deformities can, of course, occur for other reasons. Allan Virtue recalls fishing for Brown Trout on the Eden Water (Berwickshire) in 1975. He hooked what felt like a good sized fish but when he struck, to his surprise, the fish flew over his head on the end of his line. When he retrieved the trout he found that it had a good deep body but that it appeared to be stunted. Several more fish like this were caught and it transpired that considerable numbers of fish had suffered spinal deformity following a spillage of Treflan the previous year which had killed 100 fish outright.

No other fish has commanded such popularity and admiration among anglers and naturalists. Arthur Johnson was certainly one of the enthusiasts: 'The trout is a prince among fishes. He will not deign to live in waters that are contaminated by the haunts of man. The streams that are born in the rocks of the hill-tops, which wander through moors of peat and heather, are his own ... There is a common instinct - the all-powerful instinct, urging each one forward, and go they must. On, on, on, fighting the foaming flood, far up the roaring river. ... Half-way up the hillside the water off the moorlands beyond falls over a perpendicular wall of rock into a ravine below ... Silvery flashes in the sunlight, the fish follow one another in quick succession, leaping against an impossible barrier, charging the force of the flood that ever hurls them back, returning again and again with renewed vigour, burning with that mad desire to go they know not where.'

James Hogg was likewise enthusiastic in this verse from *A Boy's Song*:

> *'Where the pools are bright and deep,*
> *Where the grey trout lies asleep,*
> *Up the river and over the lea,*
> *That's the way for Billy and me.'*

As was Thomas Stoddart in his poem *The Yellow-fins of Yarrow Dale*:

> *'The yellow-fins o'Yarrow dale!*
> *I kenna whare they're gane tae;*
> *Were ever troots in Border vale*
> *Sae comely or sae dainty!*

> *They had baith gowd and spanglit rings,*
> *Wi' walth o'pearl amang them;*
> *An'for sweet luve o' the bonny things,*
> *The heart was laith to wrang them.'*

The many common names for this single species still give rise to confusion, as Peter Malloch pointed out: 'It is thus most confusing that in almost every locality the same fish should be called by different names, at different periods, simply because they change their colour with the seasons.' Thus, in different parts of the country the Brown Trout is known as Breck, Brook Trout, Brownie, Bull Trout, Burn Trout, Ferox, Gillaroo, Lake Trout, Slob Trout, Sonachen, Yellow Trout, whilst the Sea Trout has an equally confusing range of names: Blacktail, Bull Trout, Covichie, Finnock, Fordwich Trout, Grey Trout, Herling, Lammasman, Mort, Peal, Phinnock, Round Tail, Salmon Trout, Scurf, Sewen, Sprod, Truff, Whengs, White Trout, Whiting, Whitling, Yellowfin. A literal translation of the scientific name is 'Leaping Trout'.

Size

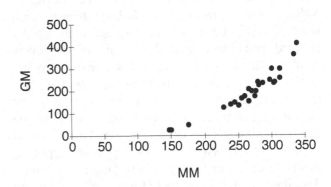

Length / weight relationship in Brown Trout from South Uist.

The length of fully grown wild Trout of both sexes can range from a few centimetres to over 95 cm, but the size encountered most commonly is around 26-42 cm, the weight equivalents being approximately 226-900 gm. The British record rod-caught wild Brown Trout caught in Loch Awe in 1996 by A. Finlay, weighed 11.502 kg (25 lb 5 oz 12 dm). The British record rod-caught cultivated Brown Trout was caught by D. Taylor in 1995 at the Dever Springs Trout Fishery in Hampshire and weighed 9.623 kg (21 lb 3 oz).

However, it is certain that at least one Trout larger than the present record has been caught in Scotland. A Slob Trout of 13.6 kg was taken in the outflow from the Loch of Stenness in Orkney in 1889, but as it was caught on a set line it is not considered to be an angling record.

Sea Trout too can attain a large size. The Scottish rod-caught record stands at 9.071 kg for a fish caught in the River Tweed in 1983. The British rod-caught record stands at 12.85 kg (28 lb 5 oz), caught in the River Test, Hampshire in 1992. However in recent years a number of large Sea Trout have been taken around the British Isles in nets, or found dead, and weights of 10.9-12.7 kg have been recorded, for example one of 28 lb 9 oz was caught by the commercial nets in the River Tweed some years ago.

Appearance

Trout are one of the most common of Scottish native freshwater fishes and to many, its shape, elongate, stream-lined and laterally compressed, typifies the Salmon family. Brown Trout are not likely to be confused with any other fish except that occasionally 'red' Salmon have been taken for large Ferox Trout, and large Sea Trout for Atlantic Salmon. In Trout the maxillary bone extends backwards at least to the level of the posterior margin of the eye. The head and shaft of the vomer bone are toothed and the palatine bone and tongue also carry teeth. There are strong teeth on the upper and lower jaws. Trout scales are noticeably smaller than those of Salmon, so there can be up to 130 scales along the lateral line. As noted in the discussion on Salmon, there are occasional occurrences of hybrid fish with characteristics intermediate between Trout and Salmon.

The colouration and marking of Brown Trout is subject to great variation, even within the same population. Young individuals may often be quite silvery, but after becoming mature, spotting usually increases progressively with age in both sexes, while the background colour becomes darker and may be brown, yellow or nearly black. Spotting is very variable too, both in form and density, and

there are innumerable combinations of large and small spots, red and black ones, some of which may be encircled by light to dark haloes. The dorsal and adipose fins may bear spots, but if any spotting is present on the caudal fin this is usually restricted to the uppermost part. In some populations individuals are found with the whole of the caudal fin heavily spotted in such individuals the whole head may be well spotted too. Occasionally, loch populations are encountered where all the individuals of both sexes are virtually identically marked regardless of size or age. At least two such populations occur in the north of Scotland and are of considerable genetic interest.

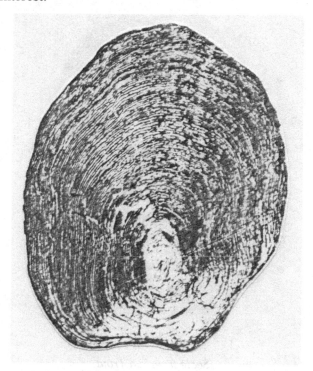

Scale from Brown Trout.

St John noted that 'Many and varied are the streams and lakes of Scotland, and scarcely any two of them contain trout of exactly similar appearance. Although of the same species, and alike in all the essential parts of the anatomy, &c., in outward appearance, shape, and colour, trout vary more than any other fish.' James Maitland put it another way: 'All the Scotch trout, with two exceptions (Islay and Orkney), pass from one form to another, so as to render, after a few generations, identification very difficult'.

At spawning time the markings and background colouration of the male become intensified. The belly may become sooty grey merging into amber orange flanks. Most male Ferox Trout develop this colouration. Sea Trout, 'bars of silver' when young and fresh in from the sea, also develop black and brown spotting with age. The female Sea Trout at spawning time assumes a dull leaden colour, but her straight edged tail and typical squared shape distinguishes her from a hen Salmon. After spawning, Sea Trout kelts quickly regain their silvery appearance and may be so 'well-mended' that anglers catching them on their way back to the sea have difficulty in distinguishing them from fresh run fish.

throughout the world, including the southern hemisphere, where anadromous populations have become established in Tasmania, New Zealand, the Falkland Islands and southern South America. It has been established, too, in some tropical countries where high altitude streams and lakes offer suitably cool conditions. It is well established in many parts of North America.

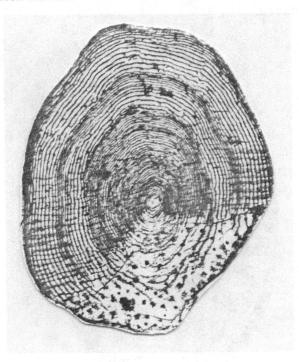

Scale from Sea Trout.

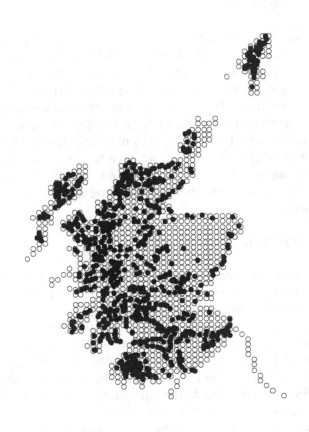

Recorded distribution of Brown Trout in Scotland.

The flesh colour of Trout ranges from white through champagne to a rich orange pink. The flesh of young freshly run Sea Trout is deeply-coloured but as the fish ages it may become much paler. That of Brown Trout from habitats where there is a good supply of crustacean food is also deeply-coloured, as is the flesh of well conditioned Ferox Trout - although the latter are almost exclusively piscivorous. The essential crustacean pigment is obtained from the flesh and stomach contents of their plankton-eating prey - small Trout and Charr.

Distribution

The Brown Trout, which is native to Europe and western Asia, has been introduced widely

All native Brown Trout populations in Scotland are descended from early, post-glacial colonisation by anadromous Trout which established populations that now appear quite distinct - some of them being still largely anadromous, others completely non-anadromous. A range of intermediate populations exist. They are probably the commonest fish in Scotland and found wherever the water quality is high and there are not too many predacious fish. Where there are no spawning facilities, populations are often maintained by anglers - by introductions from fish farms or from the wild. Sea Trout occupy all accessible rivers and often the estuaries of inaccessible ones too, which they use as feeding grounds.

Ecology

Some time after hatching, when the yolk sac has been almost completely resorbed, trout fry emerge from the gravel and can be seen hovering over the site of their redd. The author clearly remembers his excitement, at age 6+, at his first sight of a Brown Trout fry seen in Tannoch Burn, Milngavie! The time between hatching and emergence from the redd varies according to the ambient temperature (e.g. 97 days at 4.7°C, 148 days at 2°C) so that fry may be emerging from mid-March to early May, depending on the region. As with salmon fry, trout fry then gradually disperse, the stronger individuals occupying the best feeding stations while the weaker ones drop downstream, the great majority to perish during the first few months, from starvation and predation by larger fish and birds. In general, both Sea Trout and Brown Trout may spend up to five years (but usually only 1-3 years) in their nursery streams before migrating down to the sea or to a loch or large river respectively.

In river systems with Sea Trout populations, a downstream migration of silvery smolts takes place each spring. The smolts have usually spent two to three years in fresh water (although smolts of one, five and six years of age are also known). Throughout Scotland, smolt age appears to increase with latitude, and to some extent, altitude. Sea Trout smolts are usually 15-22 cm in length and more robust than salmon smolts. A similar downstream migration of Brown Trout 'smolts' takes place where substantial streams flow into large lochs but these young remain in the loch, often growing to a very large size. Unlike Atlantic Salmon smolts, not all Sea Trout smolts migrate out into the open sea: some stay in the estuary, some move into shallow coastal waters, while Sea Trout from east coast rivers may cross the North Sea to Scandinavia, and others set off on long migrations far out to the open sea. In fact, there is no real demarcation between the distribution of Sea Trout and Brown Trout, for a continuum of trout types occupy habitats from the source of a river through lochs, through estuaries into coastal waters and out into the open sea.

A proportion of Sea Trout smolts spend only a few weeks or months in the estuary or nearby coastal waters before returning to fresh water, and after midsummer, large shoals of these young Sea Trout, known by many local names (e.g. Finnock, Whitling, Peal, Harling, etc) are found in the lower reaches of their natal rivers and lochs. They are very vulnerable to anglers; large inroads can be made into their numbers at this stage by greedy fishermen. For example, Bridgett recorded in 1929 'The fact that sea-trout are so easily taken with the worm in a flood - I have seen a basket of fifteen dozen taken in a day by one rod, whose I will not say - shows that the fish retain both the power and the desire to eat in fresh water.' Peter Malloch, discussing Whitling, recorded that 'On one occasion on the 10th March, I caught with fly eighty of them, averaging three-quarters of a pound in weight.'

Food

Trout are catholic feeders and their voraciousness is well known. The great bulk of their diet throughout the seasons, however, consists of aquatic bottom-dwelling and midwater invertebrates, although from about midsummer until autumn invertebrates of terrestrial origin which fall or are blown on to the water surface may dominate. In lochs, Brown Trout patrol territories in search of food while in streams their feeding strategy is to occupy and hold the best feeding station that their rank in the local hierarchy allows, waiting for food organisms to drift past in the water column or on the surface. They also pick up nearby food off the bottom. In spite of their well-known cannibal tendencies (much exploited by anglers using fish-like lures or dead fish baits), Trout other than Ferox take surprisingly few small fish, even though the latter may be in abundance. Generally, the larger the Trout the more likely it is to feed on other fish. Certain species of small fishes appear to be more easily caught by Trout than others. Sticklebacks and Perch fry, for instance, can be important, while Minnows appear mainly to be available during early summer, when they gather into tight shoals to spawn in shallow water.

There are many observations of other vertebrates being eaten by Trout. Keith Ritchie, fishing a loch beyond Tomich, Inverness-shire in July 1994, landed a Brown Trout of two pounds. When taking out the fly, it was discovered that the fish had just eaten a newt. Willie Miller, in the late 1940s was at the Tillan Burn in Lanarkshire when he disturbed a Water Shrew which swam under water towards a large stone. 'A trout of about half a pound in weight darted out and chased it away before returning to its lair under the stone. I have to confess that I took off my boots and socks and "guddled" the trout which provided me with my tea.' Robin Ade recalls 'One summer afternoon on the upper Ken I was surprised to see a large trout doing something under the overhanging grasses of a deep pool. When I caught it, I found three well grown voles in various stages of digestion inside. The fish was 18 inches in length and weighed 2 lb. 10 oz.'

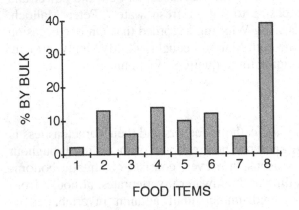

Food of Brown Trout in the Altquhur Burn (1: worms; 2: crustaceans; 3: stonefly larvae; 4: mayfly larvae; 5: caddis larvae; 6: midge larvae; 7: others; 8: aerial) (Maitland 1965).

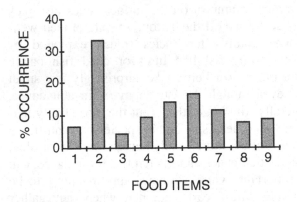

Food of Sea Trout in Lochs Etive and Eil (1: annelid worms; 2: amphipods; 3: mysids; 4: isopods; 5: insects; 6: clupeids; 7: sand eels; 8: others; 9: aerial) (Pemberton 1976).

Other, more unusual observations on trout diet include that of Derek Mills, who 'remembers watching Trout following thistle 'down' being blown across the water surface.' Alan Joyce, in Sutherland, 'remembers his pupils having a trap for Eels in 1967, near the back door of a local hotel. Some Trout were taken in this trap and on examining the gut contents of these they were found to be full of a white stodgy mass. Closer examination, with the help of iodine, proved this to be rice pudding, which (discreet enquiries confirmed) had been thrown out of the hotel into the burn!'

Tate Regan noted in 1911 that 'The Trout smolts, when they finally leave the estuaries, do not go so far out to sea as the Salmon, but feed near the coasts,...' Their food there seems to consist mainly of small fish such as sandeels and members of the cod and herring families; locally, Smelt and elvers are eaten as well as marine worms, molluscs and crustaceans. Trout caught at sea or in estuaries usually have their stomachs packed with these items. Finnock and small Sea Trout feed like Brown Trout in fresh water, but larger individuals behave more like Salmon - though unlike in Salmon, it is common to find a few small invertebrates in their stomachs (usually terrestrial insects taken from the surface and small aquatic invertebrates such as midge larvae and pupae). Occasionally fish are found, but the small bulk taken cannot be of much value and Sea Trout in fresh water lose condition.

Reproduction

Trout spawn mainly from mid-October to mid-December, but earlier and later dates have been recorded. In many places a rise in the water level of the tributary streams is necessary to entice and allow ripe fish to reach their spawning grounds: thus the time of spawning can vary considerably from year to year. The ideal situation for redd excavation is clean deep gravel in pure fast flowing water, usually at the downstream exit of a pool. Occasionally Trout are found spawning in the shallow water of highland lochs, particularly off gravelly spits where, when there is a strong wind, river conditions may be closely simulated. Some Brown Trout populations inhabiting high altitude lochs where there are no inflowing streams (e.g. Loch Vrotachan) survive in this way. Trout can be very faithful to their spawning streams - Willie Shearer recalls 'the recapture in November 1954 of tagged Sea Trout in the same spawning burns of the Aberdeenshire Dee and on the same stretches of gravel as those on which they had been captured almost exactly 12 months previously'.

Tom Stuart found that 'definite and regular migrations of immature trout have been found to occur at approximately the same times each year. A mass movement from the streams to the lochs takes place in the autumn, frequently just before the arrival of adult spawners. Some immature males with ripening gonads delay their departure until after the arrival of the adults and then depart with them. Comparable return movements in the spring repopulate the streams.' Evidence of a high degree of homing accuracy was found in lochs in the studied in the Pitlochry - Lochan an Daim, Dunalastair Reservoir and Loch Moraig.

The site of the Trout redd is carefully selected and excavated by the female with a male or males in attendance, and covered over again after the eggs are laid. Where the whole process of spawning has been observed in artificial stream sections it appeared that the female selected the site for the redd at a point where water flowed through the gravel at the tail of a pool. The requirements have been well described by Tom Stuart, who noted that the spawning gravels 'are composed of stones up to 3 in. in diameter with a large proportion of smaller materials, the effect of which is to consolidate the mass while leaving it permeable to the water. Stones embedded in fine sands or silts which form a hard bed are avoided, as are also uniform gravels and shingles of small size which move easily in a flood. ... The location of the redds in an ideal pool is towards the tail, where the gravel slopes gently upward, spreading more or less evenly from bank to bank by the slackening of the current. The depth of water here is at a minimum ...'. The presence of currents through the gravels actually chosen by Brown Trout was demonstrated by Tom by the passage of dyes through them. No doubt in a busy spawning stream, the chance fertilisation by drifting sperm from males spawning upstream can take place and where both Sea Trout and Brown Trout are present cross fertilisation between Trout with these different life styles must occur.

Trout produce from 1,100 to 1,700 eggs per kg of body weight and the rate of successful fertilisation is very high. The diameter of the eggs, which are yellow to orange yellow in colour, is 4-5 mm. In females from populations of exceptionally small individuals, the eggs remain about the same size, for example, Niall Campbell found a fully ripe female only 10.7 cm in length from a roadside ditch in Argyll which contained 31 eggs, 3.9 mm in diameter. The ripe males from this site ranged from

6.7-11.2 cm in length. Sometimes Trout are unable to spawn; Lesley Crawford recalls fishing a little-visited water in Caithness in May 1996 where she caught 'two female specimens fat with spawn who had been unable to shed their eggs. The trout weighed 2.5 lb. apiece but were poor specimens with little fight.' She concluded they were old fish waiting to die - 'rather sad.'

Predators

Trout have a wide range of predators in both fresh water and in the sea. Apart from cannibalism and predation by other fishes such as Pike, and sometimes Eels, Trout are taken by sawbill ducks, Herons, Cormorants, Black-throated Divers, terns and gulls. David Carss and Keith Brockie found Brown Trout to be one of the commonest items of the diet of Ospreys in Scotland. Large and small Trout fall prey to Otters and to Mink, while at spawning time adult Trout stranded in small streams are easy prey for predators, and even Foxes seize this opportunity. At sea, especially in bays, estuaries and sea lochs, Common and Grey Seals make considerable inroads into Sea Trout stocks.

Trout too prey upon themselves. St John noted that 'I never killed a tolerably large trout without finding within him the remains of other trout, sometimes, too, of a size that must have cost him some trouble to swallow.'

Trout are subject at times to large internal and external parasite burdens. The most conspicuous of the latter are Sea Lice (*Lepeophtheirus*), crustacean copepod parasites which infest Sea and Slob Trout, and all anadromous Trout carry them during their first few days in fresh water. Brown Trout in small ponds may be carriers of a close relative, the freshwater Fish Louse (*Argulus*).

Trout usually also contain a variety of internal parasites in and around the stomach, on the gills and in the muscle of the body itself. A fluke (*Diplostomulum*) infects the eye and may affect the fishes' vision. The most common parasites of all are cestode tapeworms such as *Diphyllobothrium* and *Eubothrium*, and nematode round worms such as *Eustrongiloides*. The latter are thin, wire-like orange worms which, to the chef, have a distressing habit of emerging from their cysts in the flesh of the fish while being cooked!

Growth

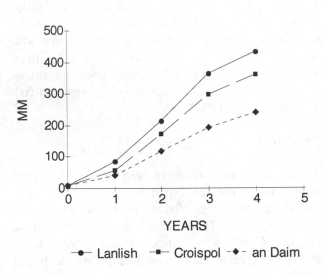

Growth of Brown Trout in three Scottish lochs (Campbell 1961).

The physical and chemical nature of lakes and rivers appears to have no direct influence on the growth rates of Trout in them, except perhaps where conditions are so acid as to prevent the hatching of eggs. Observations by Niall Campbell on the effects of a number of factors affecting fry survival have been made on a wide range of lochs, representing virtually all natural trout habitats in Scotland. High altitude lochs (i.e. those at more than 610 m above sea level), acid and alkaline lochs, all produced fast or slow growing Trout according to the relative abundance and quality of the food supply and to the productivity of the spawning grounds.

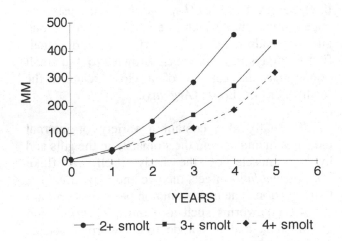

Growth of Sea Trout of different smolt ages from the North Esk (Pratten & Shearer 1985).

A special study has been made by Niall Campbell of the very large, almost wholly piscivorous 'ferox' Trout, found in the large deep oligotrophic lochs of Scotland. Characteristically, these fish grow slowly for about the first third of their life span, feeding on a predominantly invertebrate diet. On reaching a critical length, they switch to a fish diet, with a resulting spectacular increase in growth rate. In Scotland, Ferox fish occur only in lochs over a certain size (100 ha) that also contain Arctic Charr (though they also occur in Loch Lomond where Powan are the equivalent of Charr), and it is likely that the latter species is their main item of diet. Due to their large size, full grown Ferox probably breed only with each other, thus enhancing genetic characteristics such as longevity, late maturity and piscivorous feeding behaviour.

A typical water is mentioned by Andrew Lang in his *Ballade of the Tweed*:

> *'The ferox rins in rough Loch Awe*
> *A weary cry frae any toun;'*

It is from Loch Awe, in recent years, that some of the largest Scottish Brown Trout have been taken.

Populations of 'stunted' Trout, so typical of upland peaty waters, usually mature at two or three years of age (with a proportion of the males maturing at 1+ years) and have life spans of only three or four years. However, early maturity and a short life span are also typical of populations of Trout inhabiting rich alkaline waters. Slow-growing, late maturing Trout (i.e. spawning for the first time at 4+ years or more) are characteristic of large oligotrophic lochs. In such lochs however, a wide range of relationships between age and size occur, e.g. Trout that have passed one year in the nursery stream before migrating to the loch often grow faster and mature sooner than Trout that have spent two, three or more years in their natal stream.

Value

The Brown Trout is one of the most important fish to anglers in Scotland and economically must rival the Atlantic Salmon. Peter Malloch believed that 'To the angler the common yellow trout is by far the most important fish that swims, for more people capture it than any other living thing.' He also considered that 'The sea-trout is one of the best sporting fish we have in Great Britain, and is, besides, of greater value than any others to those who are fortunate enough to possess waters that contain them.' Thus he agreed with Herbert Maxwell, who had earlier recorded that 'There be some men who aver that fishing for sea-trout is finer sport than can be had with salmon.' Sea Trout angling is an important local money earner in the more isolated parts of the country while the profitability of some former marginal salmon netting stations hinged on the number of Sea Trout also taken. Due to the fact that, unlike Salmon, Sea Trout may return to fresh water to spawn many times, their population structure is particularly vulnerable to heavy cropping.

Brown Trout were once farmed on a large scale, primarily for stocking angling waters, but in recent years they have been largely displaced by Rainbow Trout both for this purpose and for the table. A few farms, however, still produce substantial numbers for stocking purposes - often for 'put and take' fisheries. Stocking Scottish lochs and rivers with Brown Trout has been a widespread practice since the mid-18th century; a search by the author of the Howietoun fishery archives, held by the University of Stirling, revealed hundreds of destinations in Scotland to which fish were sent from just this one hatchery. Many landowners also moved fish about. In 1878, John Colquhoun dedicated a revised edition of *The Moor and the Loch* to his brother who had drowned in Loch Lomond a few years earlier. 'My late brother imported ova of Geneva trout ... About 200 fry were introduced into Loch Sloy ... but have not made any appearance yet. ... My sons, in four hours' fishing, captured twenty dozen last summer, no bigger than I remember them when a boy - proof sufficient that the foreign intruders have as yet made no impression on the Arrochar Highlanders.'

Brown Trout were once netted commercially and sent long distances to markets. Robert Burns-Begg, writing about Loch Leven, recorded that 'In those days [ca 1840], owing to the difficulty and delay attending transmission of the trout from Kinross to the London and other equally distant markets, it was customary to preserve the fish by potting them, and the men engaged in performing that duty were enabled to give important information as to the different varieties of food which they were in the habit of finding in the stomachs of the fish operated

upon by them. ... It was generally in March, April, May, and June that the trout were cured and potted, when the trout were at the best.'

Poaching was also rife at this time, as Robert Burns-Begg records: 'Knew two persons in Milnathort who killed 160 trouts in one night. Understands they were salted for consumption at home, particularly by those in Milnathort. Heard a tailor in Milnathort say he was eating the salted trouts so late as in the month of May. ... This kind of fishing was not thought in any way criminal, as it was so common, and no one thought he was doing ill. The practice was common in both town and country.'

Brown Trout are a favourite angling quarry and indeed are extremely important to the economy of many rural areas. Sometimes anglers catch more than they expect - Niall Campbell recalled his experience of 'foul-hooking a Brown Trout and a Mallard duck on the same cast - like playing a kite!' On another occasion, 'a Heron regurgitated a six ounce Brown Trout which was taken home, cooked and eaten.'

Conservation

At one time there was little interest in the conservation of Brown Trout, but the recent awareness of the genetic degradation caused by stocking with domestic strains and the decline of Brown and especially Sea Trout on the west coast of Scotland has raised interest in this species. Research by Andrew Ferguson has shown that 'Genetic diversity in the brown trout is, in the main, not related to geographic distance ... Thus populations a few kilometres apart or even living in the same system may be more distinct than those from, say, 500 km apart.'

There have been a number of conservation projects, including detailed genetic studies of some important populations as well as various initiatives such as that of the Institute of Fisheries Management 'Caring for the Wild Trout in Scotland', CASTAG, Wild Trout Trust, West Coast Sea Trout, etc. There is also international concern for this species, as exemplified by the European 'Trout Concert'.

Sometimes, fishery managers take unorthodox conservation measures. To protect newly released Brown Trout fry, Niall Campbell once 'took the unusual step of shooting large predatory Trout in the vicinity by firing a 12 bore at them!'

The most important conservation initiatives needed for Brown Trout in Scotland may be summarised as follows:

(a) The production of a comprehensive genetic database of important stocks of wild Brown Trout in Scotland.
(b) Conservation management of the waters holding these stocks.
(c) Further research on the relationships between Brown Trout and Sea Trout.
(d) An evaluation of the economic importance of Brown Trout (including Sea Trout) to Scotland.

References

Alabaster (1986), Allen (1938), Bagenal (1969), Bagenal *et al.* (1973), Ball & Jones (1960), Boulenger (1946), Bridgett (1929), Burns-Begg (1874), Burrough & Kennedy (1978), Buss & Wright (1957), Calderwood (1930), Campbell (1977), Campbell *et al.* (1986), Campbell (1957, 1961, 1963, 1971, 1979), Carss & Brockie (1994), Colquhoun (1878), Craig (1982), Crozier & Ferguson (1986), Egglishaw (1967, 1970), Elliott (1976, 1985, 1994), Fahy (1977, 1978), Ferguson (1989), Ferguson & Mason (1981), Frost & Brown (1967), Gardiner (1974), Gerrish (1935), Harvie-Brown & Buckley (1896), Holmes (1960), Houghton (1879), Hunt & Jones (1972), Hynd (1964), Johnson (1907), Kennedy & Strange (1978), Le Cren (1973, 1985), Lyle & Maitland (1992), MacCrimmon & Marshall (1968), Maitland (1887), Maitland (1965, 1990, 1999), Malloch (1910), Maxwell (1904), Menzies (1936), Milner *et al.* (1978), Munro & Balmain (1956), Nall (1930), Peach (1872). Pemberton (1976), Pratten & Shearer (1983, 1985), Slack (1934), Priede & Young (1977), Regan (1911), Stephen & McAndrew (1990), St John (1891), Stuart (1953, 1957), Thorpe (1974), Treasurer (1976), Went (1979).

'The trout within yon wimplin burn
Glides swift - a silver dart,
And, safe beneath the shady thorn
Defies the angler's art;
My life was ance that careless stream,
That wanton trout was I;
But Love, wi' unrelenting beam,
Has scorched my fountains dry.'

Robert Burns (1795) *Now Spring Has Clad the Grove in Green*

*****

Arctic Charr *Salvelinus alpinus* Tarrdhearg

*'...a kind of Fishes, called by the Countrey people Red Waimbs,
from the bloud red colour of their Belly'.*

Walter MacFarlane (1748) *Geographical Collections Relating to Scotland*

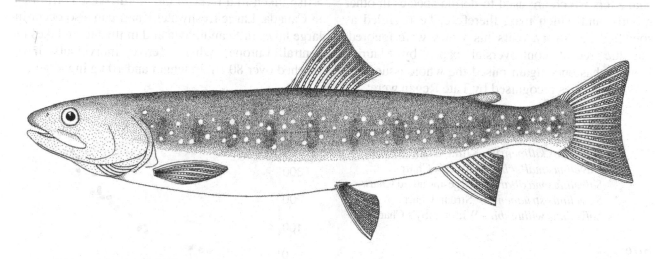

ALTHOUGH occurring quite widely in western Ireland and northern Scotland, and very locally in south west Scotland, north west England and north Wales, the Charr is not a familiar fish to most people, and when encountered, is often regarded as a curiosity - 'an Ice Age relict'. In fact, it occurs in quite large numbers in some lochs, probably outnumbering the Brown Trout population where they co-exist, as they usually do. However, due to its cryptic habits it is seldom seen, even by anglers. There has been considerable interest in the species in recent years and it is now being studied and angled for in several waters. In other parts of the charr's range it is considered to be a most valuable fish for consumption by both humans and dogs.

Charr seem to fascinate the people who become involved with them - the International Society of Arctic Charr Fanatics (ISACF) is a serious body of scientists established in 1980 to study and promote all aspects of this lovely fish. ISACF organises workshops in a different charr country every two years, where scientists from the charr nations meet to discuss current research and conservation, and to catch and eat the species. The association of Arctic Charr with some of the most isolated and extreme northerly regions of the world seems to add to their mystique. The 1992 Workshop was held in Scotland, organised by the author and Ron Greer,

and was a very successful occasion, with participants from 13 countries.

Why are people attracted to Arctic Charr? There are many reasons, but here are two quotes from just one angler, Patrick Chalmers, writing first in 1931 '... only once have I caught a char, and then he was only a very little one. But he was as beautiful as a butterfly.'

Then later, in 1938, he caught another: 'The boat drifts on. The rise for the moment is done. Then, suddenly, I am playing another, but? But? The quarter-pounder that comes fighting to the net is scarlet and green and orange - oh surely here is such a capture as never man made before! And yet, "It's just a bit char," says Chisholm dispassionately. "There's ay an odd yin in here.".'

The spellings for this species of 'Char' or 'Charr' are both acceptable, but the latter - derived from the Gaelic 'tarr' meaning belly is the more authentic. One Gaelic name for the Charr is Tarrdhearg, meaning red bellied; the Welsh equivalent is Torgoch, meaning red belly. Several common names have been used in Scotland, e.g. Alpine Charr, Cuddy, Red-bellied Trout, Red Waimb. A literal translation of the scientific name is 'Little Mountain Salmon'.

It is generally recognised that there is only one native species of charr – *Salvelinus alpinus* – in Europe. Tate Regan, however, while admitting that '... some authors contend that there is only one species of Char in our islands ...' believed that there were 15 species in the British Isles - 7 of them in Scotland. '... they differ from each other in characters which are used to define species in other groups, and which may, therefore, be regarded as specific.' For many years his views were ignored, but the recent controversial paper by Maurice Kottelat has once again raised the whole issue. The Scottish species recognised by Tate Regan were:

Salvelinus gracillimus - Shetland Charr
Salvelinus inframundus - Orkney Charr
Salvelinus killinensis - Killin Charr (Haddy)
Salvelinus mallochi - Malloch's Charr
Salvelinus maxillaris - Large-mouthed Charr
Salvelinus struanensis - Struan Charr
Salvelinus willughbii - Willoughby's Charr

Size

Although they must have done so thousands of years ago, none of the populations of Arctic Charr in Scotland migrates to the sea, and so the adults are much smaller than in places where they do (i.e. northern Norway, Iceland, Greenland and Canada). The adults vary in size in different lochs, but are typically about 20-25 cm in length (85-170 gm in weight). In some lochs, however, they may regularly reach over 35 cm and 600 gm.

Some exceptionally large Charr have been recorded recently from the vicinity of floating fish farm cages in Scottish lochs, where their diet is boosted by the large amounts of waste food pellets available to them. For a number of years the British rod-caught record fish was a natural specimen from Loch Insh, then a new record was established for a fish weighing 1.502 kg, caught in 1985 by A Robertson in Loch Earn, where there are fish cages. The record then moved up for an obese fish weighing 2.18 kg, caught in 1987 in Loch Garry, Inverness, followed by a larger fish of 3.642 kg (8 lb 8 oz) caught in Loch Arkaig, Inverness. It continued to increase for fish caught near cage systems in charr lochs, and the largest so far recorded is for a fish of 19 lb caught in Loch Arkaig in 2000. The body responsible for confirming and recording records has now said that it will no longer consider records for fish from lochs where there are fish cages.

The Arctic Charr is the most northerly freshwater fish in the world, occuring all round the northern hemisphere; in the most northerly parts of its range it forms mixed seagoing and non-seagoing populations (comparable to Sea and Brown Trout), where the seagoing Charr can reach a large size. The world rod-caught record stands at 13.46 kg for a fish taken in a river in the North West Territories in Canada. Large freshwater Charr can also occur in large lakes in Scandinavia and in the alpine lakes of central Europe, where 'ferox' individuals have reached over 80 cm in length and 10 kg in weight.

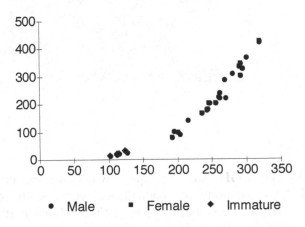

Length / weight relationship of Arctic Charr from Loch Killin (Lyle & Maitland 2003).

Formerly very large fish occurred in Loch Leven. Smith records in the *Old Statistical Account* that 'The gally-trough or charr abounds in the loch. What is remarkable of them is the size to which they often grow, some of them weighing near 2 lbs, and they are never known to rise to a fly or to be caught with a hook baited in any way whatever.' In recounting this story many years later, Robert Burns-Begg points out that 'The weight here stated by Mr Smith being the 'old pound', is equal to nearly three modern pounds.' This being the case, the Loch Leven Charr may have weighed up to 3 kg.

Appearance

In shape, Arctic Charr resemble slender Brown Trout, but their background colour and markings are quite distinctive. Like all charr (*Salvelinus*) this species does not possess any dark spots, but normally has yellow, cream, pink, red or orange spots on a background ranging from bluish grey to greenish brown. The lower half of the sides and the belly are variable in colour, ranging from dull pink to bright vermilion. The males are immediately recognisable by their more intense colouration in this area, while the females are sometimes drab; those from dark habitats may even look darkly silver - rather like mature Sea Trout finnock. The belly fins of the males are usually orange to red with a white leading edge, whereas those of females are pale yellowish. Like Brown Trout, Arctic Charr are subject to wide variation, so much so that, as noted above, the British populations were previously split into 15 different species. Though it is accepted that there are a number of distinct local races, all the populations are now considered to be one species - *Salvelinus alpinus*.

Scale from Arctic Charr.

Thomas Stoddart noted that 'The head and upper parts are of a greyish-brown colour, marked with whitish spots; the belly and lower fins pink, approaching to carmine.' He had evidently never seen this fish in October, when, as Herbert Maxwell records '... the char proclaims the approach of the spawning season by a brilliant flush overspreading

its underparts, varying in different lakes from brick red to flaming vermilion.'

The body is slim and streamlined with some lateral flattening. One of the main diagnostic features of the Arctic Charr is the presence of teeth on the head of the vomer bone on the roof of the mouth but none on the shaft of this bone. Some hyoid teeth are present, but not on the centre of the palate. The gill rakers are well-developed - as is appropriate to their commonly planktonic diet. The body is well covered with very small scales, which range in number along the lateral line from 190-240.

Young Arctic Charr are a silvery monochrome with 10-18 large grey parr marks along each side. Young and adults of this species can readily be distinguished from the young of American Brook Charr, the only other charr occurring in the British Isles, by the lack of the conspicuous dark vermiculate patterning in Arctic Charr which Brook Charr has along the back and sides as well as on the upper and lower lobes of the tail fin.

Distribution

Scotland is one of the strongholds of this species in Europe and, though declining, there are still probably about 200 populations, mainly in the north and west of the country. Some have undoubtedly still to be discovered. Arctic Charr have the reputation of being found only in large deep oligotrophic lakes lying in glaciated basins - which fits in with their Ice Age image. Though this is often the case (e.g. James Murray describes a small charr 'dredged from a depth of over 500 ft.'), there are also many populations living in shallow biologically rich habitats, as is commonly the case in Ireland. In Scotland, of 85 charr lochs investigated by the author, 43 (52%) had an average depth of 15 m and 7 (8%) less than 5 m. Arctic Charr are not limited in their distribution by having to live in particularly cold lakes and appear to thrive in much the same upper ranges as Brown Trout. There is, however, some experimental evidence that they are able to thrive in very cold conditions relatively better than other salmonids. The pelagic behaviour of Charr in lakes is probably due more to having to compete with the more aggressive Brown Trout and other competitor and predator species than to the need to find deep cool water.

There are no true river populations of Arctic Charr in Scotland, but they are found in streams connecting Charr lakes, and at times some fish swim or are washed over dams and weirs into flowing water at spawning time. In north-west Scotland though, there is one instance at least of young 0+ Charr being found in a stream flowing into the loch in which the parents had spawned.

Recorded distribution of Arctic Charr in Scotland. Note that some populations are extinct.

There are only a few situations where Charr have been intentionally re-distributed by humans. This is not perhaps surprising as to begin with it would be difficult to obtain and transport living Charr, and secondly there would be little material value in establishing them in new waters. Yet, in at least two waters in Scotland (e.g. Cruachan Reservoir) new charr populations have been unintentionally established by local hydro-electric developments. Prior to these, though, Mabie Moss noted in 1889: 'Shortly since I had the pleasure of receiving a fine specimen of the Char from a friend, who had caught it in Loch Dungeon with a fly when angling for trout. A few are caught each year in this way in the

loch named, as well as in Loch Doon. Another loch that holds char is Loch Grannoch. A few years ago char were introduced into Lochenbreck, but whether they have succeeded in establishing themselves therein I am ignorant. Can any reader tell me of any other locality for these interesting fish in the southern counties?' Harvie-Brown & Buckley noted that they were 'Introduced into the Mull lochs by Lochbuie', whilst Herbert Maxwell recorded that 'I remember having sent a hundred young char from a hatchery to be turned into Loch Ossian.' This was done in 1908, and the fish were established by 1911.

Ecology

Thomas Stoddart wrote in 1887 that: 'The most curious production of Loch Achilty is its char. This beautiful fish is indeed discovered in a number of our Scottish lakes, but nowhere have we found it so eager in its approaches towards the fly as in this loch. On a calm, warm day, the whole surface is alive with its bellings, which one would imagine proceeded from so many springs at the bottom. It rises to any colour and size of insect employed .... At table it is a perfect dainty, having a fine, delicate flavour, superior to that of any trout I ever tasted.'

Food

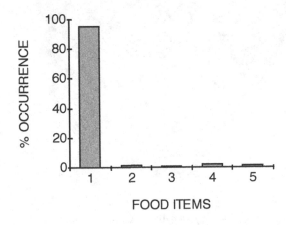

Food of pelagic Arctic Charr in Loch Rannoch (1: crustaceans (zooplankton); 2: stonefly larvae; 3: beetles; 4: midge larvae; 5: others) (Walker et al. 1988).

Arctic Charr have the reputation of living an exclusively plankton-feeding pelagic life in lochs: they are certainly well equipped to do so with their well developed gill rakers. These help Charr to feed rapidly on plankton where it is thick and can be taken in easily. This is in contrast to Trout, which have to use up much more energy in taking in each plankter individually. They are, however, by no means entirely pelagic.

Out of 15 Scottish Charr populations investigated a number of years ago, six were found at the time of sampling to be feeding exclusively on planktonic crustaceans, while in a further six lochs these food organisms dominated or co-dominated along with benthic organisms. In three, Charr were found to be feeding almost exclusively on benthic invertebrates, especially molluscs and caddis larvae. All the fish examined were adults.

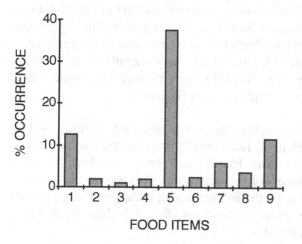

Food of benthic Arctic Charr in Loch Rannoch (1: molluscs; 2: worms; 3: crustaceans; 4: mayfly larvae; 5: caddis larvae; 6: beetles; 7: midge larvae; 8: fish; 9: others) (Walker et al. 1988).

Some Charr are piscivorous - in Loch Meallt Ronald Campbell found that they eat not only adult Three-spined Sticklebacks but also prey upon the egg clumps in stickleback nests. Here, these two species are the only fish present - one of the very few sites in the British Isles where Charr exist in the absence of Trout - but behave like them, feeding on the Sticklebacks and benthos and apparently ignoring the plankton there.

In Loch Rannoch, with its three, possibly four, races of Arctic Charr, small Charr have been found in the stomachs of the benthic group, a situation reminiscent of some charr populations elsewhere in Europe and in North America, where 'ferox' type Charr prey on populations of small stunted Charr. In some lochs, Charr eat their own eggs at spawning time. Virtually nothing is known about the food or

feeding habits of juvenile Charr in Scotland, but plankton is assumed to be a major part of their diet. Certainly this was found by the author to be the case in a few juvenile fish from Loch Killin.

Reproduction

Spawning normally takes place from late September to December, but in England there is also the well-known instance of the charr population in Windermere, one part of which spawns in the autumn, the other in the spring (February and March). So little research has been carried out on Charr in Scotland and other parts of the British Isles that there could well be other localities where there are both autumn and spring spawners. In Scottish lochs, spawning takes place over gravel and stones, normally in fairly shallow water near the shore or on a submerged reef, but in a few populations, spawning fish migrate up an inflowing river to spawn. The water temperatures at the time of spawning are usually about 6-9°C.

Andy Walker has made a special study of stream-spawning Arctic Charr. Traditionally the large Charr of Loch Insh ascend the Dunachton Burn and the River Tromie to spawn and for this reason, a special fish ladder was installed in a culvert on the Dunachton Burn when the A9 was improved and widened. The benthic race of Charr in Loch Rannoch spawn in the lower reaches of the River Gaur, a major tributary entering the west end of the loch. Charr also spawn in running water in the river-like section of Loch Coulin, above Loch Maree, where they clear silt from the gravels in mid-November and are easily seen spawning in identifiable redds. On one occasion several hundred were seen ascending the Coulin River from Loch Clair, presumably on their way upstream to spawn. Ripe Charr also ascend the relatively sluggish burn that connects Loch Clair to Loch Bharranch and ripe females are still evident there as late as December. There are also reliable reports that Charr from Loch na Sheallag, at the head of the River Gruinard in Wester Ross, spawn in a small tributary stream connecting to the main inflow.

Spawning in running water is probably more common than supposed, but rarely seen. In October 2005, the author and Andy Walker had the privilege of watching large Arctic Charr (from Loch Insh) spawning in the River Spey near Grantown.

The spawning behaviour is typically salmonid, whether in still or flowing water: the female clears the bottom of loose debris and excavates a redd at a selected site by turning on her side and sweeping her tail fin up and down. This is done within a territory that a mature male has established and is defending. In Wales, Nigel Milner found that 'Charr were found to be catholic in their selection of spawning areas. They spawned in lake sediments .. and also used artificial recent bed material (e.g. the slate spoil tip in Cwellyn and dredged channel sediments in Padarn). Redd structure was diffuse and variable in appearance with no evidence of the discrete structures typically formed by salmonids spawning in flowing water.'

Males usually mature at 2+ years of age, but occasionally at 1+ years, whereas the females are usually 3+ years old at first spawning. Data on fecundity from one population in a Scottish loch suggested that the relationship between the size (length or weight) of the female and the number of eggs produced was poor, but that a fish of 250 gm might be expected to produce 400-600 eggs. The eggs are amber in colour and comparatively large - some 4-5 mm in diameter.

Thomas Stoddart recorded in 1887 that 'The spawning of the Loch Achilty char seemed to me, in several instances, if not in all, a subterraneous operation, carried on among the roots of springs, and in the oozy and caverned outlet of its waters. The fish, I am credibly informed, have been caught repeatedly by means of a creel, during winter, in places where the effluent current, after finding its way some hundred yards under ground, emerged again into daylight, before discharging itself by other subterraneous channels into the Rasay or Black-water ...'

Predators

Little is known about the predators of Arctic Charr but the fact that, like Vendace and Powan, they spend much of their lives in deep water, means that they are not generally available to terrestrial piscivores like Herons and Otters. One known predator is Pike and it is certain that, in those lochs where both species occur, Pike prey on Arctic Charr during the spawning period. Another notable predator is the Ferox Trout; Arctic Charr is certainly a major component in its diet wherever these two fish are found together.

It appears now that benthic Charr may also be significant predators of young Charr in some habitats. Brown Trout and Eels take the eggs of spawning Charr - in Loch Doon, the author found that the stomachs of all the Brown Trout caught along with spawning Arctic Charr were full of charr eggs. Golden Eye ducks are also recorded as predators of charr eggs.

Charr seem particularly susceptible to predation by Pike, a recent species in many charr waters, and a number of charr populations may have become extinct following the introduction of this voracious species, which is still being thoughtlessly moved about by humans. Charr are also very sensitive to pollution, including both acidification, another factor that has led to their extinction in some waters - for example in Loch Grannoch .

An unusual situation was recorded by Andy Walker 'Fishing in Loch Meallt in Skye, the catch of Arctic Charr was found have been attacked by large numbers of leeches. As the fish were removed from the nets, the leeches accumulated in a large pile on the ground. Most of the fish had been badly damaged.'

Growth

Arctic Charr can grow rapidly during their first year of life, growth usually equalling or surpassing that of Trout in the same system. Growth is noticeably slower after this, however. The optimum temperature for Charr growth appears to be from 12-16°C, which is also the optimum for Brown Trout. Most Charr populations are dominated by fish of the 3, 4 and 5 year classes and the maximum age is usually 7-10 years.

In Loch Rannoch, where it has been shown by Andy Walker and colleagues that there are at least two races of Charr - one a benthic form, the other pelagic in habit - the growth of the pelagic form is fastest initially, but after the fourth year the benthic form starts to grow faster and most of the larger Charr in the loch belong to this group. These morphs show several differences in spawning site and behaviour as well as other features.

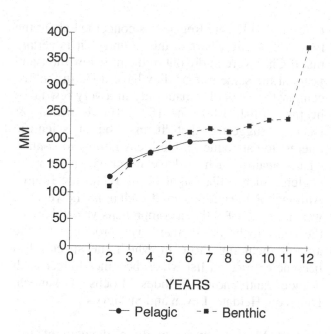

Growth of pelagic and benthic Arctic Charr in Loch Rannoch (Walker et al. 1988).

Value

Compared to many other parts of the northern hemisphere, Arctic Charr have been relatively little exploited by humans in Scotland. Peter Malloch noted in 1910 'Altogether it is a very beautiful fish. Little use is made of them in Scotland, although they are excellent for the table.' It is well known that, in the English Lake District, fish were caught and used as a basis for a small-scale local luxury food market. Daniel Defoe wrote that 'It is a curious fish, and as a dainty, is potted and sent far and near by way of present.'

Much less has been published about the situation in Scotland where, however, they were also important locally and indeed potted, as they were further south. Most fish were harvested at their spawning time, when they are most vulnerable, and sold locally. In Sutherland, Alan Joyce writes that Arctic Charr were netted regularly in the 1700s in the Badenloch area and salted away in barrels for the winter as what was called 'freshwater herring.'

The population in St Mary's Loch, now extinct, was important locally. MacFarlane records that 'In this loch are Trouts, Eels, Pearches, Pikes and a kind of Fishes, called by the Countrey people Red Waimbs, from the bloud red colour of their Belly. The fish itself is about the bigness of a herring with a forked

tail. The Herds about Michaelmass use to take great numbers of them in their blankets at a litle Rivulet that comes from the Loch of the Lowes into this, the two lochs being almost one and divided by a very small nick of ground.'

Pennant, in 1769, wrote of Loch Leven: 'The fish of this lake are pike, small perch, fine eels, and most excellent trouts, the best and the reddest I ever saw, the largest about six pounds in weight. The fishermen gave me an account of a species they call the Gally trout, which are only caught from October to January; are split, salted and dried for winter provision. By the description they certainly were our Charr, only of a larger size than any we have in England or Wales, some being two feet and a half long.'

William Thompson writes in 1856 of 'Loch Killin Charr ... So different indeed is this fish from the Charr of neighbouring localities, that it is believed by the people resident about Loch Killin to be a species peculiar to their lake, and hence bears another name - 'Haddy'. ... This fish is only taken when spawning, but then in great quantities, either with nets or a number of fish-hooks tied together, with their points directed different ways. Those, unbaited, are drawn through the water where the fish are congregated in such numbers that they are brought up impaled on the hooks.'

Charles St John, travelling in Sutherland in the 1880s, recorded that 'There is also in many of these lakes plenty of char, a fish of mysterious habits, never or seldom taking the fly or any other bait, but at a certain season (about the middle of October, as far as my experience goes) migrating in great shoals from the deepest recesses of the lake, where they spend the rest of the year, to the shallows near the shore. During this short migration they are caught in nets, and frequently in great numbers.'

Service in 1907, discussing initially Pike and Perch in southwest Scotland, recalled that 'These came from the deeper depths of the lochs they frequented to the shallows for spawning purposes at certain seasons. And so also did the vendace and more especially the charrs of the deep Galloway lochs, which could only be secured in sufficient quantity for potting purposes when they came to the margins of gravel in the autumn months.' As in Cumbria, potted charr were important in this part of Scotland.

Few anglers have bothered to fish for Charr until recently, but due to the pressures on other species, Charr are becoming increasingly regarded as an attractive alternative quarry, particularly since the British Record (Rod-caught) Fish Committee has accepted Charr as a trophy fish. It is likely therefore that in the future, the presence of Charr may be a significant attraction to angling tourists in northern Scotland. Some years ago unsuccessful attempts were made in Scotland to establish a system of cropping the Charr in some large lochs, as is commonly done in Norway, as a crofter industry in a region where other natural resources are few.

Herbert Maxwell noted that 'British char do not take high rank as a sporting fish owing to their chronically abysmal habits; nevertheless they are highly prized by sportsmen on account of their beauty and mystery. To put it colloquially, they have all the makings of a game fish, but their mode of life makes them difficult of access.' However, Arctic Charr are caught regularly in some lochs, for example Loch Lee in Angus, where the author has made a particular study of catches. Robin Ade notes that 'The char fishing in Loch Doon has long been regarded as some of the best in Britain, for unlike those from other large char waters these fish can be caught all season on fly from the shore.'

Rarely thought of as an aquarium or pond fish, the Arctic Charr, as well as being spectacularly beautiful, is a peaceful fish and, in spite of frequently living in habitats which are extremely remote from humans, can be tamed rapidly and may be taking food from the hand within a few weeks of captivity. In contrast to Trout, they do not easily panic in a tank or pond (where they will co-exist with Goldfish and other pond species) but swim calmly around in shoals. The author has persuaded them to spawn in quite a small pond by giving them a large tray of coarse gravel in which to build a redd.

Conservation

In 1873, Frank Buckland wrote that 'It is proposed, under the new Salmon Bill of 1873, to include charr under the definition of salmon, so that this fish for the future, I trust, be protected, especially in spawning time, by the provisions of the Salmon Acts.' This never seems to have happened.

Later, in 1911, Tate Regan was concerned about the future of Arctic Charr in this country: 'it is certain that if Char were to die out in the next few thousand years at the same rate as they have done in the last century they would remain only in a very few lakes in the British Isles.' In 1887, Harvie-Brown & Buckley suggested that 'It may be of scientific interest to catalogue all the lochs known to contain Charr, against such as have never yielded any to anglers. Such a list might be useful in the future.' Although R.P. Hardie started such a list, however, it was not until 1970 that a comprehensive database of the charr lochs of Scotland was produced by the author, including those in which the species has become extinct - a list which becomes longer each decade and now includes Lochs Grannoch, Dungeon, Heldale, Leven and St Mary's.

Herbert Maxwell was clearly a devotee of this species '... one of the most conspicuous survivals of the reign of ice in our islands; a slowly vanishing race, still maintaining existence in a few scattered and profound meres and lakes. ... their strict segregation for thousands of years, prohibiting all admixture of different colonies, are quite enough to account for some variation in colour, form, and even structure among fish of a single species inhabiting widely separate sheets of water.' In 1919, he was one of the first people to point out that Arctic Charr in Scotland needed protection: 'As the close season for char on Windermere is the same as for trout - from 15 September till 10th March - the fish of this lake enjoy a salutary protection which is withheld from Scottish char. These are netted, sad to say, when they enter the shallows to spawn, and the best that can be said for so barbarous a practice is that it is a beautiful sight on a bright October morning when the net is drawn ashore and the fish are seen flashing their ruby sides in the clear water.'

In spite of the fact that Scotland has considerable responsibility for the conservation of this fish, because it is a stronghold for this species in Europe, Arctic Charr are given no special protection in Scotland. Over the last few years there has been a distinct increase in interest in this species from anglers, commercial interests and from fish farmers. It is important that the resource is now looked at closely in relation to these new developments, so that there is no danger to the stocks, and that any utilisation which takes place is carried out in a sustainable fashion. Apart from the pressures from land usage, acidification and overfishing which have caused extinctions in the past, this amazing fish is now facing a new threat - charr farming.

In the last decade of the 20th Century several stocks of Arctic Charr were imported from North America: some of these have been reared in Scottish charr lochs with inevitable escapes and inbreeding with native fish. In a letter to the author from the Scottish Executive, dated 17 April, 2000, it was noted that 'Details of imports of Arctic Charr are commercially confidential ... However, ... three Scottish companies have imported charr eggs for farming in the last few years - two imports last June and one last month. 120,000 eggs came from a company in Canada last June and a further 60,000 eggs have recently come from another company in Iceland.' Further enquiries revealed that '... two imports were made in 1994 for farming purposes (totalling 125,000 eggs) and an import of eggs was made in 1997 for research purposes.' The fate of these eggs is not known. This means that our charr stocks, isolated for 10,000 years since the last Ice Age, during which each stock has developed its own genetic integrity, are now faced, not only with the usual threats from fish farming (disease, parasites, over-enrichment of lochs) but also with the loss of their unique gene pools. The recent studies by Colin Adams and Gavin Alexander have clearly shown that there there are intriguing differences between different populations of Arctic Charr in Scottish lochs (Loch Rannoch has three distinct races) and that this is indeed a species which is undergoing evolution on our own doorstep.

In spite of frequent pleas to the former Scottish Office, no action has been taken to solve this problem, and the situation remains the same with the new Scottish Executive, stocks of alien Arctic Charr from both North America and Iceland having been introduced since the Scottish Parliament started in May, 1999.

The following points summarise conservation measures which require to be implemented to prevent the decline of this valuable fish.

(a) Legislation to give protection to this species which is declining in Scotland.
(b) A complete ban on the introduction of alien stocks from abroad.
(c) Control by licence of the movement of charr stocks from one water to another.
(d) Monitoring of charr stocks using a five-year rolling programme.
(e) Further research on the distribution and status of charr populations in Scotland.

References

Formerly a rather neglected species, there have been several very valuable studies of the Arctic Charr in Scotland in recent years.

Ade (1985), Alexander & Adams (2000), Andrews & Lear (1976), Barbour (1984), Barbour & Einarsson (1987), Baroudy & Elliott (1994), Buckland (1873), Burns-Begg (1874), Campbell (1979a), Campbell (1982, 1984), Chalmers (1931, 1938), Frost (1951, 1965, 1977), Frost & Kipling (1980), Gardner et al. (1988), Greer (1995), Hardie (1940), Harvie-Brown & Buckley (1887), Hartley et al. (1992), Kottelat (1997), Lyle & Maitland (1992, 1998, 2003), Maitland (1983, 1984, 1985, 1987, 1990, 1992, 1995, 1998, 1999), Maitland et al. (1984, 1991), Malloch (1910), Maxwell (1904, 1919), MacFarlane (1748), Moss (1889), Murray (1910), Nilsson (1967), Partington & Mills (1988), Pennant (1761), Regan (1911), Robson (1986), Service (1907), Smith (1793), St John (1845), Stoddart (1867), Thompson (1856), Walker et al. (1988), Went (1971).

'To put it colloquially, they have all the makings of a game fish, but their mode of life makes them difficult of access.'

Herbert Maxwell (1919) *Memories of the Months, Sixth Series*

*****

Brook Charr *Salvelinus fontinalis* Tarragan sruthan

'He is faultless ... he is the favourite among fishes and deserves to be so.'

Livingstone Stone (1877) *Domesticated Trout: How to Breed and Grow Them*

FOR a long time after its introduction to Scotland, towards the end of the 19th Century, this colourful salmonid from North America was called the American Brook Trout, or sometimes by its specific name *fontinalis*. It is really a charr, however, and this fact has become increasingly recognised in recent years by the use of the common name Brook Charr, suggested by the author in 1972. It is still known by a variety of names in North America, mostly misleading,including Brook Trout, Aurora Trout, Brookie, Fontinalis, Speckled Trout, Mud Trout. Interestingly, although there is no Gaelic name for this species in Scotland, it was recognised as a salmonid by emigrant Gaels in Canada and there called breac (= trout). A literal translation of the scientific name is 'Little Stream Salmon'.

Though popular, the optimism of those making the introductions to Scotland was ill-founded, and well summarised by Herbert Maxwell in 1919: 'The beautiful brook trout of North America (*Salvelinus fontinalis*) is another true char, and bright were the hopes of British trout-fishers when it was introduced to this country in the last quarter of the nineteenth century. It was reared in hatcheries by hundreds of thousands, and distributed so widely that probably there is not a county in Great Britain where some attempt has not been made to establish so desirable a species. But all has proved in vain. ... these dainty aliens have refused all offers of naturalisation. They invariably disappear in a mysterious manner in the second or third year after being turned loose.' The Brook Charr is still an uncommon fish in Scotland.

Size

Brook Charr can grow well in Scotland; they are able to reach weights of 340-450 gm in three years, even in poor waters, as long as the stocking density is not too high. They seem particularly suitable for stocking lochs or ponds that are not suitable for Brown or Rainbow Trout, thriving in small shallow sheltered muddy or peaty ponds. Wild Brook Charr can grow up to 45 cm, and the British rod-caught record stood at 1.1 kg until specially fattened fish were available and placed in 'put and take' fisheries. The present British rod-caught record for a natural wild fish is vacant; the record for a cultivated specimen stands at 3.713 kg (8 lb 3 oz) for a fish caught in 1998 at Fontburn Reservoir in Northumberland. In North America, the record stands at 6.57 kg for a fish taken in the Nipigon River in Ontario. There appears to be no Scottish rod-caught record.

Appearance

The Brook Charr is typically salmonid in shape but more thick-set than the Arctic Charr or even the

Brown Trout. It can be easily distinguished from the former by the alternate light and dark wavy lines - vermiculations - forming a marbled pattern on its back and on the dorsal and tail fins. It has many more cream to greenish-yellow spots along its flanks than the Arctic Charr: these spots mix with red ones encircled by blue haloes.

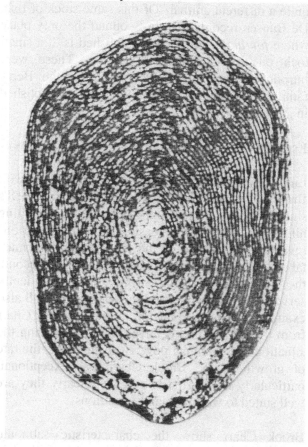

Scale from Brook Charr.

The Brook Charr is a really beautiful fish; Niall Campbell knew hardened anglers who were jubilant after landing a large brilliant male of the species, and quite unable to bring themselves to deliver the *coup de grace*! The belly fins of mature male Brook Charr are bright red-orange with white leading edges, backed by a narrow black strip. The lower flank of the fish is also a bright red-orange grading into a sooty belly. The male also develops a pronounced kype. The females are much paler, but very delicately marked. Sea-running Brook Charr (the 'Sea Trout' of eastern Canada) are overall silvery, but with some red spots showing through the silver. Young Brook Charr are greenish, well vermiculated and are not likely to be confused with any other young salmonids in Scotland.

Brook Charr have well-developed teeth on the upper and lower jaws, but elsewhere only on the head of the vomer bone on the roof of the mouth. The streamlined body is well covered in small cycloid scales of which there are some 122-240 along the lateral line. The flesh colour ranges from white to a deep pink.

Distribution

A native of eastern Canada and the east coast of the USA as far south as the State of Georgia, it is found in river systems draining to the Atlantic Ocean. It has now, however, been widely re-distributed throughout temperate North America, and to many other parts of the world also. It was introduced to the British Isles in 1868, some time before its compatriot the Rainbow Trout, as well as to most of Europe, including Scandinavia, before the end of the last century. As Hugh MacCrimmon has shown, there are now established populations in Europe, Asia, Australasia, southern Africa and South America.

In its native environment there are anadromous and non-anadromous races of the Brook Charr, and not long after it was introduced to Scotland there were reports of specimens being caught in bays and estuaries around the coast. There have been no such reports in recent times, however, probably because it is no longer so widely distributed, its popularity as a sport fish apparently declining in the face of competition from the more dashing Rainbow Trout. Gradually though, more self-propagating populations of Brook Charr have become established, and although some of these populations have easy access to the sea, there is no evidence that there is any anadromous tendency nowadays.

Although there are more established populations of Brook Charr in Scotland than of Rainbow Trout (which has never managed to establish itself permanently), it cannot be claimed that the former species naturalises successfully in this country. Of the eight established populations known to the author, five are in lochs with no stream spawning grounds, the remaining three are in lochs where the only streams are small muddy ones. None of these lochs contain Brown Trout, although one of them does have Arctic Charr and some non-predatory coarse fish. It appears that Brook Charr can maintain populations better than Brown Trout in lochs with poor spawning facilities, and also that they are more likely to become established in the absence of Trout and other fish (although there are, of course situations where the two species live together).

A variety of writers have recorded introductions to Scotland. In *The Dumfries Courier* of 30 July 1889, Mabie Moss wrote: 'An American species of char, the brook trout - perhaps better known under its scientific name of *Salmo fontinalis* - has been widely introduced in this country. It has been put into several waters in Dumfriesshire and Galloway, and if it had succeeded we ought to be hearing of an occasional capture. It would be of considerable importance to have any captures recorded.' Elsewhere in Scotland, Harvey-Brown and Buckley recorded that 'These have been introduced by Mr M'Fadyen into the lochs of the Cuifail district, over many years. They have also been placed in the Lochbuie lochs in Mull. The MacLaine writes they "have done better here than (so far as I can learn) any other part of the United Kingdom"'.

Recorded distribution of Brook Charr in Scotland.

John Berry, writing to the author about one of the Tayfield ponds where Brook Charr had been successfully introduced. 'In my boyhood it was so well established in the pond and the little burn which feeds it ... that I assumed that this was our native trout, and I remember being quite surprised when I found that the common Brown Trout was quite a different animal.' Of this same stock of fish, De Bunsen records that 'In Scotland the only place where *fontinalis* are firmly established is ... a small loch on the Ochtertyre Estates ... These were introduced from the stock owned by the Berry family at Tayfield ... and are now firmly established in ... St Serf's Water ...'

Ecology

Little biological information is available on wild Brook Charr in Scotland except Niall Campbell's studies of two self-sustaining populations inhabiting hill lochs on the west coast. There, investigations revealed that in one of the lochs, spawning must take place within the loch itself while in the second, the inflowing and outflowing streams evidently serve as both spawning and nursery areas. Niall also examined the age and rate of growth of Brook Charr from these lochs and believed that, considering the situation and chemical poverty of the lochs, the rate of growth of the Brook Charr was exceptional, particularly in the first two years. Clearly they are well suited to west highland conditions.

Brook Charr show the characteristic salmonid quality of great plasticity of form and behaviour in relation to environment. They can flourish as populations of tiny individuals in small streams, or behave like Atlantic Salmon or Sea Trout, migrating downstream to large lakes or coastal waters to feed, reaching a size of several kilograms before returning to fresh water to spawn.

Edward Hewitt commented that 'The brook trout seems to have acquired the habit of getting behind obstacles in the current, and you will often catch him behind a stone of log, or some other obstruction.' Like all salmonids they require cool, well-oxygenated water, though they can survive up to 25°C. However, they are normally found well below 20°C, preferring around 13-16°C. Edward Hewitt believed that, in the wild in North America 'Brook trout can stand a water temperature up to seventy-two degrees, if continued for any length of time he dies.' They show peak activity and consumption at just above 13°C.

The hardiness of this species is indicated by this comment by Francis Day: 'The MacLaine of Lochbuie has acclimatised this fish in a moor loch about a thousand feet [304 m] above the sea, near Loch Uisk , in Mull: in 1884 one was captured 2 1/2 lbs. [1 kg] in weight, and they are said to have attained 5 lbs. [2 kg.].' Another example is the population in Loch an Eireannaich, which sits at the head of Kirkton Glen at a height of 600 m, almost on the Teith/Tay watershed, and is obviously subject to fairly extreme conditions in winter. This population has existed for many decades and was found to be still thriving when visited by the author some years ago.

Food

Their feeding habits and behaviour are similar in many ways to those of Brown Trout, and like Trout, they can locate their food even in waters of very poor visibility. In one shallow peaty loch, where visibility was only a few centimetres, Niall Campbell found that Brook Charr and Brown Trout both feed almost exclusively on the bottom on the water slater *Asellus*.

In one of the two populations inhabiting hill lochs on the west coast, Niall observed small shoals of 0+ Brook Charr cruising swiftly over the littoral area taking surface insects. A feature of this loch was the abundance of frog tadpoles (which were in an advanced state of metamorphosis) and newly metamorphosed frogs. The stomach contents of adult Brook Charr consisted almost exclusively of tadpoles and small frogs, along with a few caddis pupae. By contrast, no tadpoles were ever found in the stomachs of Brown Trout in lochs in northern Scotland during Niall's other studies covering a span of 18 years - although adult frogs were not uncommonly found in the trout stomachs each spring.

In the second loch, where the average size of the Brook Charr was much less, and the density of fish much higher, the stomachs contained benthic invertebrates, ranging from the pea cockle, *Pisidium*, to emerging midge pupae. Adult terrestrial insects were also found. In one stomach the remains of a 0+ Brook Charr, about 60 mm in

length was found. Again, by contrast, small Brown Trout are very rarely found in the stomachs of normal loch-living Brown Trout.

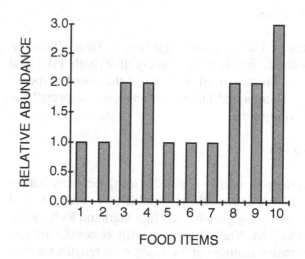

Food of Brook Charr in a loch on Mull (1: molluscs; 2: bugs; 3: caddis larvae; 4: midge pupae; 5: fish; 6: dragonflies; 7: ichneumonids; 8: beetles; 9: flies; 10: other insects) (Maitland & Campbell 1992).

Reproduction

Brook Charr spawn about the same time as most populations of Arctic Charr, beginning often at the end of September. As with many other salmonids, an area of clean gravel is selected for redd cutting; this can be in a loch (often where a spring is upwelling) or in a stream into which the fish have migrated just previously. However, many populations in this country seem to have to use quite muddy streams for spawning.

At spawning time, both sexes become very aggressive towards intruding rivals or potential egg predators. The eggs are quite large, between 3.5-5.0 mm in diameter and yellowish orange in colour. The fecundity of females varies greatly, ranging from just 100 eggs for a small female 14.5 cm in length to 500 for a large female of 55.8 cm. At 4.4°C the eggs take about 109 days to hatch, and at the same temperature the fry start to emerge from the gravel and begin feeding after a further 60 days. Scales start to appear on the fry when they reach a length of some 5 cm.

Edward Hewitt noted that, in North America, 'Brook trout prefer to spawn in small spring brooks on sand or fine gravel in fairly shallow water. This gives the fry a good chance for development when they are hatched, as the larger fish leave these small streams and spend the winter in deeper water.'

Predators

Little is known about predators of Brook Trout in Scotland, but is seems likely that both Pike and Ferox Trout as well as some of the 'usual suspects' (e.g. Herons and Otters) are involved, especially at spawning time.

Growth

Maturity is usually reached after three years but, as with other salmonids, is variable and to some extent dependent on growth rate. In a highland loch lightly stocked by Niall Campbell with Brook Charr fry, the males matured at 1+ years with lengths ranging from 18 to 21 cm., indicating a relatively fast rate of growth.

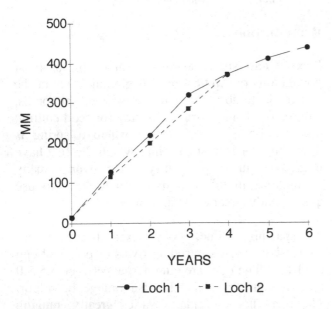

Growth of Brook Charr in two lochs on Mull (Maitland & Campbell 1992).

They are reputed to be a short-lived fish, similar to Rainbow Trout, seldom living longer than 5 years in their native habitat and apparently never beyond 8 years of age. Exceptional populations do occur, however, and in one high altitude lake in California into which the species had been introduced, which has ice cover for 8 months of the year, individuals of 24+ years were found.

Value

There is likely always to be a limited interest from anglers in Brook Charr; most of the breeding populations seem to survive partly because of their remoteness and absence of competition from Brown Trout. This is a very vulnerable species - populations could well disappear as a result of the introduction of Brown Trout, or of predatory fish such as Pike or Perch. Although not a native fish they are an interesting and attractive species, popular with some anglers, and - judging by their lack of success so far - unlikely to do any harm to our native fish fauna.

They are available from some fish farms and are stocked in limited numbers in some Scottish waters - usually stillwater fisheries or small club waters. Only rarely do they manage to breed successfully and stocks rarely last long.

Certainly anglers in the past held the Brook Charr in high repute. Viscount Grey of Fallodon wrote in 1899: 'Both the *fontinalis* and the rainbow trout are such handsome fish and have such sporting qualities that I hope efforts to establish them will be continued for some time in all sorts of water.'

J.J. Armistead took Francis Ward, a doubtful angler, fishing for *fontinalis* at the Solway fishery. Ward's account later noted '... I had more than a dozen landed in half an hour, all of which required playing, twice having two fish on at once, and I may say I never saw anything like it in my life.'

Brook Charr can also be attractive as pond fish. Tom Stuart successfully kept a small population of this species in a pond in his garden at Pitlochry and spent many hours watching their behaviour there. They were also successful for a time in Loch Dunmore near Faskally.

In his manuscript account of the species, De Bunsen recorded: '... Philip Boase ... lived at St Andrews after the First World War, and was fortunate enough to possess sufficient means to satisfy an interest in trout by building a pond in his garden fitted with an observation chamber below ground level. Here he kept *fontinalis*, as well as brown and rainbow tout, each of which was known by name. Their diet of boiled salt-water mussels suited them so well that one grew from three-quarters of a pound to four pounds between March and November.' John Berry remembers seeing one of these fish 'which had attained a weight of 14 pounds.'

Many years later, John Murray, who initiated the popular ponds for fish in the entrance hall of the Royal Scottish Museum, successfully kept Brook Charr there for several years. They became so tame that one of them once leapt out of the water and bit a visitor! At the time of writing (2006) it seems, unfortunately, that these ponds are likely to be removed.

Niall Campbell recalled the tame Brook Charr reared in an aquarium at the Freshwater Fisheries Laboratory at Faskally. It was called Charlie and jumped up out of the water for strips of liver held above. The author has successfully kept them in large aquaria, where they made attractive (and aggressive!) residents. They are easily kept, as long as live, or fresh, food is available.

Conservation

Since the species is not native to Scotland and is not notably threatened within its natural area of distribution, no special conservation measures are required here.

References

Relatively little research has been carried out on Brook Charr in Scotland and most available information is from North America.

Armistead (1895), Bridges & Mullen (1958), Day (1887), De Bunsen (1962), Grey (1899), Harvie-Brown & Buckley (1892), Hewitt (1926), Lever (1977), MacCrimmon & Campbell (1969), Maitland (1972), Maitland & Campbell (1992), Maxwell (1919), Moss (1889), Reimers (1979), Robinson et al. (1976), Scott & Crossman (1973), Stone (1877), Walker (1976), Ward (1911).

'But all has proved in vain. ...
these dainty aliens have refused all offers of naturalisation.
They invariably disappear in a mysterious manner
in the second or third year after being turned loose.'

Herbert Maxwell (1919) *Memories of the Months: Sixth Series*

*****

Grayling *Thymallus thymallus* Glasag

*'Some think that he feeds on water-thyme,
and smells of it at his first taking out of the water...'*

Izaak Walton (1653) *The Compleat Angler*

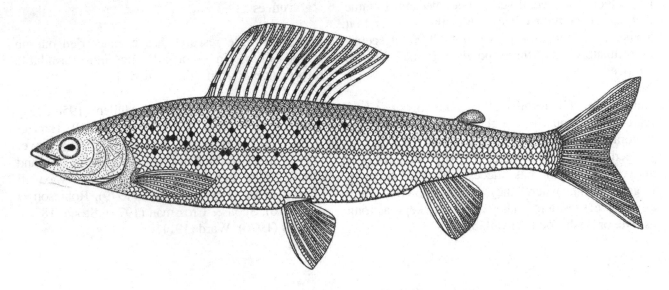

THE Grayling has an unusual position among freshwater fish in Scotland. Though generally regarded ecologically as a northern fish (its nearest relative is the Arctic Grayling *Thymallus arcticus* (Pallas) of arctic North America and Siberia), it is not a native of Scotland. Its successful introduction in the 19th century, and subsequent establishment in many rivers, has made it a favourite among some anglers, but a nuisance to others - mainly the owners of salmon fisheries. Many 'game' fishermen regard the Grayling as a 'coarse' species, whilst many coarse fishermen (and almost all grayling anglers) regard it as a game species, on a par with Brown Trout and Atlantic Salmon. Like Arctic Charr, the Grayling is a very attractive fish with many devotees. The situation was aptly described by Aflalo in 1897: 'The Grayling is an elegant fish, on the sporting qualities of which there is much difference of opinion.' Johnson mentions another enthusiast for this species: 'St Ambrose, who, with a grateful heart, called the sweet-scented grayling "the Flower of the Fishes".'

Because it is an introduced species in Scotland, there are few local traditions associated with it - for example its Gaelic name is relatively new. Local names for the Grayling in England are Umber and Umber Fish. A literal translation of the scientific name is 'The Thyme Fish'.

Size

The average adult length of Grayling is about 30 cm, with a weight of some 350 gm. However, in favourable circumstances this fish can grow up to 50 cm in length (1 kg), and there have been very large fish of about 60 cm (2-3 kg) reported. The Scottish rod-caught record is for a fish of 1.404 kg (3 lb 1 oz 8 dm) caught in the River Tweed in 1994 by J. O'Hara. The present British rod-caught record stands at 1.899 kg (4 lb 3 oz), for a fish caught in 1989 by S.R. Lanigan in the River Frome, Dorset.

Appearance

The Grayling is clearly a member of the salmonid group and is a graceful fish with a long streamlined body, spectacular dorsal and well-developed adipose fin. The mouth is small and slightly downturned, with small fine teeth present on the upper jaw and on the vomer bone on the roof of the mouth. The head is rather pointed and the slim body is laterally compressed. The pupil of the eye sometimes appears deep blue and is pear-shaped, quite different from any other fish. Apart from the smell of its flesh, the characteristic feature of the Grayling is the very large dorsal fin, which is both long and high, and is particularly large in the male. The body is well covered by medium-sized scales which have a characteristic shape with convoluted anterior edges, and are arranged in very obvious longitudinal rows along the body. There are 74-96 of these scales along the lateral line. The scales can be used for age determination.

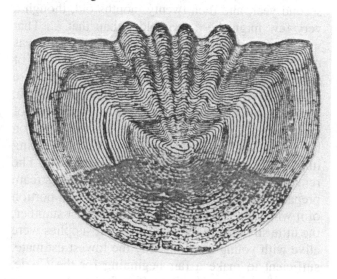

Scale from Grayling.

In colour, the Grayling has a rather silvery appearance; it is dull grey on the head and along the back, grading to a greyish green on the sides and silvery white on the belly. A number of clearly defined large black spots occur irregularly on the anterior part of the body - these spots are very variable within the species but always remain the same on a particular fish, so that experienced observers can soon identify individual fish within a river, once the detailed pattern of each has been recognised.

The sides of the body have a number of faint longitudinal violet stripes which darken at spawning time to give a purplish iridescence to the fish. The dorsal fin is well marked with 4-5 rows of reddish black spots on the fin membrane, which are particularly accentuated at spawning time. The free edge of the dorsal fin also develops a reddish margin at this time. The young fish, silvery as early post-alevins, soon develop typical salmonid parr marks along the sides, which disappear after one year. The characteristic Grayling dorsal fin is already evident in fry of around 2.5 cm.

H.C. Pennell is most enthusiastic about the appearance of this fish: 'The Grayling is certainly one of the most beautiful fish that can be imagined. ... the back is of a deep purple colour, with small dark irregular spots on the sides; the stomach is brilliantly white, with a fringe or lacing of gold; and the tail, pectoral and ventral fins, are of a rich purplish tint. The dorsal fin is very large ... and is covered with scarlet spots and wavy lines, brown or purple, and the whole body is shot with violet, copper, and blue reflections ...'

Johnson was equally ecstatic: '... the little shoal of grayling floating above the red-brown gravel. The elegant formation of their bodies, the finely modelled heads, the silvery blue of the scales, the rainbow hues of the great dorsal fins, gently waving, make a combination beautiful to look at. Presently a big fellow come sailing by ... A purple sheen infuses his steely flanks, and the softly fanning fins betray delicate tones of olive-green, yellow and red.'

Edward Boulenger, writing in 1946 agrees: 'The Grayling ... is one of, if not the most beautiful of all our fresh-water fishes. ... Its name *thymallus*, is derived from the fact that freshly-caught specimens have a faint odour of wild thyme.' Izaak Walton agrees: 'Some think that he feeds on water-thyme, and smells of it at his first taking out of the water; and they may think so as with good reason as we do that our Smelts smell of violets at their first being caught, which I think is the truth.' However, William Houghton noted: 'Mr Cholmondeley Pennell thinks that the odour rather resembles that of cucumber than of thyme.'

Distribution

The Grayling is a native of north-west Europe, including Great Britain, occuring in suitable habitats from the Arctic Ocean in the north to southern France in the south and from Wales in the west almost to the Black Sea in the east. It is absent from a number of European countries including Ireland, Iceland, Spain, Portugal and Italy. It has also disappeared from several river systems in which it was formerly abundant, for it is extremely sensitive to pollution of various kinds. In Britain, its early post-glacial distribution was limited to the river systems of eastern and south-eastern England, but now it is common over much of England and

Wales, as well as southern Scotland where it was introduced successfully during the Nineteenth Century. It now occurs as far north as the Clyde on the west coast and the Tay on the east coast. Francis Francis noted that 'there are many other rivers ... where it could be naturalised.' and to this day others agree with him. Although absent from Ireland, a clandestine attempt to introduce it there was thwarted by the quick action of the authorities.

Although it is not native to Scotland, it is certainly a fish which one might have expected to find here. The Naturalists Library in 1843 commented that 'We believe that the Grayling has not been found in Scotland, and certainly this hiatus in its distribution is not a little singular, the more especially, as being an Alpine fish, naturally fond of cool water, and abounding in much more northern countries on the Continent, the Highland rivers seem peculiarly adapted for it.'

Recorded distribution of Grayling in Scotland.

Writing in the Dumfries Courier on 30 July, 1889, 'Mabie Moss': noted that 'The facts connected with the introduction of grayling into our waters are involved in considerable obscurity, owing to insufficient records of times, places, and persons.' He had clearly not read *The Scotsman* of July, 1858: 'INTRODUCTION OF THE GRAYLING INTO SCOTLAND. The scheme for the introduction of the grayling into Scotland commenced in 1855, on the 5th December that year. Through the kind assistance of our Derby friends, three dozen healthy fish were conveyed from Rowsley to Abington, and committed to the Clyde. I judge, from remembrance of their size, that most of them were fish in their second year, and then twenty months old, though a very few might be a year older than that. ... That was the first branch of the experiment, and it was much desired to reinforce it by the importation of impregnated ova. In the spring of 1857 there were brought to Abington about 20,000 impregnated ova, of which about 2000 were given to Mr Shaw, of Drumlanrig, to make a beginning in the Nith, on behalf of Lord John Scott, who ... has this spring imported breeding fish to further it in the Nith. The remaining 18,000 ova were placed in a small stream prepared for their reception by shutting of a portion of it with zinc gauze sluices, ... During last summer, the little stream and the pond which it supplies were alive with young grayling .. by the lowest estimate, sufficient to make a fair beginning for the Clyde, and perhaps even to give off a few to friends in other districts.'

In *The Field* of March 10, 1860 there is reference to 45 brace of large breeding Grayling being transported from Yorkshire to Drumlanrig on the Nith, using a zinc container with an aerating system.

John Robertson continued the story in 1861: 'These fish are gradually spreading themselves throughout the mid and upper waters of the river and its tributaries; and many were caught in 1860 of good size and quality.'

In a letter to the author in 1985, Bill McDavid, a professional salmon netsman on the Solway commented that: 'On the Big Water of Fleet – just down from Upper Rusko Cottage – there is a beautiful pool – map reference 624 565. Carved on a rock on the west bank is "Greyling put into the Fleet 22' September 1870"

Other rivers were stocked in subsequent years and their history was reviewed by Willie Miller in 1987. 'There are conflicting versions of the stocking of the River Ayr. One gives credit to Lord Ailsa. Another states that the first batch was sent by a Mr Hunter of Coltness, in which case they would probably have been from the Clyde.... The fate of the small number sent to the Tay system in 1858 is not known. The fact that another batch was introduced about 20 years later [by Lord Breadalbane] would suggest that the original attempt must have failed. What is certain, however, is that Scotland's largest river system has provided ideal conditions for "her ladyship".... As a safeguard, a batch of grayling was sent to Matthew Brown of Greenock [in 1857] who was asked to keep them safe in a Renfrewshire pond and stream..... Could this have been the nucleus of the stock in the River Gryffe? ...The Tweed system also holds a good stock. The nucleus of this gained entry to a tributary, the Teviot, when a pond overflowed in 1860. From there they spread and in 1888 it was noted that there had been a remarkable increase in Tweed itself ... One other major Scottish river, the Annan, has contained grayling for a considerable time ... My guess would be that they came from the neighbouring Nith or the Clyde. ... In recent years ... grayling have been introduced into at least two tributaries of the Forth [Almond and Esk]'

Charles Hall writing in 1915 noted that 'Greyling. Common in River Gryffe and tributaries. Introduced many years ago into the Clyde, and has since multiplied greatly and migrated to all tributaries.'

Grayling are also now thriving in the City of Edinburgh, having been introduced successfully to the Water of Leith. In recent years too they have been introduced to lochs in the Tay catchment by Andy Walker, but the fate of these stocks is unknown.

As far a the Tweed is concerned, in *The Field* of 24th May 1862, there was a short letter to the effect that a Grayling of 'a third of a pound' was caught 'at the head of the Turnpool' by Mr. T. Todd Stoddart on 13th May. The letter also stated that the Marquis of Lothian had placed grayling in the Teviot "several years ago". On 26th June 1862, a letter from 'G.A.'. (George Anderson was the man who organised the introduction of grayling into the Clyde in 1855 and 1857. He later became M.P. for Glasgow.) confirms that T. Todd Stoddart took and killed a grayling and goes on to say that he (G.A.) was 'sorry to have to confirm the fact that Mr Stoddart was not ashamed to put it in his creel instead of immediately returning it to the stream. ...

A gentleman of such note as an angler and angling writer must have known quite well that May was not a fair month for the killing of grayling."

G.A. goes on to say that 'About the First of April this year the Earl of Shrewsbury sent from England to the Marquis of Lothian at Monteviot twelve grayling. They were put into a pond on a stream tributary to the Teviot and ... an unfortunate flood caused such overflow that they all escaped into the Teviot and some days after that Mr. Stoddart caught a grayling about six miles down the river. ... The Marquis of Lothian "intends making a further attempt next year".

George Bolam also describes the introduction and spread of Grayling in the Tweed, and the subsequent extensive attempts to eradicate it. --- 'In about ten days' time two anglers, fishing in the Teviot at Roxburgh, killed just over 300 Grayling. Another member of the Kelso Angling Association, in the course of two weeks during the month of March, killed about the same number. Two fishermen angling on Tweed and Teviot in the months of December, January and February, caught 500 Grayling between them; while three rods fishing on the Tweed at Sharpitlaw Mill, below Kelso, during six days in January, accounted for 123 of the fish. Two years ago some members of the Kelso Angling Association killed 70 lbs. of Grayling in a single day.' George Bolam continues: 'Systematic netting for the destruction of coarse fish, undertaken by the Tweed Commissioners, resulted in the capture of 5791 Grayling in 1913, and 7178 in 1914; ... it is not without interest to note that the largest catches of Grayling each year took place in the neighbourhood of the original introduction, viz., at Monteviot, where 1186 were taken in 1913 and 3269 in the following year.'

As with most other attempts to eradicate introduced species, once the Grayling had gained a foothold, these control measures failed.

Ecology

Essentially a river fish, though it does occur in some lakes (e.g. Llyn Tegid in Wales), the Grayling occurs mainly in clean well-oxygenated fast-flowing streams and rivers, but it can thrive at oxygen concentrations at which Trout are beginning to feel uncomfortable. In an experimental situation where Grayling were presented with a temperature gradient, they selected water at 18°C. However, Herbert Maxwell noted that 'It is essentially a fish of clear water, being in that respect far more fastidious than the trout, which does not quarrel with a moderate amount of pollution.' In some parts of Europe, it is so characteristic of such waters, where the bottom is stony and gravelly in the riffles and the pools are clean and deep, that they are defined in river classification terms as 'grayling reaches'. In some parts of the Baltic - as with several other species which are normally found only in fresh water - it occurs in brackish water. Yet it has shown no tendency to move around the British Isles via brackish water along the coasts.

Where it is common, the Grayling may occur in small shoals. Those fish living in lakes or in the slower flowing parts of some rivers migrate into faster flowing tributaries near spawning time. The display of the males to the females (and in aggression to other males) at spawning time is very colourful and spectacular.

Robert Service, who emigrated from Scotland to North America in the 19th century, probably never saw Grayling here. He was, however, clearly aware of them by the time he wrote this verse from *The spell of the Yukon*:

> *'The summer - no sweeter was ever;*
> *The sunshiny woods all athrill;*
> *The greyling aleap in the river,*
> *The bighorn asleep on the hill.'*

Food

When the yolk sac has been absorbed, the young fry feed in shoals (sometimes along with Minnows of similar size) on very small bottom invertebrates (e.g. crustaceans and midge larvae). They continue to rely on a wide range of bottom fauna throughout life, but particularly crustaceans, midge and caddis larvae. Stomachs nearly always contain some sand, gravel and vegetable debris.

In some situations they may take terrestrial insects which have blown on to the surface and during summer they may concentrate on surface food including adult aquatic and terrestrial insects. Unlike Trout, which usually lie just below the surface when taking adult insects, Grayling

characteristically swim steeply up from the bottom each time to take the insect, returning steeply to the bottom again afterwards. Occasionally they eat the eggs of other fish including salmonids at spawning time and even small fish, in these situations Grayling grow very large. They feed actively during the winter and can be angled during cold frosty weather.

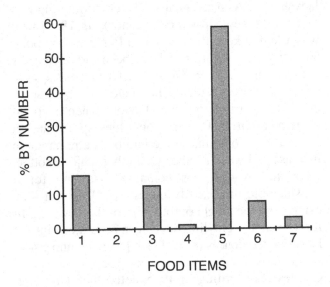

Food of Grayling in the River Tweed (1: mayfly larvae; 2: stonefly larvae; 3: beetles; 4: caddis larvae; 5: gravel; 6: others; 7: aerial) (Radforth 1940).

When in captivity, Grayling take some time to learn to eat large food items such as whole earthworms. Because stone-cased caddis larvae often figure largely in their diet in the wild, the stomach and alimentary tract may be stuffed with rough grit. This may possibly be the explanation of the ancient belief that Grayling fed on gold. Isaac Walton passes on the information that Grayling fat mixed with honey has medicinal properties 'against redness or swarthyness, or anything that breeds in the eye'.

Willie Miller has revealed a strange aspect of the Grayling's diet. 'In the late 1950s and early 1960s I was an associate member of a works club which was given permission to fish for Grayling on the Strathallan Estate water on the River Earn. Since the proprietor wished as many Grayling as possible removed from the river, we frequently killed large numbers. I have always been interested in the diet of fish and as a result I have come across a few unusual items. On 19 December 1959 a couple of

Grayling had eaten large quantities of wheat. Some of the club members were heavy smokers and threw the cigarette ends and filter tips into the river. On 16 December 1961 I found that one of my Grayling appeared to have developed a liking for such morsels.'

Reproduction

Spawning usually takes place just when river temperatures reach 4°C and are starting to rise in the spring, usually from March to May. The eggs are laid in gravelly shallows where there is some current, and spawning occurs in pairs immediately after the female has cut a redd in the gravel, in territory guarded by the male, in typical salmonid fashion.

Fred Buller and Hugh Falkus note that 'The grayling's large dorsal fin, reminiscent of the fin of the sailfish, is used in a unique way. At spawning time - usually May or June - the cock fish wraps his fin over the back of the hen. This behaviour stimulates the hen grayling to oviposit, while keeping the cock fish 'on station' ...' These eggs are covered up by fresh gravel deposited from further digging activity just upstream. If there is no suitable bottom in the vicinity, Grayling will migrate to find it. Spawning takes place during the latter part of the day, when the water temperature is at its highest. If the water temperature drops, spawning may halt for some days until it rises again.

The fecundity of each female depends very much on her size, ranging from 600 eggs in small fish to over 10,000 eggs in large fish. The eggs are about 3-4 mm in diameter and yellowish in colour. Incubation takes some 3-4 weeks (180-200 day-degrees) depending on local temperatures. Eggs are eaten by various fish in the vicinity, including the Grayling themselves. After spawning, adults of both sexes are in poor condition until late summer or early autumn.

Predators

Little has been published about the predators of Grayling but it can be assumed that, as well as fishermen, many of the 'usual suspects' (e.g. Pike, Herons, Otters) are involved. David Carss and Keith Brockie found Grayling, surprisingly, an important item (13%) in the diet of Scottish Ospreys.

Growth

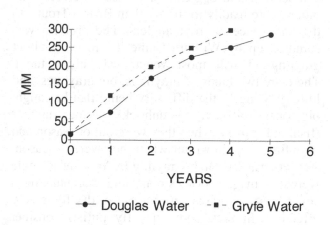

Growth of Grayling in two Scottish rivers (Mackay 1970).

The young fish hatch at about 10-12 mm and have a small yolk sac. Growth rate is variable according to habitat but young fish showing average growth are normally 8-12 cm at the end of their first year and 17-20 cm after two years - often growing more rapidly than the Trout they co-exist with. They then start to mature at 2-3 years when growth slows somewhat, most fish reaching 23-30 cm after three years and, in good conditions, 30-35 cm after four years. They live for 6-9 years on average.

Value

The Grayling has a good reputation as a table fish in some parts of Europe, but it must be eaten very fresh. Herbert Maxwell in 1904 was very enthusiastic: 'The flesh of the grayling is white, firm, and sweet, when in good condition... In my opinion small grayling furnish a more toothsome repast than any trout except the choicest, and these also should be small.' Tate Regan agreed 'The Grayling is a good sporting fish, of beautiful appearance, and as food even better than Trout, the flesh being white, firm, and of delicate flavour; it is at its best in the autumn, when Trout are out of season.' In 1910, Peter Malloch was slightly less enthusiastic 'Their flesh is white, and when in good condition is quite good for the table.'

It is not caught commercially in Scotland but is a popular fish with anglers in some areas, especially during winter when the season for Atlantic Salmon and Brown Trout is closed. Conversely, it is regarded by some game anglers as a pest in salmonid waters, mainly on the grounds of competition and egg-eating. It also tends to come rather more readily to the fly than Brown Trout - to the annoyance of trout anglers. The problem was summed up by William Caine 'I will write about graylings. I will most venemously abuse them. There are two kinds of graylings, big graylings, and little graylings' - the difficulty being that he caught big ones (difficult to unhook) during summer (fishing for trout) when they were out of season and only little ones in winter when they were in season. Anglers usually catch Grayling by 'trotting' a single weighted maggot downstream and manipulating a small float, though at times a wet or dry fly is very effective. In fact, some dry fly purists consider Grayling fishing superior to that of Trout. The two species are not mutually exclusive and an angler may take a Brown Trout followed by a Grayling with the next cast. Grayling are occasionally taken (usually accidentally) by spinning.

During the early 1970s, Grayling fry were stocked in a small hill loch near Pitlochry, lying at an elevation of 2000 feet. These fish survived for at least 20 years but there is no evidence that they spawned successfully. The last survivors must have been the oldest Grayling ever recorded in Great Britain. Andy Walker caught 'a few every year to check their progress, the largest fish weighing about 600 gm.'

The origins of the Grayling Society have been described by Ross Gardiner: '... formed in 1977/78, and now with an area network and a twice-yearly journal, has brought together those interested in grayling and in grayling fishing. An important part of its work is the collecting and recording of information on grayling.' In France, the Grayling is regarded sufficiently highly to be reared in hatcheries for subsequent stream stocking in situations where the stock has been affected by hydro-electric schemes or other developments.

Conservation

The Grayling is given no special protection under UK legislation. However it is listed in Annex IIa of the EC Habitats Directive (1992) and this new conservation obligation obliges member states to (a) designate sites to form part of the 'Natura 2000' network comprising Special Areas of Conservation (SACs), (b) protect such sites from deterioration or disturbance with a significant effect on the nature conservation interest (and take steps to conserve that interest), and (c) protect the species of Community interest listed in the Annexes to the Directive. Thus in England, where it is native, several SACs are being notified for this species.

Other than through the activities of The Grayling Society, however, it is not given any special conservation protection in Scotland, largely because it is not regarded as a native species here. Since it is still being moved into new rivers in Scotland and is likely to do well in almost all Scottish rivers, right up to the north coast, there seems no case for any conservation action.

References

Some research has been carried out on Grayling in Scotland, but most of the available literature is from studies carried out elsewhere in Europe.

Aflalo (1897), Bolam (1919), Boulenger (1946), Buller & Falkus (1988), Caine (1927), Carss & Brockie (1994), Cholmondeley-Pennell (1866), Fabricius & Gustavson (1955), Francis (1867), Gardiner (1993), Gerrish (1939), Hall (1915), Hellawell (1969, 1971), Houghton (1879), Hutton (1923), Johnson (1907), Jones (1953), Mackay (1970), Maitland & Campbell (1992), Malloch (1910), Maxwell (1904), Miller (1987, 1998), Moss (1889), Northcote 1995, Radforth (1940), Regan (1911), Robertson (1861), Walton (1653), Woolland (1987), Woolland & Jones (1975).

'If the Trout be the gentleman of the stream,
the Grayling is certainly the lady.'

Francis Francis (1867) *A Book on Angling*

*****

Thick-lipped Grey Mullet *Chelon labrosus* Muileid na Liopan Tuigh

*'This is one of the handsomest and wariest of our sea-fish,
wandering up some of our rivers.'*

F.G. Aflalo (1897) *The Natural History of the British Islands*

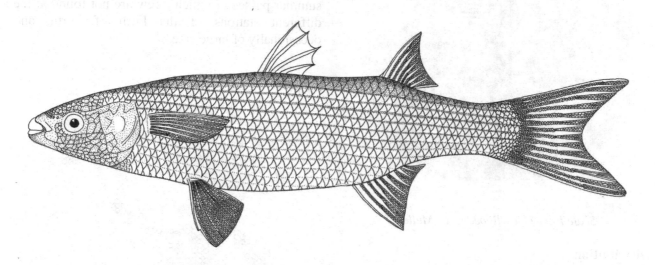

GREY Mullet are widespread around the world, but the many species behave in similar ways and occupy the same kinds of habitat. To those who know grey mullet, the description by J.W. Day and his colleagues of both fish and habitat is very familiar - even though they are describing Barataria Estuary in Louisiana and a species of grey mullet which does not occur in Scotland '... the salt marsh is visible as a thin line on the horizon ... The tide is rising ... A number of mullet (*Mugil cephalus*) are jumping and one actually lands in our boat, adding to our planned supper. Oxygen here is only 1.8 ppm, and the bottom is very soft and muddy.'

This species has a few alternative names, for example at the mouth of the Tweed, where they were formerly taken in the salmon nets, George Bolam notes that they are 'known as *The Mullet*, or *Malled*, sometimes also as *Salmon Bass*.' A literal translation of the scientific name is 'Tortoise-lipped Fish'.

Size

This attractive fast-swimming fish usually reaches a length of some 30-50 cm at maturity but can grow up to 90 cm. The Scottish boat record is for a fish caught in Girvan Harbour in 1992 by M. Kelman which weighed 1.79 kg (3 lb 15 oz 4 dm). The shore record stands at 3.25 kg (7 lb 2oz 7 dm), for a fish caught in the River Ayr by D. Smith in 1995. There are probably many fish larger than this yet to be caught. The British rod-caught record fish was caught in 1979 off Glamorgan in Wales, and weighed 6.427 kg (14 lb 5 oz).

Appearance

As its name implies, this species is characterised by its wide swollen upper lip which is quite unlike that of the other two Scottish species. Underneath this lip are 3-5 short rows of small papillae which increase in number and size as the fish grows. On this lip too are numbers of small closely-placed teeth. The edge of the eye is covered with clear adipose tissue. The body is elongate and very stream-lined in shape, with only slight lateral compression. It is covered with large cycloid scales which extend on to the head. There are 45-46 of these laterally but there is no lateral line. The two dorsal fins, which are well separated, are small, the first having 4 spiny rays the second, 9-10 soft rays. The paired pectoral and pelvic fins are both well-developed, the former being set rather high on the body just behind the gill opening.

In colour, the dorsal surface of the head and the back are a dark greenish grey-blue. This changes to a brilliant silver along the sides which also have six or seven grey bands running from head to tail. Ventrally the fish is pure white.

Scale from Thick-lipped Grey Mullet.

Distribution

The Thick-lipped Grey Mullet occurs all along the coast of Europe from western Norway to Spain, and all round the Mediterranean and Black Seas. It is probably the commonest European grey mullet. Mainly an inshore coastal and estuarine species, shoals of Thick-lipped Mullet are common in many bays and inlets, frequently coming right up estuaries almost to the limit of tidal influence. Often, very large shoals can be seen from the shore, cruising about in shallow water with their heads and backs periodically breaking the surface.

There appears to be a distinct northwards migration during spring, for there is a large increase in numbers around the British coasts in late spring and summer of fish which disappear again in the autumn. This may, however, be due in part to the fact that in some areas at least the species moves into deeper water for the winter - feeding stops at this time.

Thomas Scott and Alfred Brown noted that the Thick-lipped Mullet was 'Gregarious in estuaries of Clyde, Leven, and some Ayrshire rivers; advancing with tide to top of flood.' George Bolam remarked that 'Johnston records it as "of frequent occurrence in Berwick Bay in autumn" and R. Parnell recorded that in some seasons numbers were taken off Dunbar. This description still holds good. In some

years it is not uncommon at the mouth of the Tweed, being usually taken in the salmon nets.' Leonora Rintoul & Evelyn Baxter note that, in the Forth, it 'Occurs annually in small numbers.' Frank Buckland recorded that 'I have found it common on the west coast of Scotland, and occasionally large shoals of them appear on the east coast. Scarcely a summer passes in which a few are not found at the different stations of the Firth of Forth, and occasionally of large size.'

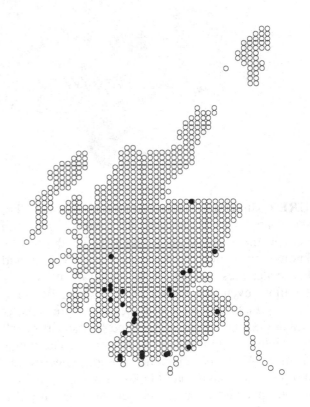

Recorded distribution of Thick-lipped Grey Mullet in Scotland.

An unusual occurrence was on July 15, 1850, when *The Glasgow Herald* reported the catch of 'a very large mullet" in Loch Lomond. A piece of paper addressed to C.C., Glasgow Herald Office was found in its stomach.' This was probably a Thick-lipped Grey Mullet, for James Lumsden and Alfred Brown recorded that this species 'Inhabits water at the mouth of the Leven, and occasionally comes up with the tide as far as Dalquhurn.'

About 1975, Rab McMath recorded a shoal of mullet in Milarrochy Bay, Loch Lomond, not long after the Leven Barrage was completed.

Ecology

As with the other grey mullet species, this is essentially a marine species which lives offshore during the winter but migrates inshore in summer. Timothy Bagenal notes that 'They are often seen swimming lazily on the surface in huge shoals'. At this time they may come into harbours, estuaries and the lower reaches of rivers. Frank Buckland records that they are 'pre-eminently, fishes of the shore, seeking shallow creeks, where they have been captured in vast quantities.'

Food

J. Hornell notes that 'The Grey Mullets are the most important of those that seldom touch animal food; their small mouths are specially well adapted to browse on the velvety growth of the smaller seaweeds that are found on rocks and harbour piling; the green varieties are their special favourites.' They have no teeth in the mouth and the pharyngeal teeth are modified to form a filter which helps to sort out those food items which they swallow from those which they spit out.

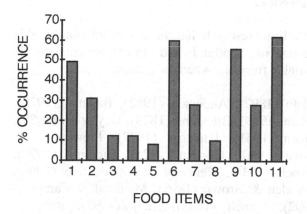

Food of Thick-lipped Grey Mullet from around the coasts of England and Wales (1: diatoms; 2: blue-green algae; 3: red algae; 4: molluscs; 5: worms; 6: copepods; 7: ostracods; 8: amphipods; 9: others; 10: fish eggs;11: detritus) (Hickling 1970).

Thus Mullet are rather unusual in their feeding habits in that, not only are they bottom feeders but they eat principally the growths of diatoms and filamentous algae which form a scum on the substrates of many of the rich shallow waters in which they occur. In browsing on these algae they also take in considerable amounts of the soft mud

on which it is growing. W.J. Gordon noted that 'It burrows in search of food, and seems to have some special fondness for the grass-wrack, *Zostera marina.*' They also take many types of invertebrates, along with the algae including many molluscs and crustaceans. The fry feed mainly on crustacean zooplankton. Very large grey mullet may feed largely on bivalve and gastropod molluscs.

Reproduction

The timing of spawning is very variable according to which part of Europe is concerned. In the Mediterranean it seems to be from December to March, whereas in the Atlantic off Spain, January to April seem to be the commonest months. Detailed information on breeding around the British Isles is scarce, but it seems likely that it takes place from June to August, as it is certainly shortly after this that very small young are found in intertidal pools in the southern parts of the British Isles.

The Thick-lipped Grey Mullet seems to spawn in shoals in shallow water and the clear eggs, 1.0 mm in diameter, are pelagic, drifting in the surface waters until hatching.

Predators

Little has been published concerning the predators of Thick-lipped Grey Mullet but, in fresh water, it can be assumed that, as well as fishermen, many of the 'usual suspects' (e.g. Pike, Herons, Otters) are involved. In the sea, of course, which is their major environment, they face additional threats from seals, sharks and other large aquatic predators.

Growth

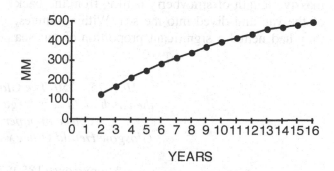

Growth of Thick-lipped Grey Mullet in Irish waters (Kennedy & Fitzmaurice 1969).

Growth varies according to latitude, but in southern waters may be very fast, fish often attaining some 15 cm after one year and 30 cm after three.

Most fish start to mature in their 3rd or 4th years and may spawn annually thereafter, often living for 10 years or longer by which time they may have reached a length of 60 cm or more.

Value

Thick-lipped Grey Mullet are an important commercial fish to various net and trap fisheries in a few places in southern Europe, on the Atlantic coast and in many places in the Mediterranean. The white flesh is flaky and most palatable, with a strong flavour of shellfish. The species is a valued sporting fish in some countries but in spite of the fact that it is often present in large shoals it is notoriously difficult to catch by rod and line, due to its food preferences. Plain white bread is a favourite bait. Sometimes, frustrated anglers cast over them repeatedly, mistaking them for Sea Trout.

Tate Regan noted that 'All the Grey Mullets are held in high estimation as food, especially in continental countries; they will thrive and fatten in freshwater ponds, and are said to be improved by this treatment.'

David Carss records that: 'in the summer of 1975, snorkelling off the beach at Barassie, my friend Grant and I occasionally encountered small shoals of quite large fish. Thinking these to be Sea Trout, we spent many hours fruitlessly trying to catch them on a fly. One day we found ourselves in what appeared to be a massive shoal of these fish. No sooner had we got to shore than a small mini van appeared and three Oriental gentlemen got out. They stripped to their underpants, pulled out a massive length of strawberry netting from the back of the van and dived into the sea. Within minutes, they had netted a significant proportion of our sea trout - which turned out to be grey mullet. Without putting their clothes on the men got back into the van. Although stunned by the spectacle, we managed to ask where they were going - they owned the Chinese restaurant in nearby Troon.'

Thick-lipped Grey Mullet can be kept successfully in aquaria. Niall Campbell once kept a 1+ specimen (caught one spring near Anglesey) in a freshwater aquarium for over a year before presenting it to the Edinburgh Zoo aquarium where it lived on for several years in a marine tank.

Conservation

There are no special protection measures for this species in Scotland; in the short-term the following would seem minimal requirements.

(a) A survey to determine the status and distribution of this species around Scotland.
(b) A standard monitoring programme covering several sites.
(c) A research study to explore the basic biology of this species in Scotland.

References

Very little research has been carried out on this species in Scotland and most information is available from elsewhere in Europe.

Aflalo (1897), Anderson (1982), Bagenal (1970), Bolam (1919), Buckland (1873), Day *et al.* (1989), Erman (1961), Farrugio (1977), Flowerdew & Grove (1980), Gordon (1921), Hickling (1970), Hornell (1934), Kennedy & Fitzmaurice (1969), Lumsden & Brown (1895), Maitland & Campbell (1992), Parnell (1838), Reay (1992), Reay & Cornell (1988), Regan (1911), Rintoul & Baxter (1935), Romer & McLachlan (1986), Scott & Brown (1901).

'*July 15, 1850. The Glasgow Herald reported
the catch of "a very large mullet" in Loch Lomond.
"A piece of paper addressed to CC,
Glasgow Herald Office was found in its stomach".*'

Anonymous (1850) *The Glasgow Herald*

*****

Golden Grey Mullet *Liza aurata* Muileid buidhe

'The fish has a golden sheen covering the head and forepart of the body.'

Pritchard (1977) *Encyclopaedia of Fish in the British Isles*

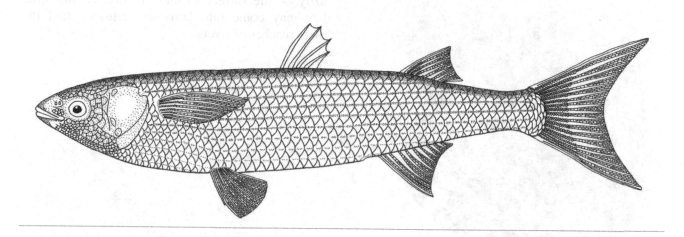

THIS is the least common of the three grey mullets found around Scotland and although Timothy Bagenal has stated that 'The golden mullet, however, is never found in fresh water.' it does move into fresh waters like the other species. The fact that it is rarer just means that it is found less commonly there than the other species. It is, however, certainly very under-recorded; the author has personally seen only one specimen in Scotland - from the upper part of the Cree Estuary.

A literal translation of the scientific name is 'Golden Lizard Fish'.

Size

As well as being the rarest, this is also the smallest grey mullet species found around Scotland. The average size of mature fish seems to be about 20-30 cm, but fish of up to 40 cm are not uncommon. Exceptionally, in very favourable habitats, it may reach lengths of up to 50 cm. The Scottish shore record (there appears to be no boat record) is for a fish caught at Fairlie in 1972 by I. McFadyen which weighed 0.322 kg (11 oz 6 dm). The British rod-caught record stands at (a) Boat: 0.992 kg (2 lb 3 oz) for a fish caught by A. Woodall in 1991 in St Peter Port Harbour, and (b) Shore: 1.368 kg (3 lb 4 dm) for two fish of the same weight, one caught by J. Reeves in 1991 on the east coast of Alderney, the other by D. Heward in 1994 at Christchurch in Dorset.

Appearance

The general body shape of the Golden Grey Mullet is close to that of other grey mullet and it is very similar in appearance to the Thin-lipped Grey Mullet. The upper lip, for example, is narrow (less than half the diameter of the eye) and there are no papillae on its lower surface. However there are teeth on this lip which are large and well spaced out. Other characteristic features of the head include the oblique truncation of the posterior end of the preorbital bone, the lower edge of which is coarsely serrated. The adipose eyelid is rudimentary. The two dorsal fins are well separated, the first with 4 spiny rays, the second with 3 spiny rays and 7-9 soft rays. The paired pectoral fins (which are long, reaching well beyond the posterior edge of the eye when folded forward), are high on the sides just behind the gill openings, while the pelvic fins are well forward on the belly. The body is well-covered by strong cycloid scales which extend well on to the head. There are 42-47 of these scales along the lateral series, but, as in other grey mullet species, there is no lateral line.

In colour, the upper part of the head and body are a dark greyish-blue, grading to a golden silver along the sides, which have a series of 6-7 length-wise greyish-gold stripes. The belly is a creamy white. There are attractive golden spots on the sides of the head and the operculum. The fins are mainly a dull greyish yellow.

Scale from Golden Grey Mullet.

Distribution

The Golden Grey Mullet occurs along the Atlantic coast of Europe from southern Norway to Spain, and along the African coast to South Africa. It occurs throughout the Mediterranean and Black Seas, and has also been successfully introduced to the Caspian Sea. It is relatively rare around the British Isles, especially in northern waters.

Like the other two grey mullet species, the Golden Grey Mullet is a coastal and estuarine species which tends to appear in its northern area of distribution during the summer months only, migrating south or into deeper waters during the autumn and spending the winter there. It often occurs in mixed shoals with either or both of the other Scottish grey mullet; like them, it will penetrate right up beyond the tidal limits of estuaries and into completely fresh water. In southern Europe it is quite common in the lower reaches of some large rivers and the lagoon lakes of their deltas (e.g. the River Danube).

Ecology

Relatively little is known about the behaviour and ecology of this species in waters around the

Scotland and it is presumed by most authorities that its biology here is similar to that of the other two grey mullet species. As with these, this is essentially a marine species which lives offshore during the winter but migrates inshore in summer. Timothy Bagenal notes that 'They are often seen swimming lazily on the surface in huge shoals'. At this time they may come into harbours, estuaries and the lower reaches of rivers.

Recorded distribution of Golden Grey Mullet in Scotland.

Food

Like other grey mullet species, after the fry stage, the Golden Grey Mullet feeds on benthic algae and the invertebrates, especially molluscs, associated with these growths. The stomachs may also contain quantities of mud. There is a tendency for larger fish to contain a high percentage of molluscs. They have no teeth in the mouth and the pharyngeal teeth are modified to form a filter which helps to sort out those food items which they swallow from those which they spit out.

Reproduction

Its reproduction has been studied in the Mediterranean and Black Seas, where it spawns from August to October, usually near the shore. Spawning is a communal process, and the shoals of adults lay large numbers of clear pelagic eggs (whose diameter is usually just under 1 mm). Juvenile fish also occur among these spawning shoals. The eggs take about 4-6 days to hatch. The females are very fecund and able to produce 1,000,000-2,100,000 eggs at each spawning.

Predators

Little has been published concerning the predators of Golden Grey Mullet but, in fresh water, it can be assumed that, as well as fishermen, many of the 'usual suspects' (e.g. Pike, Herons, Otters) are involved as predators. In the sea, of course, which is their major environment, they face additional threats from seals, sharks and other predators.

Growth

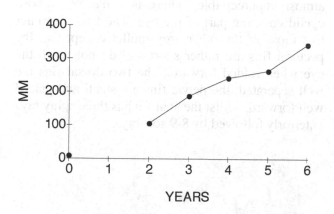

Growth of Golden Grey Mullet in Irish waters (Kennedy & Fitzmaurice 1969).

The fry are about 3 mm on hatching and remain pelagic for some time. Growth is fast however, and they soon form large shoals, sometimes near the coast but also in off-shore waters, reaching a length of 10 cm at the end of their first year and 20 cm at the end of their third year. Growth is steady but somewhat slower thereafter, individuals usually reaching their maximum sizes at around 10 years of age. The males are mature at 3-4 years and the females at 5-6 years.

Value

Though the Golden Grey Mullet has little commercial importance in northern Europe, because of its relative rarity, further south, especially in the Mediterranean and Black Seas, it is a favoured species and caught in considerable numbers in net and trap fisheries.

Tate Regan noted that 'All the Grey Mullets are held in high estimation as food, especially in continental countries; they will thrive and fatten in freshwater ponds, and are said to be improved by this treatment.'

Occasional specimens are caught by rod and line in the British Isles.

Conservation

There are no special protection measures for this species in Scotland; in the short-term the following would seem minimal requirements.

(a) A survey to determine the status and distribution of this species around Scotland.
(b) A standard monitoring programme covering several sites.
(c) A research study to explore the basic biology of this species in Scotland.

References

Very little research has been carried out on this species in Scotland and most information is available from elsewhere in Europe.

Anderson (1982), Bagenal (1972), Cihar (1976), Farrugio (1977), Kennedy & Fitzmaurice (1969), Maitland & Campbell (1992), Pritchard (1977), Reay (1992), Reay & Cornell (1988), Regan (1911).

'It lives in small shoals in the sea and likes to migrate far upriver.'

Cihar (1976) *A Colour Guide to Familiar Freshwater Fishes*

*****

Thin-lipped Grey Mullet *Liza ramada* Muileid na Liopan Tana

'The "earstones" of one are in the Robertson Museum , Millport.'

Thomas Scott & Alfred Brown (1901) *The Marine and Fresh-water Fishes*

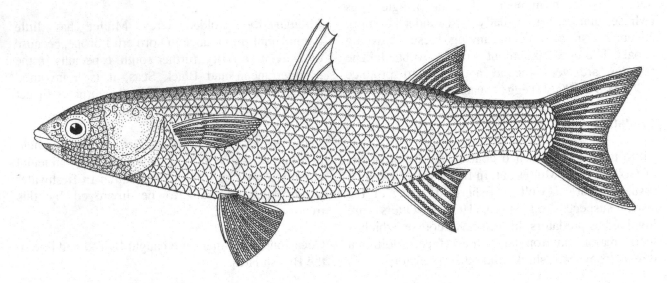

SUPERFICIALLY similar to the Thick-lipped Grey Mullet, the species are often confused - many old records simply refer to 'grey mullets'. In some areas it is called by the alternative name of Salmon Bass, but this name would often be used by fishermen for all grey mullets. A literal translation of the scientific name is 'Lizard Fish'.

Size

This species is smaller than the Thick-lipped Grey Mullet, averaging 25-40 cm, with maximum sizes up to about 60 cm and 2.5 kg. The British rod-caught record presently stands at (a) Boat: 2.693 kg (5 lb 15 oz) for a fish caught in 1991 by A. Copleym in the River Medway at Dartford, and (b) Shore: 3.175 kg (7 lb) for a fish caught in 1991 by N. Mableston in Oulton Broad near Saltside. There appears to be no Scottish record at present.

Appearance

As its name implies, this species is distinguished from the Thick-lipped Grey Mullet by its much thinner upper lip (less than half the diameter of the eye) which lacks papillae on the lower margin, and by the small bristle-like teeth which are restricted to a single row on the edge – as noted by Frank Buckland 'They have scarcely any offensive weapons, their teeth being small and delicate, often almost imperceptible.' There is a narrow adipose eyelid covering part of the eye. The fins are rather like those of the other grey mullet except that the pectoral fins are rather short and do not reach the eye when folded forward. The two dorsal fins are well separated, the pelvic fins are small and placed well forward, whilst the anal fin has three spiny rays anteriorly followed by 8-9 soft rays.

Scale from Thin-lipped Grey Mullet.

The elongate body is very stream-lined and rounded with only slight lateral compression. It is well covered with thick cycloid scales which extend on to the head even as far as the lower jaw. There are 44-46 of these scales laterally but no lateral line.

In colour, the head and back are a dark bluish-grey which grades to silver along the sides and a creamy white on the belly. Along the sides, running from head to tail, are a series of grey stripes. The fins are usually grey or a greyish yellow in colour, though the anal fin is rather darker and there is a dark spot at the base of the pectoral fin.

Distribution

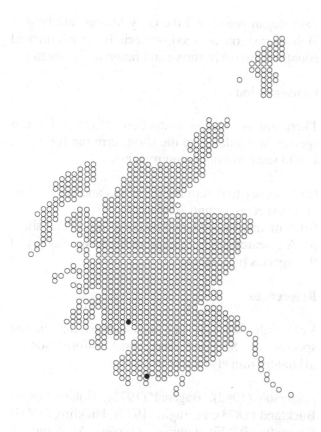

Recorded distribution of Thin-lipped Mullet in Scotland.

This species tends to be much more southern than the Thick-lipped Grey Mullet and is rare in northern Europe. Its distribution extends from southern Norway and central Scotland along the coasts of the North Sea, and the Atlantic Ocean south to north Africa, and well into the Mediterranean and Black Seas. There is a resident freshwater population in Lake Scutari in south-west Yugoslavia. The Thin-lipped Grey Mullet is common only around the southern coasts of the British Isles, where it is mainly a summer visitor to southern Ireland and the English Channel. It is rarer further north.

Mainly an inshore species, the Thin-lipped Grey Mullet occurs in a variety of shallow water habitats, including sandy and muddy bays, salt creeks, lagoons and estuaries, penetrating well above the tidal limit into pure fresh water. Where it occurs, it is usually the commonest grey mullet in fresh water. The shoals which appear along the south coasts of England in summer are assumed to have migrated north from the Bay of Biscay and other places where the species is common. These shoals disappear again in the autumn, presumably migrating south again. As with the Thick-lipped Grey Mullet, however, it seems to migrate into deep water in autumn, spending the winter resting there.

W. Yarrell mentions that, in the summer of 1834, grey mullet migrated up the River Arun in Sussex to 10 miles above Arundel. W. Thompson records them having ascended the Lagan at Belfast and into the canal there where they were shut in by the gates.

In Scotland, Leonora Rintoul & Evelyn Baxter describe it as 'An occasional visitor.' to the Firth of Forth, whilst, in the Borders, George Bolam says it is 'Rare ... The only specimen I have personally been able to identify was taken in the salmon nets at the mouth of the Tweed on 5th August 1902, and measured a little over 18 inches in length. It was called a *Salmon Bass* by its captors ...'

Ecology

As with the other grey mullet, this is essentially a marine species which lives offshore during the winter but migrates inshore in summer. Timothy Bagenal notes that 'They are often seen swimming lazily on the surface in huge shoals'. At this time they come into harbours, estuaries and the lower reaches of rivers. Frank Buckland says 'It ventures to some distance up rivers, returning with the tide.'

Food

Frank Buckland also observed that 'They come in great shoals, and keep rooting like hogs in the sand or mud, leaving their traces in the form of large round holes.' The food consists of benthic algae with some organic mud which may be ingested inadvertently. They have no proper teeth and the pharyngeal teeth are modified to form a filter which helps to sort out food items.

Reproduction

Little is known about reproduction, though it does breed in British waters. Spawning during the summer months is a communal process and the eggs (which are 1.0 mm in diameter) are pelagic. They hatch after a few days and the young are planktonic for a short period.

Predators

Little has been published concerning the predators of Thin-lipped Grey Mullet but, in fresh water, as well as fishermen, Pike, Herons, Otters, etc. are probably involved. In the sea, they face threats from seals, sharks and other large aquatic predators.

Growth

The age and growth of this species have been studied in Ireland, and are similar to other grey mullet. Maturity is reached after 3-5 years.

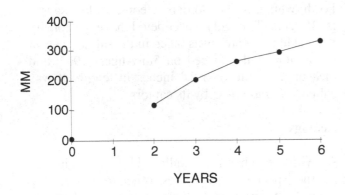

Growth of Golden Grey Mullet in Irish waters (Kennedy & Fitzmaurice 1969).

Value

Like the other grey mullet, this species is of considerable commercial and sporting value. It is caught commercially by trap nets and by seines, but it is a wary fish, skilful at avoiding nets, leaping over them at times. Anglers find it difficult to catch in many places, even although large shoals can be readily seen off-shore cruising around just below the surface. Frank Buckland records their evasiveness in this poem *The Mullet*.

'The mullet, when encircling seines enclose.
The fatal threads and treacherous bosom knows;
Instant he rallies all his vigorous powers
And faithful aid of every name implores;
The battlement of corks up-darting flies,
And finds from air th'escape which sea denies.'

Tate Regan notes: 'All the Grey Mullets are held in high estimation as food, especially in continental countries; they will thrive and fatten in freshwater.'

Conservation

There are no special protection measures for this species in Scotland; in the short-term the following would seem minimal requirements.

(a) A survey to determine the status and distribution of this species around Scotland.
(b) A monitoring programme covering several sites.
(c) A research study to explore the basic biology of this species in Scotland.

References

Very little research has been carried out on this species in Scotland and most information is available from elsewhere in Europe.

Anderson (1982), Bagenal (1972), Bolam (1919), Buckland (1873), Farrugio (1977), Hickling (1970), Kennedy & Fitzmaurice (1969), Maitland & Campbell (1992), Reay (1992), Reay & Cornell (1988), Regan (1911), Rintoul & Baxter (1935), Scott & Brown (1901), Yarrell (1836).

'It was called a Salmon Bass by its captors.'

George Bolam (1919) *The fishes of Northumberland ...*

*****

Three-spined Stickleback *Gasterosteus aculeatus* Biorag Iodain

'For voracity combined with pugnacity,
however, we shall hardly find a match for the Common Stickleback ...
It will eat almost every living thing that comes its way,
provided it is small enough to be swallowed.'

W.S. Furneaux (1896) *Life in Ponds and Streams*

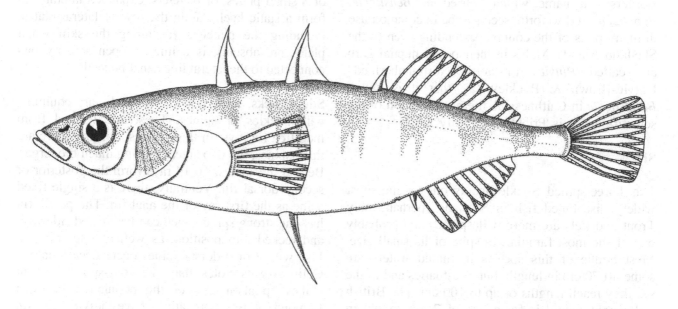

EDWARD Boulenger obviously agreed with the above comments by W.S. Furneaux: 'The stickleback is as versatile as it is greedy and pugnacious. ... At all times ready for a fight, the lovesick stickleback is a veritable tiger, battles between rival males having a fatal termination for one or more of the belligerents.'

This common species is subject to a wide variety of form. F.G. Aflalo noted that 'In place of scales, it is clad in bony plates, and the variations to which it is subject in the number of these plates, as in that of the spines, has been the basis for a number of species, which might be more properly regarded as varieties.' William Houghton seems to have disagreed for he recorded that 'The Four-spined Stickleback (*G. spinulosus*) appears to be very rare. It was first discovered by Dr J. Stark in a pond near Edinburgh in 1830. It is the smallest of all the species, being not more than one inch and a quarter in length. ... The specimens in the British Museum are from the Isle of Arran and Berwick.'

Tom Stoddart too, in 1887, noted unusual sticklebacks in Loch Achilty 'Besides the charr and trout, this beautiful lake teems with another fish of smaller dimensions, and seemingly a variety of stickleback. It swims sometimes in shoals like the minnow, and sometimes alone. ... the stickleback of Loch Achilty is itself a singular production, differing in its habits and appearance from the more common varieties of that little fish. It is thick and girthy, prefers swimming in places of considerable depth, although close to the margin, and moves at a sort of jerking but by no means rapid pace. ... There seems to be no regular season for the spawn of this diminutive animal. I observed it paired off both during summer and winter along the shallows ...'

M.A. Swan, in describing Loch Spiggie on Shetland noted that 'Sticklebacks find shelter in the reedy shallows. ... The name Spiggie is said to be derived from a Swedish word "spigg" meaning stickleback, but nowadays they are not present in such numbers as would seem to warrant this view.'

In fact, all the various forms, discussed further below, belong to the one variable species, the Three-spined Stickleback. It has a wide variety of names, most of them concerning the presence of spines: Barstickle, Doctor, Jack Sharp, Pinkeen, Prickleback, Sprickleback, Stickle, Stickling, Tiddlebat. George Bolam recorded that 'The *Bainstickle* of Berwick boys, *Boneystickle* about Alnwick, and *Baneytackle* on the Scotch side of the Borders - a name which, varied as *Benstickle*, *Banstickle*, and so forth, seems to be in common use in many parts of the country, extending even to the Shetland Islands. Males in their bright nuptial garb are called *Pinkies* in north Northumberland.' Harvie-Brown & Buckley say that it is called *Bandstickle* in Caithness. A literal translation of the scientific name is 'Prickly Bonebelly'.

Size

The Three-spined Stickleback is one of the most widely distributed fish in Scotland (perhaps only Trout and Eels are more widespread) and probably one of the most familiar, in spite of its small size. Most adults of this species in inland waters are some 40-70 cm in length, but in estuaries and in the sea they reach lengths of up to 100 cm. The British rod-caught record is for a fish of 7 gm caught in 1998 by D. Flack in High Flyer Lake, near Ely in Cambridgeshire. There is no official Scottish rod-caught record but the largest fish must surely be one of the 27 Three-spined Sticklebacks caught by rod and line by Harry and Flora Slack in a brackish quarry on Easdale Island, Argyllshire, during June 1948. The largest was, at 96 mm, the longest ever recorded in the British Isles. When Noel Hynes told Harry Slack that this was the case, Harry used to joke that he could now claim the British rod-caught record! Unfortunately, the title had to be held jointly as it was not clear whether he or Flora had caught the largest specimen.

Though no size is given, Henry Lamond recorded that 'Major Chrystal tells me that in July 1927 he hooked a stickleback through the middle, when casting in Loch Lomond with a large sea-trout fly - an odd occurrence.'

Appearance

The species is usually slender in outline and laterally compressed, though females about to spawn develop a bulging ventral profile, as do both sexes when infected with a common tapeworm (which may ultimately become as large as the host). There are no true scales on the body, which may be naked, but normally there is a variable number (1-25) of bony scutes or lateral plates which line the flanks from just behind the head to the beginning of the caudal peduncle. Along the peduncle, to the beginning of the caudal fin, there may be a further 7 or 8 small plates, or platelets, expanded laterally to form a static keel. Along the row of lateral plates, including the platelets (or along the skin when plates are absent), is a line of open sensory pits connected to the lateral line canal beneath.

Sticklebacks, as their name suggests, are equipped with stickles or spines. These are formed from modified fin rays. In this species, dorsally there are three spines, the two anterior being much the larger. Behind the spines, 10-14 rays form the posterior or second dorsal fin. Ventrally, there is a single fixed spine as the first ray of the anal fin. The pelvic fin has one strong spine which can be erected sideways and locked into position, as well as one soft ray. Fish with 2 or 4 dorsal spines occur occasionally - Colin Adams notes that 'the four-spined variant makes up about 2% of the population in Loch Lomond.' A few populations have individuals with a variable number of very weakly developed spines, from 1-3 in number or sometimes none at all. Such fish usually have few or no lateral plates and very reduced pelvic spines or none at all. So far, in Scotland, these spine deficient populations have been found only in the north and in the Western Isles, but they are also known from North America.

It is not known to what extent environmental conditions influence these features. Broadly speaking, the marine-living race possesses a full complement of well-developed lateral plates, spines and caudal keels. in a well-developed state and is known as the *trachurus* form (formerly *Gasterosteus trachurus*, the Rough-tailed Stickleback). In contrast, inland, in fresh waters, types without keels and with anterior plates only predominate - these are the *gymnurus* or *leiurus* forms (formerly *Gasterosteus gymnurus* or *G. leiurus*, the Smooth-tailed Stickleback). An intermediate form also occurs, particularly in estuaries, which has anterior plates only (usually more than *leiurus*), but also has lateral keels, although these may not be fully developed or symmetrical. This is the *semiarmatus* form (formerly *Gasterosteus semiarmatus*, the Half-armed Stickleback) which is said to be sea-going, along with *trachurus*, in the southern North Sea and the English Channel. Harvie-Brown & Buckley (1888) record the Short-spined Stickleback (*G. brachycentrus*) in North Uist in 1888. Niall Campbell believes that this is 'probably the *trachurus* morph.'

Three-spined Sticklebacks vary considerably in colour too. Marine sticklebacks have light olive to grey-green backs and bright silver flanks, while forms living inland tend to have brownish to olive backs with, often, an olive-brown to grey-brown mottling on the flanks, which may form irregular dark vertical bands. The dorsal spines and soft fin rays are pale and translucent and the pelvic spines pale to orange. At spawning time, the males assume the striking colouration that has made the species so well known: his back becomes a conspicuous pale-sandy to greenish-yellow or silvery iridescent blue-green, the iris of the eye a brilliant aquamarine and the throat and anterior flanks are suffused with a brilliant red. Males of spine-deficient populations do not colour brightly, some barely colouring at all.

Recorded distribution of Three-spined Stickleback in Scotland.

Distribution

The Three-spined Stickleback is found in the sea and in coastal waters and fresh waters throughout the temperate and polar zones of the Old and New Worlds. Its circumpolar distribution though (unlike that of the Nine-spined Stickleback) is interrupted by its absence from parts of the arctic coasts of North America and Asia.

In Scotland, where it was one of the original post-Ice Age colonisers, it is widely distributed in all types of fresh waters, from weed-choked ditches to large lakes up to about 350 m in altitude. Although impassable falls and rapids have presumably been a constraining factor, it has been further redistributed by nature and by humans, both intentionally and accidentally. It is found, too, in estuaries, tidal pools, coastal waters and far out in the open sea, where it appears to inhabit mainly surface waters - although it has also been found in the stomachs of deep water fishes.

Ecology

This species is also fairly tolerant of pollution and was one of the few species in the lower reaches of the River Clyde when the author worked there in the early 1960s. Thomas Stebbing, in his biography of David Robertson *The Naturalist of Cumbrae*, quotes Robertson: '... she led me to the place; and sure enough there was the pond, a flat hole, filled at one side with the dunghill, bounded on another by rubbish ... Still, amidst all these impurities, the sticklebacks were there in shoals, gay, fat, and healthy. I took of them with a small net what I required, and returned home, rejoicing at my good fortune.'

Sometimes this species gathers in enormous shoals of thousands, perhaps even millions, of individuals. The author has seen such shoals in Linlithgow Loch, in the River Devon at Cambus and in various other waters.

Dankers and his colleagues noted that 'The increase in numbers ... during winter indicates that the sticklebacks migrated from the shallows into deeper water. In winter they were even caught frequently in the North Sea in the surface plankton.' David Jones has caught Three-spined Sticklebacks well out at sea, off a light ship.

Food

Just as Three-spined Sticklebacks exist in a wide range of habitats, so do their modes of life vary. In weedy ditches, for instance, they may lead a rather solitary life in the thick of the vegetation, feeding on a range of small aquatic invertebrates, including small crustaceans, midge larvae and molluscs. In more open habitats they tend to feed in shoals - usually in shallow water near the shore. In large lakes, both benthic and pelagic living populations may occur, the former feeding on mayfly and caddisfly larvae, midge larvae and pupae, bottom living crustaceans, worms and small molluscs, and the latter on zooplankton (copepods and cladocerans), midge larvae and pupae, and small aerial insects on the surface film. Plant material is also taken to a limited extent, mainly diatoms, but also fragments of other algae and higher plants.

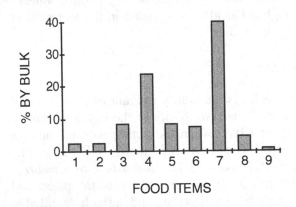

Food of Three-spined Stickleback in the Altquhur Burn (1: snails; 2: worms; 3: crustaceans; 4: mayfly larvae; 5: stonefly larvae; 6: caddis larvae; 7: midge larvae; 8: beetles; 9: others) (Maitland 1965).

Frank Buckland records a Mr Baker who wrote ' A banstickle, which I kept for some time, did, on the 4th of May, devour in five hours' time seventy-four young dace, which were about quarter of an inch long, and of the thickness of a horshair.'

Noel Hynes examined the stomach contents of the *trachurus* fish caught by Harry and Flora Slack in brackish water in a quarry on Easdale Island, in Scotland. These large fish, ranging in length from 38 to 96 mm, were feeding predominantly on higher crustaceans, especially *Gammarus*. Copepods and chironomids formed a small proportion of the diet, and cladocerans were not found at all.

Reproduction

Spawning normally takes place between March and July, with May and June being the months of peak activity. The actual spawning behaviour has been the subject of a great deal of intensive study, which has produced information basic to an understanding of vertebrate behaviour in general, particularly aspects of aggression, submission and the significance of colour and posture. At spawning time, males strong enough to hold territories select and clear an area of substrate on which to construct a nest. This is made from any conveniently available material - usually filamentous algae, plant fragments, detritus and sand. This is pushed and pulled by the male (using his mouth) until it takes the form of a low shapeless pile some 5 cm across. Where only fine sand or silt is available, the nest may actually take the form of a neatly circular low cone. The parts of the nest are glued together by sticky threads secreted by the kidneys and applied by the male rubbing his ventral region across the surface of the nest from all directions. Subsequently a hole is made through the centre of the nest by the male repeatedly poking his nose into the pile of material. Edward Boulenger noted that '... those frequenting the coast make use of such more solid material as scraps of coralline, sea mat and even the discarded spines of sea urchins.'

When the nest has been completed, the next phase of the spawning routine begins and the male embarks on a zigzag dance sequence in the presence of a ripe female. She then enters the nest, pushing her way through the entrance hole and laying her eggs on the way. The eggs are immediately fertilised by the male who has been following closely, releasing his sperm with a quivering orgasmic motion. The female then leaves, taking no further part in the procedure while the male tidies up the nest around the egg clumps. The newly-laid eggs are pale translucent yellow in colour and some 1.5 mm in diameter. The number laid by each female is usually about 50-100, but a large female is capable of producing up to 450.

The fertilisation of the eggs apparently releases the next stage in the breeding behaviour of the male, when he begins vigorous fanning of the eggs and nest to ensure that they receive an adequate amount of oxygen. Fanning is accomplished by the male going through the motions of swimming rapidly forward, but actually remaining in the same place above the nest by a rapid reversing motion of the pectoral fins. At the same time, however, he is also on guard, frequently chasing off other males or spent females, both potential nest robbers.

Niko Tinbergen, who studied this species for many years, noted that 'The male Three-spined Stickleback, while showing some hostility towards any trespassing fish, concentrates on males of its own species. Models of males release the same response, provided they are red underneath. ... Size has so little influence that all males which I observed even 'attacked' the red mail vans passing about a hundred yards away. .. When a van passed the laboratory, where a row of twenty aquaria were situated along the large windows, all males dashed towards the window side of their tanks and followed the van from one corner of their tank to the other.'

Incubation time varies according to temperature, but at a constant water temperature of 19°C hatching takes place in 7-8 days. After the young appear, the male intensifies his role as an attentive parent: as well as continuing to fan the nest he picks up in his mouth any fry which stray away, spitting them back into the nest. Gradually the family of fry shoal away from the nest, sometimes with the male in attendance, though his interest soon begins to wane.

Predators

Although there is a strong belief that the spines and lateral plates of sticklebacks are protective in function, there is actually little evidence to support this. The erectile spines may in fact be just as valuable, or more so, to the fish as a means of communication during courtship displays and defence of territory through threat, especially by the male during his nest-guarding phase. Also, in spite of the spines and body armour, sticklebacks are heavily preyed on by a number of predators - avian, mammalian and piscine. They are a valuable food item for Brown Trout and other species in many places, acting as a suitable converter of small invertebrate species into a larger item of the diet of larger fishes. Stickleback fry also suffer considerable predation from large carnivorous invertebrates such as dragonfly larvae and aquatic beetles, as well as from adult sticklebacks. Eric Hardy noted that '... four cormorants were shot and scientifically examined after feeding in Loch

Bawsinch, near Edinburgh, in response to anglers' complaints. ... One of the specimens surprised everyone by containing 82 three-spined sticklebacks. Another had six sticklebacks and a perch. A third contained one stickleback and a perch. The other had a trout's jaw in its stomach.'

Herons eat sticklebacks. Mick Marquiss recalls that: 'Nestling herons regurgitate their most recent meal when disturbed and we use this to record diet. One hot morning, my ascent to a nest near Kincardine Bridge was met with a cascade of small food items. At lunchtime, I went into the local Supermarket for a bottle of lemonade. The girl at the till was giving me that look - no not that look - the other sort. Self-consciously, I wiped my forehead with my hand and a 3-spined stickleback fell onto the counter. I knew that it had not been measured so I smiled, paid for the lemonade and put the stickleback in my pocket. The fork length was ~44 mm.'

Sticklebacks are intermediate hosts to a number of parasites. Probably the most common, and certainly the most conspicuous, of these is the large white, worm-like plerocercoid of the tapeworm *Schistocephalus solidus*. Sticklebacks become infected by eating the copepod crustaceans that themselves have been infected by ingesting the tapeworm eggs, released into the water from the faeces of a predatory bird, such as a gull, which is the final host - itself having been infected by swallowing an infected, probably disabled or dying, stickleback. In the stickleback, the parasite completely fills the body cavity of the host, eventually killing it; sometimes a very high proportion of the stickleback population is infected. In waters where the infestation level has been high, it is common to find the washed-up, dried, empty shells of stickleback carcasses with the body cavities burst open.

Growth

It is likely that in Scotland the great majority of inland and estuarine Sticklebacks spawn only once in their lives, at 1+ or 2+ years of age, dying afterwards. Their ages can usually be determined from otoliths. The larger marine-living forms may well spawn more than once and are known to live up to 4 years of age. In aquaria, 3 years is the common age limit, but a life span of 5 years has been known. Throughout the world range a maximum age of about 4 years is normal. Fast-growing individuals may reach a length of 4.5-5 cm after one year, but inland and estuarine populations nearly always include a proportion of small fish around 2.5 cm at any time of the year. These may be the progeny of late-spawning parents or just slow growers. In some populations, especially plateless and/or spine deficient ones, males may mature at around 2.5 cm, and mature specimens of less than 2 cm have been recorded from the Western Isles.

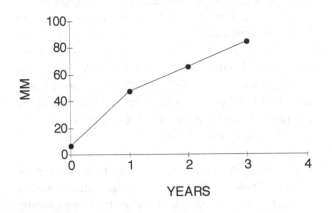

Growth of Three-spined Stickleback in Easdale Quarry (Jones & Hynes 1950).

Jack Jones and Noel Hynes, in their classic studies of sticklebacks, noted that the largest fish in the Easdale sample caught by Harry and Flora Slack were 87, 91,93 and 96 mm. 'The latter sample did not contain any of the small fish hatched in 1948, which were presumably present. ... The *G. aculeatus* from Easdale were all of race *trachura*, ... and the fish from this locality are peculiar in that they grow to a very large size. ... The large size of the Easdale fish is therefore a reflection of a higher growth rate, and not of a longer span of life.'

Value

Because of the enormous numbers in which Three-spined Sticklebacks can sometimes occur, they have been harvested by humans for a number of purposes including the extraction of oil, as an agricultural fertiliser and as fish meal. T. Pennant mentions the extraordinary number which used to appear in certain years in the River Welland at Spalding, 'where the peasantry used to scoop them out in millions, and apply them as manure to the land.' William Houghton records that 'these fish are

caught in incredible numbers in the Baltic about the middle of November, when they assemble on the coasts of that sea, and are taken by fishermen in boatloads. They are boiled, and the oil they contain is skimmed off the water; a bushel of fish is said to yield two gallons of oil. The refuse is spread over the ground for manure.'

Few people have eaten Three-spined Sticklebacks, but Frank Buckland records that 'When caught with the whitebait these fish are of an exceedingly beautiful colour, as bright as a new shilling, and the diner-out will discover sticklebacks on his plate if he will look for them; they are very good eating.' Colin Carnie has tried to eat sticklebacks: '... about six skewered on a stick and cooked over an open fire ... It was not a happy experience - not much to eat and spines seemed to be everywhere.'

Though few adults angle for sticklebacks, they may be important to young anglers. Frank Buckland notes that 'Most anglers will confess that they were first entered into the sport by fishing for sticklebacks ... In the summer-time, thousands of little boys go out fishing for 'tittlers'... I always encourage these boys, as the little urchins are thus taken out of London streets, and have a chance of learning a little natural history.'

In a recent ice-fishing tournament in Lake Vaagannet in Norway, the first prize went to a catch of two sticklebacks weighing less than 4 gm! None of the other 65 anglers caught anything at all.

They have had an important role in science for they make ideal laboratory animals. In fact, the Three-spined Stickleback may well be the most widely studied of all freshwater species of fish. Some of the most basic studies in animal behaviour have been carried out on sticklebacks - especially during their breeding period. The ability of all forms of Three-spined Sticklebacks to adapt freely to living and spawning in fresh or salt water has been a stimulus for physiological research into the mechanisms involved in osmotic control, while their great range of morphological and meristic variation is of considerable interest to evolutionary biologists - particularly as these 'adaptations' have taken place over a very short period of time in evolutionary terms and appear to illustrate 'evolution at work'. Roger Wootton noted that 'the three-spined stickleback has received more attention from zoologists than almost any other species of fish. This is not because this small fish supports an economically important commercial or sport fishery, but because it combines a fascinating biology with a hardiness that makes it an easy animal to keep and breed in the laboratory. Studies on the three-spined stickleback have made important contributions to ethology, evolutionary biology, vertebrate physiology and ecology.'

Though they make interesting aquarium inhabitants, Gregory Bateman noted that 'The Three-spined Stickleback ... though very beautiful and interesting, ought never to be confined in the same aquarium with any other fish; even in a large tank it will be a great nuisance. But if it be placed in a suitable and separate vessel during early spring, it will afford much amusement, interest, and instruction.' Chris Andrews points out that 'The ubiquitous three-spined stickleback also requires live food, and does extremely well in aquaria. Because of its territorial habits, no more than one male stickleback may be kept in a small tank during the spring-early summer breeding season.' This was one of the first species kept in aquaria (actually large jars) by the author as a boy and is an attractive fish to keep and watch - its demand for live food being the only drawback.

Conservation

This common species has no special protection in Scotland and, apart from the spineless forms which do require further study to resolve their distribution and status, does not need any at the moment.

References

Several studies have been carried out on the Three-spined Stickleback in Scotland and the species has also been well studied in other parts of the world.

Aflalo (1897), Allen & Wootton (1982), Andrews (1978), Bateman (1890), Bell (1974, 1981), Bolam (1919), Boulenger (1946), Buckland (1873), Campbell (1979b, 1980, 1985), Chappell (1969), Dankers *et al.* (1979), Furneaux (1896), Giles (1983), Hardy (1977), Harvie-Brown & Buckley (1887), Houghton (1879), Hynes (1950), Jepps (1938), Jones & Hynes (1950), Lamond (1931), Lewis *et al.* (1972), Maitland (1965), Pennant (1761), Pont *et al.* (1991), Reimchen (1979), Stebbing (1891), Stoddart (1867), Swan (1958), Tinbergen (1953), Whoriskey *et al.* (1986), Wootton (1973a,b, 1976, 1984), Wootton & Evans (1976).

'They are boiled, and the oil they contain is skimmed off the water; a bushel of fish is said to yield two gallons of oil. The refuse is spread over the ground for manure.'

William Houghton (1879) *British Fresh-water Fishes*

*****

Nine-spined Stickleback *Pungitius pungitius* Iasg deilgneach

'Regarded as a sore trouble in the trout-stream,'

F.G. Aflalo (1897) *The Natural History of the British Islands*

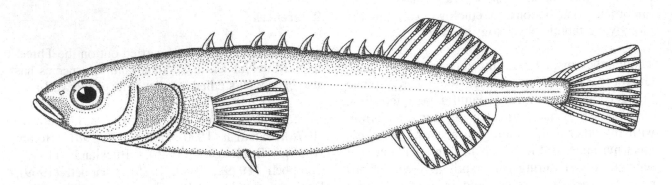

THOUGH superficially similar, close inspection of the dorsal spines will soon distinguish between the two species of sticklebacks found in fresh water in Scotland. Roger Wootton notes that 'Several sub-species of *P. pungitius* have also been recognised, though whether these should be regarded as species, sub-species or morphological variants of *P. pungitius* that are not of sub-specific status remains open to question.'

David Lewis and his colleagues noted that 'Observation ... supports the suggestion that physiologically, the nine-spined stickleback is much better adapted to life in closed, shallow, weedy, eutrophic waters, depleted in oxygen. By comparison, *Gasterosteus aculeatus* favours more open well-oxygenated water. ... This behavioural difference between the species is reflected in their reproductive behaviour. *Pungitius* builds its nest in dense weeds, off the bottom, whereas *Gasterosteus* nests on the bottom in relatively open water.' Niall Campbell suggests that 'Ten-spined sticklebacks seem to be only locally common. In waters where both species co-exist, Ten-spined sticklebacks are mostly found where the aquatic vegetation is densest.'

As with the Three-spined Stickleback, the number of spines found on the Nine-spined Stickleback sometimes can vary from that indicated by the name; the species is sometimes called the Ten-spined Stickleback. F.G. Aflalo noted that 'The Ten-spined Stickleback, or "Tinker", is widely, though

locally distributed throughout these islands, frequenting brackish as well as fresh waters.' A literal translation of the scientific name is 'The Prickle Fish'.

Size

The Nine-spined Stickleback rarely exceeds 5 cm in length, many specimens never even attaining this size. In a few exceptional habitats it is said to reach 10 cm occasionally. Apart from the Common Goby (which is mainly a brackish or marine species), it is, in fact, the smallest freshwater fish in the British Isles.

Appearance

The body is laterally compressed and more slender than that of its three-spined relative. At first, each species could be confused with the other, but the delicate, slender, caudal peduncle and general greeny-gold colour of the Nine-spined Stickleback are characteristic. Though it is also known as the Ten-spined Stickleback, in Scotland, the great majority of individuals possess nine dorsal spines - though the number can vary from 8-12. In fact, individuals with 8 dorsal spines are actually commoner than those with 10. The spines are erectile, but much shorter than those of the Three-spined Stickleback, and are off-set alternately to the left and right like the teeth of a saw. Posterior to the spines, the dorsal fin has 9-12 soft rays. The pectoral fin has 9 or 10 soft rays. The pelvic fins have 2 short erectile spines and 0-2 soft rays while the anal fin has one very small fixed spine. The caudal fin has 12 or 13 rays.

The body is scaleless and usually naked, but in Europe forms with small lateral plates - usually on the posterior body only - and inconspicuous keels along the caudal peduncle have been recorded in places. In addition, fully plated forms with up to 34 plates occur in Asia and North America. Plateless forms are sometimes classified as *Pungitius pungitius laevis* and plated forms as *Pungitius pungitius pungitius*. Both occur in Scotland, the former mostly in the north and west. A form which is deficient in pelvic spines also occurs, but mainly it seems in the central North American sector of its distribution, though it has been reported in Ireland.

In colour, the back and flanks are usually a greeny-olive gold with dark blotches or bars here and there and also many small dots. The ventral surface is pale. At spawning time, the male develops a sootiness around the head deepening to jet black on the throat and pectoral region, while the pelvic spines become a brilliant white and the rays a pale blue. Roger Wootton describes them thus: 'Reproductive mature males usually become black all over, with very prominent white pelvic spines which stand out clearly against the dark ventral surface of the fish.'

Distribution

The Nine-spined Stickleback has a circumpolar distribution, though it appears to be absent from Iceland. Thus it is found in the shallow waters of lakes and rivers and in many types of small water bodies in northern, central and eastern Europe, northern Asia and northern North America (including the Pacific coast), often in coastal brackish waters of low salinity. Roger Wootton notes that 'In marked contrast to *G. aculeatus*, *P. pungitius* is found along the Arctic Ocean coastline of the USSR, so that the species has an almost continuous circumpolar distribution, for it is also found on the northern coast of North America.'

This species is indigenous to Scotland, where it is not nearly so widespread or common as the Three-spined Stickleback, but it can be abundant locally (i.e. it is absent from most waters but common in a few). The two species often co-exist, but the Nine-

spined Stickleback tends to conceal itself in thickly vegetated shallows, often only a few centimetres deep. In captivity, the two species may hybridise but this does not appear to happen in the wild. In Scotland, the Nine-spined Stickleback is found mostly in low-lying coastal areas on the west side of the country, including the Western Isles, Inner Hebrides and the Solway. It is not as tolerant of saline conditions as the Three-spined Stickleback; it does not occur in full sea water around Scotland, though some North American populations over-winter in the sea. Under laboratory conditions, however, it can be acclimatised to full strength sea water over a period of several days. It has been suggested that, as in the Three-spined Stickleback, there is a correlation between the presence of lateral plates and keels, and a sea-going behaviour on the one hand, and between platelessness and a freshwater existence on the other.

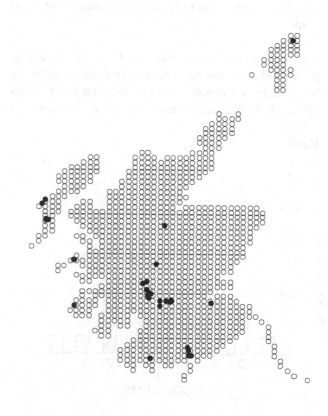

Recorded distribution of Nine-spined Stickleback in Scotland.

Leonora Rintoul & Evelyn Baxter record that, in the Forth area this species is 'Rare. Parnell found this fish near Prestonpans.' In the Loch Lomond area, Brown noted that it '... occurs sparingly in the loch and in backwaters left after floods. Spawns in May and June among stones and weeds,' The author too has found it to be uncommon in Loch Lomond, having only seen it from weed samples taken offshore and occasionally in samples from the pumping station at Ross Priory.

Ecology

A few days after hatching, as with Three-spined Stickleback, the fry swim erratically to the surface, and if they can pierce the surface film, gulp in air to inflate their swim bladders. The connection between the oesophagus and the bladder atrophies shortly afterwards. The fry are relatively small and after two weeks, they disperse individually, not remaining in feeding shoals like the fry of Three-spined Sticklebacks. This behavioural difference is very noticeable in aquaria, where young Nine-spined Sticklebacks quickly take up individual positions, only bunching when chasing after food items.

Roger Wootton noted that '... *Pungitius* tends to prefer habitats that have relatively thick vegetation, but it is by no means confined to such areas.' This agrees with the author's observations in Scotland.

Food

Food of Nine-spined Stickleback in the River Birket, England (1: worms; 2: molluscs; 3: cladocerans; 4: copepods; 5: ostracods; 6: shrimps; 7: midge larvae; 8: others; 9: aerial) (Hynes 1950).

The diet of this species is similar to that of the Three-spined Stickleback, but rarely includes plant material. It consists mainly of small invertebrates, particularly crustaceans and midge larvae, and, seasonally, the eggs and larvae of its own species.

Reproduction

Spawning takes place from spring to mid-summer in fresh or brackish water. In some waters there may be a pre-spawning migration to suitable spawning habitats. The male constructs a nest made of plant fragments suspended among vegetation, usually near the bottom, but only occasionally, actually on the bottom. In aquaria, the same male has been observed building both suspended and bottom nests.

The nest may vary in shape and is bound together by sticky secretions as with the Three-spined Stickleback, but unlike the nest of the latter, this one is also glued inside by the male. He does this by exuding a sticky thread posteriorly, catching it in his mouth and pushing his way into the nest with it. Eventually a permanent tunnel is forced through the nest (described by W.J. Gordon as '... like a muff in shape, and consists of soft-leaved water plants, the finest being confervoid filaments.') into which the female is induced to enter and lay her eggs. She is attracted by the erratic, zigzag dance executed by the male in a head down position. The male then passes through the tunnel, fertilises the eggs and drives the, now potentially dangerous, female away, for she would eat the eggs. Males have also been observed inducing females to spawn before a nest has been built. In this case the nest is constructed afterwards around the fertilised eggs.

Usually about 20-30 eggs, each 1.0-1.2 mm in diameter, are laid at a time in a single adhesive clump. Each female lays around 80 eggs altogether but not necessarily in the same male's nest. However, a nest may contain a much larger number of eggs than this, sometimes several hundreds, due to several females laying there in succession. The male guards the nest aggressively, keeping it tidy and fanning regularly to ensure a supply of oxygen to the eggs, which hatch in 6-10 days at normal water temperatures. He then removes the empty egg cases. Fry that fall out of the nest are retrieved by the male, who may then build a 'nursery' for them out of vegetation above the nest. Margaret Jepps quotes that 'Ransom also records the fact that the male stickleback tending young in the nest actually collected and conveyed to the nest young perch in the vicinity.'

Margaret Jepps also noted that 'In a charming paper written in 1865 Ransom describes the failure of a 10-spined male stickleback (*Pygosteus pungitius* L.) to persuade a 3-spined female (*Gasterosteus aculeatus* L.) to lay eggs in a nest already containing those of two females of his own species, and proceeds to remark that this "unwillingness of the female seems to be the sole cause of our finding no hybrids naturally, as artificially it is not difficult to get a cross"...'

However, Niko Tinbergen suggests that though '... the Ten-spined Stickleback's (*Pungitius pungitius*) mating behaviour is rather similar to that of the Three-spined species. It has however evolved very different nuptial colours in the male. The male of the Ten-spined species is pitchblack in spring. Just as the red colour attracts Three-spined females, so the black colour appeals to the females of the Ten-spined species. This, together with some minor behaviour differences, is sufficient to make interbreeding rare.'

Predators

Like the Three-spined Stickleback, it is a valuable food item for Brown Trout and other species in many places, acting as a suitable converter of small invertebrate species into a larger item of the diet of larger fishes. Stickleback fry also suffer considerable predation from large carnivorous invertebrates such as dragonfly larvae and aquatic beetles, as well as from adult sticklebacks. Other predators include aquatic birds such as Kingfishers and Herons.

Their behaviour when confronted by a fish predator is said to be rather different from the Three-spined species, for whereas the latter may boldly face up to the attack, the Nine-spined Stickleback will slink away into cover. It is frequently infested by parasites, though believed to be rather less vulnerable to infestations of *Schistocephalus solidus*, the tapeworm so prevalent in populations of the Three-spined Stickleback.

Growth

The Nine-spined Stickleback is more or less an annual species, usually spawning only once. Fish of 2-3.5 years of age, however (as determined from their otoliths or ear bones) have been found in the wild. The fry are relatively small, reaching a length of about 0.5-0.75 cm after two weeks. Growth virtually ceases when fish becomes mature at the end of their first year, though some may not mature until their second year. In captivity, they can live for 2-3 years, reaching lengths of 5-7 cm.

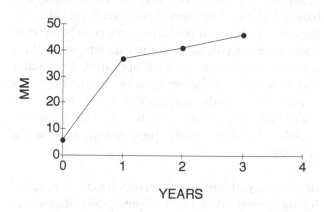

Growth of Nine-spined Stickleback in the River Birket, England (Jones & Hynes 1950).

Value

The main material value of the Nine-spined Stickleback to humans is as a forage species for larger predatory fish, which themselves are harvested commercially or caught by anglers. Like the Three-spined Stickleback, though, this species has been harvested commercially when occurring in very large numbers, yielding 'half a gallon of oil to the bushel' (i.e. 61 ml per litre).

Though rarely an item of human diet, William Houghton records that 'Sticklebacks are frequently caught with whitebait or young herrings ... and I have occasionally found some of these little fish on my plate cooked with whitebait.'

Although they make extremely interesting aquarium fish, especially as a contrast to the Three-spined Stickleback, their use as a laboratory animal has not been as extensive as the latter species, although the spine deficient form of the Nine-spined Stickleback in North America attracted much research. Gregory Bateman notes that 'The Ten-spined Stickleback, or Nine-spined Stickleback ... is often found in brackish water, and will live in either a fresh or a salt-water aquarium. ... The male fish during the breeding season ... assumes a velvety-black ' dress. ... The fish has nine spines in front of the dorsal fin: I have never found one with ten spines.'

Conservation

The Nine-spined Stickleback has no special protection in Scotland. It is by no means a common species, however, having suffered a substantial decline over the last century, disappearing from many of its former sites because of pollution and loss of habitat. Protective legislation is probably not the answer; the most useful action would be to try to restore the species to its former haunts, or, if these are not suitable or have disappeared, to suitable waters within its former area of distribution. The author has already successfully created one new 'safeguard' population and the species seems to establish relatively easily, provided the new habitat is suitable.

In summary, the main conservation actions required for the conservation of the Nine-spined Stickleback in Scotland are as follows:

(a) A comprehensive survey of the status and distribution of this species in Scotland.
(b) Preparation of a national Species Action Plan.
(c) Creation of restoration populations in those areas where the species previously occurred.

References

Little research has been carried out on this species in Scotland, most available information coming from studies carried out elsewhere in the world.

Aflalo (1897), Bateman (1890), Brown (1891), Buckland (1873), Campbell (1980, 1985), Dartnell (1973), Gordon (1950), Houghton (1879), Hynes (1950), Jepps (1938), Jones & Hynes (1950), Lewis *et al.* (1972), Mckenzie & Keenleyside (1970), Morris (1952), Parnell (1838), Rintoul & Baxter (1935), Solanki & Benjamin (1982), Tinbergen (1953), Whoriskey *et al.* (1986), Wootton (1976, 1984).

'I have occasionally found some of these little fish on my plate cooked with whitebait.'

William Houghton (1879) *British Fresh-water Fishes*

*****

Common Bullhead *Cottus gobio* Greusaiche

*'Each gill-cover carries a stout spine,
and with these weapons the fish is able to inflict severe wounds.'*

Edward Boulenger (1946) *British Anglers' Natural History*

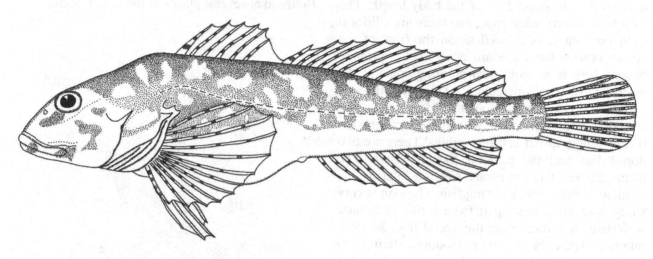

THE Common Bullhead appears to have been introduced into two catchments in Scotland quite early in the 20th Century, but it is not clear by whom. This is contrary to the advice given by William Houghton in 1879: 'However, I should not advise its introduction by any means into any waters where Trout or Salmon are found; because it is as I know a most voracious feeder, being especially fond of eggs and the newly-hatched young fry of other species of fish as well as of its own.'

The Common Bullhead is the only freshwater member of the Sculpin Family Cottidae found in the British Isles. In appearance it is typically cottid (from the Greek *Cottus* = head) and very like some of its close marine relatives which may be found commonly in estuaries.

To distinguish this species from the estuarine ones, the freshwater species was once called the River Bullhead (the term 'bull' indicating comparative greatness, i.e. Bulrush, Bull Trout, Bull Frog, etc). The alternative name of Miller's Thumb is said to have originated because of a supposed similarity of the fish to the gnarled and flattened thumb of a grain miller, due to his constant testing of the texture of the flour between his thumb and forefinger as it poured out of the meal spout. Other local names for the Bullhead were Culle (used by Isaac Walton), Chabot, Tom Culls and Tommy

Logge. In North America, freshwater bullheads are called sculpins, the Spoonhead Sculpin *Cottus ricei* being a close relation to this European species. A literal translation of the scientific name is 'Goby-like Bullhead'.

Size

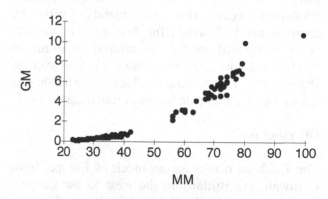

Length / weight relationship of Bullhead from the River Clyde (McAleer 1967).

Most mature Bullhead are about 5-10 cm in length, but exceptionally they may grow to over 15 cm. Females tend to grow larger than males. There is no Scottish rod-caught record. The British rod-caught record for this species is only 0.028 gm (1 oz) - for a fish caught in 1983 by R. Johnson in the Green River, near Guildford, in Surrey.

Appearance

Generally speaking, the Bullhead is not noted for its symmetry or good looks; it is often considered to be an ugly little fish, unmistakable, all head and mouth with a dorso-ventrally flattened tapering body, the widest part of which is across the head and gills and is equivalent to some 25% of the body length. The mouth has a very wide gape, and there are villiform teeth on both jaws as well as on the front of the vomer bone - but there are none on the palatine bone. There is a strong rear-pointing spine at the posterior of the operculum. This feature was remarked on by Edward Boulenger: "each gill-cover carries a stout spine, and with these weapons the fish is able to inflict severe wounds.' There are two dorsal fins and the pectoral fins are large. The comparatively large eyes are situated high on the head, as befits a bottom living fish. The skin is very slimy with no scales, apart from a row of tubular scale-like structures along the lateral line, 30-35 in number. There is no swim bladder. During the breeding season, the males have marked genital papillae. Adult fish are thought to be able to produce a sound by moving part of the operculum sharply.

The basic colouration of the Bullhead is a dark mottling over a light background, varying according to the immediate substrate on which it is resting, so that it is always well camouflaged. Edward Boulenger notes that 'It "blends" with its environment to a wonderful degree, ... its colours being readjusted to suit an altered environment much more quickly than those of the average chameleon.' The ventral surface is whitish. At spawning time the male becomes much darker.

Distribution

The Bullhead occurs across much of Europe, from Cornwall and Brittany in the west to the Caspian Sea in the east. In the British Isles it is indigenous only to south-east England, but it has been fairly widely redistributed since the last ice age, now being found throughout most of England and Wales. It is absent from Ireland and in Scotland until recently was found only in two small parts of the Clyde and Forth catchments.

The first Bullhead seen by the author in Scotland were shown to him by Hugh Gemmell, who himself had first recorded them on 6 June 1961 in the Black Loch, Renfrewshire, noting then that '... the present captures seem to confirm that *Cottus gobio* is well established in the tributaries of the Earn Water.' They were certainly abundant in the outflow from the Black Loch at the author's first visit in 1963; he subsequently persuaded Pat McAleer to study Bullhead at several places in the Earn system.

Recorded distribution of Bullhead in Scotland.

There were older records of Bullhead in Scotland, recorded by Thomas Scott and Alfred Brown: 'Upper Kelvin and tributaries', Carmel Water, Ayrshire, Dobbs Burn near Paisley' No one seems to have found them in any of these waters in recent years and the question is: Is it still in any of them?

Ken Morris records that 'In 1967 a specimen was found in the Gogar Burn near Edinburgh (the first record from East Scotland)'. This population was subsequently studied by Brian Clelland and later by Ken Morris. There is evidence that this stock is expanding: the author has found several specimens in the Union Canal in the Ratho area, from where it has no doubt the possibility of spreading to much of central Scotland. The most recent records are from the Musselburgh Esk (Susan da Prato) and from the Tweed (Ronald Campbell). In both of these systems it is likely to disperse successfully – though not if the Tweed Foundation are successful in their current attempts to eliminate it!

Ecology

The main habitat of the Bullhead has been well described by Jefferies: 'There is here some fall, and the stream is swift and bright, chafing round and bubbling over stones. Here the "miller's thumbs" are numerous. ... they hide behind stones, their heads buried in the sand, but their tails in sight. Every now and then they change positions, swimming swiftly over the bottom to another spot. Their voracity is very great ...'

Bullheads are exclusively bottom dwellers, resting on their well-developed pelvic fins which are curved outwards ventrally to lie flush with the bottom. They have no swim bladder and live a rather sedentary life under stones or thick weed beds in shallow water. When they are disturbed they move in a short dash, usually downstream, for alternative cover. They occur in a variety of streams and rivers, but usually where the current is fast and the bottom stony. They often inhabit small fast-flowing upland streams where salmonids are the only other fish present. Their oxygen requirements are much the same as those of Trout. They occur only occasionally in lochs (e.g. the Black Loch, as noted above).

They can be very numerous at times: a small hand net pushed into a bank of vegetation may catch 5-10 Bullheads of all sizes, and often other species as well. A typical Bullhead habitat might also contain high densities of Stone Loach, Minnows and Sticklebacks, and some other species, all using the same bank of vegetation for cover.

Food

At dusk, they emerge to look for food. The Bullhead has a reputation, probably much exaggerated, for eating trout eggs and alevins as well as the eggs and young of other fish. Female Bullhead are known to eat their own eggs and fry. In fact, they mainly feed on a wide range of aquatic invertebrates, including molluscs, insect larvae and

benthic crustaceans. In artificial circumstances at least, hungry Bullheads may eat quite large fish up to 4 or 5 cm in length - sometimes choking in the attempt.

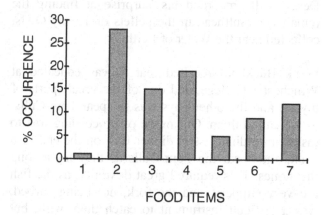

Food of Bullhead in the Gogar Burn (1: molluscs; 2: crustaceans; 3: mayfly larvae; 4: midge larvae; 5: caddis larvae; 6: vegetation; 7: others) (Morris 1978).

Ken Morris found that, in the Gogar Burn, overall, the Water Louse *Asellus* was the most important component of the diet, followed by mayfly larvae (*Baetis*). Midge larvae (chironomids) and shrimps (*Gammarus*) were also important

Reproduction

Spawning takes place from March to June, when the male excavates a small hollow under a suitable stone. A female is then attracted to spawn with him here and a clump of a few hundred pinkish yellow adhesive eggs, 2-2.5 mm in diameter, are attached to the underside of the stone. These are then guarded aggressively by the male, hatching in some 20-30 days depending on water temperature. The newly hatched fry, which the male continues to guard, are about 6-7 mm in length, well supplied with a large yolk sac which is absorbed in 10 or 11 days, by which time the fry are some 9 mm in length and start to disperse. Bullheads and sticklebacks are the only native freshwater fish which guard both their eggs and young.

Predators

Bullheads themselves are important food for Trout and many other fish eaters, especially birds such as Kingfishers and Herons. Their distended gills and opercular spines have been known to stick in the throats of these predators, causing their deaths. C. Kearton records that '...bullheads or "miller's thumbs" have proved too much for the swallowing capacity of water rails, little grebes and kingfishers.'

Derek Mills recorded his 'surprise at finding the remains of Bullhead in the pellets of Tawny Owls, collected near the Water of Leith.'

Frank Buckland recalled that 'I was educated at Winchester College, and one of the great sports of myself and the other boys was to spear Tom Culls, as we called them. Our mode of proceeding was to fasten an ordinary steel dinner fork on the end of a stick, and spear the Tom Culls as they lay among the stones. This required great dexterity, as the fish are very slippery, and the fork, not being barbed, was a difficult instrument to catch them with; but this was our sport, rather cruel I confess.'

Growth

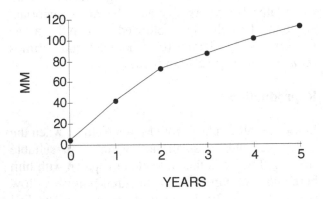

Growth of Bullhead in the Gogar Burn (Clelland 1971).

They may reach 4.5 cm in length after one year, 6 cm after two years and 7-8 cm or more after three years, by which time they are mature.

Value

Nowhere are Bullheads commercially important and they are rarely angled for. F.G. Aflalo noted that they are 'Of little account in this country, I have tasted it frequently in German towns, where they convert it into a soup, and it is also in some demand for bait.' However, they have been eaten in the past in Great Britain. At one time the 'sweet tasty' flesh of the Bullhead was much esteemed; Isaac Walton and some of the older authorities are enthusiastic

about its value as a food. William Houghton records that 'I am told by one or two persons who have eaten this fish that it is very good indeed; when the head is excluded, however, there is but little left to eat.' Leonard Mascall reported that '... Culles ... are fish holesome to be eaten of feeble persons having an ague or other sicknesse ... the Parson thereof hath tolde me he hath had so many of the said Culles and Loches to his tithe weekly, that they have found him sufficient to eat Fridays and Saturdays, wherof he was called the *Parson of Culles*.' W.B. Daniel commented that 'Notwithstanding the disgust which the form of the Bull-head creates, the largest, when the heads are cut off, are very delicious eating ...'

Fred Longrigg records that: 'When I was a boy, in England, we caught Bullheads for salmon anglers by using the stems of Ragwort. Along part of the stem the outer layer was removed and the flexible inner stem then made into a running loop, still attached to the main part of the stem. Having found a fish in the burn, we then slowly slipped the loop past the gills and snared them out of the water.'

Bullheads make interesting aquarium fish and, as Edward Boulenger has recorded, soon become tame. 'The miller's thumb changes colour under stress of its varying emotions, and in a very short while will become so tame as readily to seize worms, etc., offered in its owner's fingers.' They are easily caught with a small handnet, and F.G. Aflalo's method is not recommended: '... a prickly little fish ... it lurks beneath the stones, a favourite method of dislodging it being to strike the stone sharply, which has the effect of stunning the recluse beneath.'

They may also spawn in captivity, but successful hatching is unlikely unless special conditions of circulating, cool, well-oxygenated water are provided. However, Edward Boulenger recommends that 'If a pair of bull-heads be partnered in a well gravelled tank, they will spawn in March or April, the large eggs being laid in a compact mass placed in a hollow roughly scooped out beneath some sheltering boulder. Here the male fish incubates them, aerating the pinkish mass of about a thousand ova by movements of his tail and pectoral fins. For about a month he mounts guard, showing a bold front to all intruders by spreading his side fins, and distending his spine-clad gill covers.'

Gregory Bateman commented that 'The River Bullhead, or Miller's Thumb ... is a very interesting little fish, and will live well in an aquarium. ... In confinement it will be satisfied with small garden worms or pieces of meat, and will soon learn to take its food from its owner's fingers.' Chris Andrews agrees: 'This rather toadlike fish does well in cool aquaria. Because of its voracious appetite and large mouth, it should not be mixed with smaller fish. Plenty of live food is essential, and this fish will soon become tame enough to feed from the hand.'

Conservation

The Bullhead is given no special protection under UK legislation. However it is listed in Annex IIa of the EC Habitats Directive (1992) and this new conservation obligation obliges member states to (a) designate sites to form part of the 'Natura 2000' network comprising Special Areas of Conservation (SACs), (b) protect such sites from deterioration or disturbance with a significant effect on the nature conservation interest (and take steps to conserve that interest), and (c) protect the species of Community interest listed in the Annexes to the Directive. Thus in England, where it is native, several SACs are being notified for this species.

The question arises therefore (as it does to a lesser extent with the Grayling) that, because it has a relatively small geographic distribution in Scotland, and is therefore quite rare, does it merit any conservation action? The author believes that it does not, being an introduced species which is slowly expanding its range and is still common over much its native range.

References

Only a few studies have so far been made of the ecology of Bullhead in Scotland, but there are numerous accounts from other parts of Europe.

Aflalo (1897), Andreasson (1971, 1972), Bateman (1890), Boulenger (1946), Buckland (1879), Clelland (1971), Crisp *et al.* (1975), Daniel (1812), Fox (1978), Gemmell (1962), Houghton (1879), Hyslop (1982), Kearton (1922), Mascall (1590), McAleer (1967), Morris, (1954), Morris (1978), Rumpus (1975), Scott & Brown (1901), Smyly (1957), Walton (1653), Western (1971).

*'I am told by one or two persons who have eaten this fish
that it is very good indeed;
when the head is excluded, however,
there is but little left to eat.'*

William Houghton (1879) *British Fresh-water Fishes*

*****

Sea Bass *Dicentrarchus labrax* Beus cuan

'shoals frequently enter our rivers above tidal waters,
thus offering sport to the angler from the river-bank.'

Edward Boulenger (1946) *British Anglers' Natural History*

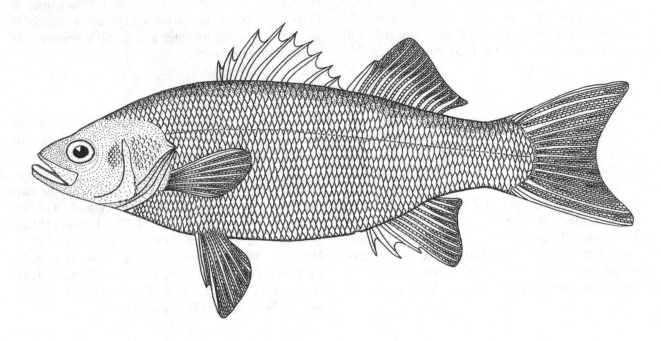

OFTEN regarded as a purely marine and largely southern species, the Sea Bass is not uncommon around the coasts of Scotland, often moving into the freshwater sections of estuaries and sometimes larger rivers themselves. Tate Regan noted that '... in the summer months the shoals may journey up the rivers for considerable distances ...', and Edward Boulenger that: '.. shoals frequently enter our rivers above tidal waters, thus offering sport to the angler from the river-bank.' With global warming, it is likely to be an increasingly common species in coastal waters, estuaries and the lower reaches of our larger rivers.

It has a variety of alternative common names, including Bass, King of the Mullet, Salmon Bass, School Bass, Sea Perch, White Mullet. George Bolam noted that 'By Tweed fishermen it is generally called *Salmon Bass*; sometimes *Pike Bass*, in order to distinguish it from the Sea Bream, which frequently shares the former appellation. *Steam Bass* is another name in occasional use, no doubt a corruption of *Stream Bass*.' Frank Buckland noted that it is 'locally called "White Salmon".'

Size

This attractive large fish matures at a length of around 30-40 cm, but specimens much larger than this are common and it can grow up to a length of 100 cm with a weight of about 9 kg. The present Scottish rod-caught boat record is for a fish caught in 1998 by A.H. McKenzie in Luce Bay, which weighed 7.162 kg (15 lb 12 oz 10 dm). The Scottish rod-caught shore record is 6.01 kg (13 lb 4 oz) for a fish caught by G. Stewart at Almorness Point in 1978. The current British rod-caught boat record is 8.877 kg (19 lb 9 oz 2 dm), for a fish caught by P. McEwan in 1987 off Reculver, Herne Bay; the rod-caught shore record of 8.618 kg (19 lb) was caught by D.L. Bourne in 1988 off South Breakwater, Dover.

George Bolam records that 'One sent into Berwick from Dunbar in June 1888, weighed 6 3/4 lbs., an exceptional weight for our district.' Herbert Maxwell has pointed out that many of the older records of Perch, caught in the lower reaches of rivers are probably Sea Bass, for example the specimen of 5 lbs 13 oz caught in the River Lunan in 1894 which was claimed to be a Perch.

Appearance

The Sea Bass is a strongly built streamlined fish, characterised by its two dorsal fins which are quite separate though one is directly behind the other, sometimes almost touching. The head is well-developed with a large mouth and eyes, and is largely covered by scales. The lower edge of the preoperculum has a series of downward and forward pointing teeth: Frank Buckland comments on the fact that there is 'an abundance of small teeth on the tongue'.

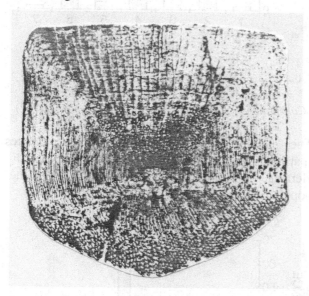

Scale from Sea Bass.

The large ctenoid scales are well-developed over the whole body and there is a strong lateral line along which are some 66-70 scales. Curiously enough, the scales between the eyes are cycloid. The ctenoid scales can be used for ageing. The first dorsal fin has 8 or 9 strong spiny rays, while the second fin is softer with one spine followed by 12 or 13 soft rays. The pectoral and pelvic fins are fairly normal with soft rays, but the anal fin has 3 spines followed by 10 or 11 soft rays.

Much of the head and back is greyish-green or bluish in colour shading to bright silver along the sides and silvery white on the belly. Small fish are often slightly spotted but these spots completely disappear in adults except for a dusky patch on each gill cover.

Distribution

The Sea Bass is mainly a marine species, found all along the coastal waters of Europe from southern Norway to Spain and well into the Mediterranean Sea. It occurs all round the British Isles but is most common in southern and western areas. As well as inhabiting coastal waters, they come into brackish areas, often moving well up large estuaries into fresh water, and even into the lower reaches of large rivers. 0+ Bass may be found in estuaries and tidal marsh pools, moving into slightly deeper water during the winter.

Along the coasts they are found in a variety of habitats ranging from sandy and rocky shores to muddy, often highly silted, areas. When they are found offshore at any time it is usually in the vicinity of submerged reefs; they have been caught at depths down to 60 m.

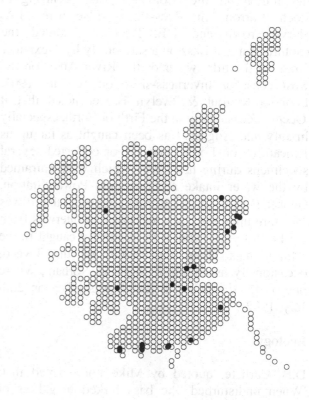

Recorded distribution of Sea Bass in Scotland.

They seem to come into very shallow seas and into fresh waters during the summer months; it is interesting that the species is also attracted to any warm water effluents which may occur in the vicinity. Sea Bass often occur in large schools which swim very actively during summer, seeking out their prey. To some extent the species is a migratory one, often appearing in the North Sea during summer, having moved in from the south west on a feeding migration. Frank Buckland observed that 'They generally arrive from the deep sea in the month of May, and remain until October.' Data from tagging experiments suggest that adults may return to the same summer feeding areas in successive years. Usually the smaller and younger fish appear first, to be followed later by larger ones. A reverse migration takes place southwards prior to spawning in the autumn. This takes the fish away from shallow waters, possibly involving journeys of several hundred kilometres.

W. J. Gordon says 'It is a summer fish, returning from the deep sea to the coast from May to November, occasionally ascending some distance up the rivers, ...' F.G. Aflalo notes that '... it is not uncommon off the Scottish coasts, occurring in Loch Carron, in Ross-shire.' Elsewhere, J.A. Harvie-Brown and T.E. Buckley recorded the capture of a Sea Bass on a salmon fly by Alexander Grant in the tidal water of the River Abort on the west coast of Inverness-shire on 16 July 1901. Leonora Rintoul & Evelyn Baxter noted that it 'Occurs occasionally in the Firth of Forth, especially in July and August. Has been caught as far up as Kincardine on Forth.' The author collected several specimens during the 1990s which were entrained by the water intake at Longannet Power Station. George Bolam, for the Borders, reported that it was 'Not rare upon the coast and taken at intervals from the Forth to the Tyne. It is frequently caught in the salmon nets about the mouth of the Tweed, occasionally ascending as high as Norham, where one of 17 3/4 inches in length was taken on 26th July, 1902.'

Ecology

D.B. Carlisle, quoted by Mike Ladle noted that: 'When undisturbed, the bass lurked in gullies or around the waving fronds of wrack, waiting to pounce out upon passing sand eels or to take shrimps moving on the sandy patches.' Sea Bass are territorial and if a territory is invade by another Bass, it will 'be driven off by the owner which 'attacked with fins spread and mouth agape in characteristic threat display.'

Food

The Sea Bass is entirely carnivorous, the larvae feeding on protozoans and small planktonic crustaceans at first.

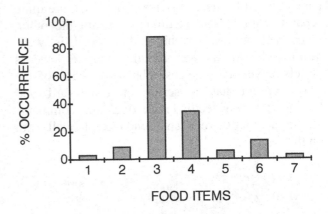

Food of juvenile Sea Bass in Irish waters (1: molluscs; 2: annelids; 3: shrimps; 4: mysids; 5: gammarids; 6: Corophium; 7: others) (Kennedy & Fitzmaurice 1972).

The adults feed mainly on invertebrates (particularly crustaceans) and small fish (especially Herring, Sprat and Sand Eels). Squids are another popular item of diet.

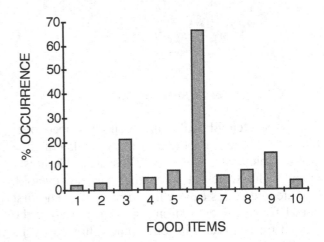

Food of adult Sea Bass around Ireland (1: molluscs; 2: annelids; 3: shrimps; 4: prawns; 5: isopods; 6: crabs; 7: sand eels; 8: herring; 9: flounders; 10: others) (Kennedy & Fitzmaurice 1972).

Reproduction

The main spawning period of the Sea Bass appears to be from February to May off southern England,

and perhaps as late as June further north. Ripe fish gather in deep inshore waters and lay large numbers of eggs (200,000-600,000 per kg of female body weight). These are clear (about 1.25-1.5 mm in diameter) and pelagic, containing 2-3 yellow oil globules. The early larvae, which are also pelagic, hatch in 4-8 days at a length of about 2 mm. Frank Buckland notes that 'when young they are bright as a new shilling and very beautiful'.

Predators

Little has been published concerning the predators of Sea Bass but, in fresh water, it can be assumed that, as well as fishermen, many of the 'usual suspects' (e.g. Pike, Herons, Otters) are involved. In the sea, of course, which is their major environment, they face additional threats from seals, sharks and other large aquatic predators. Recent correspondence with Richard Lord has revealed that Sea Bass are regularly attacked by juvenile Sea Lampreys in the sea off Guernsey.

Growth

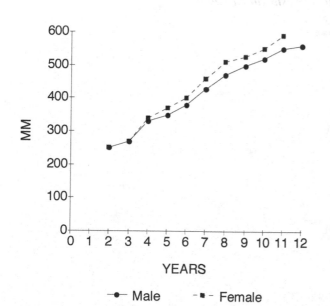

Growth of Sea Bass around the United Kingdom (Pawson & Pickett 1987).

The first post-larvae can be found in estuaries in July and August, when they are 15-30 mm in length. They reach about 3-5 cm by their first autumn and some 16-18 cm by the second. They probably start to breed at 4-6 years of age at lengths of about 35-40 cm, and they may continue to breed

up to the age of 20 years or more. Some fish are known to be long-lived - specimens of 30 years have been recorded. Females grow slightly faster than males and live longer.

Value

The Sea Bass is an important commercial species, usually exploited wherever available, although many previously abundant stocks have now declined. It has increased in value in recent years; small catches have been made by commercial fishermen in the Solway and other parts of southern Scotland.

F.G. Aflalo noted that '... is a fish of great repute with the sea-angler, chiefly on account of its extreme and capricious movements, wariness...' Thus it is a very popular fish with sea anglers, affording exciting angling in estuaries and other places around the coasts of the British Isles. W.G. Harris Cass agrees: 'The one that is outstanding and the one for which every sea angler makes a bid, is the bass, both for its fighting qualities and its table value. ... It evinces a game spirit even when only a few ounces in weight.'

It is an extremely good fish to eat with an abundance of firm white flaky flesh with a good flavour. Tate Regan noted that 'As food they have been held in much esteem since the days of the Romans, who kept them in freshwater aquaria, a practice which is said to improve the flavour.' Increasingly popular in restaurants and on fishmonger's slabs, many of the fish now available have been cage-reared, mostly in southern Europe, especially Mediterranan waters.

Young fish usually do well in captivity, in fresh or salt water and are very attractive in aquaria, fed on a diet of live foods. It is a popular fish in public aquaria and can readily be seen at Deep Sea World and other establishments.

Conservation

In the south-west of Britain and in southern Ireland the Sea Bass is heavily exploited, both commercially and by sport anglers, with the result that a minimum size limit of 32 cm has been imposed through an EC regulation in 1987. Recent research, however, has indicated that there is good evidence for raising this to 38 cm, and for the establishment of designated nursery areas where fishing will be illegal or restricted. Such areas could often be part of more general estuarine or marine nature reserves. In summary, the main conservation and research needs of Sea Bass in Scotland are:

(a) A survey of its status and distribution in Scottish waters.
(b) A monitoring scheme for catches by anglers and commercial fishermen.
(c) Protection for any areas which are known to be nursery grounds for this species.

References

Little research has been carried out in Scotland on the Sea Bass which has long been regarded as a southern species of little value here. With the advent of global warming, however, this situation may well change and it would be useful to initiate research now. Fortunately a reasonable amount of literature is available from England and other parts of Europe.

Aflalo (1897), Aprahamian & Barr (1985), Bolam (1919), Boulenger (1946), Buckland (1879), Cass (1940), Dando & Demir (1985), Gordon (1950), Harvie-Brown & Buckley (1887), Holden & Williams (1974), Jackman (1954), Kelley (1979, 1986, 1987, 1988), Kennedy & Fitzmaurice (1972), Maitland & Campbell (1992), Maxwell (1904), Pawson & Pickett (1987), Pawson *et al.* (1987), Regan (1911), Rintoul & Baxter (1935), Thompson & Harrop (1987).

'As food they have been held in much esteem since the days of the Romans,
who kept them in freshwater aquaria,
a practice which is said to improve the flavour.'

Tate Regan (1911) *British Freshwater Fishes*

*****

Ruffe *Gymnocephalus cernuus* Gibeag-muineil

'every Pope caught had a cork impaled on its dorsal spines and was set at liberty, until the surface of the canal for miles was covered with bobbing corks.'

Frank Buckland (1881) *Natural History of British Fishes*

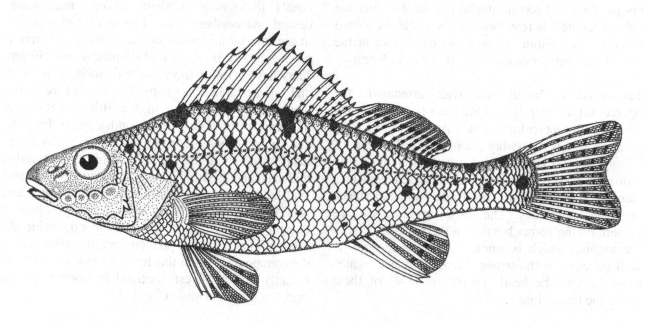

THE name of Pope for this fish provided a cruel destiny, for in Protestant England Frank Buckland described how the people of Sheffield and other towns used to go in hundreds to a place 'well named Crewell Bridge' in Lincolnshire where the 'sport' of 'corking the Pope' meant that every Pope caught had a cork impaled on its dorsal spines and was set at liberty, until the surface of the canal for miles was covered with bobbing corks. 'Of course, all these luckless fish were doomed to a lingering death.'

This species has only a couple of alternative names. W.J. Gordon refers to both of them when he points out that 'The Ruff - occasionally spelt with a useless "e" - is the same fish as the Pope.' In fact, the use of the term Ruffe is now much commoner than Ruff, and is certainly the name preferred by the author. William Houghton notes that 'The meaning of the word Pope as given to this fish is not at all clear.' A literal translation of the scientific name is 'Head Down Naked Head Fish'.

Size

The Ruffe is the smallest member of the perch family, seldom reaching a length of over 15 cm except in a few specialised habitats abroad, such as the Baltic Sea where, exceptionally, it may grow up to 50 cm in length.

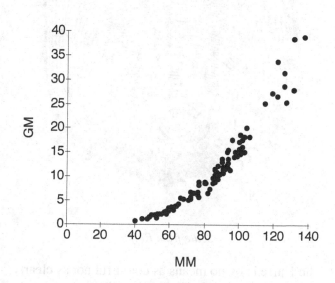

Length / weight relationship of Ruffe in Loch Lomond in 1991.

There is no Scottish record at present, the current qualifying weight being 4 oz. The present British rod-caught record is for a fish of 0.148 kg (5 oz 4 dm) caught in 1980 by R.J. Jenkins in a pond at West View Farm in Cumbria.

Appearance

Although not as high-backed, Ruffe could be initially confused with small faintly coloured Perch, but the first and second dorsal fins of the latter are quite separate whereas those of the Ruffe are joined together. In addition, there is no black spot at the rear of the first dorsal fin in the Ruffe as in Perch.

The head is broad and well armoured, the operculum ending in a long rear-pointing spine, while the preoperculum bears a 'ruffe' of 10-12 rear-pointing spines - possibly the origin of the name of this species (although apparently Pope is the older term). These spines make it painful to remove Ruffe from nets, as the author can testify from experience on many occasions. The villiform teeth are are present on the vomer but not on the palatine bone or the tongue, which is smooth. The whole body is well-covered with strong ctenoid scales which extend on to the head. There are 35-40 of these along the lateral line.

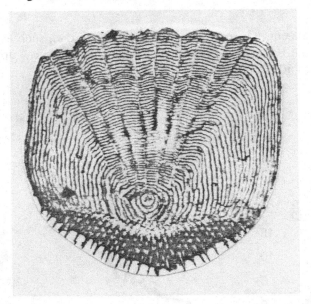

Scale from Ruffe.

The Ruffe is by no means as colourful nor as clearly marked as the Perch. The back and sides are sandy to pale brownish green with irregular dark olive brown blotches (reminiscent of those of the Gudgeon), whilst the belly is a pale yellow. The dorsal and caudal fins have broken rows of dark spots on the membranes between the rays. The paired pectoral and pelvic fins are usually a pale brown, but sometimes the former may have a rosy hue.

Distribution

Ruffe are found in lakes, slow flowing rivers and canals throughout northern Europe and across central and northern Asia. The species also occurs in the low salinity areas of the Baltic Sea where it reaches its maximum size of 45-50 cm and 750 gm. They have a lower oxygen requirement than Perch and so are able to occupy habitats where Perch would be under stress. In the British Isles it is indigenous to eastern and south-eastern England and is common locally, but it has been re-distributed to some extent within the English midlands and has now reached the lower Severn and Welsh Dee systems. It is absent from Ireland and was from Scotland also until recently when it became established in Loch Lomond, evidently brought in as a bait by pike anglers. It is now one of the commonest fish in this loch. It has also appeared recently in south-west Scotland in Loch Ken and occurs too in the Union Canal.

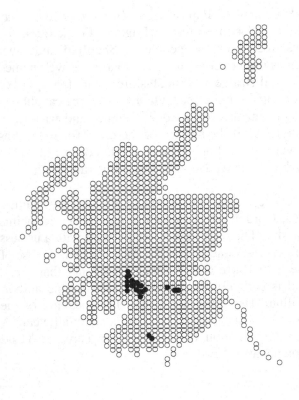

Recorded distribution of Ruffe in Scotland.

The Ruffe was first discovered in Loch Lomond by the author in 1982, when two small specimens were gill-netted at Camus an Losgainn, near the University Field Station at Rowardennan. Subsequently that year, 17 other specimens were taken on the screening system of the pumping station at Ross Priory, which the author had started monitoring in 1981. Numbers collected there increased logarithmically until, 10 years later, many thousands were being collected each year. The numbers have now stabilised at several thousand a year, and the Ruffe is one of the commonest fish in Loch Lomond, where it has initiated several major changes in the ecology of the loch.

A century ago Herbert Maxwell was very clear that this species is a northern one: consequently it is not surprising that, having been introduced to Scotland it is now thriving here. 'Probably the ruffe is of more northerly origin than the perch, having had its dimensions restricted by more rigorous climatic conditions, which have a stamp upon the little creature so permanent as not to have yielded to more temperate surroundings. Yet, if that be so, it is perplexing to note that the ruffe is not found in those parts of Great Britain where the fauna retain most traces of Arctic experience, for it is unknown in Scotland and the north of England, neither has it a home in Ireland. Yet is its range far more northerly than that of the perch, for the ruffe is not found in southern Europe, but abounds in Scandinavia, Russia and Siberia.'

Ecology

The ecology of Ruffe is similar to that of its relative the Perch, both favouring rather still waters - either lochs or the lower, slow-flowing reaches of rivers. The Ruffe, however, has been much slower in moving from the original area in southeast England in which both were found immediately after the last Ice Age, the recent advance north to Loch Lomond being due to transport by anglers. Prior to this the distribution of Ruffe appeared to reflect the canal system of the English Midlands, with none known north of the River Tees prior to 1982. In contrast, Perch are now found well north of the Great Glen.

Food

Ruffe are gregarious fish, often feeding in shoals - though these are rarely as large as some Perch

shoals. They are exclusively carnivorous, feeding on bottom living invertebrates, especially molluscs, crustaceans and insect larvae (notably midge larvae, which form a significant proportion of their diet). They take approximately the same range of organisms as Perch, but because of their ability to penetrate deeper into the mud than Perch, Ruffe appear to take a greater proportion of mud-dwelling invertebrates.

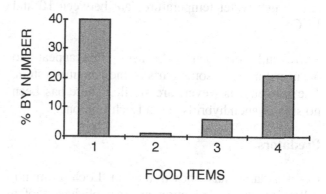

Summer diet of Ruffe in Loch Lomond in 1985 (1: crustaceans; 2: caddis larvae; 3: midge larvae; 4: beetles).

They also take fish eggs and small fish, and it has been shown in some large Russian lakes that, where Ruffe and Powan occur together, the Ruffe exert a significant control over the production of the latter, because of the enormous number of Powan eggs which they consume. Feeding takes place mainly during the day, while at night the Ruffe may lie concealed on the bottom. Unlike Perch, they feed throughout the winter but at a reduced level.

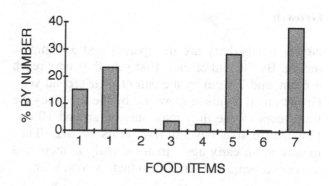

Winter diet of Ruffe in Loch Lomond in 1989 (1: crustaceans; 2: molluscs; 3: stonefly larvae; 4: caddis larvae; 5: midge larvae; 6: beetles; 7: powan eggs).

Reproduction

Spawning takes place from March to May (usually the latter month in Loch Lomond), when shoals of Ruffe move into shallow water to lay their eggs in sticky strands which adhere to rocks and vegetation. The eggs are yellowish white in colour (each yolk sac contains a large oil globule) and 0.5-1.0 mm in diameter. Each female can produce 4,000-100,000 depending on her size. The eggs hatch in some 8-12 days when water temperatures are between 10 and 15°C.

Ruffe and Perch can hybridise - this appears to occur naturally in some parts of the Danube system. The progeny, however, are sterile. There has been no sign of such hybrids yet in Loch Lomond.

Predators

Colin Adams has shown that in Loch Lomond, following the introduction and establishment of a large population of Ruffe, the predation pressure of three of the principal fish predators there (Pike, Herons and Cormorants) was deflected from native fish species there on to Ruffe. In a separate study, Dominic McCafferty showed that the main diet of another fish predator around Loch Lomond, the Otter, is now Ruffe.

Though Ruffe are known to harbour a number of parasites within their natural range, only two species were found by Thomas Shaw in a study of 254 Ruffe from Loch Lomond - a tapeworm *Eubothrium crassum* and an acanthocephalan *Acanthocephalus lucii*.

Growth

Newly hatched fry are transparent and 3-4 mm in length. By the end of their first year they may reach 3-6 cm, and 7-9 cm by the end of their second year. Growth then seems to slow and by the time they are four years of age they may only be around 10 cm. Few Ruffe live beyond five years of age. They mature at an early age - males usually in their first or second years and females in their second year.

Value

Houghton remarked of the Ruffe that 'It is a bold biter, and affords good sport to the young angler,

being taken readily with a small worm.' Certainly Ruffe are easily caught by angling with bait such as maggots or worms, but they are not considered to be an important sport fish. Sheringham agrees: 'The pope annoys the perch-fisher more than the bleak annoys the roach-fisher. Pope are chiefly remarkable for an appetite which cannot be appeased and for never growing any bigger. They could not be any smaller.' Often, small Ruffe and small Perch may be caught at the same time at the same spot. Niall Campbell reported that, when fishing for Ruffe in the River Neva in Russia, he 'baited a three-hooked cast with red caviar, black caviar and worm respectively, but the Ruffe only took the worm!' Many anglers consider them to be a great nuisance, taking the bait before a more acceptable species can get to it. As a dead bait they are favoured by Pike anglers.

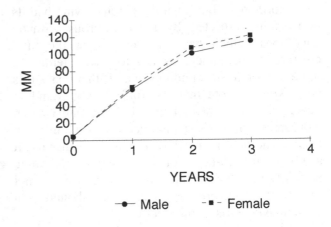

Growth of Ruffe in Loch Lomond (Murphy 1988).

Their flesh is well flavoured, like that of Perch, and there was originally an extensive fishery for Ruffe in the lagoons of the Baltic Sea. This has declined greatly in recent years owing to lack of demand.

Dame Juliana Berners says that "it is a right holsom fysshe." and Izaak Walton vouches for its excellence on the table declaring that 'no fish that swims is of a pleasanter taste.' Houghton obviously agrees 'The Ruffe is excellent food for the table, resembling that of the Perch, but of shorter texture; but Herbert Maxwell, ever one to clarify a problem, pointed out that 'The greater the pity, then, that there is so little flesh in the diminutive carcase.' Years ago, a Russian friend explained to the author how this dilemma is solved: fish are placed in a net bag and kept in boiling water for some time. The bag (and its bony fish) are then thrown away, and the remaining stock is used to make soup - apparently delicious!

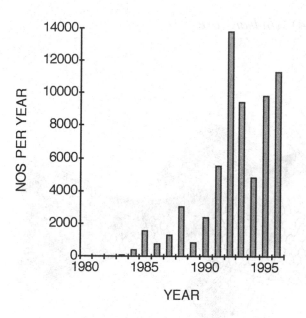

The explosive growth of alien Ruffe in the first 15 years after they were found in Loch Lomond in 1982, almost certainly introduced there by coarse anglers.

Gregory Bateman regarded it as a difficult fish to keep in an aquarium: 'Though a very handsome fish, the Pope or Ruffe ... is not so easy to keep alive in confinement as its near relative the common Perch. The reason of this may be that it is nearly always found in running water, and hardly ever in stagnant.

... Though he is so handsome, he has, however, such an inexpressibly sad looking face as almost to make one miserable to look at him.' In fact, the author has found this a relatively easy fish to keep in a large aquarium, provided sufficient live food is available.

Conservation

The Ruffe has no special protection in Scotland and, as recent alien which has expanded explosively in Loch Lomond, is now found in several other waters in Scotland and is likely to appear elsewhere, needs none.

References

Several studies of this species, new to Scotland, have been carried out since its introduction about 1980 and there is also a wealth of information from other work carried out elsewhere in Europe.

Adams (1991), Adams, *et al.* (1994), Adams & Maitland (1998), Adams & Mitchell (1995), Adams & Tippett (1991), Bateman (1890), Berners (1496), Buckland (1881), Devine *et al.* (2000), Gordon (1902), Hartley (1948), Houghton (1879), Kalas (1995), Maitland *et al.* (1983), Maitland & East (1989), Maxwell (1904), McCafferty (2005), O'Gorman (1845), Oliva & Vostradovsky (1960), Pokrovskii (1961), Salter (1833), Selgeby (1993), Sheringham (1925), Wheeler (1969).

'Though he is so handsome, he has, however,
such an inexpressibly sad looking face
as almost to make one miserable to look at him.'

Gregory Bateman (1890) *Freshwater Aquaria*

*****

Perch *Perca fluviatilis* Creagag-uisge

'The bright-eyed perch, with fins of Tyrian dye'

Alexander Pope (1688-1744) *Windsor Forest*

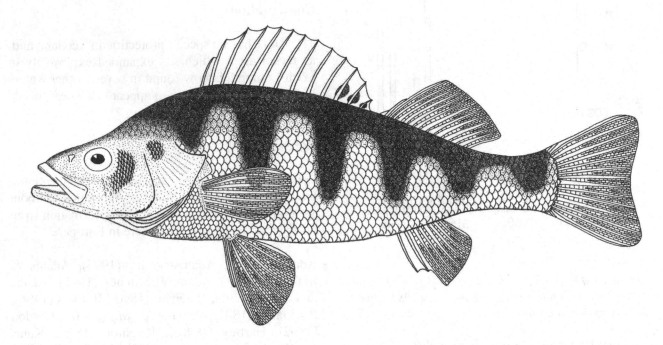

THE first fish ever caught on rod and line by the author as a boy was a Perch. Tannoch Loch in Milngavie was a Mecca for small boys with primitive fishing equipment, and there, a close packed shoal of brightly coloured Perch was a most impressive sight, when seen through clear water under sunny conditions. T.F. Salter, however, issues a warning to the unwary: 'Note the strong prickly dorsal spines. As many a young angler has learnt through painful experience, the proper way to hold a perch is to stroke back these dorsal spines with one hand before securing the fish with the other.' Unfortunately, this fact is usually only learned through painful experience - as the author found out early in his fishing career!

James O'Gorman believed that 'Perch like noise, and are fond of music, which attracts them to the surface. One of my sons (now I hope happy) assured me that he saw a vast shoal of them appear over water, attracted by the sound of bag-pipes when a Scotch regiment were marching over a neighbouring bridge, and that they remained there until the sound died away in the distance. ... How much superior the ear of perch to that of Paganini, who, on hearing the Scotch pipe, prostrated himself on the floor, declaring that it must have been invented by the devil.'

Thomas Stoddart, writing in 1866, had a high opinion of Perch: 'As an article of food the flesh of the perch is in good esteem - very superior to that of the carp, the tench, the bream, and the chub, and is held preferable by many to the flesh of the common trout. When in season, its firmness and curdy whiteness rival the same qualities as they are displayed in some of our most highly-prized salt-water fishes ... One of the drawbacks to its more frequent appearance at the table, results from the supposed necessity, before cooking it, of removing the tough coat of scales and spiny fins with which, for defensive purposes it is accoutred. ... On Lentrathan Loch, and other famed perch-yielding expanses, it is considered barbarous to subject this fish to any other process, before brandering, than that of simply wiping it.'

An alternative common name for the Perch, but in very infrequent use, is Barse. A literal translation of the scientific name is 'River Perch'.

Size

The size of adult Perch varies greatly from place to place. In some populations, individuals never attain more than 15 cm in length, even after 10 or more years of life, while shoals of younger fish 25-30 cm may occur in nearby waters. Populations of stunted Perch are usually those not subjected to pressure from large predatory fish such as Pike, but may also be the result of the presence of an exceptionally numerous year class which completely dominates the whole population for a number of years, causing a relative shortage of food and suppressing younger fish.

David Le Cren has analysed the length and growth data for 137 Perch from Windermere, all of them over 300 mm (total length). The largest was 463 mm.

The Scottish rod-caught record stands at 2.21 kg (4 lb 14 oz) for a fish caught by J. Walker in 1989 in Loch Ard, but Henry Lamond describes a larger Scottish fish - a famous catch when 'Colonel Thornton, in 1757, had a great day among the perch, capturing 97 with worm, and a giant of 7 lb. 3 oz. when trolling back to Luss through "The Straits"'. The British rod-caught record stands at 2.523 kg (5 lb 9 oz) for a fish caught in 1985 in a private lake in Kent. The Irish record stands at 2.75 kg (caught in Lough Erne in 1946), but even larger fish have been found; some of up to 4.75 kg have been recorded from mainland Europe.

T. Thornton's own description of his large fish gives further details: '... passed through the straits of Loch Lomond ... in turning gently round a dark bay, I felt a fish strike, ... After much trouble he was secured in the landing-net, and proved to be a perch of about seven pounds and a half. I never saw so fine a fed fellow, ... at Luss ... Got my portable stillards to weigh my perch, about which there were various conjectures. ... His precise weight, however, was seven pounds three ounces ...'

Appearance

Most enthusiasts would agree with Edward Boulenger that 'The Common Perch, *Perca fluviatilis*, is a strikingly handsome fish ...' It has a deep and laterally compressed body, the maximum depth being just below the beginning of the first dorsal fin. They become progressively deeper with increasing size and a really large Perch looks almost 'top heavy'. The term 'fat as Perch' is used by anglers to describe a particularly well conditioned and deep-bodied Trout. Tate Regan recorded that 'In some localities abnormal Perch are not uncommon; A hunch-backed form with a very deep body, resembling a Crucian Carp in shape ... occurs in Loch Arthur in Kirkcudbrightshire ...'

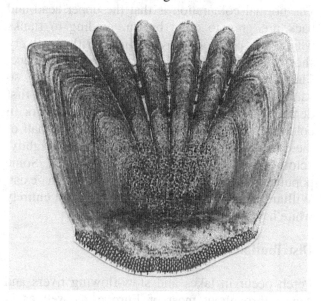

Scale from Perch.

Normally, the head is small and neat but the gape of the mouth is large. The operculum, which bears scales, ends in a very sharp rear-facing point. The mouth, which does not extend back as far as the rear of the eye pupil, is edged with many small villiform teeth which occur also on the dentary, vomer and palatine bones, though the tongue is smooth. The first dorsal fin has very long sharp 'defensive' spines, some 13-17 in number. These make the fish difficult to handle and can inflict quite a deep puncture in the unwary human. Like the spines of sticklebacks and Ruffe, however, they do not prevent Perch being a common prey of predatory fish and birds. The body is well covered with ctenoid scales which are large and strong, edged with fine spines which give the fish a rough feeling. There are some 58-68 of these scales along the lateral line; many of them are perforated by the mucu- filled offshoots of the lateral line canal.

The colouration and markings of Perch are variable in that they are subject to rapid changes in intensity according to the emotional state of the individuals.

For example, those newly introduced to an aquarium become very pale with their vertical bands hardly discernible, while at the other extreme, wild Perch in the act of pursuing small fish display striking vertical black bands and bright red-orange pelvic, anal and lower caudal fins. This full colouration, unfortunately, rarely seems to persist in captivity.

The normal colouration is that the upper head and back are grey to olive brown, grading to flanks which are pale olive (sometimes with a silvery sheen), shading to a yellow white ventral surface. The vertical bands on the sides, which are so variable in intensity, are 5 to 8 in number. The first dorsal fin, with its posterior dark spot, is dark in colour, as is the second dorsal and the upper half of the caudal fin - in striking contrast to the brightly-coloured pelvic, anal and lower caudal fins. Some populations of rather pale silvery Perch exist: William Houghton notes that 'Perch almost entirely white have occasionally been found.'

Distribution

Perch occur in lakes and slow-flowing rivers and canals throughout most of Europe, as well as in brackish lagoons in the Baltic Sea and in Asia. George Bolam recorded that 'In the Tweed ... Perch occasionally descend to within the influence of the tide.' They have been introduced successfully to Australia and New Zealand. The North American Yellow Perch (*Perca flavescens*) is very closely related and considered by some to be exactly the same species.

The Perch is indigenous to south-east England but has been widely redistributed throughout much of the British Isles, partly by natural dispersion and partly by translocations by humans. However they have not yet reached the extreme north of Scotland (there is one population in south-east Sutherland), any of the northern isles (Orkney and Shetland), nor the Hebrides, except Islay where there is one population. No doubt it will disperse slowly into many of these areas.

It is one of several species found normally only in fresh water which occurs in the Baltic Sea, confirming Frank Buckland's view that 'It rather likes than shuns a dash of salt water.' This may have been helpful in its dispersal after the last Ice Age.

Ecology

The growth and survival of young Perch is much affected by temperature. After a long warm summer, the current year class may be very strong: the reverse is true for cold summers. There are therefore great variations in year class strength within Perch populations, and the numbers involved may vary by a factor of several hundred from one year to another. As autumn approaches and the water temperatures drop, the young fish and adults move offshore into water up to 12-15 m, if this is available. However, solitary individuals are sometimes encountered sheltering under vegetation or stones in shallow water.

Recorded distribution of Perch in Scotland.

Food

Very young Perch feed mainly on zooplankton. Later, they forage for food during the day in roving shoals of fish all about the same size. Although they are accomplished piscivores from an early age, their diet comprises mainly aquatic invertebrates which they pick off the bottom or from the stems and leaves of aquatic plants. Their main food organisms include leeches, freshwater shrimps and lice, midge and caddis larvae and pupae, and small fish. Foods of lesser or only seasonal importance are mayfly larvae and, in oligotrophic waters, stonefly larvae, water boatmen and small bivalve molluscs. Snails do not generally seem to be taken and, in a seasonal study of Perch food in Loch Tummel by Niall Campbell, they were found in only a small proportion of the fish, and then only in winter.

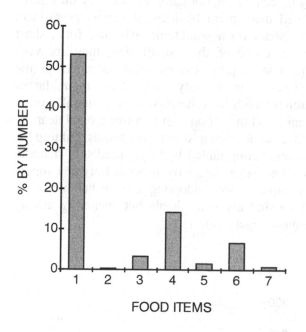

Food of Perch in Loch Lomond (1: crustaceans; 2: stonefly larvae; 3: mayfly larvae; 4: midge larvae; 5: caddis larvae; 6: sticklebacks; 7: others) (Shafi & Maitland 1971).

When attacking smaller fish, Perch show surprising speed and mobility in rounding up and out-manoeuvring their prey - the very reverse of the stalk or ambush technique employed by Pike, or the wild open-mouthed charge of Trout. Niall Campbell has watched Perch in a clear stream preying on Minnows which were shoaling densely in shallow water prior to spawning. A shoal of Perch was hovering in the current some metres away, and every now and then two or three Perch would detach themselves from their shoal and attack the Minnows in a pincer movement, herding the later into an even denser shoal before pouncing. In captivity, Perch are extremely greedy feeders and seem to have insatiable appetites. On one occasion, at Loch Leven, a 1+ Perch of 5.3 cm in length was found dead with a 2.5 cm Three-spined Stickleback in its gullet.

Reproduction

Perch spawn in the late spring, when the water temperature is between 10 and 15°C, usually from mid-March until June, with a peak of activity in late April and May. The eggs, which adhere to each other and are enveloped in slime, are laid, usually in shallow water, in long white ribbons up to a metre in length, draped over water plants or other submerged objects. As William Houghton describes it: 'a most curious and beautiful object is the spawn which they produce; it consists of a broad band of network of pearl-like eggs, a foot or considerably more in length.' George Boulenger is more lyrical: 'Often in spring one may see delicate lace-like structures, suggesting a lady's scarf, a yard or more in length, lying tangled among waterside vegetation ...' Frank Buckland notes that 'These eggs are dispersed to form a pretty pattern.' Water-logged tree trunks and submerged branches are favourite sites if weeds are absent.

Wire mesh traps set for Perch during the spawning season down to depths of 5 m may become covered with spawn. At spawning time the males arrive at the spawning area well before the females. When a ripe female arrives, one or more of the males chases her, fertilising her egg ribbon as it is laid.

Each female lays about 45,000 eggs per kg of body weight. Frank Buckland said 'that he and his secretary counted 155,620 eggs in a fish which weighed 3 lb. 2 oz. They are whitish, some 1.5-2.5 mm in diameter and take about 120-160 day-degrees to hatch (i.e. 8-16 days at normal water temperatures). Jim Treasurer found that, in the lochs of north-east Scotland, fry '... rapidly dispersed from inshore spawning areas to open water. The planktonic larval period was followed by the inshore migration of metamorphosed 0+ perch. Some juveniles remained in shallow water through the summer and all moved offshore in autumn.'

Predators

Perch eggs, like those of other fish, are vulnerable and often favoured by predators. Frank Buckland records that 'The swans in the Thames, I am sorry to say, eat large quantities of perch spawn.' David Carss and Keith Brockie found Perch to be one of the commonest items (18%) of the diet of Ospreys in Scotland. David notes that he has 'recorded them in heron and cormorant stomachs and in heron nestling regurgitations in Argyll and cormorant regurgitations in Essex, as well as specimens dropped by Ospreys in Deeside.'

Perch are hosts to a considerable number of parasites, some of which are host specific, the rest being shared with other fish species. These parasites include species that infest the exterior of the body, the mouth, gills, eye, intestine and other viscera. Perch are subject to an infectious ulcer disease which may almost eliminate a population in a short time and which can spread rapidly to other populations. During an outbreak of this disease in Windermere a number of years ago it was estimated that over 90% of the population died. This disease, and sometimes other conditions associated with the stress caused at spawning time, can result in the familiar sight of large Perch being washed up dead on the shores of lochs, as described below.

A recurring feature of the ecology of Perch is that they seem to be very prone to mass mortalities. Frank Buckland mentions 'a curious epidemic disease as occasionally occurring among Perch. In 1867 a Perch-plague is said to have destroyed hundreds of thousands of these fish in the Lake of Geneva.' Even before this, an anonymous writer had written of Hoselaw Loch: 'The present proprietor of Hoselaw, Mr James A. Somervail, informed me that the perch which at one time swarmed in the Loch had entirely disappeared. This fact strikingly confirms the report made by Rev. John Baird in his description of the parish of Yetholm in the New Statistical Account (1845), namely "Sometimes, during the most sultry period of summer, shoals of perch are, from some cause unknown, cast out dead upon the margin of the lake."' At Loch Leven, Robert Burns-Begg found that ' ... when the wind is unusually strong they have been actually blown out of the lake, and found in great numbers strewed along the beach. During a blowy afternoon towards the end of 1872, a considerable quantity, "enough to form a cartload", as stated by an eyewitness, was blown on to the east shore of the loch at Grahamston. 'Peter Malloch also wrote that 'Shortly after the great storm which blew down the Tay Bridge I had occasion to be shooting on Loch Leven, and found the shore to the width of about 100 yards, and extending for several miles, strewn with dead perch, undeniable evidence of the destruction caused by wind and waves. For years

after this they were not nearly so plentiful as formerly.' In the late 1960s, the author recorded a similar massive mortality among Perch at Loch Leven, when hundreds of large fish were washed up on the east shore.

Growth

On hatching, the fry, which are about 6.4 mm in length, can swim normally as soon as they have inflated their swim bladders at the water surface, and embark on a planktonic existence for a short time. Because of their small size, they are very vulnerable to predation by other members of the plankton community, including their larger brethren which have hatched earlier. Initial growth is rapid, and they soon start to move around near the surface, sometimes in dense grey shoals, hunting for food and being hunted by larger predatory fish and birds. Once their fins have become fully developed, they move inshore, adopting a more benthic habit, still moving about in shoals but sheltering among marginal vegetation.

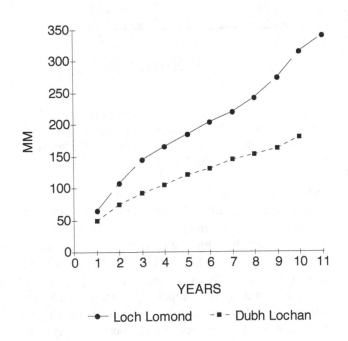

Growth of Perch in two Scottish waters (Shafi & Maitland 1971).

Subsequent growth is very variable, according to habitat and other factors. At the end of their first year they may reach 7 cm and after two years, 10-12 cm. After three years most fish are at least 12-18 cm and starting to become mature. Males mature sexually earlier than females, sometimes after one year, but usually after two or three. Most females do not mature until their third or even fourth years and by this time they have started to grow faster than males. A maximum age of about 10-13 years is normal in the wild but, in the safe environment afforded by captivity, ages of up to 28 years have been recorded for both sexes.

Value

Highly regarded in many parts of Europe, Perch are almost ignored in Scotland and elsewhere in Britain as an item of diet. Formerly, however, Perch were caught commercially at Loch Leven: Robert Burns-Begg records that 'In 1855 ... one haul of perch actually filled eleven barrows, and would have formed two ordinary cartloads.' Perch can be caught in very large numbers around the spawning season because of their habit of entering non-baited traps. These traps, widely used for this species, are made from wire mesh stretched over a heavy gauge wire frame with a funnel entrance. The largest catches are of the pre-spawning shoals of males. At other times of the year, small numbers, including females, may be taken in this way. During World War II, a small fishery was established at Windermere using such traps. The Perch caught were canned and marketed as 'Perchines' to augment home produced food.

Perch are of major commercial importance in parts of Europe; the author has eaten many excellent dishes of this fish in Switzerland and France, especially around Lac Leman, where they are a speciality and one of the top-priced dishes on the menu of most restaurants. The demand is so high that new supplies are always needed. In the early 1990s, the author was contacted by a Swiss firm, asking advice on the possibility of supplies from Scotland. They were willing to pay a high price for filleted Perch and prepared to take as many tonnes as could be supplied on a sustainable basis. Unfortunately, it was not possible to find anyone in Scotland who was willing to start up a fishery for this purpose - in spite of the enthusiasm of the Swiss firm in one of their letters: 'We are ready to flight again to Scotland at any time and for any reason to start a "perch-fishing".'

The Perch is an important match and specimen fish for anglers. William Houghton notes that 'There is no fish, perhaps, that gives better sport to the youthful angler than the Perch; bold, and always ready for a worm, minnow , or other food, they fall easy victims to the baited hook.' Colonel Thornton's sporting tour through Scotland in 1784 describes his perch fishing and the capture of the large Perch described above. Perch fishing was at one time popular at Loch Leven and Robert Burns-Begg notes that 'Of late the angling for perch has been entirely from boats, and very considerable numbers have frequently been taken ... in 1873 a party of eight anglers from Kinnesswood having caught not less than 722.'

Some anglers, however, are only interested in Brown Trout at Loch Leven, as Iain Gunn relates: 'Seizing the rod I landed a fish which proved to be a perch of about a pound weight. There are a lot of perch in Loch Leven, but this proved to be the last straw and we retired ...'

Filleted and well cooked, the flesh of the Perch is tasty, comparing very favourably with that of the best sea fish. The flesh is white, flaky, free from bones and very well-flavoured. In other parts of Europe this is well recognised and there are many varied continental recipes for its preparation. William Houghton is enthusiastic about this species 'The flesh of the Perch is excellent, being in my opinion unsurpassed by any non-migratory fresh-water species with the exception of the eel.' Herbert Maxwell is in near agreement: 'I give the palm to the perch among British fresh-water fish, always excepting the Salmon. ... Clean, firm and white, destitute of that insidious earthy flavour to which fresh-water fish are so prone, the flesh of a well nourished perch taken in pure water any time between the middle of June and the end of February is equal to all but the best of haddocks.'

Others in agreement are Tate Regan: 'The Perch is an excellent food-fish, with white, firm, and well-flavoured flesh.' and Wilfred Morris: 'Have you ever had perch, skinned (not merely scaled), and then baked with new farm butter and seasonings?' Frank Buckland too is an enthusiast 'A perch cooked immediately after it is taken out of the water is magnificent eating.' He recommends that 'The angler should transfix the perch with a sharp stick, and hold him over a fire made with sticks, &c., such as are found by the riverside. He should be eaten bit by bit off the impromptu spit.'

Harry Slack taught the author a simple method of filleting Perch which does not involve cutting into the body cavity. 'The fish is held firmly by the belly and the skin is carefully cut from the back of the head to the tail, on both sides of the dorsal fin. The skin is then eased back behind the head on either side. The head is then held firmly in one hand and the anterior part of the fillet in the other. The hands are jerked apart and one is left with a fillet in one hand and the rest of the fish, to be discarded, in the other hand.' It works!

Thomas Huxley, 'boating up the River Endrick in the mid-60s, saw a Cormorant in the water trying to swallow a fish which was then dropped when the frightened bird flew off.' The fish turned out to be a large Perch, which Thomas took home and ate - a unique experience outside China where Cormorants are tamed and trained to catch fish for humans!

Ross Gardiner, 'diving for deep water benthos samples in the River Tay in the 1970s, found Perch coming right into the area he was working to feed on the disturbed bottom material.' He also 'found inquisitive Perch peering into his eyes while diving in the Rivers Tummel and Tay - presumably attracted by their reflections in the face plate.'

Perch are good aquarium fish, becoming very tame and bold, but any other species sharing their tank should be approximately the same size! W. Furneaux says that 'The Perch is certainly one of the prettiest, if not the prettiest, of our freshwater fishes and is a perfect ornament in the aquarium;' Gregory Bateman agrees: 'The Common Perch ... is an extremely handsome fish and very suitable for the aquarium. It is hardy, and if properly cared for readily adapts itself to life in confinement. The only drawbacks to the Perch as an inmate of the tank are its voracity and a proneness to devour its companions.' ... ' The Perch is one of the most intelligent of all the freshwater fishes, soon learning to know its owner and its feeding time. In a very little time it will become tame enough to take food form the fingers.'

Derek Mills records that he 'used to keep a Perch in a tank in his room and feed it on live Minnows. It was always very noticeable that when the Perch swam round the tank, the Minnows always kept their heads away from it so that they could not be swallowed 'head on' - which is just what the Perch was trying to do!'

Conservation

The Perch, though regarded by the author as a native species, has no special protection in Scotland at the moment and requires none as it is a widespread and common species which is still, 10,000 years after the Ice Age, slowly invading new waters in the north and west of the country.

References

Substantial research has been carried out on the Perch in Scotland and there is a significant body of literature on its ecology here. Numerous studies have also been carried out in other parts of Europe.

Anonymous (1907), Andrews (1979), Bateman (1890), Bolam (1919), Boulenger (1946), Bregazzi & Kennedy (1982), Buckland (1881), Burns-Begg (1874), Burrough & Kennedy (1978), Campbell (1955), Carss & Brockie (1994), Coles (1981), Craig (1977, 1978, 1980, 1987), Craig & Kipling (1983), Furneaux (1896), Goldspink & Goodwin (1979), Gunn (1981), Houghton (1879), Jones (1953), Lamond (1931), Lang (1987), Le Cren (1992), Le Cren *et al.* (1967), Maitland & Campbell (1992), Malloch (1910), Maxwell (1904), Morris (1929), O'Gorman (1845), Regan (1911), Salter (1833), Shafi & Maitland (1971), Stoddart (1866), Thornton (1804), Thorpe (1974, 1977), Treasurer (1981, 1983, 1988), Williams (1965), Willoughby (1970), Worthington (1950).

'Prince of the prickly cohort, bred in lakes
To fest out boards, what sapid boneless flakes
Thy solid flesh supplies.'

Ausonius (undated) *Carmen de Mosella*

*****

Common Goby *Pomatoschistus microps* Buidhleis

'One of the commonest small fishes of the eastern Atlantic estuaries.'

Peter Miller (1997) *Fish of Britain and Europe*

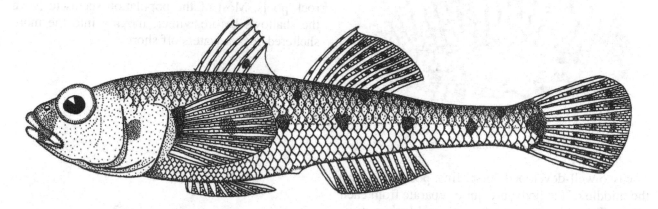

GEORGE Bolam pointed out in 1919 that 'The gobies are small insignificant fishes, which are easily overlooked, and are not very readily determined. ... All the family would repay closer attention by anyone with the necessary leisure.' Maurice Yonge agreed, noting that 'All are adapted in form, habit and coloration for life on or within sand and play their part largely as carnivores in the economy of life on sandy shore.'

Working in the west of Scotland in the 1970s, Jack Gibson noted that 'During the past few years I have been able to examine a good many of these intertidal gobies from various parts of the Clyde. These were collected for me from shore pools by my small daughter and her friends, and thus was "armchair natural history" at its very best, with the least possible exertion on the part of the naturalist! ... I have identified the Common Goby *P. microps* from places on both the east and west shores of the Clyde and on the Clyde islands, ... There is little doubt that the Common Goby *P. microps* is fairly widely distributed throughout all suitable areas of the Clyde; this will usually mean where rivers and burns come down to sandy shores.'

The Common Goby is sometimes known as Snag.

Size

This is one of the smallest of the fishes to be found in fresh water in the British Isles and is mature at a length of 4-5 cm. Its maximum length is about 6.5 cm. Perhaps not surprisingly, there are no rod-caught records for this species.

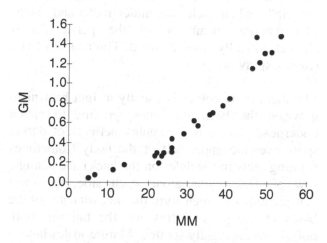

Length / weight relationship in the Common Goby in the Faveron Burn (Healey 1972).

Appearance

There are many species of goby and as pointed out by Timothy Bagenal 'It is easy to know if a fish is a goby; the difficulty is in telling which goby'. The Common Goby has the typical family shape - a broad strong head, a short sturdy body and a rounded tapering caudal region. The caudal peduncle in fact, though narrow, is almost equal in length to the head. Apart from the throat and just behind the head, the body is well covered in scales, which are much smaller at the front than they are behind. There are some 43-50 scales laterally, although there is no sensory lateral line.

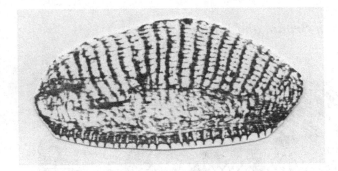

Scale from Common Goby.

The two well-developed dorsal fins, positioned near the middle of the body, are quite separate from each other. The anterior one has several spiny rays, whilst only the first ray of the posterior fin is spiny. The pectoral fins are large and held laterally, while the pelvic fins, like those of other members of the family, are typically fused to form the sucking disc which it uses to hold on to smooth surfaces, especially when sheltering under rocks and shells. The anterior membrane of the pelvic fin is characteristically smooth-edged. The anal fin has a leading spiny ray.

The main body colour is usually a light brownish-grey on the back and sides, grading to cream underneath. There is a complex network of darker spots over the upper half of the body (sometimes forming indistinct saddles on the back) and a single row of dark marks along each side, the most conspicuous of which form distinct triangles at the bases of the pectoral fins and the tail fin. Both dorsal fins are lightly spotted. Mature males have a dark spot on the first dorsal fin membrane, a dark throat and dark bars across the body.

Distribution

The Common Goby is widely distributed along the western coast of Europe, from southern Norway to Portugal, including the whole of the Baltic Sea. It occurs all round the shores of the British Isles, from Shetland to Cornwall.

As its name implies, this is one of the commonest gobies found around the coasts of Scotland, and though it is basically a marine species, it is the only goby found regularly in fresh water in the upper reaches of estuaries and in fresh/brackish situations, occurring in small ditches and trickles along the shore. It occurs in a wide range of habitats; though it is most abundant in shallow muddy estuarine situations, it is also found along sandy shores and in rock pools. Most of the population seems to leave the shallows before winter, moving into the more sheltered deeper waters off shore.

Recorded distribution of Common Goby in Scotland.

Maurice Yonge noted that the Common Goby 'may extend for considerable distances within the freshening waters of estuaries.' George Bolam found that 'The Two-spotted Goby is common about the mouth of the Tweed, and in the Forth ... no doubt it would be found equally common elsewhere if looked for.' The author has found it to be abundant in suitable habitats all along the Solway coast, in some places in association with the Nine-spined Stickleback, especially in small brackish creeks and pools. Although widely distributed, the actual recorded distribution, shown above, is poor.

Ecology

Peter Miller records the Common Goby as common around the British Isles in estuaries, salt marshes, and high shore pools. Jack Gibson, one of the few people to have examined this species in Scotland, found that 'These tiny gobies frequently inhabit very shallow pools, sometimes only a few inches deep, where they occasionally partly burrow in the sand, frequently in association with the even more abundant tiny Flounders. ... *P. microps*, the Common Goby, is typical of sandy intertidal shore pools, often only an inch or so deep, tolerates water of moderately low salinity, and is usually regarded as a freshwater fish; '

N. Dankers and his colleagues found that in the North Sea 'It spawns in June to August in the tidal gullies under *Mya arenaria* shells. The pelagic larvae are usually very abundant in summer at 0.5-1 metre depth. They start demersal life in August at a size of about 13-15 mm. ... Common gobies migrate to deeper water up to 10 metres in winter when the water temperatures decrease below 5-7oC. They cannot survive temperatures below 0°C and they were completely absent from the Dutch Wadden Sea after the severe winter of 1962-63.'

Food

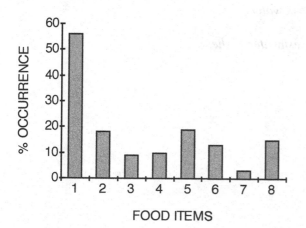

Food of Common Goby in the Faveron Burn (1: Corophium; 2: amphipods; 3: mysids; 4: copepods; 5: isopods; 6: ostracods; 7: annelids; 8: others) (Healey 1972).

The young fry feed on protozoans and larval crustaceans, later depending heavily on larger crustaceans of various sorts such as copepods, isopods, gammarids, mysids, etc.

Reproduction

Spawning can occur from April to August, but takes place mostly during May and June. The eggs are normally laid in clumps within the old shell of a bivalve mollusc such as a clam or cockle, though sometimes just underneath suitable stones, and are guarded there by the male until they hatch. The eggs are rather unusual in being oval and are about 0.7-0.9 mm in length. Both sexes may spawn several times over the summer months: in favourable conditions the population can build up enormously by the autumn.

The biology of the Common Goby was studied in some detail by Peter Miller in the estuary of the Silver Burn, at Castletown. Isle of Man. There, 'sexual maturity is reached in the spring or summer following the first winter, lengths at maturity being 31.75 mm (males) and 29.0 mm (females). The breeding season lasts from May to the end of August. Most adults die in their second autumn, few survive beyond their second winter, the oldest being 21-26 months. Maximum total length recorded was 64 mm. Fish may spawn up to eight times during the breeding season - the autumn mortality apparently being a consequence of this protracted spawning activity.'

In another study, Jones & Miller found that in intertidal pools on mussel beds at Morecambe, during winter, the shore was largely vacated, fish returning in the spring for breeding. Spawning took place in the upper shore pools from April to August. After a planktonic life of 6-8 weeks, young fish became bottom dwelling from June to September.

Predators

As well as the 'usual suspects' (e.g. Herons, Otters), a variety of other, smaller, predators are likely to prey on this common fish. These would include terns and gulls.

Growth

Hatching at lengths of 3-4 mm, the young fish grow rapidly and may reach lengths of 3-4 cm by the end of the summer. They are mostly well-grown, often maturing by the spring of the following year and usually living for only 1.5-2.5 years.

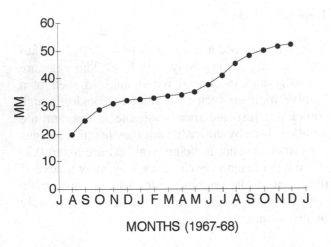

MONTHS (1967-68)

Growth of Common Goby in the Faveron Burn (Healey 1972).

Value

Like most of the small gobies, the Common Goby is of little direct commercial or angling significance, but it is important indirectly as a food source for many larger fish of importance (e.g. flatfish) and is also preyed on by seabirds such as terns. It can therefore be regarded as an important part of the food web of shallow water brackish systems. It tames readily and is an attractive fish for small aquaria, in which it can sometimes be induced to breed if plenty of live food is available.

Roger Phillips & Martyn Rix point out that 'Gobies have no known uses, though they are doubtless valuable as food for larger fishes ... They would be a suitable quarry for children with nets in pools along the shore'

Conservation

The Common Goby has no special protection in Scotland and, as its name implies, being widespread and abundant in most suitable habitats, it does not at the moment appear to need any conservation measures.

References

A few studies have been made of this species in Scotland and additional literature is available from other parts of Europe.

Al Hassan *et al.* (1987), Bagenal (1972), Bolam (1919), Dankers *et al.* (1979), Fouda (1979), Fouda & Miller (1979), Gibson (1976), Healey (1972), Jones & Miller (1966), Maitland (1970, 1879, 1997), Maitland *et al.* (1980), Miller (1964, 1975, 1997), Phillips & Rix (1985), Yonge (1949).

'It is even found in fresh water.'

Timothy Bagenal (1972) *Freshwater Fishes*

*****

Flounder *Platichthys flesus* Leabag

'Flounders are quite at home in fresh water.'

Henry Lamond (1931) *Loch Lomond*

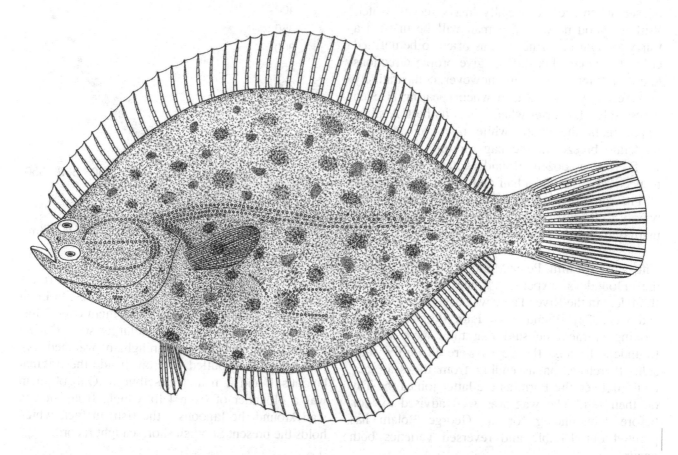

MANY people think of the Flounder only as a sea fish, but, as pointed out by Henry Lamond, it is quite common in many fresh waters during the growing phase. In the Clyde area, Thomas Scott and Alfred Brown describe it as 'Common and generally distributed; ascends most tidal rivers, and frequents some lakes as Loch Lomond, where it is not uncommon.' W.J. Gordon noted that 'For a flat fish it is singularly active and enterprising, and it can climb and cling by means of its ventral fins.' This is the only flatfish in Europe which is regularly found in fresh water; it is also abundant in estuaries and around the coasts where it may well be confused with Plaice (*Pleuronectes platessa*), Dab (*Limanda limanda*) and various other common marine flatfish.

Its unusual early development, resulting in both eyes being on the same side of the flattened body and the fish lying on its left side, has made it quite unique among freshwater fish in Scotland. Although well camouflaged on the sandy bottoms which it inhabits, its flattened shape makes it very vulnerable to an unusual method of capture - spear fishing. St John described a spear fishermen he met in the Moray Firth area in the late 1800s '... he had a singular kind of creel slung to his neck, and a long, clumsy-looking kind of trident in his hand. Walking slowly backwards, but still keeping in two-foot water, with poised weapon and steady eye, he watches for the flounders which come in at every tide. When he sees one, down goes his spear; and the unlucky fish is hoisted into the air, and then deposited in the creel.'

John Colquhoun gives a first hand account from the other side of the country, fishing in Loch Fyne: 'The shallows, however, yield abundant amusement with the spear. ... Of these methods of fishing the spear is certainly the most exciting, requiring the three qualities needful for excelling in all outdoor sports - viz., a quick eye, a steady hand, and a cool head. ... Except by a keen eye, few flat-fish (the exact colour of the sand or mud on which they rest) will be discovered. If the hand is not true, they will be missed when seen, especially in six feet of water. Without good nerves, ... a man will be in such a flurry at sight of a fine fish, as often to be unfitted either to manage his skiff or give proper directions to an assistant. ...There are, however, comparatively very few days in a season when spearing is even practicable - far fewer when it is at its best. A slight ripple mars the sport, while it is ruined by a moderate breeze. ... the bag for one Saturday in 1865 was a dozen flounders and two plaice, respectively four and a half and five pounds weight. ... The same expert "spear" has often landed from twenty to thirty flounders in a forenoon, several of them two or three pounds in weight.'

Unusually for him, Peter Malloch enjoyed a joke at the Flounder's expense: 'Once when catching flounders in the River Tay a wag asked me if those I had were Tay flounders or Earn flounders. After looking at them he said that they were all Tay flounders, because their eyes were all on the right side; therefore, on ascending from the sea, they could not see the Earn, as the latter joined the Tay on their left.' The wag was well advised to look before pronouncing for, as George Bolam has pointed out 'Double and reversed varieties both occur.'

Alternative names for this species are Butt and Fluke. Frank Buckland noted that 'In Scotland, they are known by the names of fleuke and maycock-fleuke.' A literal translation of the scientific name is 'Flounder Flatfish'.

Size

Most mature Flounders are some 20-30 cm in length but the species can exceptionally attain a length of 50 cm and a weight of 3.5 kg. The Scottish boat rod-caught record was taken at Portnockie by K.F. MacKay in 1985, weighing 1.244 kg (2 lb 11 oz 13 dm). The shore rod-caught record was for a fish caught by R. Armstrong at Musselburgh in 1970 which weighed 2.14 kg (4 lb 11 oz 8 dm). The British rod-caught boat record stands at 2.593 kg (5 lb 11 oz 8 dm) for a fish caught by A.G.L

Cobbledick in 1956 at Fowey, Cornwall; the shore rod-caught record is 2.466 (5 lb 7 oz) for a fish caught by Barry Sokell in 1994 in the River Teign, Devon.

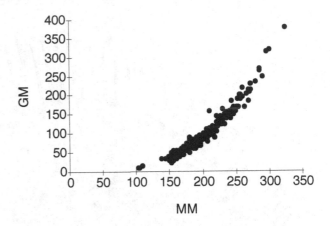

Length / weight relationship of Flounders from the Clyde Estuary.

In 1912 Peter Malloch noted that 'In rivers flounders are seldom caught over 1 1/2 lbs. In Loch Speggie in Shetland I have caught them over 2 lbs, while in the sea they grow to a larger size.' During mid-late 1970's Peter Cunningham watched big Flounders being pulled up from inside the original lagoons 1 and 2 near Musselburgh. One of them was a Flounder of over 4 lb. caught from the sea wall around the lagoons - the fish, in fact, which holds the present Scottish shore-caught record.

Appearance

The Flounder, like other members of the family, is extremely flattened and lies on one side of the body, the eyes and some other organs having migrated to the other side so that they are uppermost. Most Flounders lie on the left side with the right side uppermost, but reversed animals are quite common, where the opposite is true. In normal Flounders, the small terminal mouth is positioned to the right of the eyes. The head is large and flattened with the well-developed lateral line curving from it, round the operculum and pectoral fin, then straight down the middle of the body. The body is covered with small irregularly distributed and shaped scales which are characteristically rough on either side of the lateral line, especially between it and the pectoral fin, making this area very rough to the touch. The bases of both the dorsal and the ventral fins are also rough: this is one of the simplest ways to identify this species in the field and distinguish it from other flatfish. The dorsal fin has a sharp spine at its base and the anal fin also has a spine.

Scale from Flounder.

In colour, the upper side (usually the right) is a dull brown with indistinct blotches of greenish brown or grey and some dull red spots. The underside (usually the left) is always a pale greyish-white. In life, colour can be changed very rapidly to suit the shade and pattern of the substrate on which the fish is lying. They are also able to submerge themselves almost completely in the sand with only their eyes protruding.

Distribution

The Flounder is a very widely distributed species, occurring in Europe along coastal waters from the White Sea round northern Norway and throughout the North and Baltic Seas to Spain and north Africa, and into the Mediterranean and Black Seas. It occurs all round the British Isles in a wide range of habitats but especially on silty, sandy and gravelly bottoms. It is especially common in estuaries, from which young fish in particular often migrate upstream into fresh water: they may be found in rivers and lochs, sometimes many kilometres from

the sea. F.G. Aflalo believed that 'The Flounder may be regarded as a sea-fish that has made its way up rivers or taken to a partially fresh-water habitat.'

The author has caught this species in Loch Lomond 50 km from the sea, living at depths of up to 30 m. Its history in the area has varied, for John Colquhoun commented in 1880: 'In my early days the loch was full of flounders, some of large size. They, too, leave the fresh for salt water at certain times; but since the pollution of the Leven by turkey-red dyers, the flounders disappeared, refusing to return by such an unsavoury route.' Alfred Brown, however, noted in 1891 that 'It ascends the Leven from the sea in summer and passes up at least as far as Luss, near which village it has been taken on set lines within the past few years.' Henry Lamond commented that 'Flounders are quite at home in fresh water.' and that 'Mr John MacFarlane, Balmaha, ... netted a flounder in his salmon net measuring eight inches in length.'

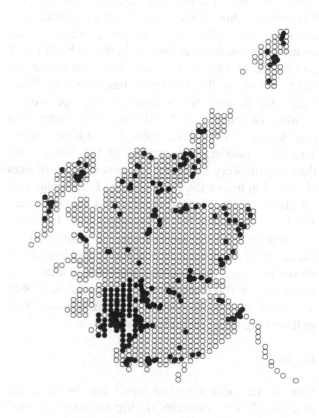

Recorded distribution of Flounder in Scotland.

Elsewhere in Scotland, George Bolam noted that in the Borders area: 'The common Fluke is well known along the coast, and ascend our burns and rivers till stopped by some unsurmountable obstruction, considerable weirs being frequently successfully passed.'

Fred Buller 'caught several flounders twenty-five miles from saltwater when roach-fishing the Tweed at Sprouston.' Hugh Falkus 'has seen them in the Tweed even further upstream - just below the cauld at Kelso.' Herbert Nall noted that 'it has been recorded .. in the Aberdeenshire Dee, a rapid and rocky river, it has been found above Banchory, twenty miles from salt water. But these long journeys are in quest of food; it returns to the sea to spawn.' Not far away, the author found them to be abundant in the shallow sandy Loch of Strathbeg in the 1970s. Even in Loch Leven, before the mills on the River Leven created physical and pollution obstacles, Robert Burns-Begg recorded that about 1900 'besides a great variety of trout of bright red colour, and all of them highly flavoured, eels, pikes, perches, and flounders are found'.

Andy Walker records an unusual habitat for Flounders: 'This species would not be expected to occur in an upland hill resevoir, yet they are common in Glenfarg Reservoir in the Ochil Hills, at an altitude of 180 m. This reservoir was formed in 1915 by impounding the upper reaches of the River Farg, which then flows through a gorge with a number of barriers and with torrential current in places, to reach the River Earn, 13 km downstream. Though Flounders are present in the lower Farg they are unlikely to be able to penetrate more than about 1 km above the head of tide. The explanation for their presence in the reservoir is simple. Since 1984, the reservoir has been augmented by water pumped up from the River Earn, via a gravel-filled basket filter within the bed of the river about 3 km above Bridge of Earn, where the river is tidal fresh water where there are plenty of Flounders. Roach are also found in Glenfarg Reservoir and are likely to have originated from the same source.'

Ecology

One of the most abundant small fish in the upper reaches of many estuaries during summer are small Flounders which have just started their migration from the sea, and may be present in thousands. Certainly this was the case in all the main rivers of the Solway when Alex Lyle and the author were carrying out surveys in the early 1990s. These

freshwater Flounders may stay in fresh water for several years before migrating back to sea as adults.

Food

The fry rely on their small yolk sacs for about two days and then feed on zooplankton, especially crustaceans such as copepods and the larvae of crabs. On the bottom they feed on a variety of benthic invertebrates, including worms, crustaceans and molluscs. The young fish often move into fresh water for a year or so where they feed on tubificid worms, insect larvae and molluscs.

It is principally a night-time feeder, and in the sea often undertakes a daily feeding migration up the shore, closely following the incoming tide up and then back down again as it retreats. It may sometimes swim well off the bottom to take an item of food.

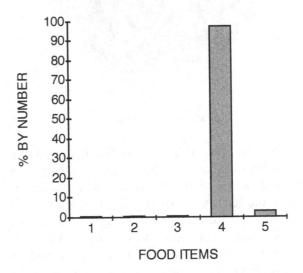

Food of Flounder in the River Tweed (1: molluscs; 2: worms; 3: beetles; 4: midge larvae; 5: others) (Radforth 1940).

Reproduction

At spawning time, adults in fresh water move to the sea; mature fish migrating off-shore to suitable spawning grounds in April and May, laying their eggs in 25-50 m of water. The eggs (about 1 mm in diameter) are pelagic initially but lose this buoyancy during development, which takes about 5-8 days at temperatures of 9-12°C. Fecundity is high and females are capable of producing some 400,000-1,000,000 eggs each depending on size.

R.W. Summers studied Flounders in the Ythan Estuary in the late 1970s. He found that 0+, 1+ and 2+ Flounders showed regular increases in length and weight between April and September, but did not grow during the winter months. Males matured toward the end of their third year and females their fourth. There were various regular migrations: in spring (April May) there is an immigration of immature fish; 0+ fish enter the estuary in late June/early July; many 3+ an 4+ leave for the sea in July; winter emigration of all age classes started in October and continued through the winter. The total population in the Ythan Estuary at high tide during August and September was 265,000.

Predators

The habit of following the tide in and out on shallow sandy shores during daylight makes it a valuable (and available) prey for Ospreys in Scotland, partly leading to their concentration as breeding birds around some of the large east coast estuaries (e.g. Findhorn Bay). In Loch Lomond, John Mitchell has watched Ospreys fishing successfully for Flounders in the shallow sandy waters near the mouth of the River Endrick. 'When the water level was low, in late summer, Ospreys successfully caught Flounders, which were quite distinctive as they flapped in the Osprey's talons.' However, David Carss and Keith Brockie found Flounder to be one of the least common items (1%) of the diet of Ospreys in Scotland. Sometimes, fish-eating birds have difficulty swallowing large Flounders, which are very wide: Peach noted that 'a cormorant with a large flounder is even a more ridiculous object than one with an eel.'

J. Cunningham records '... a curious belief current among fishermen, to the effect that the Flounder carries its eggs on its back, the so-called eggs being in reality tumours.'

Growth

The fry are some 2.5-3 mm at hatching and are perfectly symmetrical pelagic little fish. As they grow, however, they start to develop asymmetrically, metamorphosing rapidly at 20-30 mm to become flattened and bottom living. They grow fairly rapidly in good conditions and can reach a length of 10 cm after one year and 15 cm after two. Males mature before females at a length of ca

12 cm (18 cm for females). They usually reach about 25 cm after four years and mostly live for 8-10 years, occasionally surviving up to 20 years.

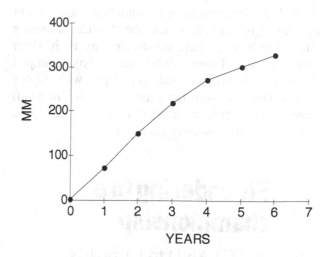

Growth of Flounder in the Ythan Estuary (Summers 1979).

Value

The Flounder is an important commercial species caught in trap nets and seines in several parts of Europe, though it is not rated highly in Scotland, where it is usually a by-catch of some other type of bottom trawl fishery.

Charles St John, commenting on the situation in Morayshire in 1891 noted that 'Among the available products of the sandy creeks and bays on this coast are immense quantities of excellent flounders. ... Notwithstanding the abundance and excellence of the flounders, left, as it were, for any person to pick up, with scarcely any exertion, the country people very seldom take the trouble to catch them, excepting now and then by the line, in a lazy, inefficient way.' Tate Regan obviously agreed: 'Although very abundant, the Flounder is generally not held in much estimation for the table, and is considered much inferior to its relative, the Plaice.'

It is, however, a popular sport fish, caught in large numbers by sea anglers all round the soft coasts of Scotland. In fact, the author believes that large specimens, freshly caught, are delicious fish to eat when fresh, especially when hot smoked. One of the finest meals of fish he has ever eaten was in Finland, of fish caught by the author and colleagues in the Baltic Sea (Flounder, Perch and Haddock) and hot smoked almost immediately on the shore.

Alan Ayre was surprised recently (October 2006), 'when fly fishing on the River Tay at Stanley, to catch a large (11") Flounder.'

As well as being speared, Flounders are also caught by 'tramping' and each year the flounder tramping championships are held on the Solway Firth. Peter Cunningham's father, (who, as a boy, tramped flounders elsewhere, and set lines with hooks baited with lugworm for Flounder buried in sand) remembers catching a sea gull - which was unhooked and apparently survived

Floundering to a championship

SCOTLAND today boasts a new world champion after Drew McDougall, from Dalbeattie, won the Annual World Flounder Fish Tramping Championship at Glenisle on the Solway Firth. His fish – 92mm from head to tail – was so small it couldn't be weighed.

Report of the 2006 Championship (The Herald).

Frank Buckland notes that 'They have been successfully transferred to fresh-water ponds.' Though not often kept in aquaria by amateurs, small specimens tame readily, making very attractive additions to suitable freshwater or marine tanks. They have the attractive habit of adhering to the glass side by means of the flattened body (thus allowing the pale underside to be examined closely), or almost disappearing on the sandy bottom by a combination of camouflage colouration and burying in the sand. They require live food and eventually must be returned to an appropriate river from which they have easy access to the sea.

Conservation

The Flounder has no special protection in fresh water, but in the sea, fishing is subject to normal sea fishery regulations.

References

Several studies have been carried out of the Flounder in Scotland as well as in other parts of Europe.

Beaumont & Mann (1983), Bolam (1919), Brown (1891), Buckland (1873), Buller & Falkus 1988), Burns-Begg (1874), Carss & Brockie (1994), Colquhoun (1896), Gibson (1972), Gibson (1984), Gordon (1902), Jones (1952), Kennedy (1984), Kislalioglu & Gibson (1977), Lamond (1931), Malloch (1910), Moore & Moore (1976), Mulicki (1947), Nall (1930), Radforth (1940), Rae (1970), Regan (1911), Scott & Brown (1901), St John (1891), Summers (1979), Van Den Broek (1979).

'The Flounder is generally not held in much estimation for the table,
and is considered much inferior to its relative, the Plaice.'

Tate Regan (1911) *British Freshwater Fishes*

*****

FISH WHICH DID NOT STAY

*'In November a consignment of one hundred yearling bass
was received from Germany, where these fish are artificially reared.
They measured from three to four inches long ...'*

Herbert Maxwell (1897) *Memories of the Months, First Series*

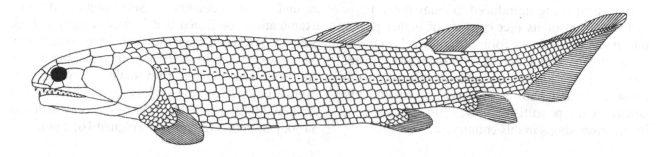

HUNDREDS of species of fish which were once native to Scotland became extinct millions of years ago and are now known only from fossils. Much more recently, several contemporary species have been introduced to waters in the wild in Scotland, but have never managed to establish populations. One native species, the Vendace, became extinct during the last century. All of these are fish which did not stay - for one reason or another.

The reasons for these failures are often uncertain. The Rainbow Trout, which is considered in full elsewhere, has been introduced widely and often over the last 120 years, but has never yet managed to breed successfully over a number of years and establish a population in any of the hundreds of waters into which it has been released. Similarly, the Chinese Grass Carp, which is considered below, has been introduced to a number of waters in Scotland but will never establish a permanent population because climatic conditions are not suitable for its reproduction.

Fossil fish

Many geological deposits in Scotland are rich in fossil fish, and over the geological time periods a wide variety of species occurred here, most of them apparently marine rather than freshwater species. R.H. Traquhair noted that 'The Carboniferous rocks of the west of Scotland are rich in fish-remains, although entire specimens of fishes are not of such frequent occurrence as in Midlothian ...'

These were among our oldest large animals, for as Hugh Miller remarked: 'Of all the vertebrata, fishes rank lowest, and in geological history appear first.' Many of them were first discovered in Scotland, as noted by A.S. Alexander: 'The fish banks consist of hard, grey, flaggy shale ... *Lanarkia* was named after Lanark ... *Birkenia* was named after Birkenhead Burn, where this genus was first found. It is, as its specific name (*elegans*) implies, an elegant little fish, fully two inches in length, completely clad in plate armour ...' *Lanarkia spinosa* and *Birkenia elegans* were found in the Upper Silurian Shales. *Jamoytius* was a freshwater fish of the freshwater shales of Lesmahagow with no paired fins or jaws.

Ever one to bring the past to life, Hugh Miller noted that 'The old Devonian species of fishes, like those of the present day, seem to have had their favourite haunts and feeding or spawning grounds, and must now be sought for where they congregated of old.' As with modern fish, these old species had their tolerance limits; eventually these were exceeded and entire fish communities disappeared.'

Failed introductions

Chinese Grass Carp *Ctenopharyngodon idella* - also known the White Amur - have been introduced widely in the British Isles, including Scotland, and is now accepted as an angling species here. The present British rod-caught record is for a fish of 7.257 kg caught in 1986 in a private lake near

Canterbury. There is no Scottish record. Nowhere has it succeeded in establishing itself, however, and all populations are maintained by stocking. The Grass Carp is native to the middle and lower reaches of the River Amur in Asia. In many temperate countries, due to low water temperatures, it never matures sexually and so has never been known to breed in Scotland, nor is it likely to. One of the important features of this species, and a major reason for it being introduced to many areas, is the fact that most of its diet consists of higher plants; thus it can be used as a form of biological control to keep down unwanted and troublesome weed growths: for example it has been introduced for this purpose at Sandyknowes Fishery. It is an attractive aquarium and pond fish and was formerly available in aquarium shops in this country.

Pink Salmon *Oncorhynchus gorbuscha* - sometimes called Humpback Salmon - is one of the five species of salmon native to the Pacific rivers of North America and Asia. Three of these species have been introduced by Russian fisheries workers to White Sea rivers, and thus given access to the North Atlantic. Pink Salmon are now established in the White Sea area, providing a useful commercial fishery for the Russia. A few breeding populations have also become established in some Norwegian rivers, while stray fish have turned up in Iceland.

The first specimen in Scottish waters was caught in July 1960 in coastal nets set just south of Aberdeen: it was a fully mature male, 52.1 cm in length and 1.8 kg in weight. Its appearance caused some consternation, for its heavily spotted tail indicated that it was a Pacific salmonid, but its other features did not fit with the provisional identification as a Steelhead Trout - the most likely species to turn up, since so many Rainbow Trout were being moved around the country. Subsequent detailed examination confirmed that it was actually a Pink Salmon, presumed from the White Sea area. A few others were taken in the following years, but a long gap followed until a fine male specimen was captured by two Renton anglers, William Miller and Andrew McElwee, in the River Leven. This fish had a fork length of 532 mm and weighed 1612 gm.

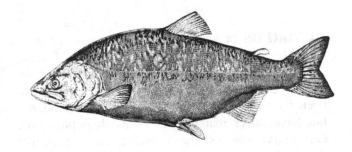

Pink Salmon (from Maitland 1972).

Pink Salmon recorded from Scottish waters			
DATE	PLACE	CAPTURE	SEX
16.07.60	Altens, Kincardineshire	bag net	male
19.07.65	Norham, River Tweed	sweep net	male
07.07.67	Bonar Bridge, Sutherland	sweep net	male
29.08.67	Stromness Voe, Shetland	sweep net	male
12.07.73	Portlethan, Kincardineshire	bag net	female
12.07.73	Delnies, Nairn	bag net	male
01.08.73	Loch Duich, Ross-shire	sweep net	male
02.08.73	Macduff, Banffshire	bag net	female
25.09.73	River Naver, Sutherland	landing net	male
19.08.03	River Leven, Dumbarton	angling	male

Bob Williamson noted that 'Pink salmon are not the only exotic salmonids that might be found in Scottish waters. Coho and chum salmon (*Oncorhynchus kisutch* and *O. keta*) have also been used in acclimatization trials. Sea-run rainbow trout (*Salmo gairdneri*), which have spots on their tails like pink salmon, have been recorded in Scotland.'

Unilever's attempts to rear Coho Salmon in Scotland have been described by A.L.S. Munro;

some 26,000 eggs of this species were introduced from British Columbia. Growth in the artificial system was slow and fish smolted as yearlings. After sea water adaptation, growth was better, fish reaching 2.2-2.4 kg in 18 months. Maturation was slow but it was possible to strip the adult fish and obtain F1 stock. Legislation established following this import prohibited the rearing of Coho Salmon in outside cages, and allowed the F1 fish to be used for experimental purposes only.

Two other species of Pacific salmon have been successfully introduced to waters in Europe outside the British Isles - the Chum Salmon, *Oncorhynchus keta* and the Coho Salmon, *Oncorhynchus kisutch* - and it is possible that both of them could make their way to Scottish waters in future years - indeed Coho Salmon have already been recorded in the Channel Islands. Chum Salmon have been introduced to the rivers of the White Sea area by the USSR but so far none have followed the Pink Salmon to the shores of Scotland. Some Chum Salmon have been caught, however, by commercial netsmen off the Norwegian coast; their status has been difficult to ascertain, for the fresh run fish can easily be mistaken for Atlantic Salmon.

Largemouth Bass *Micropterus salmoides* is one of the six species of this family now established in Europe. It is native to North America, where it is found streams and rivers, ponds and lakes. Because of its fine sporting and eating qualities, the species has been widely introduced to the waters of many temperate and tropical countries where it has often succeeded in establishing itself. It was first introduced to Europe towards the end of the 19th century and is now firmly established locally in a number of countries in the south. In Scotland, however, in spite of several introductions (some along with Smallmouth Bass) it has failed to establish. Only in the extreme south of England did it become established, at two sites in Dorset where it seems now to be extinct.

Herbert Maxwell describes its introduction to Scotland. '... in 1892 it was decided to introduce them to Scottish waters. In the spring of that year a pond about one hundred and fifty yards long was cleaned out for their reception ... In November a consignment of one hundred yearling bass was received from Germany, where these fish are artificially reared. They measured from three to four

inches long ... The pond was drawn in April 1895 ... six bass in lusty condition with fine olive green backs and white bellies, were drawn ashore. ... A year later, in April 1896, we drew the pond again, and this time ... seven bass were drawn to land .. they were transferred to the lake ... During the thirty-five years since these bass were liberated, nothing has been seen or heard of them.'

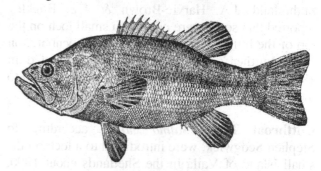

Largemouth Bass (from Maitland 1972).

In fact, apparently unknown to Herbert Maxwell, J.A. Harvie-Brown & T.E. Buckley recorded in 1892 that Largemouth Bass had been previously released in Scotland: 'These fish were imported by the Marquis of Exeter in 1879 - "So far as I recollect, I imported through my pisciculturist, Mr. Silk, the Bass which were turned into Loch Baa by the Duke of Argyll in 1881 or 1882. These were, I think, mostly small-mouthed Bass, but there were a good many of the big-mouthed among them."'

Smallmouth Bass *Micropterus dolomieu* are similar in many ways to Largemouth Bass and have similar ecological requirements. The Smallmouth Bass is native to clear lakes and rivers in eastern central North America, but has been introduced to Europe where it is now established in several countries. Reproduction is from May-July, spawn being laid in a nest excavated by the male among sand or gravel. The male guards eggs and fry, the former hatching after 4-10 days. The young mature in 3-4 years and may live to 15 years. The main food is invertebrates (mainly crustaceans and insect larvae) when young, large invertebrates and fish when older. Though of little significance in Europe, it was formerly important commercially in North America and is still a prized sport species there.

Frank Buckland records that Smallmouth Bass first appear to have been introduced to Scotland in waters at Dunrobin Castle in Sutherland in December 1878. In 1881, as noted above, they were introduced (along with Largemouth Bass) to Loch Baa by the Duke of Argyll: 'Only one of the Bass which were put into Loch Baa has been caught. The old keeper there thought he saw some fry in one of the bays; but he is not certain that they were Bass.' Richard Fitter recorded that in 1890 more were released in Loch Baa. In 1882, 10 large fish from Baltimore were put into a loch at Golspie in Sutherland. J.A. Harvie-Brown & T.E. Buckley reported that some were put into 'a small loch on the top of the hill in Laggan Forest.' near Lochbuie, on Mull. Further south, H.S. Gladstone recorded in 1912 that this species was introduced about 1890 to the upper River Annan.

Cutthroat Trout *Salmo clarki*, according to Stephen Sedgwick, were introduced to a loch on the small island of Vaila in the Shetlands about 1890, but little is known about this introduction.

The following species were all introduced between 1920 and 1930 by John Berry and his father to various ponds at Tayfield in Fife.

American Lake Charr *Salvelinus namaycush*, native to large lakes in North America.

Dolly Varden *Salvelinus malma*, an exciting riverine charr native to north-west North America.

European Mudminnow *Umbra krameri*, a small 'freshwater dogfish', native to central Europe.

Bleak *Alburnus alburnus*, native to waters in southeast England and continental Europe.

Bitterling *Rhodeus sericeus*, native to continental Europe and Asia, but established in England.

Danube Catfish *Silurus glanis*, native to central Europe, but established in England.

Brown Bullhead *Ameiurus nebulosus*, native to North America, but established in Europe.

Pumpkinseed *Lepomis gibbosus*, native to North America, but established in southern England.

None of these species apparently was successful in establishing a permanent population there, which is perhaps not surprising in view of the habitat requirements of some of them. Most were destroyed

during a very serious pollution incident which occurred in 1934.

There are undoubtedly many other cases where fish were introduced from abroad, but there is no record of the introduction. Where there is information, it is often very vague. For example, an anonymous writer in *The Field* in 1885 recorded the shipment to Liverpool for 'the Marquis of Lorne, a consignment of ova of the celebrated bass and whitefish from the inland lakes of the Dominion (of Canada)'. A previous consignment of ova had been sent to Inveraray, but the lot referred to were sent to London. The results of these importations are not known, but there must have been some success, for a year later in *The Field* of 1866 is was noted that 'in consequence of the success attending the introduction of the whitefish in this country last year special attention has been given to their culture with a view to their distribution in some of our lakes.' There appear to be no further records of such distributions , nor of populations which could have resulted from them.

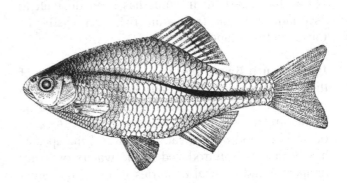

Bitterling (from Maitland 1972).

Other uncertain introductions include the 'catfish sp.' which J.A. Harvie-Brown and T.E. Buckley record as having been introduced to 'Loch Uisg; Lochbuie, island of Mull.' Gudgeon and Tench were also released to the same water, but none of these species seem to have been recovered since.

Close relatives

As well as species new to Scotland, it has been common practice among fishery managers and fish farmers to bring in foreign stocks of species which are actually native to Scotland. The belief that 'new blood' and 'hybrid vigour' were desirable attributes, held sway among many folk during the 19th and

20th Centuries and, unfortunately, still holds true with some fishery managers and fish farmers today who are legally, but irresponsibly, importing stocks of several species which cannot benefit, and are likely to damage, our native populations.

Although Scotland is one of the world strongholds of Atlantic Salmon, with over 400 different populations each originally with a unique genetic identity, when salmon farming started here in the 1970s, it was somehow felt essential by fish farmers to bring in foreign strains to cross breed with Scottish stocks and thereby create fast-growing domestic strains. Thus, stocks of fish from Norway, Iceland and elsewhere were imported to form part of a farmed population of very mixed genetic identity. Many complaints were raised at the time by the author and others, concerned that escapees, diseases and parasites from the many salmon farms developing in and around Scotland would constitute a threat to native stocks. These concerns were ridiculed as scaremongering by the fish farming community and ignored by the Scottish Office. In fact, the reality has proved even worse than envisaged: many millions of farmed Atlantic Salmon of variable genetic background have escaped from cage farms and are known to have interbred with native fish in many of our rivers.

Introductions of other races of Atlantic Salmon have been proposed in the past. A letter, dated 1 November 1991, from the Scottish Office to the Nature Conservancy Council for Scotland (now Scottish Natural Heritage) concerned the possible introduction of landlocked salmon into Scotland: 'Since 1989, there have been several queries from fish farmers and salmon fishery interests about the possible introduction to Scotland of "landlocked salmon" for controlled fisheries. We have been considerably embarrassed to deal with correspondence on this issue dating from last year. We are responding to this now but we will need to establish a policy and procedures for licenses to release such fish in Scotland under the Wildlife and Countryside Act 1981.

The background is as follows. A fish farmer, Mr Hugh Horrex, has imported to a hatchery in Wales eggs of a North American strain of *Salmo Salar* (sic) .... to breed smoults (sic) from his stock for sale to "put and take" fisheries. ... The first such proposal (in Scotland is) ... to stock a loch in

Kintyre. ... The policy issues we need to resolve are whether licences are required at all under the 1981 Act for releases of this kind of "landlocked salmon" ... and, if so, whether it is desirable to agree to any such licences; ...' It is not clear how this issue was finally resolved but none of these salmon (*Salmo salar sebago*) seem ever to have been introduced to Scotland.

Brown Trout (from Maitland 1972).

Scotland has always been famous for its Brown Trout, especially the population at Loch Leven, where the stock was regarded by anglers as ideal - silvery, pink-fleshed and fast growing. As a result of this reputation, stocks of Loch Leven fish were transferred, not only to waters all over Scotland, but also all over the world, including Australia and New Zealand, where they quickly established themselves in suitable waters.

However, not content with having one of the most celebrated stocks of this species, many landowners and fishery managers wished to import to Scotland foreign stocks of Brown Trout, some of them regarded at the time as separate species. Notable among these were the so-called Lake Trout of some of the large lakes of Europe, and several different stocks of these were imported during the 19th and 20th centuries and released into Scottish lochs. John Colquhoun records that: 'My late brother imported ova of Geneva trout ... About 200 fry were introduced into Loch Sloy ... but have not made any appearance yet. ... My sons, in four hours' fishing, captured twenty dozen last summer, no bigger than I remember them when a boy - proof sufficient that the foreign intruders have as yet made no impression on the Arrochar Highlanders.'

J.A. Harvie-Brown & T.E. Buckley record that 'The "Dog Loch" in Mull - a small mountain tarn, has been stocked by Lochbuie with Great Lake Trout, and from a clear shallow creek on the margin of the loch we saw a trout of at least ten pounds weight dart out towards the middle of the loch, raising a wave on either side of him, like a salmon.' It was also noted that 'Loch Dhu, in Benderloch, was partially cleared of its native stock by dynamite, and stocked with Loch Levens, but with what result is unknown, as it is isolated and unattended to.'

The latest episode in this saga of salmonid imports concerns our third native salmonid, the Arctic Charr, for this lovely fish is now being farmed at various sites in Scotland. Again, in spite of Scotland being a stronghold in Europe for this species, with over 200 populations, fish farmers felt compelled to bring in foreign stocks. So far, several different alien strains have been brought in from Canada and Iceland and it is likely that problems similar to those with Atlantic Salmon will arise. This is nothing short of genetic vandalism; in spite of approaches by the author and others to the Scottish Office and more recently to the Scottish Executive, nothing is being done to stop these imports. Our populations of Arctic Charr, with few exceptions, are pristine, unlike those of Atlantic Salmon and Brown Trout where there have been extensive transfers and mixing of stocks. They have been isolated for many thousands of years and it is clear from current research that many of them have developed individual genetic identities within that time. This 'evolution in progress' about which we still have much to learn, is now going to be destroyed by escapes from charr farms.

Arctic Charr (from Maitland 1972).

The need for studies of these native fish stocks is nothing new. In 1887, J.A. Harvie-Brown & T.E. Buckley commented: 'The isolation of the charr in certain lochs on the shoulder of Ben Hope, in Sutherland, is a subject distinctly deserving of the attention of naturalists, as indeed is the isolation of charr and trout in several other localities ...' Only now is attention being paid to the ecological and genetic differences among the still pristine populations of these species.

Casual visitors

Occasionally, strange fish can be found in fresh waters - especially those closely connected to the sea. For example, John Colquhoun noted in 1888 that 'We made a curious discovery last autumn - viz., that seithe frequent fresh water. When trouting in Arrochar Burn, my son caught eighteen "cuddies" nearly 100 yards from the sea. The most experienced Loch Fyne fishermen would not believe it until they saw his basket. They were taken with brandling worm.' The Saithe *Pollachius virens* - also known as Coalfish - is a common fish in coastal waters all round Scotland. Similarly the Sand Smelt *Atherina presbyter* on the west coast moves into fresh water for short periods, as J.A. Harvie-Brown & T.E. Buckley mention: 'I have captured it from shoals at the mouth of a stream in Loch Creran'. This species is much commoner further south in Britain, as Frank Buckland records 'on the Hampshire coast ... they come into the river in the summer, and are caught by thousands in the fresh or brackish water. The method of taking them is this way: a man from a boat drops a large round shallow hoop-net, fastened by a rope to the end of a pole, into the water to the bottom, and then strews crumbs of bread over it, and by leverage of the pole on the side of the boat sometimes brings up thousands at a time; they have the same cucumber scent, as the sparlings and smelts sold in the London market.'

Sometimes dead fish appear in strange places and several instances of this have been reported to the author at various times: a Skate (*Raja*) in Loch Lomond and a South American Armoured Catfish (*Loricaria*) in the Cree Estuary are two examples. Dead Tiger Barb *Puntius paripentazona* reported in 1996 to Colin Adams as having been found in Loch Lomond 'turned out to be a practical joke by a post-graduate student - the fish were mortalities from his tropical aquarium.'

Though the Vendace has been extinct in the Castle Loch at Lochmaben for about 90 years, people are still on the lookout for it there and from time to time there are reports that a dead specimen has turned up. On the two occasions when these specimens have been available to the author for inspection, one turned out to be a Sparling, the other a Grayling. It appeared that both these species are used as bait by pike fishermen who regularly fish the loch.

Visitors to expect

It is likely that other alien species will be introduced to Scotland in coming years, especially if the legislation and management structure for freshwater fisheries is not improved. A wide range of species are possible immigrants likely to be brought in by humans at some time or another. A few of them have already been introduced unsuccessfully, as described above.

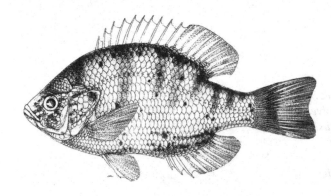

Pumpkinseed (from Maitland 1972).

Several species native to southeast England are contenders, for example Silver Bream *Abramis bjoerkna*, Bleak *Alburnus alburnus* and Spined Loach *Cobitis taenia*. A number of foreign species are already established in England, for example the Black Bullhead *Ameiurus melas*, Pumpkinseed *Lepomis gibbosus*, Largemouth Bass *Micropterus dolomieu* (from North America) and the Sunbleak *Leucaspius delineatus*, Danube Catfish (Wels) *Silurus glanis*, Pikeperch (Zander), *Sander lucioperca*, False Harlequin *Pseudorasbora parva* and Bitterling *Rhodeus sericeus* (from Europe).

Other species, not yet established anywhere in the British Isles, but readily available in aquarium shops until recently, include Brown Bullhead *Ameiurus nebulosus*, Channel Catfish *Ictalurus punctatus*, Fathead Minnow *Pimephales promelas* and the Japanese Weatherfish *Misgurnus anguillicaudatus*.

Some of these species, but not all, are included in The Prohibition of Keeping or Release of Live Fish (specified species) (Scotland) Order 2003, introduced recently by the Scottish Executive. The list of species in this Order is given at the end of this chapter. It is to be hoped that a number of other, potentially damaging, alien species will be added to this list in due course - including those Engish species not yet found in Scotland.

Solutions

The solution to the problems which have been outlined above is to have a national policy and management structure for all fish species across the whole of Scotland, and it is not until we have this in full that we can expect the uniqueness and integrity of out native fish populations to be protected. This topic is fully explored elsewhere in this book.

References

Alexander (1925), Buckland (1873, 1881), Colquhoun (1887), Cross (1969, 1970), Fitter (1959), Harvie-Brown & Buckley (1887, 1892), Lever (1977), Maitland (1972, 1990), Maitland & Campbell (1992), Maitland & Price (1969), Maxwell (1897), Miller (1841), Muckle (1988), Munro (1979), Pentelow & Stott (1965), Shearer (1961), Stott & Cross (1973), Traquhair (1901), Wheeler & Maitland (1973), Williamson (1974).

*'My late brother imported ova of Geneva trout
... About 200 fry were introduced into Loch Sloy ...'*

John Colquhoun (1887) *The Moor and the Loch*

*****

Fish species listed in 'The Prohibition of Keeping or Release of Live Fish (specified species) (Scotland) Order 2003'.

Asp	*Aspius aspius*
Barbel	all *Barbus* (except *Barbus barbus*)
Bass	all *Morone*
Bighead Carp	*Hypophthalmichthys nobilis*
Bitterling	*Rhodeus sericeus*
Blacknose Dace	*Rhinichthys atratulus*
Blageon	*Leuciscus souffia*
Blue Bream	*Abramis ballerus*
Blue Sucker	*Cycleptus elongatus*
Burbot	*Lota lota*
Catfish	all *Ictalurus, Ameiurus & Silurus*
Charr	all *Salvelinus* (except *Salvelinus alpinus*)
Chinese Black Carp	*Myxopharyngodon piceus*
Chinese Sucker	*Myxocyprinus asiaticus*
Common White Sucker	*Catostomus commersoni*
Danubian Bleak	*Chalcalburnus chalcoides*
Danube Salmon	all *Hucho*
Eastern Mudminnow	*Umbra pygmaea*
European Mudminnow	*Umbra krameri*
Fathead Minnow	*Pimephales promelas*
Dragon Fish	*Zacco platypus*
Grass Carp	*Ctenopharyngodon idella*
Land-locked Salmon	all non-anadromous *Salmo salar*
Largemouth Bass	*Micropterus salmoides*
Marbled Trout	*Salmo marmoratus*
Nase	*Chondrostoma nasus*
Northern Redbelly Dace	*Chrosomus eos*
Pacific Salmon and Trout	all *Oncorhynchus* (except *Oncorhynchus mykiss*)
Paddlefish	all *Polyodon & Psephurus*
Perch	all *Perca* (except *Perca fluviatilis*)
Pike	all *Esox* (except *Esox lucius*)
Pikeperch	*Stizostedion lucioperca*
Red Shiner	*Notropis lutrensis*
Rock Bass	*Ambloplites rupestris*
Ruffe	*Gymnocephalus cernuus*
Schneider	*Alburnoides bipunctatus*
Silver Carp	*Hypophthalmichthys molitrix*
Smallmouth Bass	*Micropterus dolomieu*
Snakehead	all *Channa*
Southern Redbelly Dace	*Chrosomus erythrogaster*
Sturgeon	all *Acipenser, Huso, Pseudoscaphirhynchus & Scaphirhynchus*
Sunbleak	*Leucaspius delineatus*
Sunfish	all *Lepomis*
Topmouth Gudgeon	*Pseudorasbora parva*
French Nase	*Chondrostoma toxostoma*
Vimba	*Vimba vimba*
Weatherfish	*Misgurnus fossilis*
Whitefish	all *Coregonus* (except *Coregonus albula & lavaretus*)

PART 3

THE ISSUES

'Altogether the prospect was not pleasing.
There was no sign of a stream, a loch or a pond,
one or other of which was essential to my happiness.'

Halliday Sutherland (1934) *A time to keep*

FISH COMMUNITIES

'Our plenteous Streams a various Race supply;
The bright-ey'd Perch, with fins of Tyrian Dye,
The silver Eel, in shining Volumes roll'd,
The yellow Carp, in scales bedrop'd with Gold
Swift Trouts, diversfy'd with Crimson Stains,
And Pykes the Tyrants of the wat'ry Plains.'

Alexander Pope (1688-1744) *Windsor Forest*

A lowland fish community (drawing by Robin Ade, from Maitland & Crivelli 1996).

ONE of the most important parts of fish habitat is other fish. A good example is the relationship, unique in Great Britain, between Powan and River Lampreys in Loch Lomond, where many of the former bear scars from attacks by the latter, the River Lampreys dependent on Powan as a source of food in the form of blood and muscle tissue. Although Scotland has a relatively impoverished freshwater fish fauna compared to many other parts of Europe, the fish communities found in fresh waters here do vary according to habitat and, perhaps surprisingly, to latitude, for there is somewhat of a north/south divide in the present fish world. Continental and some English fish biologists have categorised river reaches according to their fish communities (e.g. trout, grayling, barbel, bream zones) but these are not really relevant to most of Scotland where many of the fish species concerned are absent.

Within each species, freshwater fish are probably more variable in their form than most other vertebrates. This is because they are much better adapted at responding to varying conditions than, say, warm-blooded birds or mammals. The latter, when food is very short, die. Fish simply stop growing: there are many populations of stunted fish as evidence of this. Henry Thoreau was well aware of this when he pointed out that '... indeed all the fishes which inhabit this pond, are much cleaner, handsomer, and firmer fleshed than those in the river and most other ponds, as the water is purer, and they can easily be distinguished from them.'

Because of this adaptability, the same species can behave quite differently in one fish community, compared to another. Nils Arvid Nilsson was one of the first people to observe this, and he showed that, in the Swedish lakes which he studied, when Arctic

Charr live on their own in a lake, they occupy most of the available habitat, especially the pelagic and littoral areas. However, when the more dominant Brown Trout is present, it drives the Arctic Charr away from the more favoured littoral areas, and the latter become an almost entirely pelagic species.

Thus, as well as knowing something about the ecology and habits of individual species of fish, it is important to be aware of the different fish communities which occur in Scotland, and the various ways in which our freshwater fish behave within them. As a boy, the author first became familiar with common native species in quite simple fish communities, in a local river and loch - rather like C. Braithwaite: 'My earliest experiences of fishing are those of many an urchin - chiefly concerned with sticklebacks, minnows, loaches, eels and sometimes perch and roach.'

Native fish communities

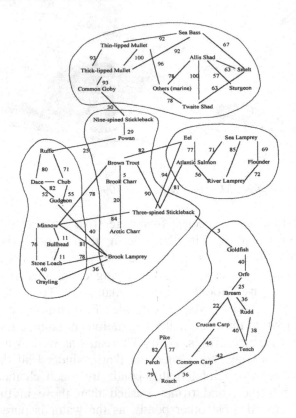

An association diagram of freshwater fish species in Scotland suggesting five main communities. Each species is linked to the two others with which it is most commonly associated. Numbers represent percentages of maximum possible associations. Because of their rarity, Barbel, Vendace and Golden Grey Mullet are not included in the analysis.

There are several recognisable native fish communities associated with the main habitats found in Scotland's lochs and rivers. It must not be forgotten, however, that there are now many introduced species which occur also in these habitats and which are affecting the local communities in various ways - many as yet undetermined and few understood.

Estuaries

Conditions here can be very changeable with considerable variations in salinity, flow and water level. The 'freshwater' element of the fish community might commonly comprise River Lamprey, Eel, Sea Trout, Thick-lipped Mullet, Three-spined Stickleback, Sea Bass, Common Goby and Flounder. Fish passing through on their way to and from the sea would include several of these species and also Sea Lamprey and Atlantic Salmon. A few estuaries have populations of Sparling and several have regular visits from Allis Shad, Twaite Shad and Thin-lipped Mullet and, increasingly rarely now, from Sturgeon.

Many Scottish rivers have little in the way of a proper estuary before entering the sea, but a few rivers have large and important estuaries, among them the Rivers Clyde, Cree, Forth and Tay. The Moray and Solway Firths must also be included here. Many smaller estuaries are of significance and have fish communities of interest, but there is little information available on them. We need to learn more about these lesser known systems.

Depositing rivers

These are mostly the lower sections of our larger rivers where the gradient is low, leading to a meandering channel and slow current, with deposition of the sands and silts which have been eroded by the faster currents and rolling stones further upstream. The water tends to be turbid and the river bed is soft and silty. The fish community commonly includes the larvae of River, Brook and Sea Lampreys, Eel, Roach, Minnow, Pike, Brown Trout, Three-spined Stickleback, Perch and Flounder. Fish passing through on their way to and from the sea would include River and Sea Lampreys, Atlantic Salmon and Sea Trout, and there may be occasional forays from the estuary by Thick-lipped Mullet and Sea Bass.

Because a significant gradient in many Scottish rivers continues right down to the sea or estuary (e.g. the Rivers Dee, North Esk, Tay, Awe), such systems do not have a significant depositing section but are stony right into their estuaries. Notable systems with important lowland sections include the Rivers Clyde, Endrick and Forth. The River Spey is unusual in having a typical depositing section on the level land halfway along its course, where it runs through the Insh Marshes. Below this, the river reverts to its previous eroding condition.

Eroding rivers

Here, the gradient is steeper and the current faster, consequently eroding any fine sands and silts present. Thus the river bed is mostly made of clean gravels and stones and the channel comprises a series of riffles and pools. The native fish community includes River, Brook and Sea Lamprey (which spawn in these reaches), Eel, Minnow, Stone Loach, Atlantic Salmon, Brown Trout and Three-spined Stickleback. The larger pools can sometimes hold Pike or Perch, and Flounders may occur in numbers in some rivers. In the north of Scotland, some of these species may be absent.

There are many excellent examples of eroding river systems in Scotland and, as indicated above, some of these continue with the same habitat right down to the coast. The larger systems include the Rivers Tay, Dee, Clyde, Tweed, Beauly and North Esk.

Highland burns

The head waters of most rivers are small, steep in gradient with fast currents and clean stony or rocky beds. They are often inhospitable places for fish, especially in winter, and not uncommonly have no fish at all, especially small burns at high altitudes near their sources. The most common fish is Brown Trout, but in accessible burns the native community can also include Eel, Minnow, and Atlantic Salmon and, in the south of Scotland, Brook Lamprey and Stone Loach.

There are hundreds of such burns all over the upland parts of Scotland. Though small individually, collectively they make up an impressive amount of habitat, much of which is important as nursery area for salmonids, which always dominate the community.

Rich lowland lochs

There are many naturally rich eutrophic lochs in the lowland areas of Scotland; and some of these have been further enriched by nutrients washed in from various land use practices in the catchment (e.g. agriculture and forestry). Though such waters are very productive, associated hazards for fish include the possibility of toxic algal blooms and deoxygenation in deep water during summer and under ice in winter. Typically, the native fish community includes Eel, Minnow, Roach, Pike, Brown Trout, Three-spined Stickleback and Perch. Often, especially where stream spawning is unavailable, the community may be dominated by Pike and Perch.

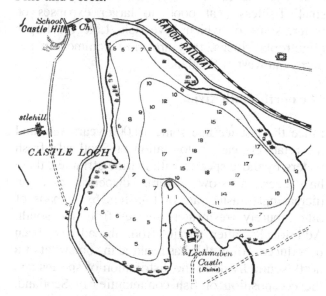

The Castle Loch, Lochmaben (Murray & Puller 1910).

Notable examples of this type of habitat include Loch Leven, Castle Loch (Lochmaben), Loch Flemington, and the Loch of Menteith.

Poor highland lochs

In highland areas of Scotland, these oligotrophic lochs are the commonest fresh waters and characterised by clear water, low in nutrients. They are often deep and rarely very productive. The native fish community characteristically includes Eel, Minnow, Brown Trout, Arctic Charr and Three-spined Stickleback and commonly has also Atlantic Salmon, where it is possible for this species to gain access from the sea. Larvae of Brook Lamprey may occur in the southern part of Scotland where there are suitable soft sediments.

These are the typical attractive lochs of Scotland, often long and narrow and set amid magnificent mountain scenery. They are especially common in the north and west, fine examples including Lochs Tay, Rannoch, Morar, Shiel and Einich.

Highland lochans

These small standing waters are mostly found at high altitudes and are commonly peat-stained and low in nutrients. Frequently they have no fish at all, but sometimes there are Brown Trout and occasionally Eel, Arctic Charr and Three-spined Stickleback.

Such systems can be very variable, ranging from small fishless peat pools to larger expanses of water, some of which have fish. Like the larger oligotrophic lochs, they are most common in the north and west of Scotland.

The north/south divide

Since the last ice age some 10,000 years ago, and the relatively easy colonisation of Scotland's fresh waters by those species with marine affinities, there has been a slow natural dispersal and an increasingly faster rate of transfers by humans of other, purely freshwater, species from the south. Additionally, alien species from abroad have been introduced. All of this has resulted in somewhat of a north/south divide in the distribution of species and the composition of fish communities in Scotland. Whilst the situation is more complex than a single dividing line on the map, it is certainly true to say that the fish communities of waters north of the Great Glen have many fewer species than comparable systems south of the Highland Boundary Fault. Moreover, most fish communities of waters north of the Great Glen have only native species, whilst many of those further south have an alien component.

For example, typical northern running waters like the Rivers Naver, Thurso and Helmsdale have only about six or seven species of freshwater fish, all of them native, whereas comparable waters in the south, for example the Rivers Nith, Annan and Tweed have sixteen to nineteen species. Similarly, standing waters in the north, such as Lochs Stack, Shin and Calder have only six or seven species, whilst Lochs Ken, Lomond and St Mary's Lochs in

the south have nine (Ken) to nineteen (Lomond). In terms of species composition, many of the northern communities have probably remained the same for thousands of years and are relatively stable, whereas those in the south have changed substantially over the last two centuries, and continue to do so with introductions by anglers and others. Each new introduction can produce instability in its new community, as studies by the author and Colin Adams at Loch Lomond have shown in recent years.

Rare species

As well as fish communities which have only their complement of common native species, and those where there have been additions due to transfers by humans, there are also a number of fish communities which include much rarer species. Such communities are of particular interest, not only because they are more complex, but also because of their conservation value for the rare species concerned. Unfortunately, several of the waters concerned, which were previously of considerable interest, are now less so because of the extinction of various species. Examples are the demise of Sparling from the Rivers Almond, Annan, Bladnoch, Clyde, Cree, Dee, Esk, Fleet, Forth, Girvan, Lochar, Nith, Stinchar, Tay and Urr, Vendace from the Castle and Mill Lochs, and Arctic Charr from Lochs Dungeon, Grannoch, St Mary's, Leven and Heldale. The great decline in other species, such as Sturgeon, and Allis and Twaite Shad, has reduced the biodiversity of many of our estuarine fish communities.

Of particular importance, therefore, are those waters which still retain all their more common native species, but also have stocks of our rarer fish. Obvious waters in this category include the following:

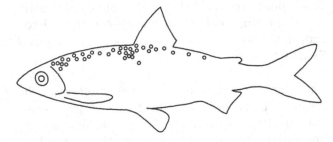

Powan showing typical positions of lamprey scars.

Loch Lomond. Unique for its freshwater River Lamprey, this is one of only two natural sites for Powan, the two species making a unique inter-related combination found nowhere else in the British Isles. Loch Lomond has the highest fish community diversity in Scotland, with 15 native species. Unfortunately, there are now at least five alien species established there, all of them introduced since 1970.

Loch Eck. One of only two lochs for Powan, Loch Eck also has Arctic Charr, this being a unique combination in Scotland and found elsewhere in the British Isles only in Haweswater in the English Lake District (now a water supply reservoir). The fish community as a whole is pristine, with no introduced species.

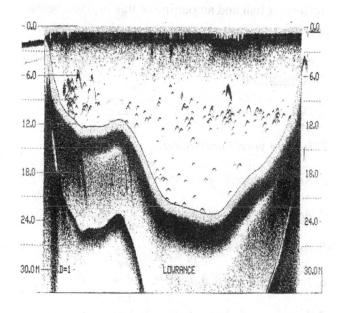

Echo sounding taken by Alex Lyle at Loch Eck which has a fish community unique in Scotland.

Loch Rannoch. Three distinct morphs of Arctic Charr are found here, something which has not yet been found in any other Scottish loch.

Loch Skene. This was judged by the author and Alex Lyle to be the most favoured new site in southwest Scotland for the (then) extinct Vendace. Introductions were carried out in the 1990s. Although an introduced species there, it is now established, and this loch is considered to have a high conservation priority with just three native species - Vendace, Minnow and Brown Trout.

Loch Meallt. Here, there is a very simple community of only Arctic Charr and Three-spined Stickleback, apparently unique in the British Isles.

Loch Crocach. The unusual race of Brown Trout found here has always been of exceptional interest. It is believed that the whole population may have descended from just one female and male.

River Cree. The only population of Sparling left on the west coast of Scotland is found here in the lower river and estuary. In addition, Twaite Shad, which are not known to breed anywhere else in the country, apparently spawn in the upper estuary, thus giving a unique combination found nowhere else in Scotland.

River Forth. One of the only three surviving populations of Sparling in Scotland occurs here in the lower river and estuary. Most of the main stem of the Teith and lower Forth is now a Special Area of Conservation (SAC) for River, Brook and Sea Lampreys.

River Tay. One of the only three surviving populations of Sparling in Scotland occurs here in the lower river and estuary. The Tay is also an SAC for Sea Lampreys.

River Endrick. As the main, and probably the only, spawning and nursery river for the unusual freshwater River Lamprey of Loch Lomond, this system is unique in the British Isles. Much of the main stem is now an SAC for River and Brook Lampreys.

As has been indicated, some of these sites already have some form of protection but what is essential, as far as their native fish communities are concerned, is that each has an individual management plan which is acted upon as soon as it has been agreed by all the stakeholders involved. In addition, there are also many other waters which merit further research and conservation (e.g. Lochs Einich, Harray, Obisary and Druidibeg).

Restoration of fish communities

Considerable research and practical effort is going into the restoration of some fresh waters, especially rivers which have been canalised and whose banks have been degraded (e.g. River Tweed), and lochs which have become eutrophic (e.g. Loch Leven) or acidified (e.g. Loch Enoch). Less consideration and effort has been given to the problems of restoring the original fish communities - perhaps rightly so since there is no point in attempting this until the habitat has been returned to normal. However, as indicated above, there are a number of waters where fish have become extinct through human activities and it would be pleasant to think that the original fish communities might be restored there by the end of the 21st Century. Unfortunately, such restorations are often expensive and this is usually the main, but not the only, barrier to action.

One of the main losses across Scotland, which has been little discussed because it is apparently unimportant as far as angling species are concerned, has been that of lowland ponds and streams. Many of the former have been filled in and the latter ditched. This has lead to the disappearance of many communities of small fish; one species in particular - the Nine-spined Stickleback – appears to have declined substantially in Scotland as a result. Many of these systems, especially small ponds, could be restored or recreated and the appropriate species reintroduced successfully.

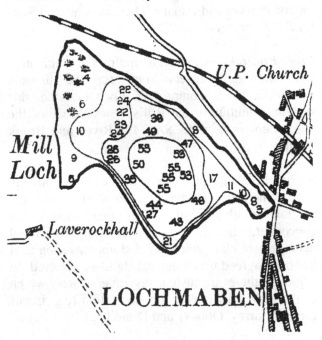

The Mill Loch, Lochmaben (Murray & Pullar 1910).

In an ideal world, high among such restoration projects would be to restore the Mill Loch at Lochmaben to its original status and re-establish the

Vendace there. This would not only be an admirable conservation project, but would offer the opportunity for local people to reclaim an important part of their local culture and traditions. Not too far away, the restoration of the Sparling to one or more of the Solway rivers and their estuaries, say the Nith or the Annan, would not only be an important conservation initiative, but would also provide some local employment by restoring the fishery for this fish which used to exist there. These and other fish restoration projects are certainly worthy of consideration in the future: many have social and economic values as well as conservation ones.

Selection of important sites

Some years ago the author developed a protocol for the selection of sites which were important for freshwater fish and an outline of this is given below.

Algorithm for the selection of important sites for freshwater fish.

		See
1. Does the site have fish?	Yes	2
	No	A
2. Is there good information?	Yes	3
	No	B
3. Are rare species present?	Yes	C
	No	4
4. Are there unusual races?	Yes	C
	No	5
5. Are any stocks pristine?	Yes	C
	No	6
6. Is the community unusual?	Yes	C
	No	7
7. Is native fish diversity high?	Yes	C
	No	D
8. Is the site representative?	Yes	C
	No	D

A. If fishless, then the site is of potential value and attention should be given to other fauna and flora to determine value.

B. Survey is needed to assess status.

C. The site is important and should be notified.

D. The site is probably not important as far as freshwater fish are concerned.

It should be noted that sites can be of importance under several different criteria and that the more of these criteria which are relevant to a single site, the greater is its conservation significance. For example, Loch Lomond is of particular significance in Scotland because it scores highly under all the criteria used – but unfortunately, this high conservation value has now been reduced by the number of alien species which have been introduced in recent decades and are now well established there.

The use of fish in establishing the status of a water body is an important part of the assessments required under the EC Water Framework Directive. High status is only achieved when the original fish community is present and not affected in any major way by pressures such as alien species.

Unfortunately, the question of whether any species is native to a water can be of special significance in relation to the importance of its fish community. Thus Nick Giles considers that the River Tweed, with sixteen established fish species, is of importance in terms of its fish diversity, whereas Ronald Campbell rejects eight of these species as introduced, and would claim that the native community included only eight species, thus giving it a lower diversity than, say, the River Cree. The

author's opinion lies somewhere between these two stances; the arguments concerning native and introduced species are considered in detail elsewhere in this book.

The distribution of native freshwater fish in the UK is uneven, reflecting the effect of the last Ice Age and more recent translocations by man. In general, there is a decreasing number of species from south to north and, to a lesser extent, from east to west. Thus, in order to know how natural any fish community is, geographic information on the complement of native and alien fish species in any water body under assessment must be available. This can then be compared with what is believed to be the natural distribution of each of the native species. It must be remembered that there is a variance of opinion as to the extent to which UK native species have moved naturally.

References

Adams (1991, 1994), Adams & Maitland (1991, 1998, 2002), Braithwaite (1923), Campbell (1963), Giles (1994), Maitland (1966, 1976, 1983, 1985, 1987, 1989), Maitland & Crivelli (1996), Marwick (1909), Murray & Pullar (1910), Nilsson (1967), Thoreau (1886).

'1884: Besides herring, we find from an old list that there were not fewer than fifty-seven different kinds of fish then caught in the Clyde at Greenock, including pellocks or porpoises, salmon, sea-trout, cod, haddocks, whitings, gudgeons, sturgeons and others of the piscatorial genus.'

J.J. Marwick (1909) The River Clyde and the Clyde Burghs

*****

A summary of some of the waters in Scotland which are (or were) important for their fish.

SITE	INTEREST	COMMENT
Standing		
Loch Crocach	Brown Trout	unique gene pool
Loch Eck	Powan & Arctic Charr	unique & pristine community
Loch Lomond	River Lamprey & Powan	unique comunity
Loch Maree	Arctic Charr & Brown Trout	important & pristine community
Loch Meallt	Arctic Charr & 3-spined Stickleback	unique community
Loch Skene	Vendace	'safeguard' stock
Running		
River Cree	Twaite Shad & Sparling	unique & pristine community
River Endrick	River Lamprey	unique stock
River Forth	Sparling	one of three surviving stocks
River Tay	Sparling	one of three surviving stocks
Engineered		
Carron Valley Reservoir	Powan	'safeguard' stock
Daer Reservoir	Vendace	'safeguard' stock
Meggat Reservoir	Arctic Charr	'safeguard' stock
Loch Sloy	Powan	'safeguard' stock
Talla Reservoir	Arctic Charr	'safeguard' stock
Lost		
Castle Loch	Vendace	extinct
Loch Dungeon	Arctic Charr	extinct
Loch Grannoch	Arctic Charr	extinct
Loch Heldale	Arctic Charr	extinct
Loch Leven	Arctic Charr	extinct
Mill Loch	Vendace	extinct
St Mary's Loch	Arctic Charr	extinct
River Annan	Sparling	extinct
River Nith	Sparling	extinct
River Clyde	Sparling	extinct

THREATS TO FISH

'Salmon roe, snatching, guddling (or gunnling or tickling), ottering,
set lines, netting, burning the watter, dynamite.'

J.A. Harvie-Brown & T.E. Buckley (1889) *A Vertebrate Fauna of the Outer Hebrides*

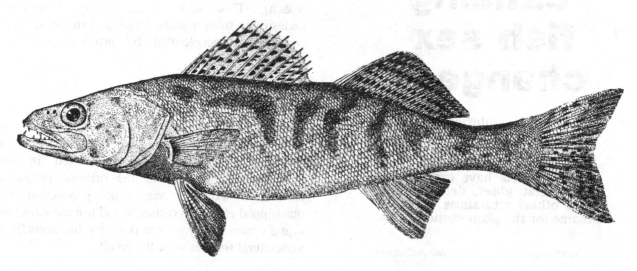

FISH populations have been utilised by man for many thousands of years, and it is often difficult to separate the effect of his impact from changes due to more natural processes. New threats are a continual challenge, for example aggressive alien species such as the Pikeperch *Sander lucioperca* (above) rumoured to have been introduced to Scotland in recent years. Over the last two hundred years, particularly the last few decades, many new and intense pressures have been applied, and in most cases these have been detrimental to fish conservation as a whole. Inevitably, many of the threats are interlinked, the final combination often resulting in a complex and unpredictable situation.

The main conclusion of a recent review by the author of trends in freshwater fish populations in Scotland was that most native species were declining whilst most introduced species were expanding. The reasons for this are various but there is no doubt that native fish in Scotland are facing a variety of threats, some of which they are unable to withstand. Many of these threats are not new. In 1891, Andrew Lang in his *Angling Sketches* recorded that 'It was worth while to be a boy in the South of Scotland, and to fish the waters haunted by old legends, musical with old songs, and renowned in the sporting essays of Christopher North and Thomas Stoddart. Even then, thirty long years ago

(1860), the old stagers used to tell us that "the watter was ower sair fished", and they grumbled about the system of draining the land, which makes a river a roaring torrent in floods, and a bed of grey stones with a few clear pools and shallows during the rest of the year.'

Pollution

Pollution of fresh waters in Scotland comes mainly from domestic, agricultural, or industrial wastes and it can be either toxic, thereby eliminating all the fish species present, or selective, killing off only a few sensitive species, or altering the environment so that some species are favoured and others are not. Eutrophication is sometimes thought of as mild pollution. Because of the large numbers of rivers in Scotland and proximity to the sea, there is relatively little direct pollution of standing waters compared to running waters. Most effluents are directed into running waters and estuaries (or the sea), rather than into lochs, and it is the species concerned with such systems which are mostly influenced by pollution. Species with marine affinities are especially affected, for it is normally the lowest reaches of rivers and their estuaries that are most seriously polluted, but through which such fish must pass at two stages in their life history. Thus one extreme belt of pollution in a river system can affect the fish population of that whole system by acting as a barrier to these fish.

Pollution 'causing fish sex changes'

CHEMICAL pollution in the Solway Firth could be causing sex changes in fish, two scientific reports claim.

The reports have expressed fears that glues, detergents, and other substances are to blame for the phenomenon.

A worrying side-effect of some pollutants.

Polluting discharges have affected some of Scotland's rivers for hundreds of years. The Bonhill Parochial Board noted in 1879, however, that the discharges from local mills into the River Leven (Lomond) were apparently 'not so prejudicial to fish life as is generally imagined. This is proved by a series of elaborate experiments by Professor Penny of Glasgow about 7 or 8 years ago.'

The process of eutrophication in a wide array of waters has received considerable attention in many parts of the world. Essentially, the phenomenon is due to increasing amounts of nutrients entering the water from geological, domestic, industrial and agricultural processes. The general changes which take place in a body of water that is undergoing eutrophication involve large increases in the abundance of algae (associated with which is a change in the species composition, and subsequent changes in the invertebrate and fish populations).

As far as fish are concerned, one common response to increased eutrophication is a marked increase in growth rate. There appears also to be a direct relationship between accelerated 'cultural' eutrophication and an increased rate of parasitism in fish. With the change from clean oxygenated substrates to silted, sometimes poorly oxygenated, substrates in waters undergoing eutrophication,

many fish lose valuable areas of spawning ground. This has a serious effect on species such as Powan which spawn in over clean gravels. In lakes that are undergoing natural eutrophication, there appears to be a succession of species and dominance from salmonids (e.g. Arctic Charr) to coregonids (e.g. Powan) to percids (e.g. Perch) to cyprinids (e.g. Roach). This succession, usually considered undesirable from a game angling point of view, is considerably accelerated by artificial (cultural) eutrophication.

Agriculture

Many of the problems arising from agriculture have been reviewed by Ron Allcock and Duncan Buchanan. Post-war changes resulted in the agricultural equivalent of the original industrial revolution. For a long time production far outstripped efforts to conserve and use manures and liquid wastes; many serious pollution incidents from agricultural sources were the result.

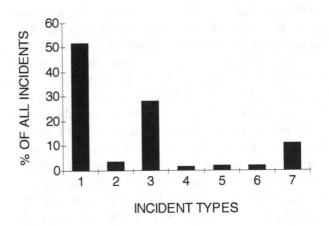

Reported pollution incidents due to various agricultural activities: 1982-90 (1: silage; 2: chemicals; 3: livestock; 4: sheep dip; 5: fuel oil; 6: farm tip; 7: others) (Allcock & Duncan 1994).

During the summer months, the main problems arise from silage liquor discharges and, in a few places, water abstraction. In the winter months a major problem is the run-off from overloaded waterlogged fields following the spreading of slurry. Other problems arise from the careless use of chemicals and fertilisers, sheep dips, fuel oils and the general tipping of waste materials.

The situation has improved greatly in recent years with improved legislation and more responsible behaviour on the part of farmers. Nevertheless there are still regular problems in some waters due to agricultural activities in their catchments.

Acidification

Acid precipitation has caused major damage to fish stocks in various parts of the world including Canada, Norway and Scotland. Salmonid fish are particularly vulnerable, but many other fish species have been affected. A survey of fish populations in 50 lakes in southwest Sweden showed that, in acidified lakes, Brown Trout, Arctic Charr, Roach and Minnow were all affected. Atlantic Salmon have been completely eliminated from several Swedish rivers over recent decades. Acid precipitation has also devastated fish populations in southern Norway where the principal fish species affected are Atlantic Salmon, Brown Trout, Arctic Charr and Brook Charr.

One of the most characteristic effects of acidification on fish populations is the failure of recruitment of new age classes into the population. This is manifest in an altered age structure and reduction in population size, with decreased intra-specific competition for food and increased growth or condition of survivors. To anglers, the fish stock then appears to 'improve' because larger fish are caught, albeit in lower numbers. However, with no recruitment, year by year, the population contains fewer and fewer fish until eventually there are none. This was the situation in Loch Doon, when the author studied it in the 1980s and, because this was the last remaining population of Arctic Charr in southern Scotland, stock were successfully transferred to Talla Reservoir, where a safeguard stock is now established.

Afforestation

The impact of coniferous afforestation and forestry practice on freshwater habitats in Europe and North America caused much concern in the second half of the 20th Century. The effects of each stage of the forestry cycle - ground preparation, tree planting to canopy closure, the maturing crop and felling - may have an impact on local fresh waters. The physical aspects of afforestation affect: (i) the hydrology of streams, as shown by (a) increased loss of water through interception and evaporation from the coniferous forest canopy, and (b) a tendency to

higher flood peaks and lower water levels during droughts; (ii) the release of sediments to streams because of erosion following ploughing and weathering of exposed soils; (iii) reduced summer water temperatures in afforested streams where the channel is shaded.

The principal chemical changes in fresh waters in afforested catchments include: (i) increased nutrient levels from the leaching of exposed soils and applied fertilisers (thus leading to eutrophication), and (ii) the acidifying effect of air pollutants (especially sulphates) which are intercepted by conifers (as airborne particles) and then transferred to the ground (by rain - itself often very acid) and eventually to adjacent water courses. The base-richness of local soils and rocks is of major importance here, for it is only in areas lacking basic ions that acidification occurs. Much of the biological damage is due to the high amounts of aluminium (leached from the acid soils).

These physical and chemical effects combine in various ways to affect the plants, invertebrates and fish of fresh waters in afforested areas. Changes in hydrology and ambient water temperatures tend to make conditions in streams more extreme for most biota. Turbidity decreases plant growth through reduced light penetration and physical siltation. Increased nutrients alter plant communities and cause problem crops of algae in streams and lochs. Acidification of water courses affects the composition of their plant and invertebrate communities, completely eliminating fish in some cases. Other vertebrates, such as amphibians and birds, may also disappear.

Engineering schemes

The creation of reservoirs for hydroelectricity or water supply, where shallow streams or small ponds are transformed to large, often deep, reservoirs can have deleterious effects on local fish populations; this is certainly the case with Atlantic Salmon in certain waters. In some cases these reservoirs have highly fluctuating water-levels: this can have a significant influence on littoral populations, including fish, which either live or spawn in shallow water or inundated tributaries. The dams of such reservoirs normally act as barriers to migratory fish. Remedial measures in such situations have previously been directed towards the conservation of anadromous species of sporting value (i.e Atlantic Salmon and Sea Trout).

Arctic Charr seem to be less affected than other salmonids, and there is evidence that some stocks (which are mainly plankton feeders) may be favoured by the fluctuating water level, which adversely affect Brown Trout in hydro reservoirs. It is believed that, because the fluctuating levels often devastate the littoral flora and fauna, the trout population, which mainly feeds in the littoral area, is also adversely affected. The plankton, on the other hand, is not affected and so Arctic Charr still have their main food source.

However, Hugh Elder has pointed out that there has been considerable mitigation: 'Anadromous fish conservation measures include guarantee of minimum river flow, provision of freshets, construction of fish passes, erection of smolt screens, passage of smolts through turbines, creation of artificial spawning gravels, use of hatcheries and the opening up of previously unutilised natural spawning gravels.'

Water supply schemes may also have significant effects on fish populations, especially where large volumes of water are transferred from one catchment to another. In England, for instance, numbers of Arctic Charr and Schelly (Powan) are pumped out of Haweswater each year as part of the water supply system supplying Manchester. The impact can be particularly serious where no account is taken of local fish ecology when the engineering works are being designed. This can lead to serious damage to fish stocks as, for example, at Loch Lee in Scotland, where substantial numbers of adult Arctic Charr are washed out of the loch in some years at spawning time, due to construction of a spillway near charr spawning grounds.

Although some impoundments, artificial pools and engineering schemes may improve the quality of sport fishing, practically never is thought given to species other than those of sporting or commercial value.

General land use

The impact of different forms of land use can be important to fish populations in various parts of Scotland and much of the variation depends on soil and climate.

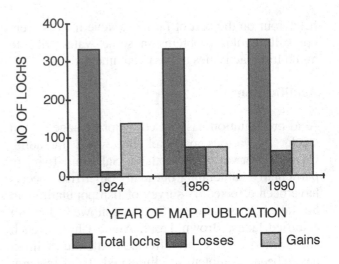

Chronology of loch numbers in central Scotland during the 20th Century: the first map 1896 (Lassiere 1993).

One serious problem related to land use is the widespread destruction by draining or filling of many thousands of small ponds, ox-bow lochans, and other minor waters in many parts of Scotland - particularly in lowland areas. Such habitats are of major importance to populations of smaller fish species (as well as being important aquatic communities in their own right), but rarely is there any outcry about their destruction. The economic and social claims of those destroying them are well known (cheap dumping grounds for garbage, reclamation of land for agriculture or industry, and removal of a potential 'nuisance' and danger to children), and apparently outweigh all else in the minds of those concerned.

In a few areas the trend is the reverse, where new waters are dug out for water shortage, for agriculture or the winning of gravel. Although the original aim is an economic one, such waters can be important in conservation terms if constructed and managed intelligently, and they are becoming an increasingly important amenity in some areas.

Fisheries management

The impact of fisheries (both commercial and sport) on the fish populations which they exploit can range from total extinction of the populations to the more or less stable relationship of recruitment, growth and cropping which exists in many old-established fisheries. The essence of success in fisheries is to have a well-regulated fishery where statistics on the catch are monitored continuously and used as a basis for future management of the stock. It is clearly in their own interests for those concerned with fisheries to adopt sound policies in relation to the close season, the type of gear used, and the numbers and sizes of fish caught, etc., so that the stocks will survive. Although there are considerable gaps in our knowledge of the population dynamics of many fish species, we have sufficient information to suppose that well-regulated fisheries, whose management is based on existing scientific information, should prove sustainable for the species of fish concerned. Yet all too often freshwater fishery management is still not based on sound ecological information and is consequently unsustainable.

Stocking was once considered by many (and is still thought by some) to be the main management tool to be used in fisheries; a number of derelict fish hatcheries and ponds in various parts of the country provide evidence of the faith of past generations in this procedure. There is no doubt that stocking is a valuable part of the management policy of some fisheries, but it is only necessary in waters where spawning and nursery areas are absent or inadequate to provide recruitment for the losses due to the angling or other pressures involved. The numbers of waters where this is true are increasing, and are likely to continue to do so, resulting in the creation of more and more 'put and take' fisheries. Fish food production in these is largely irrelevant, as the fish introduced are of a catchable size and likely to be caught before they starve. There are very few aquatic ecosystems into which a species of fish could be introduced and established without some major alteration within the system itself. This being the case, all proposed introductions of fish, either as species that do not occur in Scotland or that would be new to a water or catchment system within it, should be given the most careful consideration before action is taken.

Overfishing

Whilst most responsible anglers will decry poaching, there have been many cases in the past where more damage was probably done by legal angling than by local poaching. Thomas Stoddart, a well known and respected angler, recorded in 1866 that 'The Water of Leith, however, was my favourite resort on these occasions. ... the prime portion of this stream lay above Malleny, on which

range I have taken, in the course of a day's fishing, as many as five or six dozens of yellow trout, ...' His companions seemed no better: '... my friend Wilson was directed by our host to a small lake and its connecting rivulet, on the opposite side of Loch Laggan, ... He came safely back, however, with not only his creel, but his handkerchief and pockets crammed with trout, twenty-six dozens in all, which, averaging them at one fourth of a pound each, and the average I think was heavier, speaks to seventy-eight pounds.'

The Fishing Gazette of 8 December, 1883, gives details of Thomas Stoddart's diary which records that, over 50 years, he had in 4,150 days fishing caught 928 salmon, 1,540 sea trout, 61,573 yellow trout and 378 pike a total of 67,419 fish in total - other fish are not mentioned!' However even this does not match up to Adam Dryden, who, in his *Hints to Anglers* in 1862 boasts 'caught nearly 5,000 trout weighing 860 pounds in February-June 1858.' Another example is given by C. MacDonald: 'Between nine and midnight (midges permitting) we fished for trout in the burn and set snares for hares, to augment our provender. One evening, in three hours, four of us caught a total of 365 trout - mostly small but all of eatable size.'

Atlantic Salmon too have suffered from overfishing. The Highland and Islands Development Board reported that 'On the memorable night of June 7th, 1648 a famous pool on the Findhorn gave up no less than 1300 fish.' ' In 1778, over 700 fish were taken in a single haul of the net on the Tweed,'
Fish stocks can, however, also be damaged by regular illegal fishing. MacVine recorded in 1891 that 'These poachers were fairly civil when spoken to in a pleasant manner, and it was no use remonstrating with them ... and, as they asserted that these fish were sent by Providence to feed hungry people in winter, when they could not work, they considered it no disgrace to be convicted and fined, or go to prison ... and a poacher when he came out of gaol was looked upon as a sort of martyr - a victim to the unjust laws of the country! The lower classes might be actuated by sinister motives. They belonged to their class, and, better still, these poachers furnished them with an abundant fresh supply of fresh salmon all the winter through at the low price of three-halfpence or twopence per pound. Before the author became known in the district he was frequently accosted in

the streets of Peebles: "Will ye buy a nice clean salmon, sir? I'll sell ye ane cheap."'

At the other end of Scotland, poaching is more in the nature of subsistence fishing, as Ratcliffe Barnett recorded: 'Benbecula ... a lonely little world of its own with its one rounded hill of Rueval. We passed many tinkers' tents by the roadside. These are generally pitched most cunningly by the side of a good trout loch with a peat stack close by. For tinkers' morals are easy when it comes to fish or a fire, and nobody in these islands grudges them their perquisites. A wedge of wild geese flew above us. White swans with their little cygnets floated on the blue waters of a loch.'

In 1899, Hamish Stuart bemoaned 'the peculiar character of the Scottish law as to the right to take trout ... A few years ago one could take baskets of from 33 lb. to 40 lb. on Loch Harray. Its deterioration began with the institution of poaching by set lines and nets, and has been steadily maintained. In one season a single Orcadian sent 1000 lb. of trout to Billingsgate, and there are some hundreds of nets - the number is by some placed at 1000 - on this lake alone!'

Fish farming

As widely predicted by the author and others at the beginning of salmon farming in Scotland, it has become evident over recent years that such farms do pose a number of environmental problems.

The Highlands and Island Development Board (HIDB) noted in 1985 that 'With the development of salmon farming 'The Scottish salmon resource - now a combination of the traditional and the new - has formidable economic potential in both commercial and sporting terms. As in the past, continued success will depend ... upon new and effective measures to protect this traditional and prized resource.' Few such measures have ever been implemented.

The effects of solids and nutrients originating from waste feed, fish faeces and urine can be potentially serious, especially in the long-term. Solids from waste food and faeces are known to build up significantly on the bed below cages in both freshwater and marine environments and eventually these can completely envelope the bed to produce

an anoxic layer of sludge, with local deoxygenation of the water. For this reason, it is now common practice to move cages at least every second year to allow the original sites to recover. Other effects of eutrophication are various, including increased algal growths, deoxygenation of water during stratification in summer and under ice in winter, and a tendency for the fish community to change from one dominated by salmonids to one where coarse fish predominate.

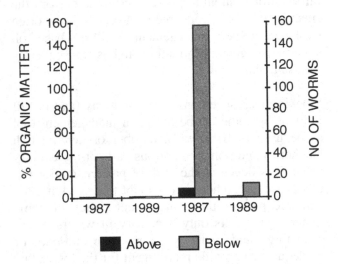

Improvements in discharge quality from a fish farm by the River Polly with installation of settlement ponds (Allcock & Buchanan 1994).

Fish farming activities can be an important source of disease and parasites to wild fish populations (and vice versa). Although many fish farmers claim that farms are not a source of disease, they themselves are frequently the first objectors, on disease grounds, to proposals for new fish farms or hatcheries above or near them in a catchment or sea loch. One of the most notable problems in recent years has been that of fish lice, which have not only caused considerable damage to coarse fish but have almost certainly been responsible for the decline and near-extinction of both Atlantic Salmon and Sea Trout in many rivers along the northwest coast of Scotland. In fresh water, several outbreaks of the fish louse *Argulus* have originated from fish farms.

In order to combat disease, the use of disinfectants, antibiotics and other control measures is commonplace in modern fish farming practice. For example, farmers add a chemicals to their cages to treat fish lice - a major pest to fish in sea cages. As well as direct chemical treatment, one of the principal ways of treating fish is by enteral administration of pharmaceutical products incorporated within the feed, and antibiotics are commonly administered in this way. The use of chemicals has risen dramatically in recent years as fish farmers try to control diseases among their fish.

Another feature of cage systems is that they soon attract predators, notably (in Scotland) Herons, Cormorants, Mink and Otters. These can be controlled using anti-predator devices such as nets and scaring devices, but there is little doubt that, in the past at least, many have been shot or trapped.

One common feature of all fish farms, and especially floating cages, is the fact that many fish find their way into the local waters by intentional release or escapes during handling or from accidents. Even in land-based systems fish may escape with ease. The numbers involved are unpredictable, but an escape from one salmon rearing system of 1.6 million salmon fry into one small river in Scotland has been reported. It can reasonably be expected that escaped Salmon from cage farms will interact with native fish species thereby competing for space and food and by actively predating the eggs and young of native fish.

It is now widely accepted that the numerous stocks of Atlantic Salmon in Scotland and other parts of Europe and eastern North America have been largely separated since the last Ice age and have some genetic individuality. In Scotland there are about 400 individual rivers with salmon populations. The salmon farming industry has imported a number of strains of salmon from abroad (especially Norway). These are being cross-bred with Scottish stocks to develop a domestic race with characteristics desired by the industry (fast growth, disease resistance, placid nature, etc.) but unlikely to be advantageous in the wild. There is strong evidence that farmed salmonids are much less fit in the wild than native fish. However, farmed fish are introduced into the wild so frequently and in such large numbers (either intentionally through stocking or unintentionally through escapes and accidents) that there is considerable concern that they may upset the genetic integrity of native stocks. Already, an appreciable proportion of the adult Salmon appearing in some rivers is of farmed origin. In 2005, over 30 years after salmon farming started, hundreds of thousands of fish were still escaping

It is of interest to note that, in the US State of Washington, it was ruled in 1997 that Atlantic Salmon (an alien species there) which manage to escape from floating pens are a 'living pollutant' which can be regulated by the State just as sewage discharges.

From the beginning of salmon farming in Scotland there has always been concern that escaped domestic fish would find their way into rivers and breed with wild fish. Some of the first evidence of this was provided by John Webb on November 1989 at Strabeg, on the River Polla, Sutherland when he made the first observation of (radiotagged) male escaped farmed salmon spawning with a wild female grilse. Radio tagging proved to be a useful technique for tracking salmon movement: John Webb recalls November 1996 by the River Clunie on Braemar Golf course seeing a radiotagged male spring salmon spawning. The fish had been tagged and released as part of the new catch-and-release programme on the Dee on 1st March at Kincardine O'Neil.

After a large escape of fish from a cage farm in the Clyde Estuary, widely reported in the press, Colin Adams found two adult farmed salmon in November 1998 (one male and one female) which had migrated into Loch Lomond and up the very shallow burn into the Dubh Lochan, and then up an equally shallow burn running into the Dubh Lochan.

Fish introductions

With native species of restricted distribution, or foreign species already well established in Europe, careful consideration should be given to any proposals to introduce them outside their present areas of occurrence. Many native species appear to be gradually distributing themselves by natural means, for example Perch and Pike, but even in these instances the effect may be drastic. For example, several cases are known of Pike gaining access to waters containing populations of Brown Trout, and completely eliminating them.

Many species, too, have been dispersed by man. The introduction of Dace and Roach to Ireland, and their influence on other fish there, has been carefully documented, while recently the Barbel has been moved well beyond its former area of occurrence in Great Britain and most recently into the River Clyde. Many hundreds of introductions take place annually, mostly within the existing distribution boundaries of the species involved.

Danube Catfish, now widely introduced in England. An aggressive predator growing to 3 m, this is a fish which is not wanted in Scotland (from Maitland 1972).

In the case of fish species that are new to Europe, there are several points which should be taken into account before introduction is contemplated: (a) It should be quite clear that there is some real purpose behind the introduction, such as sport or aquaculture. (b) It is undesirable to introduce species which are likely to have a major influence on ecosystems through predation, competition, or feeding habits (such as destruction of vegetation). (c) Consideration should be given as to whether or not the species could be controlled readily. Some species may not breed because of climatic conditions, while others may have very specialised requirements for spawning, etc.

Most of the introductions to sensitive waters which take place within Scotland are transfers from fish hatcheries or other waters. Such transfers are rarely needed or justifiable, and can be potentially damaging. Further measures are needed to control them, and in general it should be illegal to introduce any fish into a water body without a permit from a national controlling authority. Such permits would only be issued if a good case had been made for the introduction.

A recent news report on a new fishery on the River Devon is typical of the unscientific and ambiguous approach taken by some anglers to the question of stocking. 'Initial stocking of Trout will take place later this year - probably in August - to supplement an already adequate stock of well-conditioned brownies. The club will attempt to establish an indigenous strain to the river, which is lacking because of cross breeding of Trout, with the resultant effect that the major proportion of the stock may be unsuited to the type of water they live in.'

In spite of existing legislation, there are obviously still potential dangers from disease introduced with ornamental fish. The bulk of these fish are of tropical origin and destined for private aquaria and there is probably very little risk here. With temperate species, however, there are definite disease and parasite risks associated with their introduction to the country. Both fish and parasites are adapted to the climate and could potentially become established in the wild. Moreover, even if the host fish with which a disease or parasite was introduced was unable to establish itself in the wild, the parasite might well do so by transferring to native species.

There are obvious potential dangers from temperate fish which are brought into the country and sold for aquaria or ponds. Until recently, it was possible to purchase a variety of Eurasian and North American species through the aquarium and pond trade and a number of these could establish here and pose a threat to our native stocks.

One other loophole in the legislation which potentially allows even prohibited species (and their parasites) into the country lies with the problems of identification. It has been suggested, for instance, that Rainbow Trout could accidentally become mixed with batches of ornamental species imported regularly from Denmark and so bring in diseases which could prove a major threat to salmonid farming in the British Isles. The question of the identification of imported batches of fish is a very critical one, for there is little point in having legislation banning the import of certain fish if the controlling authorities are actually unable to carry out accurate identifications.

Predators

Natural predators are rarely a threat to wild fish - something which many anglers find difficult to accept. Almost all the 'problems' which have arisen in relation to freshwater fish in Scotland due to predators, have been related to fish farming, introduced predators (e.g. Mink), or stocked fish. The recent controversy concerning Cormorants at Loch Leven is a good example of the latter.

Wild fish do have many predators (and parasites) but it is normally the young, diseased or displaced fish which are killed, and rarely are there any serious effects on the stock as a whole. Thus, fishery managers should look towards their own practices and stop blaming wildlife for their own inadequate and unsustainable attempts at management.

Tweed salmon losses blamed on birds

Typical newspaper headline about predators.

Climate change

There is increasing evidence that human activities of various kinds are altering the atmosphere to such an extent that global warming may create major climatic changes over the next few centuries. The most certain changes seem to be a rising sea level and a general rise in atmospheric temperatures, especially at high latitudes. Changes in precipitation, wind and water circulation patterns are also likely but their nature is uncertain. Scotland will undoubtedly be affected by these changes.

These changes are highly likely to affect fish: a number of scenarios are possible. Everywhere there is likely to be a shift of southern species to the north and a retreat northwards of northern species. In the open sea, changing temperature and circulation patterns are likely to affect pelagic, demersal and migratory species. Along the coast and in estuaries, increased sea levels will create many changes to shallow water systems and produce problems for humans in low-lying areas.

In 1991, the author pointed out that in fresh waters, as well as latitudinal changes, there are also likely to be parallel changes related to altitude, with coldwater species moving into higher cooler waters and their place being taken in the lowlands by warmwater species. In rich lochs in summer there will be an increasing tendency to low oxygen conditions in deep water, with 'summer kill'; there will, however, be less freezing in winter and so a lesser tendency to 'winter kill' in these lochs.

One of the major concerns for the fish fauna of Scotland is that many of our native species thrive only in cool conditions. These include Atlantic Salmon, Brown Trout, Arctic Charr, Vendace and Powan, all of which must be considered as under threat as global warming proceeds. In contrast, alien species from the south, most of them introduced by man, favour warmer conditions and so are likely to do well, outcompeting and eliminating native stocks. Some of these species, which previously could not breed in Scotland because of low summer temperatures, will now be able to reproduce and disperse. These fish include many of the cyprinids – Common Bream, Barbel, Common Carp, Goldfish, Crucian Carp and others.

Thus the most likely scenario for freshwater fish in Scotland is a loss of coldwater fish species in their southern ranges and a reciprocal extension of southern species to the north. Similar changes will take place in relation to altitude, especially in stream situations, where migration is easier than in lochs. Such changes are likely to lead both to a loss of stocks and to the isolation of previously contiguous populations.

Apart from political and practical measures to reduce the amounts of greenhouse gases entering the atmosphere, aquatic biologists can improve understanding and subsequent planning by developing existing monitoring programmes, devising ecological experiments, implementing conservation plans for threatened species, controlling northward movement of undesirable species, analysing existing data and developing better models. Sadly, it seems likely that mankind is incapable of making adequate sacrifices to reverse the increasing trends towards what will be a climate disaster for many parts of the world.

References

Allcock & Buchanan (1994), Anonymous (1997), Bailey-Watts & Maitland (1984), Barnett (1927),

Dryden (1862), Elder (1965), Forestry Commission (1993), Harriman *et al.* (1987), Harriman & Morrison (1980, 1981, 1982), Harvie-Brown & Buckley (1889), HIDB (1983, 1985), Lang (1891), Lassiere (1993), MacVine (1891), Maitland (1969, 1972, 1977, 1985, 1987, 1990, 1991, 1994, 1998), Maitland *et al.* (1980, 1985, 1987, 1990), Maitland & Price (1969), Maitland & Turner (1987), Muniz (1984), Stoddart (1847, 1866, 1867), Stuart (1899).

'One day I killed in the Esk thirteen dozen of trout which weighed 23 pounds.
The water of Leith is a good stream above Balerno Bridge.
Four dozen and a half trout taken in this water weigh on an average 14 pounds.'

A. Dryden (1862) *Hints to Anglers*

*****

FISH AND PEOPLE

'Fishing next to prayer is the most personal relationship of man,
and a constant reminder of the democracy of life, and of human frailty,
for all men are equal in the eyes of fish.'

U.S. President H.C. Hoover

SADLY, the only reminder of the extinction of the Vendace from Scotland is the ironic dead-end road sign in a housing estate in the village of Lochmaben beside the Mill Loch, the last known site for this important fish. Many aspects of our past treatment of Scotland's valuable freshwater fish resource have been shameful and only recently have more positive attitudes started.

No part of Scotland is far from the sea and so sea fish and fishing have played a much greater part in the lives of most Scottish folk than freshwater fish, especially compared to, say, landlocked countries elsewhere in Europe such as Austria and Switzerland, where freshwater fish are of major significance. Nevertheless, freshwater fish have been important to many Scottish communities in the past and are still valuable in many places, particularly for sport.

Unfortunately, many of the traditions associated with most freshwater fish in Scotland, as well as a recognition of their value as food, have been lost. This is in contrast to the situation a few hundred years ago. Having visited Scotland at the end of the 15th century, Don Pedro de Ayala wrote in 1498 that 'It is impossible to describe the immense quantity of fish. The old proverb says already "piscinata Scotia". Great quantities of salmon, herring, and a kind of dried fish ... are exported. The

quantity is so great that it suffices for Italy, France, Flanders and England.' In 1618, Taylor, The Water-Poet wrote that 'In the river of Tweed, which runnes by Barwicke, are taken by fishermen that dwell there, infinite number of fresh salmons, so that many households and families are relieved by the profit of fishing.'

Marian McNeill, however, records that quite some time prior to this, 'During the early Celtic period, when adoration was paid to the waters, fish as food was taboo, and even after the introduction of Christianity it continued for a time to be considered dangerous to the purity of the soul. When, in the eleventh century, the Roman Church superseded the Celtic one, her fast-days and fastings encouraged the development of the fisheries, which at an early period became a source of national wealth.'

One obvious relationship between fish and people lies in surnames: many people in Scotland have names which are the same as those of our fish. A scan of the telephone directories for Edinburgh and for North Glasgow by the author in 2001 revealed the following impressive list: Sturgeon (30 entries), Shad (2), Bream (2), Gudgeon (1), Roach (25), Rudd (10), Pike (10), Smelt (1), Trout (1), Salmon (17), Mullet, (1), Bass (7) and Ruff (1). In addition to these there were other fish-related names such as Lamprecht (1), Eeles (2), Goldie (a local name for

Arctic Charr: 130), Carper (1), Minnis (2), Stickle (1), Gobey (1) and Flook (a local name for Flounder: 1). No doubt a search of other directories would provide interesting additions to this list.

Food

In Britain as a whole, there is little doubt that both freshwater and marine fish were an important and popular diet in times past. Arthur Bryant records that in the age of chivalry, 'In 1248, John Pecham, an Oxford friar, when created Archbishop of Canterbury had, among other items, to provide for his enthronment feast: '... 300 ling, 600 cod, 40 fresh salmon, 7 barrels of salt salmon, 5 barrels of salt sturgeon, 600 fresh eels, 8,000 whelks, 100 pike, 400 tench, 100 carp, 800 bream, 1,400 lampreys, 200 large roach ...' In Scotland, Herbert Maxwell reminds us that: 'We read in Barbour's famous poem, *The Brus* (ca 1375), that Lord James Douglas acted as caterer to the King of Scots during his wandering in the Highland hills after the battle of Methven.'

> 'Bot worthy James of Douglas
> Ay travaland and besy was
> For to purchas the ladyis met,
> And it on mony wis wald get.
> For quhile he venesoun than brocht,
> And with his handis quhile he wrocht
> Gynnis to tak geddis and salmounis,
> Troutis, elis, and als menounis.'

Three centuries later, in 1656, Richard Franck describes of Edinburgh: 'She has fish and flesh in abundance ... salmon, trout, pike, perch, eel, &c. ...'

Despite this evidence, results from archaeological excavations carried out at several Scottish burghs so far have not revealed bones of Pike or Perch, although bones of other species were present. At Perth, for instance, bones of these two species were not found during excavations covering the period 1200-1400 AD, a time when they were commonly used as food in English towns. Pike were certainly used as food during the reign of James VI, as the palace accounts in 1649 reveal. These fish probably came from the Loch of Menteith.

By the 20th Century, the situation was completely different; Edward Boulenger states that 'To those whose interest in our freshwater fishes extends beyond their mere capture with rod and line, their present day neglect as a food is a matter for regret. At one time every country house of any size had its private ponds, or stews, in which fish were kept and fattened for the table. Every castle in the Middle Ages kept its moat well stocked with fish, not merely as a source of sustenance, but also as a standby in time of siege.'

SPECIES	PRICE £ / lb
River Lamprey	31.9
Sea Lamprey	4.19
Eel	7.83
Allis Shad	3.07
Twaite Shad	1.50
Pike	5.12
Vendace	3.29
Powan	6.25
Rainbow Trout	3.37
Atlantic Salmon	4.42
Brown Trout	7.45
Thick-lipped Grey Mullet	0.84
Thin-lipped Grey Mullet 2.79	
Sea Bass	7.45
Perch	2.84

Some market prices for fish species in Finland and France. For particular stages some prices may be much higher - e.g. Glass Eels can fetch £140 / lb.

Commercial fishing

Much of the fishing in Scotland was carried out on a commercial basis, mostly for Atlantic Salmon, which were caught in large numbers in various places. In 1636, William Brereton recorded that at '... Barwicke [on the Tweed] ... They speak of three hundred and sixty salmons taken at one draught, and ordinarily about eighty, and one hundred, or one hundred and twenty at one draught.'

Further north, Thomas Tucker mentions in 1655: '... Garmouth and Findhorne in Murray-land, two small places, from whence some 60 lasts [= loads, generally estimated at 4,000 lbs.] of salmon in a yeare are sent out, for which salt is brought in from France ...' In central Scotland, Richard Franck, in 1656 noted that 'The Firth [of Forth] runs here [Stirling] that washeth and melts the foundations of the city, but relieves the country with her plenty of salmon; where the burgomasters (as in many other parts of Scotland) are compell'd to reinforce an ancient statute, that commands all masters and others, not to force or compel any servant, or an apprentice, to feed upon salmon more than thrice a week.'

Hundreds of years later, commercial fishing for Atlantic Salmon and Sea Trout continued successfully, as John Robertson described in 1859: 'I was walking up a hill towards Berwick in the evening. A horse was pulling a heavy cart up the road in front of me ... loaded to the brim with the kind of salmon that men tell lies about in every public-house in the Highlands! They were enormous. The life was hardly out of them. ... I went down to a strip of shore at Tweedmouth. ...

Nothing could be simpler than the methods of the Berwick fishery. ... The fishers prepared to pull the net. ... Then as the men pulled, narrowing the circle every minute, we saw an odd commotion out at sea as if the water were boiling. "Yon's a big fush! " shouted a boy. ... The crowd began to shout excitedly. The dogs began to bark. As the great net came into shallower water we could see that the bag, or bosom, at the end of it, was full of salmon, lashing out, leaping, fighting for life. The desperate struggle of the silver-flashing life was terrible. They fought like sharks. ... When the net was pulled in thirty-two salmon lay on the shore. They ranged from fifteen to thirty pounds in weight. ... "We shall fish right on until midnight." said one of the men.'

During the latter half of the 20th century, commercial fishing for Salmon suffered a substantial decline. The advent of salmon farming (with a subsequent reduction in the market value of Salmon), a decline in stocks in many places, and the increased value of Salmon to rod fisheries (leading to many commercial fisheries being bought out or closed down) have all contributed to this reduction.

Commercial fishing for other freshwater species has always played a minor role compared to that for Atlantic Salmon. Nevertheless, at various times and places, there have been small commercial or subsistence fisheries for Eels, Roach, Pike, Sparling, Vendace, Powan, Brown Trout, Arctic Charr, Sea Bass, Perch and Flounder. A few of these continue to this day, notably for Eels, Sparling, Sea Bass and Flounder.

Sea Bass (from Wood 1863).

Angling

Angling is a well-established pastime in Scotland, which has been famed for game fishing for hundreds of years. As well as being popular with local folk, the fisheries available in some of Scotland's waters have attracted enthusiastic anglers from far and wide. Richard Franck in 1656, in an account of the Sanquhar area, recorded that '... yet are their rivers and rivulets replenished with trout, because undisturb'd with the noosey net, which augments the angler's, if not the artizan's entertainment.' A few years later, in 1661, another visitor, Jorevin de Rochford arrived '... at Arington [Haddington] on a river [Tyne] ... I followed the river, full of good fish, particularly trout of a delicious taste...' He would not find the same situation in the Tyne now.

Coarse fishing, though practised by large numbers of people in England for centuries, was not popular in Scotland until the second half of the 20th century. One of the reasons for this was the abundance of game fish in Scotland, originally the main quarry of most anglers; the other reason was the absence of most species of coarse fish other than Roach, Pike and Perch. However, with the transfer and introduction by coarse fishermen of a variety of coarse species to sites in Scotland, especially in south-west Scotland and across the central belt, coarse fishing has become much more popular and is now a well organised sport here, with economic implications in some areas. The damage done to native fish communities by the hundreds of introductions of coarse species, some of them new to Scotland, though difficult to assess accurately, is probably considerable.

Angling is not of course, just about the economy. It is an important and long-established pastime which has given pleasure and relaxation to many thousands of Scots and others over the centuries. For this reason alone it must be sustained and fostered, something which is only feasible if our fish species and populations are protected and managed in a sustainable way.

GENTLEMEN at a distance are respectfully informed that they may procure any of the

TACKLING, &c.,

RECOMMENDED IN THE

"HAND-BOOK OF ANGLING,"

OF THE

BEST QUALITIES,

AND AT THE

LOWEST PRICES,

BY ADDRESSING ORDERS TO

JOHN ROBERTSON,

Care of Messrs GRANT BROTHERS, Printers, 14 St James' Square, Edinburgh.

_____

Price Lists, and Sample Parcels, from 2s. 6d. upwards, forwarded to all parts of the Country, Free by Post, on a remittance of the amount.

1859 advert in John Robertson's Hand-book of Angling.

The challenge, the camaraderie, and sometimes the solitude are all important to anglers: during the wars of the last century the chance to fish during periods of leave, and even the thought of getting back to the wild after the war, sustained many servicemen. Many did not get back, of course, as this abbreviated version of John Buchan's emotive 1917 poem *Home Thoughts from Abroad* indicates:

Aifter the war,
says the papers,
they'll no be content at hame

No me!
By God! No me!
Aince we hae lickit oor faes
And aince I get oot o' this hell,
For the rest o' my leevin' days
I'll make a pet o' mysel'.

'Whiles, when the sun blinks aifter rain,
I'll tak my rod and gang up the glen;
Me and Davie, we ken the pules
Whaur the troot grow great in the howes o' the hills;
And, wanderin' back when the gloamin'fa's
And the midges dance in the hazel shaws,

But Davie's deid!
Nae mair gude nor ill can betide him.
We happit him doun by Beaumont toun,
And half o' my hert's in the mools aside him.

The therapeutic value of fishing is emphasised by Henry Thoreau: 'Sometimes, after staying in a village parlour till all the family had retired, I have returned to the woods, and, partly with a view to the next day's dinner, spent the hours of midnight fishing from a boat by moonlight, serenaded by owls and foxes, and hearing, from time to time, the creaking note of some unknown bird close at hand. These experiences were very memorable and valuable to me, ...'

Few writers have considered anglers as an important part of the landscape, but Robert Louis Stevenson in *On the Willebroek Canal* appeared to think so: 'I do not affect fishes unless when cooked in sauce; whereas an angler is an important piece of river scenery, and hence deserves some recognition among canoeists. He can always tell you where you are after a mild fashion; and his quiet presence serves to accentuate the solitude and stillness, and remind you of the glittering citizens below your boat.'

Fish farming

Once regarded as a truly wild creature associated only with pure northern waters, the Atlantic Salmon, now readily available on supermarket shelves, can be quite a different 'kettle of fish'. Following success with the rearing of Salmon in the sea in Norway during the 1960s, the advantages of the sea lochs on the west coast of Scotland were realised, and after initial trials in the early 1970s, salmon farming developed very rapidly. It is now so successful that the production of farmed salmon in 2000 was about 100 times that of wild fish when salmon farming started. There has been a parallel development in freshwater units to supply the large numbers of smolts which the industry needs for rearing in the sea. Parr are grown rapidly in freshwater cages until they smolt and are ready to be moved to the sea cages, where they grow rapidly to marketable size. Selected broodstock are retained for stripping to produce the next generation, and it is at this point that stock selection begins.

The success of cage farms has been due to a number of factors, mainly because, for the farmers, they have advantages over land-based systems. The investment involved is relatively small and the principle of unit systems means that each can be added to quite simply. There is an uninterrupted water supply of relatively constant temperature. Probably most significant of all, however, up until recently it has been relatively easy to obtain planning permission to install cages, and there is no requirement to treat the 'effluent' emanating from them as there would be with land-based systems.

It has become evident over recent years, however, that such farms do pose a number of environmental problems: these are discussed elsewhere.

Atlantic Salmon and a fish cage (drawing by Robin Ade, from Maitland 1989).

Thus, although salmon farming has brought significant social benefits with it - notably a substantial number of worthwhile jobs to many of the remotest parts of Scotland - it has also created problems. Lack of planning and foresight by government and the industry created a 'boom and bust' situation and many of the smaller farms have sold up. Environmental problems - many of them foreseeable - have added to the industry's economic difficulties. However, the increasing awareness of the importance of a sustainable relationship with the environment, the Code of Conduct which has been proposed by the aquaculture industry, and the impending aquaculture and fisheries Bill in Scotland, all give hope for the future.

Culinary aspects

As a perishable commodity, fish has to be eaten fresh or preserved in some way. Robertson recorded that: 'In early times, before the advent of preserving fish in ice, the Highlander usually boiled and pickled salmon and sea trout, which rendered them every bit as edible a provision as bacon is to the Sassenach. In other words, they were "kippered", by splitting them open and salting them, after which they were dried over a wood or turf fire.'

Salmon may be eaten at many fine restaurants all over Scotland. Formerly, all the Salmon were wild caught, usually from the nearby river, and very fresh and tasty. The same was true of smoked Salmon. In good years in the past, up to 1,000 tonnes of wild Salmon were caught in Scotland, much of it exported south of the border and to other parts of the world. With the advent of salmon farming in the 1970s, however, much has changed. The numbers of wild fish have declined substantially (some suggest that this is more than a coincidence), while salmon farming has boomed, now producing over 100,000 tonnes each year. Not only is the fish on your plate now most likely to be of farmed origin, but the Scottish scenery, in many previously remote places, has changed also - with most sea lochs on the west coast and many freshwater lochs in the highlands, having extensive rafts of floating cages housing enormous numbers of *Salmo salar* - this previously completely wild fish.

As well as their scientific importance, the economic and social value of many of our native species is often not realised. For example, Arctic Charr is an important species in some northern countries. Although there are no fisheries in Scotland for Sea Lamprey or River Lamprey, both these species are highly valued as commercial species in Portugal and Finland respectively, where they fetch high prices. Similarly, both Pike and Perch, which many anglers and fishery managers regard as vermin, command high prices in parts of Europe (e.g. Switzerland). Eels too are an extremely valuable species, not only in other parts of Europe, but also in Japan and other parts of Asia. Thus the total worth of Scotland's freshwater fish resource, assuming that any fisheries were managed in a sustainable way, is much higher than normally appreciated.

Tourism

Atlantic Salmon and Brown Trout are an important part of the Scottish economy, especially on the larger salmon rivers and in some of the more remote parts of the country. Although salmon fishing can be relatively cheap on some waters, e.g. less than £100 for a whole season on parts of the River Annan, on other rivers it is expensive, for example £3,000 for one week on some beats on the River Spey. Because of this, access to many of the best salmon rivers is very restricted, both because the owners are concerned about poaching (which is a serious problem in some areas), and employ bailiffs to counteract this, but also because they do not wish their paying customers to be disturbed by other people straying along the river banks. This has led to conflicts over public access in some areas, something which it is hoped may now be resolved with the recent Bill on access to the countryside.

The tourist in Scotland can see many aspects of Scotland's freshwater fish, especially the Atlantic Salmon, and of the many attractive and wild lochs and rivers in which they occur. For those with an especial interest in Salmon, its life cycle is depicted in many places, such as the Salmon Centre near Oban or more simply on displays such as that by the riverside at Bonar Bridge. Wild fish and their aquatic companions may be seen at many places - public and private, and farmed Salmon at appropriate visitor centres. After dark, a more intimate acquaintance is possible with the fish on your dinner plate.

Many years ago the author suggested the idea of a 'Fish Trail' to the Scottish Tourist Board but the idea was rejected. This seems a pity, for there is a wide range of fishy places of interest in all parts of the country which could be incorporated into local trails, themselves combining to make an extensive national trail, for those with a special interest in fish and fishing. The following could all have a part to play in such a proposal: local public aquaria and ponds, fishing ports and fish markets, commercial salmon netting stations, waters open to the fishing public, angling demonstrations, salmon leaps and fish passes with viewing facilities, the 'weigh in' at fishing competitions, fish spawning sites, natural history and fishery museums, special events such as the flounder tramping championships on the Solway.

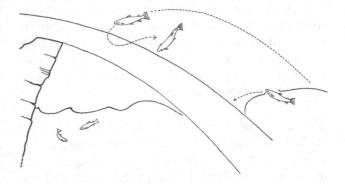

Atlantic Salmon leaping at the Pot of Gartness (from Stuart 1962).

Salmon rivers are exciting places to visit and there are more than 400 of them in Scotland. In fact, the majority of Scottish rivers offer a home to Salmon, unless they are polluted or have an obstacle such as a high weir or waterfall near the mouth. They offer an amazing variety of habitat, from tiny west highland waters which are only a few metres across where they enter the sea, to the mighty River Tay, the largest river in Great Britain with twice the flow of the River Thames. The salmon populations themselves can differ from river to river: in the smaller waters most of the Salmon are small (about 2 kg) and adults come in from the sea in late summer; in larger rivers, in addition to a large late summer run of fish around 2-4 kg, there may be a spring run of very large fish, sometimes as heavy as 20 kg. These are the stuff that salmon anglers' dreams are made of!

Artefacts

Some of the earliest Scottish artefacts involving fish were left to us by the Picts, who inhabited much of Scotland before they were dominated by the Scots. The Picts left little written record, but, as the indigenous inhabitants of Scotland living mainly to the north of the Clyde-Forth valleys, they left many carved stone ornaments which give considerable insight to their interests. The earliest Pictish stone, on Orkney, comes from the 6th century and an excellent illustrated gazeteer to the major Pictish stones has been provided by the Royal Commission on the Ancient and Historical Monuments of Scotland (RCAHMS). Fish, a sign of Christianity since early times, are the second most common animal carved on stones: most of these carvings seem to represent Salmon. As Duncan Jones has pointed out, many folk-stories involve Salmon and they frequently were believed to represent knowledge and the power of prophecy.

Stones with fish carvings are found over a wide area of Scotland, but especially in the north and east. They are found in Aberdeenshire (Dunnicaer, Keith Hall, Kintore Churchyard, Hillhead of Clatt, Rhynie), Angus (Glamis), Highland (Ackergill, Drumbuie, Dunrobin, Edderton, Gairloch, Garbeg, Golspie, Latheron, Ulbster, Easterton of Roseisle), and Perth & Kinross (Inchyra). Fish carvings are also found in some caves, notably those at East Wemyss in Fife and Covesea in Moray.

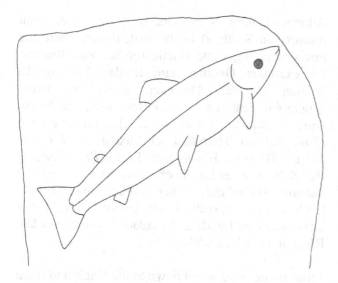

Outline of an Atlantic Salmon on a Pictish stone at Rhynie.

Dick Shelton has noted 'their striking symbol stones are well known to historians as one of the principal artistic legacies of the Dark Ages. The standing stone in the manse garden at Glamis in Angus is renowned for its accurate representation of an Atlantic salmon incised between that of a serpent (a viper) above and a mirror below. We have no way of knowing how the 7th century (or probably much earlier) sculptor incised the salmon. The accuracy of its proportions suggests that the initial outline was made with an iron spike with the fish in situ ... Other details ... would then have been added to complete the representation.'

In a few Scottish cemeteries there are gravestones to fishermen. These may have engravings of boats, fish, oars and nets. The term 'lax fisher', sometimes found on inscriptions on such stones, is the traditional old name for a salmon fisherman. Few such stones are ever erected now.

Scotland has a number of fine statues depicting fish or fishermen. One of the most notable is the memorial to Neil Gunn at Dunbeath, of a boy with a giant Salmon on his back, recalling a scene from the author's *Return to the River*. In Ayr, there is a striking statue of a fisherman holding a fine Salmon in his hands. The forecourt to the nuclear power station at Hunterston has a series of heads rising out of the water with a fish above each.

Fish appear in some places in Scottish heraldry. If placed in the normal swimming (horizontal) position they are defined as *naiant*, if vertical with the head up, as *hauriant*, if vertical with the head down, as *urinant*, and if arched, as though jumping, as *embowed*. The colours used for fish are usually either the natural colours, *proper*, or silvery - *argent*. Examples of arms are those of Hobbs, which include a Salmon, and an excellent one of three Pike or geds for the Geddes or Ged family. The Chief of the Clan Donald, Lord MacDonald, has a Salmon in one of the quarters of his coat-of-arms.

Several towns in Scotland regarded fish as so important in past times that they were incorporated into the town coat-of-arms. One of the best known examples is Glasgow, in which Salmon are a prominent feature. The coat-of-arms always shows the fish with a ring held in its mouth. This, according to legend is because a King of Strathclyde had given his wife a ring as a present. But the Queen gave it to a knight who promptly lost it. Some versions of the story say that the King took the ring while the knight was asleep and threw it in the river. The King then demanded to see the ring - threatening death to the Queen if she could not do so. The knight confessed to St. Mungo who sent a monk to catch a fish in the River Clyde. When it was brought back (presumably catching Salmon in the Clyde in those days was a lot easier than now!) St. Mungo cut open the fish and found the ring. When the Bishop of Glasgow designed his own seal around 1271, he used a Salmon with a ring in its mouth and this has come down to us in today's coat-of-arms.

The attributes of St Kentigern (Mungo) are remembered in a nonsense rhyme taught to

Glasgow school children about the city's coat-of-arms:

> *'This is the bird that never flew,*
> *This is the tree that never grew,*
> *This is the bell that never rang,*
> *This is the fish that never swam.'*

The Glasgow coat-of-arms is incorporated into a number of artefacts around Glasgow, for example there is an ornate lamp-post, located near Glasgow Cathedral – itself founded nearly 1500 years ago by St Mungo. The same symbolism has also been used by the University of Glasgow. Other Scottish towns which have fish in their coats-of-arms include Port Glasgow, Peebles and Lanark. The Burgh Arms of Peebles, for example, have three Salmon - one going upstream and two downstream. The associated motto is *Contra Nando Incrementum*, meaning 'I swim against the current and increase'. South of the Scottish border, Kenworthy notes that 'three pike in a red field were the arms of the ancient baronial families of Lucie, of Cockermouth and Egremont.'

Official crest of the University of Glasgow.

Atlantic Salmon are also a symbol of the Clan Campbell and the Duke of Argyll's banner, usually on show in Inverary Castle, is spangled with Salmon.

Place Names

An inspection of many Ordnance Survey maps of Scotland will reveal that, perhaps curiously, there are few places or geographical features which are named in English after fish. In contrast, towards the north and west of the country there are many lochs and rivers named after fish - but in Gaelic or derived from Norse. Some of the names used are very general, for example: Loch an Iasgair *Loch of the Fisherman* (Gairloch), Loch an Iasgair *Loch of the Fisherman* (Eddrach), Loch Eisg-bhrachaidh *Loch of the Speckled Fish* (Inverpolly). The latter, presumably refers to Brown Trout.

Atlantic Salmon were clearly a very important resource in Scotland in the past, though there are few waters using the Gaelic for Salmon: *Bradan*, for example: Geodha nam Bradan *Voe of the Salmon* (Lewis). However, there are many geographic features derived from *Lax*, the Norse word for Salmon, for example: Glen Laxdale *Glen of the Salmon* (Harris), Loch Laxdale *Loch of the Salmon* (Harris), Eilean Mor Laxay *Big Island of the Salmon* (Lewis), Loch Laxford *Loch of the Salmon* (Sutherland). Other similar names include Loch Laxavat Iorach, Loch Laxavat Ard, River Laxdale, Glen Laxdale, Laxadale Lochs, Laxadale River and Loch Laxdale.

Trout *Breac* were well known to the Gaels and there are numerous lochs named after this species, especially in Sutherland and the Western Isles, for example: Loch nam Breac *Loch of the Trout* (Sutherland and Eddrach), Loch nam Breac Buidhe *Loch of the Yellow Trout* (Ross & Cromarty), Loch nam Breac Dubha *Loch of the Black Trout* (Western Isles), Loch nam Breac Dearga *Loch of the Spotted Trout* (Inverness), Loch Dubh nam Breac *Black Loch of the Trout* (Argyll)

Sea Trout were distinguished by the Gaels as *Gealag*, for example: Loch na Gealag *Loch of the Sea Trout* (Western Isles).

The Gaels were also very familiar with Pike *Gead*, for example: Loch a Gheadais *Loch of the Pike* (Perthshire), Loch nan Geadas *Loch of the Pike* (Ross & Cromarty), An Gead Loch *The Pike Loch* (Ross & Cromarty), Poll na Geadas *Pool of the Pike* (Dunkeld).

The name Lochan Allt an Sgadain *Loch of the Burn of the Herring* (Argyll) is a curious one, for though it mentions Sgadain, which usually means Herring, it implies that they are associated with fresh water. Also, the lochan concerned is in the hills to the east of Loch Awe. An explanation may be that not Herring, but Arctic Charr (occasionally known in the highlands as 'Freshwater Herring') may be intended. Curiously, though the Gaels knew about Arctic Charr *Tarrdheargan*, and fished for them in places, there do not seem to be any lochs or rivers named after them. For further information, W.F.H. Nicolaisen has given an excellent review of the origin of place names in Scotland.

In a few places streets are named after fish. In Lochmaben, for example, 'Vendace Drive' is named after the now extinct Vendace which occurred in the nearby Mill Loch. In Polmont there is 'Salmon Inn Road' to the west of Inchyra roundabout.

Hobbies

A number of hobbies, all relating to fish, are popular in Scotland and all have beneficial effects on local economies. As well as angling, considered above and elsewhere in this book, these include fly-tying, aquarium and pondkeeping (also considered elsewhere), art, the use of fish skins in making purses and other objects, collecting fish stamps (enjoyed in the past by the author) and so on.

Damage

Humans have inflicted enormous damage on the freshwater fish populations of Scotland and we have irretrievably lost some parts of their value and diversity. The kind of incident described by Herbert Maxwell '... in 1894, not for the first time, the sudden release of mineral waste from Dalmellington pits swept havoc down the channel of Burns's Bonny Doon. Thousands of salmon and trout were destroyed in a single night.' has been repeated in one form or another hundreds of time in many parts of the country. The range of threats to which our fish are still exposed in many parts of the country is considered in more detail in a separate chapter.

Research

As can be seen from the bibliography at the end of this book, there has been extensive research on the freshwater fish of Scotland, though many areas are still to be explored. Past research has been carried out for many reasons and has had many facets. The emphasis has been on species which are important in economic terms - notably Atlantic Salmon, on which more effort has been expended than all other species together, though some studies have been made of the other larger species which are important to angling - Brown Trout, Pike and Perch.

Towards the end of the 20th Century, rather too late unfortunately, fish conservation came to the fore and studies were carried out on rare and threatened species, such as Allis and Twaite Shad, Sparling, Powan and Vendace, Arctic Charr and the three lamprey species. Small species such as Three-spined Stickleback and Minnow have also claimed some attention, mainly because of interest in their behaviour and physiology.

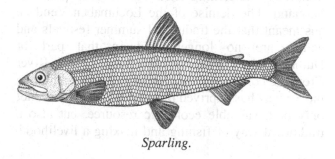

Sparling.

All of this research has created a useful bank of information for future research and for teaching in both schools and universities where ecology and other disciplines can now be based on factual information and experiments in the field. Perhaps even more important is the fact that the information is essential if we are to manage our fish populations adequately, so that the sustainable use of this valuable resource in the future is based on a firm foundation of science.

Education

Some years ago, based on a project which had initially been successful in the United States, Alistair Stephen started an educational initiative using fish in schools in southwest Scotland. This has now spread to several other areas of Scotland, via the Fisheries Trusts, and, with titles like 'Trout in the classroom', is proving to be very successful in educating young children about the aquatic environment and its problems.

Culture

Andrew Lang, in his *Angling Sketches*, recalls that 'Thirty years ago the burns that feed St. Mary's Loch were almost unfished, and rare sport *we* had in them, as boys, staying at Tibbie Shiel's famous cottage, ... not in our time will any man, like the Ettrick Shepherd, need a cart to carry the trout he has slain in Meggat Water.... Nothing can be so good as what is old, and, so far as angling goes, is practically ruined, the alternate pool and stream of the Border waters, ...' One wonders if the change is not, in fact, due to the fact that the Meggat trout population could not sustain its members being taken away in 'cartloads', though the point is now academic since most of the burn concerned now lies under Meggat Reservoir.

Nevertheless there have been losses of culture and traditions concerning fish in several parts of Scotland. The demise of the Lochmaben Vendace has meant that the traditional summer festivals and games are no longer held in that part of Dumfriesshire. With the exception of the River Cree, the collapse of the sparling fisheries all along the Solway has deprived that part of the country, not only of a valuable economic resource, but also a traditional way of fishing and making a livelihood.

The collapse of stocks of Sea Trout and Atlantic Salmon along parts of the west coast of Scotland has meant great losses in these areas, not only to the local economies, but also in terms of traditions and expertise as ghillies, etc.

Though some of these elements of our culture are probably lost forever, there could be a future for them again if there was a proper concern for all our freshwater fish, and if there was a national structure for their management and conservation - something which Scotland has desperately needed for hundreds of years. This important matter is considered in more detail in a separate chapter.

References

Barnett (1927), Barrett *et al.* (1999), Boulenger (1946), Brereton (1636), de Ayala (1498), Dennis (1999), Franck (1694), Fraser & Ritchie (1999), Grant (1812), Gunn (1931), Jackson (1989), Jones (1998), Lang (1891), Maitland (1989), Maitland & Campbell (1992), Maxwell (1897), McNeill (1929), Nicolaisen (1957), RCAHMS (1999), Robertson (1859, 1861), Shelton (1993), Stevenson (1913), Stuart (1962), Taylor (1618), Thoreau (1889), Tucker (1655), Wood (1863).

'At this time in the Highlands,
we were so remote from markets that
we had to depend very much on our own produce ...
Yet we lived in luxury; game was plentiful, ...
the rivers provided trout and salmon, the different lochs pike and char.'

Elizabeth Grant (1812) *Memoirs of a highland lady*

*****

WATCHING FISH

'The fish lay on the bottom at a depth of some four feet, and was visible from tip to tail.
Crouching over the water, Hugh saw its delicately shaped head, its lithe form,
its desirable beauty. He saw them too clearly, so clearly that obviously the fish saw him.
It was watching him. ...'

Neil Gunn (1931) *Morning Tide*

FISH watching is not yet a popular pastime - certainly compared with bird watching, which has many devoted followers (including the author). Yet, if the weather and the light are right it is possible to spend many hours by a loch or river and get great pleasure and satisfaction from watching fish go about their daily business. A good vantage point is a bridge, such as the one at Lamington, depicted above, by L.F. Brotchie in *Glasgow Rivers and Streams*. A helpful aid in this pastime is a good pair of polaroid sunglasses. Many anglers are happy to admit that, even if they catch no fish, they find great pleasure just being beside water and close to nature. Often, special moments occur, such as that recalled in Hamish Hendry's verses from *Pax sine pace*:

> *'Lo! from its silver lair*
> *Hunting the lazy fly,*
> *Gleaming a curve on high*
> *Glances a trout in air:*
>
> *Lo! from the blossomed thorn*
> *Swiftly on wings a-whirr*
> *Flashes the kingfisher:-*
> *Straightway the trout is torn.'*

Neill Gunn was one of the many Scottish authors fascinated by the Atlantic Salmon; there is a wonderful statue dedicated to him which holds pride of place at Dunbeath in Sutherland. The statue is of a boy carrying an enormous Salmon, and depicts an exciting scene from his novel *Highland River*. In another novel, *Morning Tide*, he faithfully conjures up a scene involving Salmon, familiar to many naturalists: 'The fish lay on the bottom at a depth of some four feet, and was visible from tip to tail. Crouching over the water, Hugh saw its delicately shaped head, its lithe form, its desirable beauty. He saw them too clearly, so clearly that obviously the fish saw him. It was watching him. ...'.

The problem

Fish are difficult to observe in their natural environment - this is probably the main reason why they are one of the neglected elements of our native wildlife. Only in recent times has it been possible, thanks to sophisticated diving equipment, to go beneath the surface and observe fish in their natural surroundings. In the restricted environment of our lochs and rivers, however, most native fishes react nervously to the presence of divers (and their air bubbles) who, because of poor light penetration and the lack of clarity in our fresh waters (compared to those of the sea along our west coast), are usually too close to fish for the latter to behave normally.

One very useful aid in watching fish and underwater life is a glass-bottomed box - an essential tool of the former pearl fishermen of Scottish rivers. Pearl fishing is now an illegal occupation because of the threat to Pearl Mussels, which have disappeared from many waters because of overfishing, pollution and other threats. Robert Gibbings wisely kept such a box on board: 'When the water cleared I put my glass-bottomed box over the side of the boat, and watched the life that circulated among the roots of the water lilies.'

Trout and lilies (a decoration by Louis Rhead, from Crandall 1914).

There are some exceptions to this, of course. Some fish are very obvious in shallow clear waters, especially when they are engrossed in spawning activities. Lampreys, sticklebacks and Minnows are good examples of this; the author has spent many enjoyable hours sitting quietly in the sunshine on river banks watching the fascinating natural behaviour of these and various other species. The amount of light and the clarity of the water is obviously important, as S.E. White implies: 'Through the green water the bottom was plainly to be seen, with all its hills and vales, its old snags, its rock, and the clean white sand. ... Trout lay singly, in twos and threes. Some were close to the bottom in plain sight, their gill-covers moving slowly. Others could be made out dimly as shadows in the shadows. All were big.'

In many places a quiet approach is essential if fish are not to be disturbed, for most species are very wary of vibrations and shadows. The problem is treated amusingly by John Wolcot in his poem *To a Fish of the Brooke.*

'Why flyest thou away with fear?
Trust me there's naught of danger near,
I have no wicked hooke
All covered with snaring bait,
Alas, to tempt thee to thy fate,
And drag thee from the brooke.

Enjoy thy stream, O harmless fish;
And when an angler for his dish,
Through gluttony's vile sin,
Attempts, a wretch, to pull thee out,
God give thee strength, O gentle trout,
To pull the raskall in!'

Richard Jefferies describes the best tactic: '... any one may see them who may stroll by the water's edge on a bright, warm day, taking care to walk slowly and not to jar the ground or let his shadow fall on the water before he can glance round the willows and bushes. Jack may then be seen basking by the weeds. ... a willow tree overhangs the pool there. By lying on the grass and quietly looking over the brink, the roach may be seen swimming in the deeper part, and where it shallows upstream is a perch waiting for what may come down. Where the water runs slowly on account of a little bay, there, in semi-darkness under the banks on the mud, are a few tench.'

The perception

There are four million anglers in Great Britain and under a million bird watchers, but the differences in attitude between them are astonishing. Relatively few anglers show any interest in fish other than those they wish to catch, and few fish are protected by conservation laws. In contrast, there is an intense interest in birds by ornithologists and the public in general, and most, even very common species, are protected under the Wildlife and Countryside Act and other legislation.

Yet some anglers are prepared to admit that, sometimes at least, they are as interested in the fish as in the fishing. Arthur Johnson notes that 'The angler, whose curiosity for observing the ways of the fishes in their natural element is as keen as his delight in trying to outwit them with rod and line, creeps cautiously to the bank.' George Dewar, too confesses to his attention wandering: 'There are intervals, I grant, when the trout is at the back of the thoughts, and the scene and delicious influences of Nature by the river come to the front ... Whilst trout are moving, or even whilst dun and Mayfly are appearing at the surface, and a rise of trout is hoped for, we can give little attention to wild life. ... But in the waiting hours of the dry fly angler's day, we can be very sensitive to the beauty and interest of the riverside.' Anatole France is certainly among the most philosophical of anglers: 'The careful watch for fish particularly suits my meditative spirit, holding the insidious line, and pondering on the eternal verities.'

It is perhaps because fish are often difficult to watch in the wild that they have no popular following. Instead of being warm and feathery, they are cold and wet. People see fish mostly as dead objects on fishmongers slabs, or served in batter with chips. Yet fish do have a fascination for the public. A well filled slab in a fishmonger's window always attracts little knots of curious people, and some fishmongers (e.g. S.L. Neill's in Glasgow) used to make a point of buying any unusual fish at the market and displaying them in the window.

Clearly too, there is an interest among some anglers and aquarists, but the former are mainly interested in quarry species and the latter in exotic ones (sometimes paying very high prices for them). There is sadly little interest in our less common native fish and their conservation, although some of them are in dire need of support and are every bit as fascinating as birds. If the interest in catching and eating fish could be transferred to watching and conserving them - enjoying them as the delightful wild creatures they are - the future for our native fishes would be brighter.

The reward

Thus, although watching fish is not a popular pastime, yet it can be both rewarding and relaxing. As mentioned, a prime reason for this poor following is that fish in the wild are usually hard to see, so the prospective fish-watcher is limited to certain waters and favourable weather. Also one must know something about fish before going out to

watch them - which fish can be found in an area, what they are likely to be doing during the daytime and at that particular season, and which are the most likely places to see them. Finally, fish are often thought of as being rather drab, with few of the interesting habits of birds - a profound misconception if ever there was one. Something of the pleasure of watching fish is captured in this quote from an anonymous author: 'Even the iris blades are not so coolly green as the backs of minnows at midsummer; and, like small children, they delight to caper in the sun. One flicks his tail on the shallow, and the whole shoal flashes its armour; they are away and back again with a speed that diverts the idle eye.'

Nine-spined and Three-spined Sticklebacks - easy fish to watch in the wild (from Wood 1863).

Fish-watching in the wild combines well with watching birds, but is generally only feasible in bright weather when there is little wind to ruffle the water surface. Then fish can be watched by carefully approaching the edge of a loch or river and keeping still. It is surprising what can be seen with a little patience, and there is something to be seen at most times of the year. The exception, weatherwise, is at waterfalls, often after heavy rainfall, when Salmon and Sea Trout may be very evident and people gather to watch them leaping.

Of course, salmonids are not the only occupants of these rivers they rub shoulders with a variety of other aquatic creatures - some of which they eat, others which may eat them. Typically, salmon rivers have few submerged plants, other than mosses, due to the high flows and the constantly eroding rocky beds. In the more barren northern rivers of Scotland, such as the River Naver, there are only a few other fish species present, such as

Brown and Sea Trout, Eels and sticklebacks. Further south, however, in rivers such as the River Tweed many more fish species occur, perhaps 20 altogether, including coarse fish like Roach and Dace. Obvious predators of young Salmon (and other fish), such as the attractive Goosanders and Red-breasted Mergansers, can often be seen on most salmon rivers. Less obvious, but equally widespread, are Otters, and that less welcome introduction from North America, the Mink.

What, where and when

Bridges make particularly good observation posts for fish watchers. From such a vantage point an anonymous writer notes that it may be possible to 'Watch some plump roach as it lets the current slide by it; it seems gently fanning its cool throat with its fore-fins.' Clear streams are the best places to see Brown Trout with their varied colours and beautiful patterns of rings and spots. Small flies or other insects which drop on to the surface upstream may bring them up to the surface for a closer look. In some of the same streams, the poise and elegance of Grayling can be admired also.

Walkinshaw Brig (from Brotchie 1914).

A boat is also a useful vantage point for seeing fish, as Henry Thoreau indicates: 'In warm evenings I frequently sat in the boat playing the flute, and saw the perch, which I seemed to have charmed, hovering around me, and the moon travelling over the ribbed bottom, which was strewed with the wrecks of the forest.'

Richard Jefferies knows that the river bank is also a useful place to stand quietly and look down into the water. The jack lies in shallower water and keeps close to the shore under shelter of the flags, or concealed behind the weeds. It is as if he

understood that every now and then the shoal of roach will pass round the curve ...'

Although many commercial fishermen have given up in recent years, either because they were bought out by salmon 'conservation' organisations or because of declining catches, there are still some parts of the country where it is possible to watch commercial netsmen at work. Net and coble fisheries still exist in some estuaries and, on different parts of the coast, stake and bag net fisheries are still operated. At these places, it is possible to see good catches of Atlantic Salmon and Sea Trout and, on occasion, bycatches of various other species, including Allis and Twaite Shad, Thin-lipped Mullet, Sea Bass and Flounder.

Some fish farms are keen to have visitors, and at places such as the Salmon Centre near Oban, or trout farms such as that at Beecraigs Country Park, near Linlithgow it is possible to see Atlantic Salmon and Rainbow Trout respectively being reared for the table. Such centres have a very commercial eye for visitors and they hope that the latter will not only pay to get in but also buy some fish food to feed the fish and finally buy some fish on the way out. As one fish farm puts it: 'See them, feed them, eat them!

Spring

In spring, River and Brook Lampreys build nests and spawn in shallow streams in April and May, (the much larger Sea Lamprey does not appear until June or July) and can be readily observed spawning in large and small rivers. The brilliant turquoise and red Three-spined Stickleback males build nests in shallow water between April and June and are readily seen nest-building, courting females towards their nest or chasing off virtually anything which moves near their nests. They can subsequently be observed watching over a batch of young. Richard Jefferies describes the scene: 'In quiet, sheltered places, where the water is clear but does not run too swiftly, the "minnie", as the stickleback is locally called, makes its nest beside the bank. ... He guards the nest with the greatest of care, and if he is tempted away for a minute by some morsel of food he is back again immediately.'

Many fish are migrants and several species are very obvious when travelling in large numbers. In the early months of the year, young Eels (elvers) move upriver and can sometimes be seen in thousands, often attracting dozens of gulls and other predators.

As the water warms up in spring, fish become hungry, as Robert Gibbings found out on one of his boat trips. 'While at this same anchorage I found in my larder a hard-boiled egg, whose shell had been broken. It only needed the evidence of my nose to tell me that it had seen its best days, so I broke it into a few pieces and held these under water. Keeping my hand absolutely still it was soon surrounded by hundreds of small fish, first the little ones, then larger and larger. Minnows, dace, gudgeon, they came and ate the egg out of my hand. While the larger ones of six or eight inches were feeding from my fingers the little ones went right inside my half-closed palm to see what they could find there. Of course I had to remain very still and to show as little as possible of myself over the side of the boat, but this was an occasion where I found my glass-bottomed box of use, for through it I could see everything as clearly as if it was all happening in the open air. At the same time it was difficult for the fish to see me.'

Summer

During the warmer summer months the members of the carp family become active and shoals of Minnows, Roach and Dace can be seen swimming and feeding in shallow water or perhaps spawning among weed (Roach) or in shallow running water over clean pebbles (Minnow). 'But there is a mystery about the thinnest layer of that alien element; and the fishes that swim there are never so full of happy suggestion as in June sunshine. Those that greet us as we pass above or beside them are mostly roach and dace and minnows; their scales of green and silver are the perfection of midsummer attire.' (Anonymous).

By watching from a raised vantage point over a secret pool, it may be possible to observe some of our introduced species, perhaps giant old Carp or the mysterious shapes of dark, thickset Tench, which send streams of bubbles up from the mud among which they are feeding. The scene is well described by A.C. Williams: 'The branches of many hollow-stumped willows gently sweep the surface of the water, which itself is liberally decked with beds of snow-white lilies. And it holds carp; carp that one can see. Carp that lazily swim round and

round all day long. Carp that have the appearance of well-fed alderman. Carp that despise the angler who dares to pit his skill against them.'

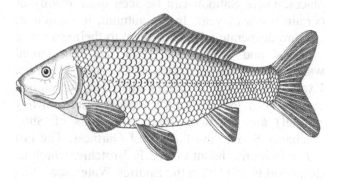

Common Carp.

In the few waters where they have been introduced, groups of blunt-nosed Chub may be found, and on hot airless days Bream may be seen rolling at the surface, sometimes creating a turbulent wash at the edge of a lake or river. Dull green shapes flash silver as Roach and Dace twist and turn, while the long slender forms of Pike may be picked out as they rest motionless, suspended among weed. Richard Jefferies describes watching Pike: 'After a long look I began to examine the stream near at hand: the rushes and flags had forced the clear sweet current away from the meadow, so that it ran just under the bank. ... Under a large dead bough ... across ... the stream I saw the long slender fish lying a few feet from the bank, motionless save for the gentle curving wave of the tail edges.'

Occasionally, one has a once-in-a-lifetime sighting of the kind recounted by W.H. Hudson. 'I stood on the margin, looking across the sheet of glassy water at a heron on the farther side, ... Between me and the heron scores of swallows and martins were hawking for flies, ... and all at once, ... there was a great splash ... later out of the falling spray and rocking water rose a swallow, struggling laboriously up, its plumage drenched, and flew slowly away. A big pike had dashed at it and tried to seize it at the moment of dipping in the water, and the swallow had escaped as by a miracle.'

Autumn

Although juvenile Atlantic Salmon and Brown Trout are relatively easy to see in the crystal waters of their nursery streams, in many rivers it is difficult to see adult Salmon, their presence being revealed only by the occasional fish leaping from the water trying to free itself of parasites or from its capture by anglers. However, there are quite a number of places where Salmon can be seen quite readily at certain times of year. In the autumn, in particular, fish are desperate to get upstream to their spawning grounds and are prepared to leap substantial waterfalls, making themselves very visible. Examples of the many places where leaping Salmon can be seen in October (especially after a heavy rainfall) are the Rogie Falls, the Falls of Shin, Buchanty Spout, and the Pot of Gartness. The last was a favourite haunt of T.C.F. Brotchie, which he described in 1911: 'Up the Endrick Water, sparkling in the sunlight, an unfrequented pathway takes us to the famous Pot of Gartness - a deep and romantic linn shaped like a caldron or pot, hence the name. Vast shoals of salmon and trout from Loch Lomond make their way up the Endrick, and afford no little entertainment in their efforts to force their way over the falls at the Pot of Gartness. We saw dozens of the boldest salmon conquering "the leap" in magnificent fashion.'

Sometimes migratory fish are held up by river conditions, as described by Dugald MacIntyre: 'Once, when the run was delayed by drought, I lay on a shingle bank at the mouth of the Glen Barr River and saw the sea-bottom simply paved with salmon and sea trout. So closely packed were the fish as to look like masses of seaweed, and they lay still on the bottom, in the brackish water, the only signs that they were aware of my presence being an occasional silvery gleam as an individual fish made a nervous movement.'

Salmon may also be seen during the summer and autumn at the viewing chambers on fish passes, which have been constructed to allow them to circumnavigate hydroelectric dams. Good places are the viewing chambers at the Pitlochry Dam on the River Tummel and the Borland lift at the Aigas Dam on the River Beauly. At such places, especially in early summer, it is also possible to see the young Salmon passing downriver to the sea as silvery smolts.

Again, one can be lucky enough to see something exceptional, if in the right place at the right time. Edward Hewitt obviously was. 'There were only one or two small salmon but a number of good-sized sea trout in water six to eight feet deep. ... a red squirrel jumped into the water from the opposite bank and began to swim across the river. The largest trout made for him at once and coming behind him caught his tail and pulled him below the surface. The squirrel broke loose and struggled to the surface again with part of the hair of his tail gone. Again the trout caught him and a second time the squirrel got away with more of his hair stripped off. The third try was near our bank, and when the squirrel finally came ashore ... his tail was stripped like a rat's and he was completely exhausted.'

Winter

During autumn and winter, Salmon and Trout spawn. Each female digs out a nest in gravel while her male partner chases off other suitors. Spawning can be seen in rivers and in surprisingly small burns in many parts of Scotland and, if the approach is a cautious one, and the light is favourable, the full spawning act can be observed.

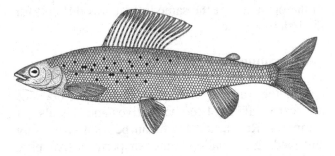

Grayling.

Other fish, however, are generally not easy to watch in the wild in winter. The weather is cold and may be wet (or snowy) and windy. Days are short and often the light is poor. Many fish (especially members of the carp family) are hiding in deep water or under stones or weed. Nevertheless, some fish (especially salmonids) are active at this time, as are those who angle for them. Grayling fishing is popular during the winter and most fishermen, if approached cautiously and respectfully are willing to show you these attractive fish as they are caught - and often then released.

During cold weather, young Salmon and Trout continue to feed - they have been found to have food in their stomachs even when the water temperature is near 0°C. Though more normally a summer occupation, what H.T.Sheringham describes as 'The Bridge Habit' can still be carried out in winter. '... the said bridge generally gives shelter to the largest trout in the neighbourhood. If he is a well-known trout, & respected by the inhabitants, he may be seen lying a foot or so below the bridge waiting for the worms which are thrown to him from time to time by his admirers. There is a bridge over another river ... below which are half-a-dozen trout constantly in waiting for pellets of bread, and I have there seen as many stalwart anglers, each with his slice of bread, solemnly making votive offerings.' Richard Jefferies also had the bridge habit: 'The trout I watched so long, and with such pleasure, was always on the other side, at the tail of the arch, waiting for whatever might come through to him. There in perpetual shadow he lay in wait, a little at the side of the arch, scarcely ever varying his position except to dart a yard up under the bridge to seize anything he fancied, and drifting out again to bring up at his anchorage.'

Yet another fish watcher from a bridge was J.H.B. Lockhart ' ... there was a wooden bridge over the burn, and here I would stand for long intervals, keeping very still ... It fascinated me to see how the trout took up their line in order of physical superiority, with the biggest fish commanding the best feeding. If he left his place for one second on some predatory expedition, the next biggest would automatically move up. Then the big fellow would return and fiercely drive away his weaker rival. Occasionally a huge eel, for the burn contained some monsters, would wind his way out of the weeds. Then all but the biggest trout would dart under the weeds ...'

An alternative

To many people the word 'aquarium' conjures up little more than the idea of a sterile goldfish bowl and its prisoner, or a brightly lit tank in a restaurant with multi-coloured plastic gravel, plastic plants and sometimes even plastic fish! Aquaria mean much more than this, however, and, to many millions of people in Europe and elsewhere, form the basis of an absorbing hobby. In other spheres aquaria offer a valuable educational tool, or the basis for exciting work on processes ranging from animal behaviour to research on cancers. With modern technology it is possible to keep almost all species of fish alive in aquaria, with the exception perhaps of those which live in very deep water. Gregory Bateman is right to be enthusiastic! 'It is

almost unnecessary to say that there are no more interesting and attractive inhabitants of an aquarium than fish. Fish are both intelligent and, to a certain degree, affectionate.'

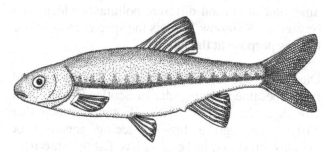

Common Minnow - one of the best native fish for an aquarium.

Aquaria are employed in two main ways: firstly, simply as containers for keeping alive fish which are required for some purpose such as food or experimentation. Secondly, as small isolated habitats replicating those found in nature, and including not only fish, but also aquatic plants and invertebrates. Such aquaria can be both educational and aesthetically pleasing, as can many types of outdoor pond. Important studies of behaviour have relied on fish in captivity, for instance, the complicated and ritualised breeding sequence of sticklebacks, from nest building to the dispersal of fry, can be watched in minute detail in a small well-lit aquarium.

The object of experimental aquaria is to test the performance of fish under various, often extreme, conditions. For example, light, temperature, water flow and quality (including oxygen content), are all variables which may be altered during experimentation. Some of the work in this field is of a very theoretical nature, but much of it is totally applied and relevant to man's needs. The study of fish genetics and the value of various types of foodstuffs are two examples of research related to priorities in fish farming at the moment. One of the obvious impacts of pollution on fresh waters is the death of fish - often in very large numbers. Indeed, were it not for the death of fish many people would not know (or even care) that a stream was suffering from severe toxic pollution. Attempts are being made at the moment with various species of fish in aquaria to develop a 'standard fish', whose precise reactions to various polluting substances, and to combinations of these, are known. This allows the possibility of biological assay of effluents, etc. by introducing them to fish in aquaria, or suspending fish in cages in the effluent channel. Such tests have a number of advantages over existing, purely chemical methods of detecting pollution, mainly in the ways in which they act as integrators both of time, and of several different pollutants which may be present. Rainbow Trout is the species most in use for this purpose at the moment.

During the 20th century, keeping fish in aquaria and ponds became very popular in Scotland. Writing in *The Aquarist* in 1977, K.L. Brown recorded that 'Fifty years ago a few far-seeing aquarists or persons interested in keeping live fish in watertight glass aquaria as they were known then, banded together, meeting in each other's houses, attracting others, until they realised that they should form a properly constituted society with a hall as a meeting place. This was the birth of the Scottish Aquarium Society.' He goes on to note that 'Looking at a show catalogue for 1936, for instance, the biggest section, by far, was the coldwater section with 10 Goldfish classes... One class devoted to domestic fish other than goldfish e.g. golden carp, orfe, rudd, tench. Four classes devoted to others e.g. bitterling, roach, dace, minnow, bream, sunfish, stickleback, pope, pike, perch. Lastly a section called "Sundry" e.g. a shoal of minnows or sticklebacks.'

Ponds too can be a source of pleasure to those who like watching fish. One famous saltwater pond in Scotland is described by H.V.Morton: 'On the way down through the Rhinns I came on the west side of the peninsula to a tiny place called Port Logan. There is a white turreted building on the shore approached by a long drive, and near this building there is a pond of pet codfish. ... a beautiful deep pool hewn out of the rock in such a way that every tide fills it without giving the fish their liberty. The young man keeps a basket of shelled mussels always ready with which to feed the fish. ... Andrew M'Dougall of Logan made the fish-pond in the year 1800 with the idea of providing his larder with a fresh supply of fresh cod and other sea fish. Somehow or other the idea, happily for the fish, was not always observed ... the water was cut with the powerful green forms of cod fish, some of them ten pounds in weight. He ... held a mussel in his hand. Five or six fish put their heads out of the water and nosed for the morsel like dogs. It was an extraordinary sight. Some of them cruised round,

but one, the canniest, just waited with his head out of the water!'

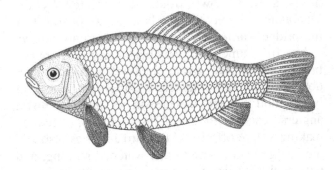

Goldfish - an excellent aquarium fish given the correct conditions.

The main attraction of the indoor aquarium is that it represents an easily observed microcosm, with conditions similar to those found in nature. The object of keeping such aquaria is to achieve a balance between the various components, so that something equivalent to natural recycling can be attained. It is rare for this to happen completely, and normally some material is added to the system in the form of fish food and some removed in the form of fish droppings and decaying or excessively growing plants. Nevertheless something near a balance can be achieved.

The plants in an aquarium are normally rooted in gravel and are able to use up quantities of waste salts and carbon dioxide produced by the fish. Under the action of light, the plants photosynthesise and produce oxygen (much of this also comes through the surface of the water) which is respired by the fish. Certain fish eat parts of the plants and also algae, which grows on all surfaces in the aquarium, and the ecosystem is often made more complex by adding invertebrates such as molluscs and crustaceans, which may help to prevent excessive plant growth and whose eggs and young form a source of food for the fish. Many very useful books have been written about aquaria, and some of these are listed in the bibliography. In addition there are aquarium societies in various parts of Scotland and anyone with an interest in this subject is recommended to join one.

The educational value of fish in aquaria is something which should not be underestimated, for if appropriate species are kept in the right conditions they may be persuaded to go through their full life cycle in one year and to demonstrate a wide range of biological principles which are relevant to the theory being taught in the classroom. This may include many aspects of behaviour, reproduction, ecology, anatomy and physiology. Common species found in Scotland which are of great value as educational aids since they may be kept and bred in aquaria include: Brook Lamprey, Minnow, Stone Loach, Three-spined and Nine-spined Sticklebacks and Bullhead. Suitable foreign species which may be obtained from aquarium shops are the various varieties of Goldfish and Carp.

The value of aquaria in teaching is well known, and they are a common feature of many educational establishments. Their value to the general public is less appreciated; it is a lamentable fact that most inhabitants of European countries have seen only a few of their native fish, and these only too often on a fishmonger's slab. Fish are among the most difficult of animals to watch in nature and many can only be fully appreciated when observed in aquaria of sufficient size, set up according to their ecological requirements. The few really good public aquaria in the world today (for instance in Amsterdam and Vancouver) show that it is quite possible to display the range of mature species found in a country, and that these are often more attractive and exciting than the standard brightly coloured exotic species so commonly featured in other public aquaria. It is very unfortunate that Edinburgh Zoo was eventually forced to close its Carnegie Aquarium, then the only major public aquarium in Scotland, a number of years ago.

Finally, some people dream of fish - and not just the large ones which got away! David Carss records that: 'Regularly since the 1970s, although I have never seen a living Ten-spined Stickleback, they appear to me in a recurring dream about once every year or so. It is always the same dream: I'm standing in a lush grassy field beside an extensive, well-vegetated ditch. By some means unknown I am catching lots of sticklebacks. Then I'm in what appears to be a farmhouse kitchen examining the catch in an aquarium on the table. Suddenly I realise that amongst my catch of three-spined fishes there are some ten-spined ones. As this realisation hits me, I wake up with a fantastic sensation of elation. Sad but true!'

References

Anonymous (1933), Bateman (1890), Brotchie (1914), Brown (1977), Caine (1927), Crandall (1914), Dewar (1908), Gibbings (1941), Gunn (1931, 1937), Jefferies (1937), Johnson (1907), Lockhart (1937), Maitland (1995), Maitland & Campbell (1992, 1994), Maitland & Evans (1986), Morton (1933), Thoreau (1886), White (1911), Williams (1928), Wood (1863).

'The trout by the bridge - you know him.
You have only to look over the parapet,
and there he is, fat, insolent, tolerant of your presence as ever.
It matters nothing what bridge or what river. He is always there,
somewhere about mid-stream, poised over a patch of gravel.'

William Caine (1927) *Fish, Fishing and Fishermen*

*****

This table includes a few examples of where to watch fish in Scotland. It should be used in relation to the distribution maps of the species concerned - there is no point in looking for Grayling in Sutherland!

WHEN?	WHAT?	WHERE?	HOW? - WATCH FOR:
January	Brown Trout	most clean burns and rivers	a bridge or high vantage point
	Goldfish	aquarium shops	some healthy specimens to buy
February	Grayling	south Scotland	a grayling fisherman to ask
	Powan	Lochs Lomond & Eck	dead fish along shore at dawn
March	Sparling	River Cree	shoals of adults in pools
	Pike	lochs & lowland rivers	spawners in shallow weedy areas
April	River Lamprey	rivers south of Great Glen	spawners over gravels
	3-spined Stickleback	the shallows of still waters	colourful males nest building
May	Brook Lamprey	most burns and rivers	spawners over gravels
	Common Minnow	ditto	shoals spawning over gravels
	European Eel	ditto	elvers, especially along edges
June	Sea Lamprey	larger southern rivers	spawners over gravels
	Stone Loach	burns and rivers south of the Dee (Aberdeen)	adults active on sandy areas and fine gravels
July	Roach	slow rivers, lochs & canals; River Tay and southwards	spawning activity among weed beds
	Perch	slow rivers, lochs and canals	shoals under boats, piers & bridges
August	Common Minnow	most lochs and rivers	shoals of fry in shallows
	Stone Loach	burns and rivers south of the Dee (Aberdeen)	fry on sandy areas, camouflaged on bottom
September	Common Goby	coastal pools and bays, near freshwater inflows	small darting fish - well camouflaged
	Rainbow Trout	fish farms	farms which welcome visitors
	Sea Trout	major waterfalls and weirs	leaping fish
October	Brown Trout	clean burns and rivers	spawning activity over gravels
	Flounder	shallow coastal waters	fish under your feet when paddling
November	Atlantic Salmon	clean rivers and burns	spawning activity over gravels
	Arctic Charr	River Spey & other waters	spawning activity over gravels
December	Atlantic Salmon	clean rivers and burns	late spawners
	A range of species	aquarium displays - Balloch, North Queensferry, etc	opening & closing times

FISH AND FISHERY MANAGEMENT

'I have heard of a Cabinet Minister who,
when asked to advance salmon fishery legislation,
promptly declined, saying he wished to die in his bed.'

William Calderwood (1907) *The Life of the Salmon*

THE existing protection, conservation and management of freshwater fish populations in Scotland is fragmented and greatly biased in favour of game species, especially Atlantic Salmon. Yet much fishery management for salmonids in the past - illustrated above by one of the fine drawings in J.R.G. Maitland's *A History of Howietoun* - has been detrimental to both fish and fish habitat. The present system is inadequate in that it does not meet all the needs of the angler, riparian owner or fish - particularly species other than salmonids. It is apparent that this system has, in the past, had serious difficulties in safeguarding the status of populations of freshwater fish (including salmonids) in many parts of the country from the numerous hazards that they face, which include the introduction of parasites and diseases, overfishing, barriers to migration and habitat loss. Currently there is concern about the status of many stocks of all native salmonids - Atlantic Salmon, Brown and

Sea Trout, and Arctic Charr. One freshwater fish, the Vendace, became extinct in Scotland during the 20th century and most other native species (e.g. Allis and Twaite Shad and Sparling) have declined. This chapter reviews the recent and current (2006) situation; the latter is in a state of flux and, it is to be hoped, there will be significant improvements in the management of fish and fisheries in a few years.

Yet an understanding of some of the principles of fishery management is not new, as Piers of Fulham described about the year 1400:

'And ete the olde fisshe, and leve the yongr.
Though they moore towgh be uppon the tonge.
But stynkkyng fisshe, and unsesonable,
Latte passe, and taak such as be able.'

Though there is a long history of legislation, stretching back as far as 1030, almost all of it concerns Atlantic Salmon and Sea Trout, and even here there are still many gaps and anomalies. Marwick records that in 'Rutherglen ... feu right granted by King Robert II on 6 February 1387-8, in favour of his faithful burgesses and community of the burgh, with the power of holding courts, and with the mills, fishings, and former customs, to be held feu for payment of a feu duty of £13 sterling yearly'. and in 'Renfrew ... King David I ... made a grant to the Abbey of Holyrood of a toft in it, with the draught of one net for salmon, and to fish there for herrings freely.' Salmon and Sea Trout are not classed as 'freshwater fish' in Scots fisheries law; this distinction has placed the latter in an invidious position with little protection over the centuries. Now that it is accepted that Sea Trout and Brown Trout are one and the same species it is ridiculous that one is given extensive protection and the other very little. Herbert Maxwell was aware of this problem and in 1897 noted that 'Unluckily, the killing of trout has been made in Scotland a matter of prize-winning. Almost daily is the bosom of Loch Leven desecrated by this ignoble rivalry ... Shameful to say there is no close-time for breeding trout in Scottish waters. In many districts the open fishings are rendered worthless to fair anglers by the miserable snigglers who destroy gravid and unseasonable fish all through the winter months.' Fortunately, shortly after this, he was able to write: 'Since these lines were penned a statutory close-time for Scottish trout has been enacted, extending from 15th October to 28th February.'

Thus, although in a few areas of Scotland there is excellent management at a local level, elsewhere there continues to be poor or even damaging management. There is not yet comprehensive monitoring of key fish stocks across the range of Scotland's fish fauna, nor even any realisation of the full economic value of this diverse resource. The educational, social and cultural importance of our fish fauna is diminishing as biodiversity is lost - as, for example, in the case of the Vendace. Many aspects of essential fish habitat (e.g. spawning grounds, riparian cover and shade, backwaters) are declining, not only through agricultural practices, etc., but also through damaging 'management' of some fisheries. The *status quo* is unacceptable.

The need for change is great, and, though comprehensive agreement among all the stakeholders may be difficult, most - anglers, fishery owners and managers, fish farmers,

conservationists and others - are dissatisfied with the present system (or, more correctly, lack of one). It is believed that, with the many changes which are taking place in relation to the management of fresh waters at the moment, a suitable opportunity exists for a complete review of the status and management of freshwater fish in Scotland.

Present fishery arrangements

There is not at present, nor has there ever been, an overall administrative framework for the management of fish populations and fisheries in Scotland. Instead, there has been a range of systems, largely relating to game fishing, the most important of which are described below.

The Scottish Executive. Though not involved directly in fishery management, the former Scottish Office and presently the Scottish Executive - through the Environment and Rural Affairs Department (SEERAD) and the Freshwater Fisheries Laboratory (FRS) at Pitlochry - has always had an important role to play. It deals with the implementation of fisheries legislation, the compilation and analysis of catches of Salmon and Sea Trout, advice to fishery managers and research on Salmon and Sea Trout and their environment.

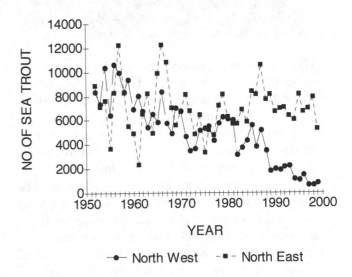

SEERAD catch statistics for Sea Trout in northern Scotland.

The validity of at least some of the reported catch statistics has always been questionable. Proprietors were formerly taxed in relation to the numbers caught - a very questionable basis for obtaining reliable statistics. The returns by anglers have also been called into question at times, as implied in 1932 by Herbert Maxwell in quoting this verse from *An Angler's prayer*:

> *'Oh grant that I may catch a fish*
> *So big, that even I,*
> *In talking of it afterwards*
> *Shall have no need to lie.'*

Henry Lamond was also aware that angler's catches were sometimes called into question: 'If he *said* a dizzen, then he *had* a dizzen, for I never kent *him* lee aboot his fish.'

District Salmon Fishery Boards. The Salmon Fisheries (Scotland) Act of 1862 established the Salmon District Fishery Boards (DSFBs) and the Fishery Districts. The main remit of the DSFBs is to regulate local matters concerning Atlantic Salmon and Sea Trout. The boards are composed of proprietors (or their representatives) who hold rights to salmon fishing. The maximum number on any board is seven and the chairman is elected by other proprietors (Salmon Act 1986). These boards have varied very much in size and, until recently, many parts of Scotland with salmon rivers were never covered. Some areas still have no board. The duties undertaken by the boards vary considerably, including bailiffing (the main responsibility at present), stocking, predator control, etc.

An exception is found on the River Tweed, where just over half of the Tweed Council is appointed by local authorities, and the rest by salmon proprietors or their representatives. Although the latter number less than half the Council, however, they provide over 95% of the income.

The Salmon Fisheries (Scotland) Act of 1868 defined and extended the powers and duties of the DSFBs. Power was given to them to purchase dam dykes, cruives and other fixed engines, for the purpose of removal; to remove any natural obstruction or waterfall by agreement and to attach fish passes to these; and generally to carry out any work which they considered necessary for the protection and improvement of fisheries under their charge. The River Tweed Council has some compulsory powers (which have never been used) to remove natural obstructions in the river at its own expense.

ANNO VICESIMO QUINTO & VICESIMO SEXTO

VICTORIÆ REGINÆ.

****

C A P. XCVII.

An Act to regulate and amend the Law respecting the Salmon Fisheries of *Scotland*.

[7th *August* 1862.]

Cover of the 1862 Act.

The DSFBs have various powers under existing legislation, and notably under the Salmon Act 1986 they are able to control the stocking of Atlantic Salmon and Sea Trout into the wild (but not into fish farms) in their board areas. In the past, however, many Boards themselves had indiscriminate stocking policies which may well have damaged the stocks in their own rivers. Andrew Ferguson has pointed out that 'One of the main threats to the maintenance of natural genetic diversity is indiscriminate stocking with hatchery fish of non-native origin.'

With the reorganisation of the DSFBs in recent years there have been many improvements, including the amalgamation of some of the smaller boards into larger ones, the creation of new Boards where there were none before, the provision of a regular newsletter and a general increase in the efficiency of their structure and operation.

Protection Orders. Historically and in law, the rights to fish for Brown Trout and other freshwater fish species have accompanied ownership of the land adjoining the water concerned. Fishing for them without permission has been a civil rather than a legal offence and owners have had to take out an interdict to protect against individual anglers returning again, except in the case of 'stanks', where the removal of Brown Trout is a criminal offence. Following the provisions of the Freshwater and Salmon Fisheries (Scotland) Act 1976, Protection Orders have been made in several areas. Their main purpose is to (1) increase the availability of fishing for freshwater fish in inland waters to which the proposals (for protection) relate, and (2) protect freshwater fishing, by prohibiting persons without legal right, or without written permission, from fishing for, or taking, freshwater fish in the inland waters of the prescribed area.

As part of the process, the Scottish Executive has to (1) receive proposals from, or on behalf of, the owner or occupier of the right of fishing in relation to the improvement of, or the giving or availability of access to, fishings. (2) consult a body representative of persons wishing to fish for freshwater fish in inland waters in Scotland (Consultative Committee on Freshwater Fisheries), (3) be satisfied that the fishings being offered are reasonable in relation to demand, and to the terms and conditions, and (4) take into consideration the need for conservation of any species of fish.

Although the Orders apply to all freshwater fish except Salmon and Sea Trout, species other than Brown Trout are neglected in most proposals. Exceptions include the Tummel/Garry Protection Order (which considers fishing for Arctic Charr, Grayling, Pike and Perch as well as Brown Trout) and both the Tweed and Tay Protection Orders (which refer also to Grayling). Nor do the Orders protect the fish as such, for there is no offence of killing or injuring fish, or of damaging their habitat. The Orders protect the right to fish for freshwater fish, which is held by riparian owners or tenants.

Fishery Trusts. Since about 1980, starting with the West Galloway Fisheries Trust, local groups in different parts of Scotland (some of them registered as official charities) have been established which can be included under the general term 'fishery trusts'. Individually, some of them have demonstrated how effective good local fishery management can be when it is based on sound scientific principles. For economic and political reasons, most trusts initially had to limit their interest to Salmon and Trout and have been concerned primarily with rivers, as opposed to lochs.

Logo of the West Galloway Fisheries Trust (now the Galloway Fisheries Trust). Drawing: Janet Stephen.

As the various trusts developed, covering more and more Scottish catchments, they became an important force in fish and fishery management in Scotland, standardising techniques and monitoring an increasing number of local fish populations. The formation in 1997 of the Scottish Fisheries Co-ordination Centre (SFCC), based at the Freshwater Fisheries Laboratory at Pitlochry, has allowed a central point for the collection and collation of trust fish data. In 2004 the Scottish Trusts became linked through a central organisation - Rivers and Fisheries Trusts of Scotland (RAFTS).

Angling Associations and Clubs. There are several hundred angling associations and clubs in Scotland (about 450), varying greatly in size and in objectives. Many of them are small groups of people who have a common interest in fishing (there are both game and coarse fishing clubs, though the majority are interested in trout angling), but own no waters and have no fishing rights or management involvement. Other clubs do own waters and/or fishing rights, and carry out various active management practices related to the fisheries concerned. In general, angling clubs are as varied as their memberships. Most neither have nor seek scientific advice but carry out various management practices, invariably dominated by stocking. Others have members who are fisheries biologists or seek scientific advice before any management is practised. The attitudes and management policies of such clubs may change quite rapidly as committee membership changes from year to year.

Private Ownership. Much of Scotland's land and fresh water is owned by private individuals (including estates and hotels) who also possess the fishing rights, which, in the case of salmon fisheries, can be extremely valuable. Again, management practices are extremely variable, ranging from no action whatsoever (and sometimes even no fishing), through occasional stocking and bank clearance or allowing the management to be carried out by one of the trusts, to intensive management which may involve draining or poisoning whole lochs to remove native species (usually Pike or Perch) and creating put-and-take fisheries for Rainbow Trout. However, the role of stocked stillwater fisheries can be a positive one, and they do give protection to wild fish stocks. It is estimated by the Association of Stillwater Fisheries that commercially managed fisheries (mainly for Rainbow Trout) cater for some 85-90 per cent of those fishing for trout in Scotland. If such fisheries did not exist there would inevitably be further greater pressure on wild Brown Trout.

Public Ownership. A considerable number of waters are owned by the state in its various guises, notably the Forestry Commission, the Scottish Executive (within its agricultural holdings) and Scottish Water (e.g. reservoirs). The policies adopted in relation to the management of these waters and their fisheries are varied. An important aspect of state ownership is that the fishing rights belong to the Crown Estate. Several other bodies have extensive ownership of freshwater systems, for example the National Trust for Scotland, the Scottish Wildlife Trust, the John Muir Trust and the Royal Society for the Protection of Birds. Though not really 'public' organisations, they have an important role to play in fish conservation and fishery management, and indeed could play a key role in the future development of some policies - for example, the conservation of rare species and the encouragement of the use of a rod licensing system.

Other responsibilities for fish

Scottish Natural Heritage. Although little attention had previously been paid to the conservation of freshwater fish in Scotland, the Nature Conservancy Council and its successor body, Scottish Natural Heritage, have initiated several studies of native freshwater fish in Scotland. Partly, this has followed from the Wildlife and Countryside Act 1981, where five species of freshwater fish are now listed in Schedule 5, and the more recent EC Habitats and Species Directive which lists several of the freshwater fish found in Scotland in Annex II. In this way, a number of native fish species are now given protection and, in addition, the introduction to Scotland of several species, which are alien to the United Kingdom but are established in England, is now illegal (Wildlife and Countryside Act 1981). Most recently, there have been requirements for the establishment of Special Areas for Conservation (SACs) and action programmes under the EC Habitats and Species Directive. Several SACs have now been established for fish species such as River, Brook and Sea Lampreys, and Atlantic Salmon.

Scottish Environment Protection Agency. The Scottish Environment Protection Agency (SEPA) and before it the River Purification Boards (RPBs) in Scotland have never had direct powers in relation to fish populations and fisheries. Nevertheless SEPA and the RPBs have done an excellent job in restoring and maintaining water quality which is such an essential part of fish habitat. In addition, they have carried out considerable research in relation to fish (as part of their assessment of water quality), and fish habitats, and have monitored fish kills, providing a wealth of valuable information for the future. The recent EC Water Framework Directive has included a requirement to use and monitor fish populations in relation to the status of water bodies, and so SEPA has now had to take a greater interest in fish populations in Scotland.

Present deficiencies

Although some aspects of the current management of fish populations and fisheries in Scotland are good, especially at a local level, there are many deficiencies, largely due to the absence of any nationwide structure or scientific control of managemnt in some places. The problems can be summed up briefly as follows:

(a) Unlike the standard Hydrometric Areas , none of the existing systems of fishery management, not even the long-standing District Salmon Fishery Boards, cover the whole of Scotland. The result for fish populations in Scotland is that in some areas they are managed well, in some areas they are managed badly, whilst in the remaining areas there is no management at all and fish stocks are subject to a variety of pressures. In addition, fish habitat has suffered abuses in various parts of the country.

(b) The organisations which do exist cover variable parts of the country (sometimes duplicating within one catchment), and have a variety of objectives, often without a scientific basis. There is therefore little opportunity for national initiatives (such as those outlined by Winstone in 1993 and the National Rivers Authority in 1995) which may benefit fish stocks, as there is in England and Wales. Nor is there any opportunity to control fish movement and translocations on a catchment to catchment basis, as there is in England and Wales.

(c) In the main, the organisations concerned do not employ fish biologists. To some extent this is now changing, largely in the interests of salmon fisheries, but ad hoc groups are necessary to try to resolve major problems (e.g. West Highland Sea Trout and Salmon Group 1995). Both angling interests and fish stocks themselves would benefit greatly from long-term national policies, implemented countrywide, as appropriate, by sound local management.

(d) Most of the objectives and management are concerned solely with Atlantic Salmon, Brown and Sea Trout and in some cases Rainbow Trout. There is no national policy related to fish habitat, as in other countries (for example, the 'no net loss of habitat' law which has operated in Canada since 1986) nor is there proper consideration given to the role of fish and fisheries in the wider countryside and to the general public.

(e) In many areas there is little contact between the fishery 'managers' (e.g. the DSFBs) and the 'users' (the anglers). This has caused unnecessary friction which could have been avoided if management aims and strategies had been explained at open meetings. In addition, as noted by Medway in 1980, there are important aspects of the public perception of angling to be addressed at national level.

Hunter Report

The Committee on Scottish Salmon and Trout Fisheries was appointed on 12 March, 1962 (under

the chairmanship of J.O.M. Hunter) 'to review the law relating to salmon and trout fisheries in Scotland, including the Tweed, and its operation, with special reference to the constitution, powers and functions of District Boards, and the responsibilities of the Secretary of State, and to consider in the light of current scientific knowledge the extent to which fishing for salmon and trout by any method, whether in inland waters or in the sea should be regulated, and to recommend such changes in the law as might be thought desirable'.

DEPARTMENT OF AGRICULTURE AND FISHERIES
FOR SCOTLAND

SCOTTISH SALMON AND TROUT FISHERIES

Second Report by the Committee
Appointed by the Secretary of State
for Scotland

*Presented to Parliament by the Secretary of State for Scotland
by Command of Her Majesty
August, 1965*

Cover of the Hunter Report.

The final report was submitted on 24 May, 1965: 'Having shown the need for regulation and management, and recommended objectives for the salmon and trout fisheries, we go on to consider how those objectives might be attained.'

At the time, this was the most thorough consideration ever given to freshwater fisheries in Scotland and, though restricted to Atlantic Salmon and Trout, is highly relevant to the present review. The 10 chapters of the final report contained 127 conclusions and recommendations. The Hunter Report caused intense interest in some quarters both during its preparation and after its publication. However, in spite of the White Paper produced in 1971 virtually no action was taken by the Government, and it was 11 years before just one of the proposals was implemented - the statutory protection of fishing for Brown Trout, which was incorporated into the Freshwater and Salmon Fisheries (Scotland) Act, 1976.

Many articles debating the Report appeared in the general press, and especially in the angling magazines. It received considerable criticism in some quarters, especially from riparian owners of salmon fisheries. For example, in 1967, Sinclair concluded a review of the report thus: 'The strength of the report is its recommendation that the salmon fishing industry should be put under sound and well-informed management. The weakness is that it goes on to recommend the wrong sort of management policy'.

Though there was widespread criticism from anglers at the suggestion of angling licences, many supported other aspects of the report. Shortly after a discussion of the Report by the Scottish Grand Committee in 1967, Williams wrote prophetically: 'One wonders whether after its airing in debate the Hunter Report will again be swept into some administrative backwater or will ultimately attain legislative fruition. There is no doubt that if anglers could forget their prejudices, adopt a forward-looking attitude, and arrive at basic essentials to achieve management and control of Scotland's fishings a worthwhile result would be attained.'

The 1971 White Paper implied that the Government proposed to implement many of the proposals in the Hunter Report. It was apparently accepted that local fisheries administration required to be reorganised, with new Area Boards replacing the existing District Salmon Fishery Boards, with additional responsibility for trout fishing and, in some cases, coarse fishing. The functions for the new Area Boards were seen in the White Paper as: (a) representing fisheries interests generally, (b) controlling methods of fishing, (c) levying fishing rates, (d) issuing rod and net licences, (e) employing staff, including technical officers and wardens, and (f) dealing with applications for statutory protection of trout fisheries.

Although some people in Scotland felt that there were several shortcomings in the White Paper, in 1972 it was accepted by many, including Drew Jamieson 'that at last something was being done about the Hunter Report'. In the event, over the four decades since its publication, very little was done

about the Hunter Report, and the major subsequent legislation relevant to freshwater fisheries in Scotland, the Freshwater and Salmon Fisheries (Scotland) Act 1976 and the Salmon Act 1986, merely reinforced the *status quo* as far as the absence of any comprehensive fish and fishery management policies in Scotland are concerned.

A

BILL

TO

Set up a Scottish Anglers' Trust to administer fresh-water A.D. 1984. fishing rights in Scotland; and for connected purposes.

BE IT ENACTED by the Queen's most Excellent Majesty, by and with the advice and consent of the Lords Spiritual and Temporal, and Commons, in this present Parliament assembled, and by the authority of the same, as follows:—

5 1. There shall be set up a corporate body called the Scottish Scottish Anglers' Trust, hereinafter referred to as the Trust.   Anglers' Trust.

An unsuccessful Bill proposed by Dennis Canavan and other Scottish MPs.

In retrospect, the Hunter Report suggested exactly what was required for salmonid fisheries in Scotland at the time and, if it had been implemented in its entirety, would have resulted in a much better situation than actually exists today. The failure to implement the sensible policies recommended in the Report must rest with both anglers (who seemed unable to reach a consensus, even among themselves) and politicians who, having set up a very powerful Committee with wide powers of consultation, failed to do anything effective in the areas which the Report covered.

WWF Report

This report was commissioned from the author by WWF Scotland in 1995 (following his critical comments (largely directed at the former Scottish Office) over many years), and published in 1996. The present lack of protection of native freshwater fish and threats to their habitat - including those from unscientific fisheries management - have led to increasing concerns in recent years, highlighted by the extinction of one species (the Vendace), the loss of major populations of others and the spread of aggressive, alien fish. The report is summarised below.

(a) The freshwater fish fauna of Scotland is a highly valuable resource of importance in economic, recreational, educational, scientific and cultural terms. It is at present under-valued and under threat.

(b) There is not at present, nor has there ever been, an overall administrative framework for the management of fish populations and fisheries in Scotland. The existing protection, conservation and management of freshwater fish populations in Scotland is fragmented, and greatly biased in favour of game species, especially Atlantic Salmon.

(c) A new structure is suggested to take the best aspects of the present systems and involve minimal change to existing organisations. Its main features are: (a) Scotland would be covered by about 25 catchment-based Regional Fishery Boards with statutory powers for scientific and sustainable fishery management. (b) In parallel with these Boards would be independent Regional Fishery Trusts, each with a similar core structure but otherwise flexible according to local conditions, (c) Both Boards and Trusts would liaise with a Central Fisheries Unit, the former on a statutory basis, the latter on a voluntary one. This unit should be based at the Freshwater Fisheries Laboratory at Pitlochry. (d) The new Boards would be funded similarly to present District Salmon Fishery Boards. The new Trusts would be funded from various sources, possibly including local initiatives, angling permits from public fisheries, a national angling licence and grants from the Scottish Executive, the Scottish Sports Council, Scottish Natural Heritage and other bodies. (e) On both Boards and Trusts, the recognised national body representing anglers and angling clubs would be the Scottish Anglers National Association (SANA) whilst fishery managers would be represented by the Institute of Fisheries Management (IFM).

(d) In addition to the range of fish species within each catchment, there is also a range of needs from human society. These include the conservation of fish species and their habitats, the requirements of new European legislation, the demands for angling,

the importance to tourism, the rights of traditional commercial fishermen and the expectation of income from fishing by riparian owners and others. The much needed restructuring of the way in which fish populations and fisheries in Scotland are managed should benefit both fish and people.

(e) The consequences of no, or little, change in the present policies and management practices are that there will be further loss of local wild stocks and even of some species and their habitats - due to variable management over the country and continued fragmentation of interests. This is bad for fish and for anglers and it represents a lost opportunity for a strong voice and dedicated resources for this important part of our natural heritage. Lack of action will also have an impact in the wider countryside in relation to continued loss of biodiversity and unsustainable use of our lochs and rivers.

Nickson Report

In 1997, the Scottish Office published the 'Report of the Scottish Salmon Advisory Task Force', initiated by the Secretary of State in 1995 and chaired by D. Nickson 'To consider the challenges and opportunities facing Scottish salmon fisheries with a view to recommending a Strategy for the management, conservation and sustainable exploitation of the stocks into the next century.' It admitted that 'The management of salmon fisheries raises many issues and, when a species excites as much interest as the Atlantic salmon, it is inevitable that divergent views are sometimes held.'

Parts of the Task Force paralleled earlier recommendations in the WWF report: (a) The formation of new fishery areas, covering all Scotland, each with a mandatory board. (b) The inclusion of a range of fish-related interests on these boards. (c) The need for state financial support for scientific expertise. (d) Improved methods of collecting fishery statistics. (e) A central body to co-ordinate monitoring and research. (f) Support for fishery trusts through the Freshwater Fisheries Laboratory, Pitlochry. (g) Strict control of the transport of freshwater fish into and within Scotland.

There were, however, several major criticisms by the author and others of other Task Force recommendations, including (a) Only Salmon (and to a lesser extent Sea Trout) were included, leaving all other native fish, as at present, in limbo. (b) The reason for the importance attributed to Salmon is the contribution it makes to the Scottish economy, yet no assessment was made of the valuable contributions made by other species, especially Brown Trout (which, of course, includes Sea Trout), which may be just as valuable as Atlantic Salmon. (c) A major flaw in the logic of the report concerned the inclusion of only part of the population of trout (Sea Trout); the remaining part (Brown Trout) was left without adequate legislation or management. (d) Proposed conservation and management of the aquatic habitat was totally dominated by salmocentric considerations. (e) All management decisions would be controlled by salmon fishery proprietors and there was no clear commitment to following scientific fishery management principles or to the production of local science-based management plans for independent scrutiny.

There were some surprising omissions from the Task Force Report. For example, there was no consideration of several organisations which are highly relevant to freshwater fisheries in Scotland. Notable among these is the Institute of Fisheries Management (IFM) and the Scottish Anglers National Association (SANA), both of which must surely have an important part to play in any future fisheries administration in Scotland?

It is clear that the Salmon Task Force Report is not a realistic blueprint for the equitable management of fish populations and fisheries in Scotland. The salmocentric nature of most proposals, with the expectation that salmon interests will guide virtually all other activities which might affect salmon, but without a corresponding gesture of commitment to science-based fishery management, is unacceptable. To base the only national structure for fisheries management in Scotland on only one species and part of another, leaving the remaining 41 species with virtually no legislation or strategic management, is an objectionable concept. Scotland's diverse and valuable native fish populations deserve a better future and a more sympathetic treatment.

At the end of the 20th Century, the Scottish Executive was consolidating fisheries legislation and reviewing Protection Orders. Surely that was the time to take account of all issues relating to the management of fish populations in Scotland, as proposed in the WWF report. In addition, following European legislation, there were strong moves at this time by government agencies (e.g. SNH and SEPA) and most NGOs towards a more integrated and holistic approach to conservation and management of the environment, which must inevitably broaden management of fish and fisheries from the narrow consideration of just two salmonids to that of all native fish and their habitats.

The inevitable conclusion is, that in order to retain the overall biodiversity and economic value of the freshwater fish populations of Scotland, a much wider revision of freshwater fish policies, along the lines of those recommended in the WWF Report, is required. Such an approach is essential if the unique nature of our fish fauna is to be saved in a sustainable way for the pleasure and economy of future generations.

Angling for Change

Following the WWF Report, a consensus was achieved among the main bodies concerned with fish conservation and angling in Scotland. This consensus group, called Angling for Change (AfC), was a unique initiative which brought together anglers, fishery proprietors, and fish conservation interests to consider the needs of freshwater fish and fisheries in Scotland, and develop proposals for the future. The AfC Group considered that Scotland's legislative and management structures for fisheries in fresh water should contain effective measures which operate throughout Scotland to provide:

(a) appropriate conservation for all fish species;
(b) science-based fishery management;
(c) well-publicised and readily available access for angling of all types;
(d) recognition for the full diversity of sporting species available;
(e) mechanisms to gather and disseminate accurate up-to-date scientific information;
(f) control and monitoring of the movement, introduction and reduction of stocks of fish, invertebrates and aquatic plants.

The present legislation covering angling, freshwater fisheries and fish conservation in Scotland fails to achieve these objectives. There is thus an urgent need for change. In considering legislation and structures for the future, it will be particularly important to:

(a) ensure that, wherever possible, decisions are taken at a local level;
(b) build on the strengths of the effective and well-established structures which currently exist in some areas, and promote the spread of best practice to all areas;
(c) provide sufficient flexibility to accommodate the diverse interests in different areas, and give scope for relationships to develop, and local structures to evolve over time;
(d) operate cost-effectively by, among other things, harnessing and maintaining the voluntary effort currently invested by individuals, clubs and other organisations.

A Consensus Statement on the need for a freshwater fish policy review in Scotland

April 1999

Urgent action is required to address the inadequacies of current policies for one of Scotland's most precious natural resources – freshwater fish.

This statement is signed by a broad range of groups and creates a powerful case for change to improve fish conservation and management in Scotland.

First publication by Angling for Change.

After considerable deliberation and debate, Angling for Change devised a model for new legislation and management structures which provides comprehensive coverage, encourages extension of access for angling, and places appropriate weight on the need for sustainable, scientifically-informed fisheries management and fish conservation. It provides a template into which the structures which already exist in large areas of Scotland can grow over time.

Angling for Change recognised that these proposals required further refinement to translate them into a set of working structures and put them into operation.

The AfC group met regularly over a number of years to develop proposals and tried, with varying degrees of success, to involve the relevant departments of the Scottish Executive. During the same period, the Scottish Executive produced a number of consultations (e.g. on Protection Orders) and reports (e.g. Protecting and Promoting). Eventually it was accepted that the only way to move forward was for the Scottish Executive to liaise directly with angling bodies and conservation organisations; this lead to the formation of the Freshwater Fisheries Forum.

Freshwater Fisheries Forum

The Freshwater Fisheries Forum (FFF) is a Scottish Executive convened process designed to develop ideas to ensure the better management and use of Scotland's freshwater fish and fisheries. It has proved to be an effective way of achieving consensus across a wide range of organisations with common and different interests and agendas, and for presenting cogent, credible, broadly supported ideas to policy makers. Recognising that consensus cannot be achieved on all matters, it was understood from the beginning that no member of the Forum should be prevented from making representations in the normal way to MSPs, The Executive, etc.

The Forum includes annual public meetings, a website and e-noticeboard written consultations, and a broadly representative steering group. The steering group, chaired by SEERAD, consists of all the major fish and fisheries management and angling representative bodies in Scotland, plus representatives from other important stakeholders, and also including representatives from FRS, SEPA and SNH. Initially, it was hoped that the FFF would develop a Bill which would cover all the requirements of freshwater fish and fisheries in Scotland, as AfC had debated for years. However, with the realisation that there was the possibility of a Bill before the next Scottish election in May 2007, the FFF Steering Group was forced to focus on a smaller number of relevant measures.

The Forum process, therefore, focussed on developing and achieving consensus on a proposed Aquaculture & Freshwater Fisheries Bill, to be put before Parliament in the summer/autumn of 2006. The Forum has concerned itself almost exclusively with the freshwater fisheries part of this bill.

One of the most controversial aspects of the proposed legislation is the banning of livebaiting, discussed in an earlier chapter. In spite of the fact that anglers can readily catch Pike and Perch on other lures, anglers are fighting the proposed ban and naming the author in newspaper articles.

ANGLING Times is this week joining forces with the Pike Anglers Club of Great Britain to fight against a ban on livebaiting.

The controversial tactic, which is already banned in Southern Ireland and the Lake District, is under further threat from new Scottish legislation that could consign its use to the history books.

The reaction of pike anglers to the proposed ban.

It is intended, however, that the Forum process will continue, and develop ideas on the future of Scottish fisheries management and access arrangements. Considerable work has previously been conducted in this area by AfC but there is still more to be done. The Forum is a credible and widely-supported vehicle for the discussion of these ideas and, whilst there is broad agreement on some of the principles required for reform of management structures, there are still a number of issues - some of them contentious - to be resolved.

Conclusions

The problem of inadequate national fisheries organisation and management in Scotland is not new; the problems have been outlined from time to time by various authors. Over a century ago, in 1887, James Maitland noted that if his studies "persuade District Fishery Boards that their sphere

of usefulness is wider than they have hitherto held, I shall be amply rewarded".

Three decades later, in 1929, Henry Lamond was mistakenly optimistic: 'By the Salmon and Freshwater Fisheries Act, 1923, which came into force on 1st January 1924, the laws of England and Wales regarding salmon, trout and coarse fish have been consolidated and amended. Similar legislation for Scotland is long overdue, and probably, although in many respects English and Scottish conditions vary in fundamentals, any fresh legislation for Scotland may be expected to follow, where appropriate, the lines of this statute. In particular, it is more than likely that any new Scottish consolidating and amending statute will deal with all species of fresh-water fish ...'

Yet, many decades later, in spite of the new Salmon Act (1986), even salmonid fisheries still have inadequate protection and management, as noted by the UK Salmon Advisory Committee in 1991: 'There is ... a need for a stronger and more comprehensive framework for Scotland and we therefore recommend that the present regulatory structure there should be urgently reviewed in this context and amended if appropriate'.

There is an obvious historical basis for many of the proposals put forward in the recent WWF and AfC reports, for it will be clear that several of them are close to those proposed by the Hunter Committee 40 years ago - for example, the creation of a nationwide series of fishery boards or trusts. It is believed also by the author that these new proposals build on the best of the existing structures, notably the DSFBs and the new Trusts, but also take on board many additional features and advantages.

The present trends towards the creation of local Fishery Trusts in parallel with existing District Salmon Fishery Boards (or groups of Boards) are already moving in the right direction. Some would favour this 'evolutionary' approach as the right one, even though some recent Trusts have been created on the basis of a specific 'firefighting' approach to local stock decline. However, nationally, a more radical approach to restructuring and fishery legislation is necessary if Scotland is not to end up with an even greater complexity of organisations and activity in some areas, and nothing in others.

Scotland's important freshwater fish resources deserve better.

The objectives of the recent WWF and AfC proposals, largely attained through the FFF, will produce an improved aquatic habitat for native fish species in Scotland, and thus one in which they will thrive. This will obviously also benefit anglers, who prefer to catch wild fish in natural surroundings where many other forms of wildlife are also an attraction. To gain maximum benefit, however, from river improvements it is obviously also important that the fish stocks themselves are adequately protected, through the sound management which would be possible through a new national structure for the management of fish and fisheries. Unfortunately, promoting better legislation and fishery management has proved to be, in the words of Thomas de Quincey 'an unceasing expectation and a perpetual disappointment.'

The broad objectives of management remain the same as in the Scottish statutes of 500 years ago - to put down practices which 'destroy the breed of fish, and hurt the commoun profite of the realme'. The much needed restructuring of the way in which fish populations and fisheries in Scotland are managed should, therefore, benefit both fish and people; it is to be hoped that the activities of the Freshwater Fisheries Forum will lead to a permanent, sustainable management structure based on sound science which will suit both fish and fishermen.

References

Angling for Change (2000), Association of District Salmon Fishery Boards (1977), Bastard (1598), Calderwood (1907), Department of Fisheries and Oceans (1986), Ferguson (1989), Hunter (1965), Jamieson (1972), Lamond (1929), Maitland (1887), Maitland (1969, 1995, 1996, 1997, 1998), Maitland & Turner (1987), Marwick (1909), Maxwell (1897, 1903), Medway (1980), National Rivers Authority (1995), Nickson (1997), Radford *et al.* (2004), Salmon Advisory Committee (1991), Scottish Executive (2000), Sinclair (1967), West Highland Sea Trout and Salmon Group (1995), Williams (1967), Winstone *et al.* (1993).

'Fishing, if I a fisher may protest,
Of pleasures is the swet'st of sports the best,
Of exercises the most excellent,
Of recreations the most innocent,
But now the sport is marde, and wott ye why?
Fishes decrease, and fishers multiply.'

Thomas Bastard (1598) *Chresteleros, Seven Books of Epigrames*

*****

FISH CONSERVATION

'A new philosophy and practice was needed in which the conservation principle operated independently between sites, species and the countryside as a whole.'

J Morton Boyd (1999) *The Song of the Sandpiper*

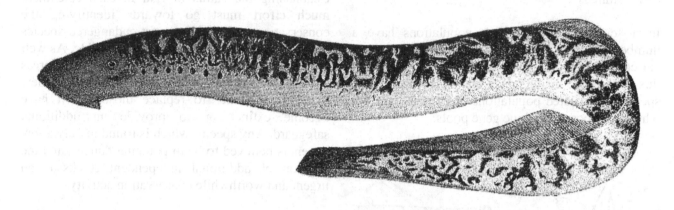

THE conservation of freshwater fish means a number of quite separate things to different people. For example, the Sea Lamprey figured above (from Goode 1884) is regarded as a pest in North America, but given conservation protection in Europe. Most of the discussion on fish conservation in the past has centred around species which are of direct importance to man, either commercially or for sporting purposes. The number of people involved in these activities is very large, and forms a significant proportion of the Scottish population. For this reason, to many, fish conservation means the conservation of Atlantic Salmon: this is undoubtedly an important issue at the start of the new Millennium, due to apparent declines in this species and, of course, Sea Trout in some rivers. Many other species deserve consideration, however; some are increasing their range of distribution, whereas others are decreasing in this respect. A few are in danger of extinction - mainly as a result of stresses imposed by man. There is a considerable amount of fundamental research still required, especially on the less common species, if all aspects of this valuable resource are to be protected and the extinction of rare species is to be prevented.

In considering the conservation of fish populations, there are a number of other important aspects which have received little attention - presumably because most freshwater fish appear to have no economic importance or, if it is admitted that they have such

importance, because it is too difficult to assess in terms of 'cash value'. Amenity is one such reason. The amenity value of water bodies is an accepted fact now, and one of the attractions of being near them is to see fish swimming and leaping in clear water. For many people it is psychologically satisfying to know (without necessarily seeing) that a water body contains healthy, stable fish populations. Fish are excellent indicators of the degree of pollution of a water body: fish deaths are often one of the first obvious indications of sudden pollution, and waters which are fishless owing to pollution are generally considered to be unsatisfactory - regardless of their visual condition. Usually among the top consumers in food webs in aquatic systems, fish often provide a very sensitive measure of pollution of the environment by insecticides and other insidious poisons which build up in the food chain and accumulate in their tissues.

The recreational value of fish populations is often equated solely with their use for sport fishing, whereas this is often far from the case. The economic element and adult participation in sport fishing often mask the fact that, as that great conservationist Frank Fraser Darling was well aware, large numbers of children (and their parents!) gain very great recreational pleasure from catching small fish and later keeping them in ponds or aquaria. The range of important species in this context is far wider than that which is appropriate to sport fishing. Related to this activity is the educational value of these species to schools and universities - especially from behavioural and anatomical viewpoints. The ecological relationships within waters can also be important as far as fish are concerned. For example, they often form the main food supply for interesting or rare predatory birds and mammals.

In a scientific context, fish populations have a number of uses, and are widely employed in a variety of research studies. One important aspect in this field is the maintenance of populations of rare species, or isolated populations of common species which may possess unique gene pools.

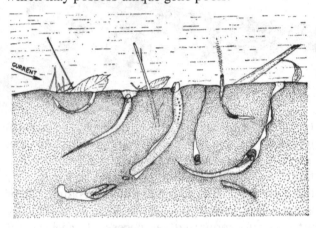

Lamprey larvae in their burrows in silt - a habitat now protected in several Scottish rivers (from Applegate 1950).

The existing protection given to most native fish species in Scotland is completely inadequate both in terms of the establishment of appropriate reserves, and of management and legislation. The exception relates to Atlantic Salmon and Sea Trout which, as discussed elsewhere, are given substantial protection both in the water and through available legislation. Even here, however, the situation is not entirely satisfactory. In Scotland, in general, no reserves have been established primarily for native fish conservation and virtually the only national legislation other than that relating to fisheries is the Wildlife & Countryside Act 1981. The 1992 EC Habitats Directive has been very beneficial for several fish species in Scotland, by forcing conservation agencies not only to take more interest in native freshwater fish but also to establish Special Areas of Conservation for some of them. The forthcoming EC Water Framework Directive

will also be of considerable help in relation to the requirements for monitoring fish and other biota.

Conservation options

There is an enormous amount of work to be done in the field of fish conservation. In addition to establishing the status of fish in each catchment much effort must go towards identifying the conservation needs of the most endangered species and implementing these as soon as possible. As well as habitat restoration, one of the most positive areas of management lies in the establishment of new populations, either to replace those which have become extinct or to provide an additional safeguard. Any species which is found in only a few waters is believed to be in potential danger, and the creation of additional independent stocks is an urgent and worthwhile conservation activity.

Habitat restoration

Obviously enormous damage has been done to many fish habitats and the situation is often not easy to reverse - especially in the short term where fish species or communities are severely threatened. In many cases, unique stocks have completely disappeared. Even where habitat restoration is contemplated, stock transfer (discussed below) could be an important interim measure. There are, however, a number of important examples of habitat restoration in temperate areas, and it should be emphasised that habitat protection and restoration are the principle long-terms means through which successful fish conservation will be achieved.

There have been enormous strides in pollution control in Scotland over the last few decades and a number of the worst rivers are now much cleaner. For example, the Rivers Clyde and Carron are now so much better than 50 years ago that migratory fish have been returning to them in increasing numbers. At their worst, both rivers were virtually fishless in their lower reaches: because of this pollution no migratory fish could pass through to reach the clean upland waters. Rehabilitation of the River Clyde has been a slow but steady process. The final arbiters of water quality are surely the fish themselves and the return of the Atlantic Salmon to this river after an absence of more than 100 years is a marvellous tribute to decades of work by the Clyde River Purification Board (now part of SEPA).

Many fresh waters in Scandinavia, North America and the British Isles have lost their fish populations over the last three decades because of acidification, altogether many thousands of individual stocks having disappeared. Various ways of ameliorating the impact of acid precipitation have been investigated, most of them involving adding calcium in some form, either direct to the water body or to the catchment of the system involved. Most of the pioneering work in this form of habitat restoration has been carried out in Scandinavia.

In Scotland, various attempts at liming to ameliorate freshwater acidification have been attempted, most notable among which has been the work at Loch Fleet. Here, the former healthy population of Brown Trout started to decline during the 1950s and became extinct during the 1970s. In 1984, a restoration project costing over £1.5 million was initiated and calcium carbonate was added to the catchment in various ways. The loch responded quickly - the pH rose from about 4.5 to 6.5 within a few weeks; at the same time the amounts of aluminium in the water decreased. Adult fish which were introduced to the system in 1986 subsequently spawned successfully.

As expected, this experiment verified the earlier work of others, but in spite of being successful, it was transitory and very expensive; though providing a possible short-term answer to the acidification problems affecting important local stocks of fish, e.g. the Arctic Charr at Loch Doon, it does not provide a satisfactory long-term form of habitat restoration.

Habitat management is of major importance to many of our native fish species. It is essential that a number of waters, both running and standing, are given high priority in this context; these systems are described in an earlier chapter. Habitat management is also of importance in relation to any new stocks which are initiated from translocation experiments. The restoration of habitats from which important stocks have disappeared is a much more difficult and expensive procedure, and in some cases may not be considered worthwhile. Important examples are discussed elsewhere and include sites where rare species are now extinct, such as the Mill Loch at Lochmaben. Unfortunately, restoration of most of these sites would seem both difficult and expensive.

Stock transfer

This can be done without any threat to the existing stocks, but it important that certain criteria are taken into account in relation to any translocation proposal. With most fish it should be possible to obtain substantial numbers of fertilised eggs by catching and stripping adult fish during their spawning period. The adults can then be returned safely to the water to spawn in future years. Fortunately, most fish produce an enormous excess of eggs, and so substantial numbers can be taken at this time without harm. Having identified an appropriate water in which to create a new population, this can be initiated by placing the eggs there, or hatching the eggs in a hatchery and introducing the young at various stages of development.

Logo of the Fish Conservation Centre which has been involved in a number of fish conservation projects.

Preliminary transfers of three of Scotland's most endangered fish have already been carried out by the author and Alex Lyle - Vendace, Powan and Arctic Charr. These are discussed in full elsewhere. However, much further work remains to be done with these and other threatened species. The results of these transfers need to be monitored, and additional transfers are required to safeguard other notable stocks of fish. Thus translocation proposals for threatened fish species in Scotland are one of the most realistic ways to help to conserve endangered stocks. In general, these are likely to be less expensive and have a greater chance of success than many habitat restoration proposals - especially in the short term. All translocation proposals should follow the guidelines given at the end of this chapter.

Captive breeding

Captive breeding is widely used throughout the world for a variety of endangered animals, including fish. For most fish, however, it can only be regarded as a short term emergency measure, for a variety of genetic and other difficulties are likely to arise if small numbers of animals are kept in captivity over several generations or more. Captive breeding in the long-term does not seem appropriate to any of the freshwater fish species at present under threat in the British Isles.

Eyed egg of Powan, a species now well established in two 'safeguard' sites.

Short-term captive breeding, however, involving only one generation, does have some advantages for a number of species, and has already been carried out by the author with Arctic Charr. It is especially relevant where translocations are desirable, but it is difficult to obtain reasonable numbers of eggs or young because of ecological or logistic constraints. In such cases, there are considerable advantages to be gained in rearing small numbers of stock in captivity, and then stripping them to obtain much larger numbers of eggs and then young for release in the wild. Because of genetic problems related to the 'bottleneck' effect and inbreeding it should not

be carried out for more than one generation from the wild stock and as many parent fish as possible should be used.

Cryopreservation

Modern techniques for rapid freezing of gametes to very low temperatures have proved successful for a variety of animals, including fish. After freezing for many years and then thawing, the material is still viable. The technique is successful only for sperm, however, and though much research is at present being carried out on eggs, no successful method of cryopreservation for them has yet been developed. The technique is therefore of only limited value in relation to the conservation of fish species.

Nevertheless, where a particular stock seemed in imminent danger of dying out it would seem worthwhile giving consideration to saving at least some of its genetic material through the cryopreservation of sperm. When it is possible to preserve female gametes in a similar way, the technique will have obvious possibilities in relation to the short-term conservation of a wide variety of fish species. For the reasons explained above, cryopreservation does not seem to be a realistic option for any of the threatened Scottish species at the moment, this is an active research area of considerable relevance to fish conservation and as such support should be given to it.

Legislation

As mentioned elsewhere, there is considerable legislation giving protection to Atlantic Salmon and Sea Trout in Scotland; this was reviewed in a consultation document produced by the Scottish Executive in 2000. Most other freshwater fish have been given little protection but some, for example our three lamprey species, have been helped by recent EC Directives. There are, however, some obvious anomalies in this legislation as far as Scotland is concerned; for example introduced species in Scotland, such as Grayling and Bullhead are given some protection by the EC Habitats Directive, whereas native species such as Sparling and Arctic Charr, whose populations are under threat, have virtually no protection of any kind.

All freshwater fish in England and Wales are given substantial protection in by the Environment

Agency in relation to capture and to threats from translocated competitors or predators; in Scotland this is not so. There is an urgent need to revise the situation so that all native species of fish in Scotland are given some protection in their waters, and also that there is some form of catchment to catchment control to prevent harmful translocations. It is hoped that the Aquaculure and Freshwater Fisheries Bill, at present under consideration by the Scottish Parliament, will go some way towards giving better protection to all our native fish.

Monitoring

It is believed that all populations which are sufficiently worthy of conservation must be monitored adequately - at the very least to prove that they still exist. Ideally, the system used, though it may not be quantitative, should at least be standardised if possible, so that comparisons can be made in time, and also perhaps in space in some cases. Monitoring programmes of this type can be expensive, however, both in terms of manpower and of finance, and any long-term monitoring programme must be designed to keep both of these to a minimum.

In spite of the fact that catches of Atlantic Salmon and Sea Trout have been monitored for many decades, the data are inadequate and better methods of assessing catch and stock are required. For centuries, writers have claimed that the demise of the Salmon is imminent! For example, St John wrote in 1891 that 'It is a matter quite beyond doubt that salmon are decreasing every year in most of the Scottish rivers. With short-sighted cupidity, these valuable fish are hunted down, trapped, and caught in every possible manner; ... Prolific as they are, fish, like all other animals, must of necessity decrease, unless allowed fair play and time to breed. ... It is not the angler who injuriously thins their number. The salmon is too capricious in rising to the fly to make this possible. ...'

In some cases it is possible to take advantage of local circumstances which make monitoring relatively easy. In others, special methods and monitoring programmes must be set up. These are not outlined in detail as each site needs to be reviewed and considered on its merits. It is important that an investigation is carried out to assess the value of non-damaging methods which

are low in man-power requirements; the author is currently (2006) carrying out research in this area.

Predators

The attitude of many anglers to predators of all kinds is usually negative, but sometimes ambiguous. In 1891, Charles St John wrote: 'Why the poor osprey should be persecuted I know not, as it is quite harmless, living wholly on fish, of which every one knows that there is too great an abundance in this country for the most rigid preserver to grudge this picturesque bird his share.'

Later: 'We reached Rhiconnich ... and immediately started for a lake some two or three miles off, where the osprey was said to build. ... In the nest he found a half-grown young bird and an unhatched egg, both of which he brought safely to land. He remained on the spot to try to shoot one of the old birds ...'

And again: 'After resting ... we again left the inn to look for another osprey's nest ... our perseverance was rewarded by presently making out that one osprey was on the nest, and the other soaring above her. Dunbar again swam off to the rock ... and found three young birds in the nest, which he brought to land in his cap. ... I saw the male osprey perch on a rock ... I shot him deliberately in cold blood as he sat. He fell down the face of the rock, and lay at the bottom perfectly dead. I then had to consider how to get at him ... In the meantime, Dunbar having shot both barrels at the hen bird, she took her flight straight off to the sea. ...'

Herbert Maxwell, a complete naturalist, had a contrary view: 'There is no more fascinating display in bird life than is presented by an osprey circling high over the dark waters of a Highland loch, when he suddenly contracts his wingspread and, falling headlong into the waves, emerges with a trout or pike in his rough talons ...' It was no surprise, however, that this elegant fish-eater became extinct in Scotland, though it is, of course, now re-establishing itself successfully in many parts of the country.

Charles St John had an equally ambivalent view of Otters: 'While the river is in this state of confusion with ice, &c., I see that the otters take themselves to the unfrozen ditches and springs to hunt for eels and flounders, which fish they fed on apparently with great perseverance. ... Altogether this is a most interesting animal, graceful in its movements, and in salmon rivers not nearly so destructive and injurious as he is supposed to be, feeding on eels, flounders and trout far more than salmon: in such situations he is most unjustly persecuted.'

Negative attitudes towards fish predators persist today among many anglers and fishery managers, who feel that any piscivore should be eliminated from their waters. Cormorants are another topical but persecuted species: the supposed reason for the recent collapse of the trout fishery at Loch Leven. Here, as at many other put-and-take fisheries around the country, it is not surprising that Cormorants and other piscivores are attracted to the easy pickings of farmed fish which have been introduced to an alien water and are easily caught. With the end of stocking at Loch Leven it is likely that the loch will return to a more natural condition, predator numbers will drop, and an excellent fishery for wild Brown Trout will emerge.

Conservation management

What can be done to further the cause of freshwater fish conservation in Scotland as a whole? An essential step must be to inform and interest that large and influential section of the public already interested in natural history and the conservation of wildlife. It must be generally realised that the conservation of the freshwater fish element of our native fauna has been, and still is, neglected in comparison with the more "popular" elements of our wildlife, and that a number of our original fish communities and individual rare species are endangered. While there is much scientific information already available on the biology and ecology of many of our freshwater fish species, this is spread widely over a range of sources, only readily obtainable by fishery scientists.

Obviously the conferring of National Nature Reserve (NNR) or SSSI status on a water body cannot guarantee that the rare fish species or community therein will be safe from the deliberate introduction of a disruptive species for selfish motives - and indeed such cases have already occurred. Thus, in certain instances it might well be safer not to confer any official status to a site, thereby avoiding the usual publicity. This is a sad commentary on the times we live in! In some cases

the re-distribution or even re-introduction from abroad of rare or newly extinct species might be justified. There are, currently, precedents involving other animals; SNH and others have re-introduced the White-tailed Sea Eagle to the west coast of Scotland by importing young birds from Norway over a period of years; other agencies are involved in re-establishing Otters in areas now devoid of them, where habitat conditions have improved over recent years. Fish re-distributions and introductions should be at or near original sites, using the nearest suitable available stocks, but as a last resort, stock from abroad could be used.

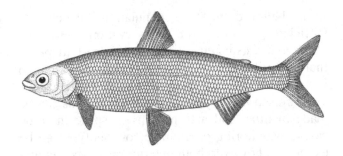

Powan - further 'safeguard' populations are proposed.

The establishment of further 'safeguard' populations of Powan using stock from Loch Eck is an example of a project which should be given serious consideration. With some threatened species or unusual stocks, captive breeding may be necessary so that a significant stock of the species is available for release. Newly-made artificial lakes (e.g. worked out gravel pits or quarries) could be used for stocking with rare species, or original communities, to increase their representation and security in relevant parts of the country, and possibly these waters could even be opened to controlled angling. On the other side of the coin, the spread of dangerous or disruptive species might be halted by their elimination through the careful use of fish poisons, or their numbers controlled by less drastic methods such as trapping or netting. For example, in this context, Ruffe and Signal Crayfish are alien species, posing a potential threat to rare species or valuable aquatic communities. The early discovery at a new site, of the presence of any of these alien species, could help greatly with their subsequent control if prompt action is taken by an appropriate authority, presumably SNH.

There are many problems related to fish conservation proposals such as these. Should any non-Scottish fish species (many of which are now established in Scotland) be included for consideration as being threatened? The Grayling and Bullhead are examples, both alien to Scotland but given protection under the EC Habitats Directive. How should the status of species native to temperate areas, but transferred outside their area of natural distribution, be regarded? For example, in Sweden the Danube Catfish (*Silurus glanis*) is an endangered native species, whereas in Scotland, in spite of the wishes of a section of the coarse fishing community to introduce it, this is definitely a fish which we do not want.

There are now many nature reserves of various types, managed by different organisations throughout Scotland. They have been established for numerous different reasons, some for a unique type of habitat or plant community, many for their ornithological interest and others for the rare flowers or butterflies which occur there. We have no reserves which have been set up especially for their fish, but it is hoped that this may be remedied

in the future. Of course, many nature reserves have water bodies on them containing fish. The exact status of such populations is now known and fish there are normally already given a good measure of protection. Such a baseline is a prime requirement before suggestions can be put forward for any new reserves which may be required specifically for fish conservation.

References

Applegate (1950), Boyd (1999), Goode (1884), Hammerton (1994), Harriman & Morrison (1981, 1982), Hicks (1855), Langton (1992), Lyle & Maitland (1992, 1994, 1998), Maitland (1974, 1976, 1979, 1982, 1985, 1987, 1991, 1992, 1995, 1998), Maitland & Campbell (1992), Maitland & Crivelli (1996), Maitland *et al.* (1997), Maitland & Evans (1986), Maitland & Lyle (1990, 1991, 1992, 1996), Maitland & Morgan (1997), Maitland & Turner (1987), Morris *et al.* (1989), Muniz (1984), Scottish Executive (2000), St John (1891), Sweetman *et al.* (1996).

'There is a burn flowing out of Loch Urigill at the north end (which we had to wade nearly to our waists), in which I took five trout in six casts, averaging above half a pound each. The sum total of our capture today, amounted to two hundred and twelve. Early in the season, on a favourable day, we should, in all probability, have had thrice that number.'

James Hicks (1855) *Wanderings by the Lochs and Streams of Assynt ...*

*****

Important criteria concerning the translocation of threatened species to create new populations (after Maitland & Lyle 1992).

1. The translocation activities must pose no threat to the parent stock.

2. The introduction proposals must pose no threat to the ecology or scientific interest of the introduction site.

3. The introduction site must be ecologically suitable. In general, sites from which the species concerned has disappeared should be considered unsuitable unless the causal factors have ameliorated.

4. Ideally, the introduction site should be in the same catchment, or the same geographic region as the parent stock; or in the same geographic region as a former stock, now extinct.

5. Permission must be obtained from riparian owners or relevant legal authorities, where appropriate.

6. Stock may be transferred as eggs, fry, juveniles or adults, but the latter may pose a threat to the parent stock.

7. Special consideration should be given to the genetic integrity of the stock to be translocated. Once the stock has been defined, maximum genetic diversity should be sought by selecting material widely in space and time. When transferring fish or stripping adults to obtain fertilised eggs, at least 30 adults of each sex should be used wherever possible.

8. Consideration must be given to avoiding the transfer of undesirable diseases or parasites. Most of these can be avoided by taking eggs only from the parent stock and checking for disease before the eggs or the resulting fry are introduced.

9. Notes of each translocation experiment should be kept and details published where relevant.

10. The fate of the translocated stock should be monitored.

THE FUTURE

'Each day a new adventure in itself ...
The plop of trout in a stretch of still water at the gloaming,
and the splash of leaping salmon below the fall.'

Ratcliffe Barnett (1927) *The Land of Locheil and the Magic West*

THERE can be little doubt that we need to take action soon if the future envisaged by Ratcliffe Barnett (and depicted above by Louis Rhead in Lathan Crandall's *Days in the Open*) is to hold true. Therefore we need to make many changes in the way we manage our freshwater fish stocks if we are to conserve them for the future. Anyone who doubts the seriousness of the situation only needs to think of the current state of salmonids in north-west Scotland and compare it with what Henry Lamond wrote in 1914: 'What an extraordinary stock of sea-trout our numberless Western streams must hold in spite of nets and natural enemies. The run in this one little stream is prodigious, and the fish are of noble proportions, not mere whitling. Three-pounders and even four, five, and six pounders, are not uncommon. Nor is it in any degree remarkable that one should secure a grilse.'

Of course, the problems lie not only with our stocks of freshwater fish but also with the disasters which have befallen the stocks of several marine species. Again, compare the present situation with what Joseph and Elizabeth Pennell recorded in 1889: '... the fishing season was almost over. So they said. But when one thousand boats came in, and twenty thousand fisher-folk were that day in Fraserburgh, to us it looked little like the end. In all this busy

place we heard no English. Only Gaelic was spoken, as if we were once more in the Western Islands.'

Yet, when thinking about the future of freshwater fish in Scotland it is surely best to be optimistic. This does not mean that we should forget the mistakes and losses of the past, which must form our experience and be the basis of, it is to be hoped, making a better job of managing our fish stocks from now on. This chapter reviews what has been elaborated in previous chapters and draws some conclusions for the future.

On too many occasions in the past, disaster has been predicted for the Atlantic Salmon. In 1932 Herbert Maxwell noted that 'The salmon-fishing seasons of 1898 and 1899 must be reckoned of culminating leanness among a number of lean years. In fact, our goose of the golden eggs seems to be approaching its last gasp. ... the combined greed and ingenuity of man is on the point of overcoming the extraordinary fecundity of the finest of our food fishes, the noblest of our sporting ones.'

On the same theme, H.V. Morton recorded that 'The salmon has had a bad time in Scotland. Every one is on the look out for him. His only hope of survival is to gain the freedom of some expensive river in the Highlands where there is nothing more dangerous than a colonel or two with a rod bought in St. James's Street ...'

It is essential, for the survival of our rare native fish species and communities, that changes are made in the way that fish, their habitats and their fisheries are managed. Apart from the aesthetic and scientific reasons for fish conservation, there are other important reasons: commercial value, sporting value (often with direct and indirect financial and social implications), reservoirs of important genes (of potential value in the future), amenity and education. Freshwater fish are also an important part of the habitat of other wildlife, for example as food for many waterside birds and waterfowl, including some rare and threatened breeding species such as Black-throated Divers, while for Otters, whose survival in much of Scotland has been difficult in the past, a good fish fauna is essential.

What can be done to further the cause of freshwater fish, especially our native species, in Scotland? An essential step must be to inform and interest that large and influential section of the public already interested in natural history and the conservation of wildlife. It must be generally realised that the conservation of the freshwater fish element of our biodiversity has been, and still is, neglected in comparison with the more "popular" elements of our wildlife, and that a number of our original fish communities and individual rare species are also endangered.

While there is much scientific information already available on the biology and ecology of many of our freshwater fish species, this is spread widely over a range of sources, only readily obtainable by fishery scientists. There is a need for a summary of such information to be found within one cover; the author hopes that this book will make some contribution towards that objective. In particular, there are a great many anglers and angling organisations in Scotland and their co-operation is essential if the future of our native fish populations is to be secured. After all, fish conservation should, in the long term at least, be in the anglers' best interests, but it is unfortunate that the majority of anglers are not primarily interested in the natural history or ecology of their quarry and even less in fish species which are not angled.

Habitat

Our native fish face a number of problems, some of them common to other forms of wildlife, others more particularly to fish. Rivers, and to a lesser extent lochs, are repositories of enormous amounts of human waste, ranging from toxic industrial chemicals, through agricultural slurries and herbicides, to domestic sewage. Even aerial pollutants such as sulphur dioxide from power station chimneys, are eventually washed into water courses as 'acid rain'. Many rivers have become completely fishless as a result, especially those in the industrial and heavily populated lowland areas of Scotland.

The River Clyde is a good example of a river which formerly had rich and diverse fish populations of some 20 species but which eventually became completely fishless in the lower reaches. Here at one time, not too many years ago, the waters were totally devoid of oxygen and comprised a lethal cocktail of various industrial chemicals. As well as eliminating all fish in this stretch, the gross pollution prevented all migratory species reaching any part of the river upstream. The river is recovering now but we must never let this happen again.

Other factors have affected fish in various ways. Barriers on rivers, such as weirs or some hydro-dams have blocked the passage of migratory fish to their spawning grounds and so eliminated them upstream. Enrichment from farm fertilisers, overfishing and the introduction of new fish species have all contributed to the decline of fish stocks - especially those of the rarer and more sensitive native species. Fish populations are limited by land boundaries to their immediate water body and so the whole population is vulnerable to a single incident of toxic spillage or acidification. Where a native species is found in a few waters only - sometimes only one or two (as was the case with the Vendace) - it is obviously extremely vulnerable and in urgent need of protection.

Over the last few decades there have been very significant advances in combating pollution in Scotland, thanks to the work of the River Purification Boards and more recently their reincarnation in the form of the Scottish Environment Protection Agency (SEPA). The trend towards increasingly polluted waters and declining fish populations has largely been reversed; many rivers are now much cleaner than they were fifty or even a hundred years ago. Some rivers which were so badly polluted that they became fishless are now clean again and supporting good stocks of fish. The Rivers Clyde and Carron are good examples.

One of the most important revolutions which is taking place in this country at present is the pressure for change in land use. This has been brought about for a variety of reasons, partly the tax or subsidy advantages in planting some crops and not others (e.g. the massive areas of conifers planted during the 20th Century), partly the problems with intensive animal husbandry, and partly the over-production of food stuffs. There are likely to be enormous changes in the countryside over the next few decades which will undoubtedly impact on native fish populations. Some of these effects may be good, others bad.

The main trend in the foreseeable future is likely to be a change away from conventional and intensive agriculture towards other forms of land use, particularly forestry and leisure activities. Overall, this will be good for the freshwater environment - especially in lowland areas, where a reduction in fertilisers, herbicides, pesticides, and less intense drainage with more ground under both deciduous and coniferous trees, will all be beneficial.

In the uplands, however, the increasing amounts of blanket coniferous afforestation has created problems for fish populations; in some base-poor heavily afforested areas the waters have become acidic and completely fishless. However, the implementation of the Forests and Water Guidelines, and the more enlightened policies of the Forestry Commission, are rapidly changing things for the better. Elsewhere, the development of estate and farm diversification projects of many kinds is resulting in the 'enhancement' of ponds and lochs through the introduction of various alien coarse and farmed salmonid species, usually Rainbow Trout, for 'put and take' fisheries. These often use extremely small, unsuitable water bodies.

Management

The problems of management, and some possible solutions, have been considered in a previous chapter. It is to be hoped that the Bill at present passing through the Scottish Parliament will soon become law, and that it will be followed eventually by Bill covering fish and fishery management.

Every year, changes take place in the distribution of freshwater fish in Scotland, for the situation has never reached equilibrium during the ten thousand years or so since the last Ice Age. In addition to the slow natural dispersal northwards by many of our native fish species, humans have caused many changes. Few of these have been beneficial to native species, and if things are to improve in the future we must learn from our mistakes. Many of the relevant aspects have been discussed in earlier chapters, but what can be done to rectify them?

Many thousands of different introductions involving most of our fish species have been carried out in various parts of the country. Rarely have any of these had a reasoned scientific basis, and practically never have the results been monitored. Fortunately, a high proportion of these introductions have been unsuccessful. Certainly some damage to native species has been done by some of them, and there can be few cases where any good has been done. In spite of existing legislation, casual introductions continue to occur - sometimes with potentially serious consequences. The introduction of Ruffe to Loch Lomond in recent years is an example of this, and the even more recent establishment there of Gudgeon, Dace and Chub has presented additional threats to the unique native fish community - especially the endangered Powan and the unusual race of River Lamprey there.

As well as tightening the legislation to prevent such introductions, we need a change of heart among those involved in carrying out such transfers - in the majority of cases, fishery managers and anglers. With many thousands of lochs and streams in Scotland is there any justification at all for moving relatively common species (in the south of the country at least) into additional waters further north, just because they do not occur there? The rare local fish populations and the valuable wild salmonid fisheries in some of these waters are under enough threats without those posed by the indiscriminate introductions of aggressive alien species.

In an era of high stress and unemployment, the national importance of the recreational value of angling to a large number of Scots must not be underestimated. In addition, as well as a scientific, there is also a socio-economic element within the new proposals, for hopefully more jobs will be created for fishery biologists and fishery wardens within Scotland as a whole. The potential value of Scotland's freshwater fisheries has been measured recently as well over 100 million pounds, much of which is spent on a local basis, not only on fishing permits and lets of beats, but also on accommodation, food, fishing tackle and so on. Fish and fisheries are patently an important element of the tourist industry.

Fisheries

One problem related to conserving fish compared to most groups of plants, amphibians, reptiles, birds and mammals, is that many more fish are exploited (commercially or by anglers). The majority of planned introductions taking place at the moment are of sport fish, and the usual objective is either to diversify and enhance the existing natural population or to provide catchable size fish which can be caught immediately - the 'put-and-take' fishery. Enhancement is very rarely justified, and in the author's opinion should never be carried out unless the following questions have been asked. "Will it do any harm to the existing fish population and fishery?", "Is it necessary?" and "Is it likely to fulfil its objectives?". If these questions were answered honestly, many fewer enhancement introductions would be carried out, and less damage would be done to the genetic integrity of our indigenous fish populations.

The popularity of 'put-and-take' fisheries is undoubtedly increasing, though many find it difficult to understand the satisfaction in catching tame, fat fish which come easily to a lure and may only have been released from the fish farm the day before. There they will have been fed on an expensive pellet diet (much of it consisting of meal made from other fish species to which flesh colouring agents have been added), and perhaps treated with antibiotics and other medications. Such fish are normally far less attractive than their wild counterparts - often having foreshortened snouts and stunted and worn fins through being kept in close confinement with hundreds of their brethren. In addition to the unreality of the situation, the introduction of such fish can damage the native

stock through competition for food and space, the introduction of diseases and the reduction of the genetic integrity and fitness of the native stock.

An additional problem related to 'put and take' fisheries is that the fish being introduced have been reared somewhere else with all the attendant problems created by fish farms. These include water abstraction, pollution from waste food and faeces, increased demand for industrial fish meal (i.e. more fish such as Sand Eels will be removed from the marine food chain to the detriment of, for example, Atlantic Salmon and Sea Trout at sea). The introduction of diseases, problems with predators (Herons, Cormorants, Otters, Mink, etc) attracted to the farm, and impacts on native fish from escapees through competition and reduction of genetic integrity are also problematic issues. All these factors must be borne in mind when the pros and cons of 'put and take' fisheries are discussed.

Perhaps the answer to these various introductions is to accept that they are going to continue to take place in some waters, especially artificial reservoirs and ponds, but to try to identify natural waters which are important for their wildlife, and make sure that no introductions take place there. This would seem to be a reasonable compromise which would ensure that the native stocks will continue in some waters, and that these truly wild fish will be available to those anglers who wish to use their skill to pursue them there.

One of the modern ways to transport fish is by helicopter and Donald Mitchell tells of a 'fair sized' Salmon caught in a small loch high in the hills above Scourie. 'I suspected that it dropped out of a helicopter bucket.'

The importance of catchments was pointed out by Charles St John in 1891: 'Fish are as dependent on the nature of the soil through which a stream runs, as oxen are on the richness of the meadows on which they pasture. The reason is obvious: a river which runs through a fertile country always abounds in flies, worms, snails, &c., on which its inhabitants feed; while a mountain stream, which flows rapidly through a barren and rocky country, has not the same supply.' Yet it is only in the last decade or so that serious attention has been given in Scotland to catchment management. This has been given further impetus by the acceptance of the EC

Water Framework Directive, currently being implemented by SEPA.

Legislation

The existing legislation in Scotland controlling fish introductions has developed on a rather piece-meal basis as a response to a number of emergencies that have arisen in the past. As a result it has a number of deficiencies which need to be remedied if we are to be in full control of the destiny of our native fish populations. Formerly, one of the biggest loopholes related to the control of fish coming into the country, for whilst it was difficult because of controls to bring in most salmonids in this way, it was relatively easy to import almost any other fish which could be classified as a 'pond or aquarium' species. Thus many pet shops around the country were until recently selling several species which could well establish themselves in Scotland if introduced, intentionally or unintentionally, into suitable waters. Examples of such fish which could well pose a threat to our native stocks are Danube Catfish *Silurus glanis*, Channel Catfish *Ictalurus punctatus*, Black Bullhead *Ameiurus melas*, and others, as discussed in and earlier chapter.

A simple improvement in legislation which has recently taken place both south and north of the Border was the production of a 'black list' of some temperate species which are likely to cause damage if released in the wild. This list, however, could be improved as there are several potentially damaging species missing from it.

Though EC bureaucracy is not one of the author's favourite institutions, there can be no doubt the several of the EC Directives which have appeared in the last two decades have been of substantial help to freshwater fish in Scotland, and forced the UK and Scottish governments to make conservation provisions which should have taken place many years ago. The EC Habitats Directive and the Water Framework Directive are forcing member states in Europe to pay attention to freshwater fish and their habitats in a way that can only be beneficial to their future conservation and sustainability.

Education

Lack of publicity and education by all concerned in the world of freshwater fish in Scotland has resulted in low 'fish appeal'; fish are just things that anglers catch while standing or sitting on stools in the rain, or commercial fishermen harvest callously in large numbers! Because of this, in recent times, the study of fishes and their conservation has been left mainly to scientists supporting commercial fisheries, or the management of sport fisheries; to biology teachers; to a small proportion of anglers; to the relatively few professional and amateur ichthyologists and to children with nets and jam jars. All this adds up to a sad lack of general awareness of an important element of our native fauna, and no effective country-wide policy for fish conservation can emerge from such relative apathy. How different is the present attitude to the conservation of many other forms of wildlife, especially plants and birds.

One of the main problems in appreciating the beauty and interest of our native fish is the difficulty of observing them, a subject explored in a previous chapter. Some species can be observed in the field from time to time, especially during their migrations (e.g. Salmon and Eels) or spawning acts (e.g. lampreys and sticklebacks), but most are difficult to observe and therefore rarely seen. Thus very few people in Scotland have ever seen Twaite or Allis Shad, Powan or Vendace.

An obvious way to remedy this is via aquaria - both public and private. Some excellent marine aquaria have been established, partly linked to aquaculture, and these are very well patronised by the public. Certainly captivity in zoos and wildlife parks is the way that most people see many of our native mammals (e.g. Wild Cats, Otters, Pine Martens and Polecats); and there is no reason why this should not also be the case with our native freshwater fish. Scotland needs a national aquarium for native fish!

Yet there are very few public aquaria or other displays concerning our native fish: several (in different parts of the country) are surely needed if we wish to demonstrate the variety and interest of our fish fauna. These need not be too expensive nor necessarily try to include all the Scottish species. They may try to emphasise or explore very local fish, or show some aspects of our commercial species, or demonstrate conservation of our rarer species and so on. It is an ironic fact that it is very much easier to see (and even purchase) a wide variety of exotic fish from many other countries, but impossible anywhere to see native fish.

In addition to aquarium displays of different kinds, there are other means of giving people pleasure and an interest in our fish. For instance, there are many places all over the country where some species can be easily seen at some times of the year. The fish viewing chamber on the salmon ladder over the hydro-electric dam on the River Tummel at Pitlochry is a good example - many thousands of people pass through this each year in the hope of seeing a wild Atlantic Salmon. Why should we not follow the excellent example of ornithologists and create hides and viewing areas in our country parks, and some wildlife reserves where the public can see fish in action? Good examples abroad are the Fairy Springs at Rotorua in New Zealand and the Capileno Dam near Vancouver in Canada. At the latter site there is a marvellous display of fish with uniformed hostesses conducting visitors around.

In Scotland, Sea, River and Brook Lampreys are all obvious at spawning time, when their nesting activities and reproductive behaviour are fascinating. At many waterfalls and obstructions Salmon and Trout can be seen on their spawning migrations, and the actual act of their spawning is much easier to observe than most people believe. Common Minnows and several other members of the carp family gather in large numbers on the spawning beds where they can be watched with ease at the right time of year.

Some of the commercial aspects of our freshwater fisheries are also of considerable interest. Of course, there are many anglers throughout the country who spend much of their time in pursuit of various species. Only in a few places can commercial netting be seen, however, and only rarely is it ever explained properly to the public. A step in the right direction are the attractive information notices provided by the Highland Council telling the public about Atlantic Salmon - one at a commercial netting station at Bonar Bridge, another at the Falls of Shin. There are many other areas of the country where this example could be followed by local councils and others.

Fish farms are another area where the interest of the public could be fostered more. Indeed many Rainbow Trout farms are already proving an attraction, and thousands of people now visit them each year, usually paying to enter and even to buy the pelleted food on which the fish would be fed anyway by the owner. Fish at various stages of their life cycle can usually be seen on such farms, and indeed it is common practice for the adults to be sold on the way out - either fresh or smoked. Such displays should be encouraged, and could surely be developed in various parts of the country and for other species.

Following on this theme, another idea is to develop 'trails' such as already exist for other interests such as castles, gardens and even whisky. There could be 'fish trails' around the country, where it would be possible to visit in a day, say, a public aquarium, a fish farm, a place where fish can be seen in the wild (a waterfall in the autumn or a viewing chamber in a river or lake), a port where fish are landed commercially, a river where commercial netting is being carried out or eel traps are being emptied. Such trails would undoubtedly boost the tourist potential of any area as well as educate the public about the beauty and value of our native fish and fisheries.

Research

In an ideal scenario, the way forward in managing our fish populations must be for each local 'fishery board' (whatever these might be in the future) to develop rational conservation-orientated policies for the management of its own fish stocks. Too often, however, the basic information for this is lacking, such as knowledge of the distribution and status of each species within each catchment area concerned. Where this information is available it is possible to produce realistic management plans which may often allow conservation and exploitation to go hand-in-hand for many of the commoner species. The real status of threatened species is also much clearer under these circumstances; here too it may be possible to proceed with effective measures for their conservation.

We know a great deal now about the natural history of most of our fish and much of this has been researched by scientists over the last few decades. Much remains to be learned, however, especially, as indicated in other chapters, about our rarer and endangered species. This remains an important area for future research. Nevertheless, we are also still extremely ignorant about the population biology of most of our species and the factors controlling their numbers. Information of this kind is vital to the management of our stocks, but unfortunately is difficult to acquire because of the problem of obtaining exact counts of entire fish populations, and the complexity of factors involved. Nonetheless, some useful studies have already been made and with the advances in modern technology related to echo sounding, revolutionary new fish counters and new methods of tagging, the future could be bright.

Conservation

The conservation of fishes has been sadly neglected until now, but there are several indications that this is changing. There is no doubt that public support and feeling behind the conservation movement has strengthened and broadened substantially over the last decade. Many readers of this book are likely to be supportive of wildlife conservation in some form, and to have his or her own image of what that means. Natural habitats like meadows and mountains, forests and seashore, rivers and lochs, wildlife like trees and flowers, butterflies and bees, birds and mammals - these are all areas of popular interest and concern. But what about fish, and why conserve them? Do they need it? Are they worth it? The author believes that the case is strong and hopes that the reader would think so too.

Fish suffer from a bad press! Many are difficult to observe in the wild and so do not have a popular following like field botanists or ornithologists. Yet there is substantial interest among several groups of people in living fish. There are reputed to be 250,000 anglers in Scotland, but sadly their main concern is with the species they wish to catch. Another major interest group is aquarists, but the great majority of these seem uninterested in our native fish, being more concerned with exotic tropical species.

Our native fish species have much of merit and interest. They were among the very first recolonisers after the last Ice Age along with the first plants and birds. Many are important commercially in some parts of the world. Others (contrary to popular opinion) are colourful and have fascinating life cycles and habits. Several are much easier to observe in the field if you can be in the right place at the right time: a number make excellent subjects for aquaria, and indeed can often be induced to breed there.

As far as our rarer species are concerned, although all of them do occur in other countries, the Scottish populations have all been isolated for at least 10,000 years, developing distinct gene pools during that period. This is in contrast to virtually all our birds, which are mostly just part of the north-west European stock. In some countries, where some of our fish species are less rare, they are valuable commercially (for example, River and Sea Lampreys, Allis and Twaite Shad, Arctic Charr and Sparling, Pike, Perch and Eels). The importance of conserving this resource to provide options for the future is clearly imperative.

The author considers that at least 10 of our native species require specific conservation measures of some kind. The conservation action needed for some species (like the Sturgeon) must involve international action, and, it is hoped, its re-establishment at least as a regular visitor to this country. The immediate concern in Scotland must be for species which are still with us, but are rare and under very significant threat. Such fish include the Vendace (which was extinct, but has recently been restored to Scotland), the Powan (which occurs naturally in only two lochs), the Allis and Twaite Shads and the Sparling (with only three stocks remaining out of at least 15 previously). Surprisingly perhaps, even the Nine-spined Stickleback requires some conservation initiatives.

Finally there are several species which are not immediately threatened, but which have declined significantly within this century. We need to learn much more about the threats to such species and about their current status. Within this category are Atlantic Salmon. Brown (and Sea) Trout, Arctic Charr, River, Brook and Sea Lampreys and even Eels and Three-spined Sticklebacks in some catchments. Some of these are known to have (or have had) quite distinct races or populations - several of which are now extinct, others of which are struggling for survival.

Conclusions

Some scenarios for the future look bright. There have been enormous strides in pollution control and a number of our worst rivers are now much cleaner. Several of our fish and parts of their habitat are now given some protection - thanks largely to the European Community. There is an admission in the

Scottish Parliament that management is not as it should be, and discussions are taking place about the future. All of this is very welcome.

Hopefully, we may eventually understand much more about the status and requirements of our native species, and we will give them much more protection in nature reserves, etc. The future in these islands of our rarest species could be assured through their re-establishment in previously occupied waters, and the creation of additional populations as a form of safeguard. The need is clearly a long-term one. Perhaps, by the end of the 21st Century, many lochs and rivers which have been lost as suitable habitat will have been restored, and our native fish species will be safe for future generations to use and enjoy in various ways.

The author hopes that this book will give pleasure to those readers already interested in the freshwater fish of Scotland, perhaps stimulating them to observe more of their natural history. It is his hope too that other readers of the book, who perhaps had only a passing interest in the subject, may have gone beyond the illustrations to delve into some aspect of the text. It is hoped also that more Scottish anglers will consider a broader view of their hobby, taking an interest, not only in the natural history of the species they fish f,or, but in the other members of our fish fauna also. The strength of the angling lobby in this country is very great, and all our native fish need support if they are to survive into future centuries.

References

Anderson (1977), Barnett (1927), Bregazzi *et al.* (1982), Cooper & Wheatley (1981), Crandall (1914), Crisp *et al.* (1975), Fitter (1959), Harden-Jones (1968), Holt & Talbot (1978), Kennedy (1975), Lamond (1931), Maitland (1966a, 1982), Maitland *et al.* (1981, 1987, 1994), Maitland & Turner (1987), Marshall (1971), Maxwell (1932), Pennell & Pennell (1889), Pitcher & Hart (1982), St John (1891), Toner (1959).

'By the Salmon and Freshwater Fisheries Act, 1923,
... the laws of England and Wales regarding salmon, trout
and coarse fish have been consolidated and amended.
Similar legislation for Scotland is long overdue, and
... it is more than likely that any new Scottish ... statute
will deal with all species of fresh-water fish ...'

Henry Lamond (1931) *Loch Lomond*

*****

BIBLIOGRAPHY

'Knowledge is of two kinds: we know a subject ourselves,
or we know where we can find information upon it.'

James Boswell (1775) *Life of Johnson*

AASS, P. 1972. Age determination and year-class fluctuations of Cisco, *Coregonus albula* L., in the Mjosa hydroelectric reservoir, Norway. *Report of the Institute of Freshwater Research, Drottningholm,* 52, 5-22.

ADAMS, C.E. 1991. Shift in pike, *Esox lucius* L., predation pressure following the introduction of ruffe, *Gymnocephalus cernuus* (L.), to Loch Lomond. *Journal of Fish Biology,* 38, 663-667.

ADAMS, C.E. 1994. The fish community of Loch Lomond, Scotland: its history and rapidly changing status. *Hydrobiologia,* 290, 91-102.

ADAMS, C.E., BROWN, D.W. & KEAY, L. 1994. Elevated predation risk associated with inshore migrations of fish in a large lake, Loch Lomond, Scotland. *Hydrobiologia,* 290, 135-138.

ADAMS, C.E., BROWN, D.W. & TIPPETT, R. 1990. Dace (*Leuciscus leuciscus* (L.)) and Chub (*Leuciscus cephalus* (L.)) new introductions to the Loch Lomond catchment. *Glasgow Naturalist,* 21, 509-513.

ADAMS, C.E., FRASER, D., HUNTINGFORD, F.A., GREER, R.B., ASKEW, C.M. & WALKER, A.F. 1998. Trophic polymorphism amongst Arctic charr from Loch Rannoch, Scotland. *Journal of Fish Biology,* 52, 1259-1271.

ADAMS, C.E. & MAITLAND, P.S. 1991. Evidence of further invasions of Loch Lomond by non-native fish species with the discovery of a roach x bream, *Rutilus rutilus* (L.) x *Abramis brama* (L.), hybrid. *Journal of Fish Biology.* 38, 961-963.

ADAMS, C.E. & MAITLAND, P.S. 1998. The Ruffe population of Loch Lomond, Scotland: its introduction, population expansion and interaction with native species. *Journal of Great Lakes Research,* 24, 249-262.

ADAMS, C.E. & MAITLAND, P.S. 2002. Invasion and establishment of freshwater fish populations in Scotland - the experience of the past and lessons for the future. *Glasgow Naturalist,* 23, 35-43.

ADAMS, C.E. & MITCHELL, J. 1992. Introduction of another non-native fish species to Loch Lomond: Crucian Carp (*Carassius carassius* (L.)). *Glasgow Naturalist,* 22, 165-168.

ADAMS, C.E. & MITCHELL, J. 1995. The response of a Grey Heron *Ardea cinerea* breeding colony to rapid change in prey species. *Bird Study,* 42, 44-49.

ADAMS, C.E., MURRAY, K.R. & HUNTINGFORD, F.A. 1992. The periodicity of primary increment formation in the otoliths of Arctic Charr *Salvelinus alpinus* (L.). *Journal of Fish Biology,* 41, 515-520.

ADAMS, C.E. & TIPPETT, R. 1990. *The status of the fish populations of Loch Lomond.* Edinburgh, Nature Conservancy Council.

ADAMS, C.E. & TIPPETT, R. 1991. Powan ova (*Coregonus lavaretus*) predation by introduced Ruffe (*Gymnocephalus cernuus*) in Loch Lomond, Scotland. *Aquaculture and Fisheries Management,* 22, 261-267.

ADE, R. 1989. *The trout and salmon handbook.* London, Croom Helm.

ADE, R. 1985. *Fisher in the hills.* London, David & Charles.

AFLALO, F.G. 1897. *A sketch of the natural history (vertebrates) of the British Islands.* Glasgow, Grant Educational Co.

AITKEN, P.L., DICKERSON, L.H. & MENZIES, W.J.M. 1966. Fish passes and screens at water power works. *Proceedings of the Institute of Civil Engineers.* 35, 29-57.

ALABASTER, J.S. 1986. An analysis of angling returns for trout, *Salmo trutta* L., in a Scottish river. *Aquaculture and Fisheries Management,* 17, 313-316.

ALABASTER, J.S. & LLOYD, R. 1980. *Water quality criteria for freshwater fish.* London, Butterworths.

ALDOORI, T.Y. 1971. Food and growth of roach (*Rutilus rutilus*) in two different environments. *Proceedings of the British Coarse Fish Conference.* 5, 72-77.

ALDOORI, T.Y. 1972. The ecology of the roach (*Rutilus rutilus* L.) in Humbie Reservoir. *PhD Thesis, University of Edinburgh.*

ALEXANDER, A.S. 1925. *Tramps across watersheds.* Glasgow, Smith.

ALEXANDER, G. & ADAMS, C.E. 2000. Phenotypic variation in Arctic Charr from Scotland and Ireland. *Aqua,* 4, 77-88.

AL-HASSAN, L.A.J., WEBB, C.J., GIAMA, M. & MILLER, P.J. 1987. Phosphoglucose isomerase polymorphism in the Common Goby, *Pomatoschistus microps* (Kroyer) (Teleostei: Gobiidae), around the British Isles. *Journal of Fish Biology,* 30, 281-298.

ALI, S.S. 1976. The food of Roach, *Rutilus rutilus* (L.) in Llyn Tegid (North Wales). *Sind University Research Journal of Science,* 9, 15-33.

ALI, S.S. 1979. Age, growth and length-weight relationship of the Roach *Rutilus rutilus* L. in Llyn Tegid, north Wales. *Pakistan Journal of Zoology*, 11, 1-19.

ALLEN, A. 1987. The olfactory world of the goldfish. *Aquarist and Pondkeeper*, 1987, 52.

ALLEN, H. 1977. Koi. *The Aquarist*, March, 454-455.

ALLEN, J.R.M. & WOOTTON, R.J. 1982. Age, growth and rate of food consumption in an upland population of the three-spined stickleback, *Gasterosteus aculeatus*. *Journal of Fish Biology*, 21, 95-106.

ALLEN, K.R. 1938. Some observations on the biology of the trout (*Salmo trutta*) in Windermere. *Journal of Animal Ecology*, 7, 333-349.

ALLEN, K.R. 1941. Studies on the biology of the early stages of the salmon (*Salmo salar*). 3. Growth in the Thurso River system. *Journal of Animal Ecology*, 10, 273-295.

ALLEN, K.R. 1944. Studies on the biology of the early stages of the salmon (*Salmo salar*). 4. The smolt migration in the Thurso River in 1938. *Journal of Animal Ecology*, 13, 63-85.

AMUNDSEN, P.A., SIIKAVUOPIO, S., SVENNING, M. & CHRISTENSEN, G. 1998. Cannibalistic responses in Arctic Charr - individual and population differences. *Proceedings of the 7th ISACF Workshop on Arctic Charr*, 1992, 5-11.

ANDERSON, J.A. 1944. Gannets fishing in fresh water. *British Birds*, 38, 17-18.

ANDERSON, L.G. 1977. *The economics of fishery management*. Baltimore, John Hopkins University.

ANDERSON, M. 1982. The identification of British grey mullets. *Journal of Fish Biology*, 20, 33-38.

ANDREASSON, S. 1971. Feeding habits of a sculpin (*Cottus gobio* L.) population. *Institute of Freshwater Research, Drottningholm, Report*, 51, 6-30.

ANDREASSON, S. 1972. Distribution of *Cottus poecilopus* Heckel and *C. gobio* L. (Pisces) in Scandinavia. *Zoologica Scripta*, 1, 69-78.

ANDREWS, C.W. & LEAR, E. 1976. The biology of Arctic char (*Salvelinus alpinus* (L.)) in northern Labrador. *Journal of the Fisheries Research Board of Canada*, 13, 843-860.

ANDREWS, C. 1979. Coldwater fish in the aquarium. *The Aquarist*, June, 106.

ANGLING FOR CHANGE. 2000. *Angling for Change proposals – July 2000*. Aberfeldy, WWF.

ANONYMOUS. 1843. *Carp*. London, New Naturalist's Library.

ANONYMOUS. 1904. Old Scottish salmon traps. *Country Life*, December 17, 901-903.

ANONYMOUS. 1907. A ramble round Yetholm. *History of the Berwickshire Naturalists' Club*. 19, 331.

ANONYMOUS 1978. Goldfish. *The Aquarist*, May, 62.

ANONYMOUS. 1933. *The Thames in June*. In: Stanley, A. (Ed.) *The Out-of-doors book*. London, Dent .

ANONYMOUS. 1883. The woods of Loch Lomond and the Gairloch. *Journal of Forestry*. 6, 318.

ANONYMOUS. undated. *Glen-Albyn, or tales of the central highlands*. Fort Augustus, Abbey Press.

ANONYMOUS. 1997. Atlantic Salmon ruled a pollutant. *Sea Wind*, 11, 31.

ANONYMOUS. 2000. A fisherman's diary. *The Link*, 12, 7.

APPLEGATE, V. 1950. Natural history of the sea lamprey, *Petromyzon marinus*, in Michigan. *Special Scientific Report, U.S. Fish and Wildlife Service*, 55, 1-237.

APRAHAMIAN, M.W. 1981. Aspects of the biology of the twaite shad (*Alosa fallax*) in the Rivers Severn and Wye. *Proceedings of the Second British Coarse Fish Conference, Liverpool*, 2, 373-381.

APRAHAMIAN, M.W. 1985. The effect of the migration of *Alosa fallax fallax* (Lacepede) into fresh water, on branchial and gut parasites. *Journal of Fish Biology*, 27, 521-532.

APRAHAMIAN, M.W. 1988. The biology of the twaite shad, *Alosa fallax fallax* (Lacepede), in the Severn Estuary. *Journal of Fish Biology*, 33A, 141-152.

APRAHAMIAN, M.W. & BARR, C.D. 1985. The growth, abundance and diet of 0-group sea bass, *Dicentrarchus labrax*, from the Severn Estuary. *Journal of the Marine Biological Association of the UK*, 65, 169-180.

ARMISTEAD, J.J. 1895. *An angler's paradise, and how to obtain it*. London, The Angler.

ARMSTRONG, J.D. 1987. Metabolism, feeding and cardiac function in pike, *Esox lucius*. *PhD Thesis, University of Aberdeen*.

ASHWORTH, A.W. & BANNERMAN, I.C.W. 1925. On the tetracotyle (*T. phoxini*) in the brain of the minnow. *Transactions of the Royal Society of Edinburgh*, 55, 159-172.

ASSOCIATION OF SCOTTISH DISTRICT SALMON FISHERY BOARDS. 1977. *Salmon fisheries of Scotland*. Farnham, Fishing New Books.

AUSEN, V. 1976. Age, growth, population size, mortality and yield in the whitefish (*Coregonus lavaretus* (L.)) of Haugatjern - a eutrophic Norwegian lake. *Norwegian Journal of Zoology*, 24, 379-405.

BAGENAL, T.B. 1966. The Ullswater Schelly. *Field Naturalist*, 2, 1-2.

BAGENAL, T.B. 1969. The relationship between food supply and fecundity in brown trout *Salmo trutta* L. *Journal of Fish Biology*, 1, 167-182.

BAGENAL, T.B. 1969. Relationship between egg size and fry survival in Brown Trout *Salmo trutta* L. *Journal of Fish Biology*, 1, 349-353.

BAGENAL, T.B. 1970. Notes on the biology of the schelly *Coregonus lavaretus* (L.) in Haweswater and Ullswater. *Journal of Fish Biology*, 2, 137-154.

BAGENAL, T.B. 1970. *Freshwater fishes*. London, Warne.

BAGENAL, T.B. 1972. *Sea fishes*. London, Warne.

BAGENAL, T.B. 1978. (Ed.) *Methods for assessment of fish production in fresh waters*. Oxford, Blackwell.

BAGENAL, T.B., MACKERETH, F.J.H. & HERON, J. 1973. The distinction between brown trout and sea trout by the strontium content of their scales. *Journal of Fish Biology*, 5, 555-557.

BAILEY-WATTS, A.E. & MAITLAND, P.S. 1984. Eutrophication and fisheries in Loch Leven, Kinross, Scotland. *Proceedings of the Institute of Fisheries Management Annual Study Course*. 15, 170-190.

BAIN, G. 1944. A list of the fish of the Clyde area. *Transactions of the Paisley Naturalists' Society*. 5, 18-23.

BALL, J.N. & JONES, J.W. 1960. On the growth of brown trout of Llyn Tegid. *Proceedings of the Zoological Society of London*, 134, 1-41.

BANKS, J.A. 1969. A review of the literature on the upstream migration of adult salmonids. *Journal of Fish Biology*, 1, 85-136.

BANKS, J.A. 1970. Observations on the fish population of Rostherne Mere, Cheshire, *Field Studies*, 3, 375-379.

BANKS, S. 1989. Land-locked salmon are no gimmick. *Fish Farmer*, July/August, 25-26.

BARBOUR, S. E. 1984. Variation in life history, ecology and resource utilisation by Arctic Charr *Salvelinus alpinus* (L.) in Scotland. *PhD Thesis, University of Edinburgh*.

BARBOUR, S.E. & EINARSSON, S.M. 1987. Ageing and growth of Charr *Salvelinus alpinus* (L.) from habitat types in Scotland. *Aquaculture and Fisheries Management*. 18, 1-13.

BAROUDY, E. & ELLIOTT, J.M. 1994. Critical thermal limits for juvenile Arctic Charr *Salvelinus alpinus*. *Journal of Fish Biology*, 45, 1041-1053.

BARNETT, T. R. 1927. *The land of Locheil and the magic west*. Edinburgh, Grant.

BARRETT, J.H., NICHOLSON, R.A. & CERON-CARRASCO, R. 1999. Archaeo-ichthyological evidence for long-term socioeconomic trends in northern Scotland: 3500 BC to AD 1500. *Journal of Archaeological Science*, 26, 353-388.

BASTARD, T. 1598. *Chresteleros, Seven Books of Epigrames*. London.

BATEMAN, G.C. 1890. *Freshwater aquaria*. London, Link House.

BAXTER, E.W. 1954. Lamprey distribution in streams and rivers. *Nature, London*, 180, 1145.

BEAMISH, F.W.H. & MEDLAND, T.E. 1988. Age determination for lampreys. *Transactions of the American Fisheries Society*, 113, 63-71.

BEAN, C. 1978. Longevity in fish. *The Aquarist*, July, 157-160.

BEAN, C.W. & WINFIELD, I.J. 1989. Biological and ecological effects of a *Ligula intestinalis* (L.) infestation of the gudgeon, *Gobio gobio* (L.), in Lough Neagh, Northern Ireland. *Journal of Fish Biology*, 34, 135-147.

BEAN, C.W. & WINFIELD, I.J. 1992. Influences of the tapeworm *Ligula intestinalis* (L.) on the spatial distributions of juvenile roach *Rutilus rutilus* (L.) and gudgeon *Gobio gobio* (L.) in Lough Neagh, Northern Ireland. *Netherlands Journal of Zoology*, 42, 416-429.

BEAN, C.W. & WINFIELD, I.J. 1995. Habitat use and activity patterns of roach (*Rutilus rutilus* (L.)), rudd (*Scardinius eythrophthalmus* (L.)), perch (*Perca fluviatilis* L.) and pike (*Esox lucius* L.) in the laboratory: the role of predation threat and structural complexity. *Ecology of Freshwater Fish*, 4, 37-46.

BEAN, C.W., WINFIELD, I.J. & FLETCHER, J.M. 1996. Stock assessment of the Arctic charr (*Salvelinus alpinus*) population in Loch Ness, U.K. Pages 206-223 in: Cowx, I.G. (Ed.) *Stock assessment in inland fisheries*. Oxford, Fishing News Books.

BEAUMONT, W.R.C. & MANN, R.H.K. 1983. The age, growth and diet of a freshwater population of Flounder, *Platichthys flesus* (L.), in southern England. *Journal of Fish Biology*, 23, 607-616.

BEIRNE, B.P. 1952. *The origin and history of the British fauna*. London, Methuen.

BELL, M.A. 1974. Reduction and loss of the pelvic girdle in *Gasterosteus* (Pisces): a case of parallel evolution. *Natural History Museum, Los Angeles, Contributions to Science*, 257, 1-36.

BELL, M.A. 1981. Lateral plate polymorphism and ontogeny of the complete plate morph of threespine sticklebacks (*Gasterosteus aculeatus*). *Evolution*, 35, 67-74.

BELYANINA, T.N. 1969. Synopsis of biological data on smelt, *Osmerus eperlanus* (Linnaeus). *Food and Agriculture Organisation Fishery Synopsis, Rome*, 78, 1-55.

BERG, L.S. 1965. *Freshwater fishes of the U.S.S.R and adjacent countries*. Jerusalem, Israel Program for Scientific Translations.

BERNERS, J. 1496. *A treatyse of fysshynge wyth an angle*. London.

BERRY, J. 1933. Notes on the migration of smolts from Loch Ness, summer, 1932. *Fisheries Scotland, Salmon Fisheries*, 1933, 1-12.

BERTIN, L. 1956. *Eels*. London, Cleaver-Hume.

BEUKEMA, J.J. 1970. Acquired hook avoidance in pike, *Esox lucius* L., fished with artificial and natural baits. *Journal of Fish Biology*, 2, 155-160.

BEUKEMA, J.J. & DE VOS, G.J. 1974. Experimental tests of a basic assumption of the capture-recapture method in pond populations of Carp *Cyprinus carpio* L. *Journal of Fish Biology*, 6, 317-329.

BEVERIDGE, M.C.M., SIKDAR, P.K., FRERICHS, G.N. & MILLAR, S. 1991. The ingestion of bacteria in suspension by the common carp *Cyprinus carpio* L. *Journal of Fish Biology*, 39, 825-832.

BIBBY, M.C. 1972. Population biology of the helminth parasites of *Phoxinus phoxinus* (L.), the minnow, in a Cardiganshire Lake. *Journal of Fish Biology*, 4, 289-300.

BIDIE, G. 1896. Char in Loch Lomond. *Annals of Scottish Natural History*, 124, 258.

BIRD, D.J. & I.C. POTTER, 1979. Metamorphosis in the paired species of lampreys, *Lampetra fluviatilis* (L.) and *Lampetra planeri* (Bloch). 1. A description of the timing and stages. *Journal of the Linnaean Society of London*, 65, 145-160.

BIRD, D.J. & POTTER, I.C. 1979. Metamorphosis in the paired species of lampreys, *Lampetra fluviatilis* (L.) and *Lampetra planeri* (Bloch). 2. Quantitative data for body proportions, weights, lengths and sex ratios. *Journal of the Linnaean Society of London*, 65, 127-143.

BIRD, D.J., POTTER, I.C., HARDISTY, M.W., & BAKER, B.I. 1994. Morphology, body size and behaviour of recently metamorphosed sea lampreys, *Petromyzon marinus*, from the lower River Severn, and their relevance to the onset of parasitic feeding. *Journal of Fish Biology*, 44, 67-74.

BLACKMORE, R.D. 1910. *Lorna Doone*. London, Sampson Low.

BOARDER, A. 1978. Breeding goldfish. *The Aquarist*, April, 18-24.

BOARDER, A. 1978. Coldwater queries. *The Aquarist*, April, 9-12.

BODALY, R.A., VUORINEN, J., WARD, R.D., LUCZYNSKI, M. & REIST, J.D. 1991. Genetic comparisons of New and Old World coregonid fishes. *Journal of Fish Biology*, 38, 37-51.

BOLAM, G. 1919. The fishes of Northumberland and the Eastern Borders. *History of the Berwickshire Naturalists' Club*, 23, 153-197; 250-304.

BORCHARDT, D. 1988. Long-term correlations between the abundance of smelt (*Osmerus eperlanus eperlanus* L.) year classes and abiotic environmental conditions during the period of spawning and larval development in the Elbe River. *Archiv fur Fischereiwissenschaft*, 38, 191-202.

BOSWELL, J. 1775. *Life of Johnson*. London.

BOULENGER, E.G. 1936. Pond and stream life. Pages 98-129 in: *Nature in Britain*. London, Batsford.

BOULENGER, E.G. 1946. *British anglers' natural history*. London, Collins.

BOYD, J.M. 1999. *The song of the sandpiper*. Grantown-on-Spey, Baxter.

BRACKEN, J.J. & KENNEDY, M.P. 1967. A key to the identification of the eggs and young stages of coarse fish in Irish waters. *Scientific Proceedings of the Royal Dublin Society*, 2B, 99-108.

BRAITHWAITE, C. 1923. *Fishing vignettes*. London, Home Words.

BRASSINGTON, R.A. & FERGUSON, A. 1975. Electrophoretic identification of Roach (*Rutilus*

rutilus L.), Rudd (*Scardinius erythrophthalmus* L.), Bream (*Abramis brama* L.) and their natural hybrids. *Journal of Fish Biology*, 9, 471-477.

BREGAZZI, P.R. & KENNEDY, C.R. 1980. The biology of Pike, *Esox lucius* L., in a southern eutrophic lake. *Journal of Fish Biology*, 17, 91-112.

BREGAZZI, P.R & KENNEDY, C.R. 1982. The responses of a perch, *Perca fluviatilis* L., population to eutrophication and associated changes in fish fauna in a small lake. *Journal of Fish Biology*, 20, 21-31.

BRERETON, W. 1636. In: Brown P H. 1891.

BRIDGES, C.H. & MULLAN, J.W. 1958. A compendium of the life history and ecology of the Eastern Brook Trout *Salvelinus fontinalis* (Mitchill). *Massachussetts Division of Fish and Game Fishery Bulletin*, 23, 1-30.

BRIDGETT, R.C. 1929. *Sea-trout fishing*. London, Jenkins.

BROOK, A.J. & HOLDEN, A.V. 1957. Fertilisation experiments in Scottish freshwater lochs. 1. Loch Kinardochy. *Freshwater and Salmon Fisheries Research, Scotland*, 17, 1-30.

BROTCHIE, T.C.F. 1911. *Scottish western holiday haunts*. Glasgow, Menzies.

BROTCHIE, T.C.F. 1914. *Glasgow rivers and streams: their legend and lore*. Glasgow, Maclehose.

BROUGHTON, N.M. & JONES, N.V. 1978. An investigation into the growth of 0-group Roach *Rutilus rutilus* (L.) with special reference to temperature. *Journal of Fish Biology*, 12, 345-358.

BROWN, A. 1891. The fishes of Loch Lomond and its tributaries. *Scottish Naturalist*, 10, 114-124.

BROWN, A. 1896. Char in Loch Lomond. *Annals of Scottish Natural History*, 123,192.

BROWN, E.A.R. 1989. Growth processes in the two Scottish populations of Powan, *Coregonus lavaretus* (L.). *PhD Thesis, University of St Andrews*.

BROWN, E.A.R., FINNIGAN, N. & SCOTT, D.B.C. 1991. A life table for powan *Coregonus lavaretus* (L.) in Loch Lomond, Scotland: a basis for conservation strategy. *Aquatic Conservation*. 1, 183-187.

BROWN, E.A.R. & SCOTT, D.B.C. 1987. Abnormal pelvic fins in Scottish powan, *Coregonus lavaretus* (L.) (Salmonidae, Coregoninae). *Journal of Fish Biology* 31, 443-444.

BROWN, E.A.R. & SCOTT, D.B.C. 1988. A second hermaphrodite specimen of *Coregonus lavaretus* (L.) (Salmonidae, Coregoninae) from Loch Lomond, Scotland. *Journal of Fish Biology*. 33, 957-958.

BROWN, E.A.R. & SCOTT, D.B.C. 1990. Anabolic adaptiveness in the two Scottish populations of powan *Coregonus lavaretus* (L.). *Journal of Fish Biology*. 37A, 251-253.

BROWN, E.A.R. & SCOTT, D.B.C. 1991. Adolescence in female powan. *Proceedings of the 4th*

International Symposium on the Reproductive Physiology of Fish, Norwich, 1991, 163.

BROWN, E.A.R. & SCOTT, D.B.C. 1994. Life histories of the powan, *Coregonus lavaretus* (L) (Salmonidae, Coregoninae) of Loch Lomond and Loch Eck. *Hydrobiologia*, 290, 121-133.

BROWN, K.L. 1977. The Scottish Aquarium Society's Golden Jubilee. *The Aquarist*, October, 292.

BROWN, P. H. (Ed.). 1891. *Early travellers in Scotland.* Edinburgh, Douglas.

BROWN, W. 1862. *The natural history of the Salmon.* Glasgow.

BUCHAN, J. (Ed.). 1921. *Great hours in sport.* Edinburgh, Nelson.

BUCHAN, J. 1917. *Poems Scots and English.* London, Jack.

BUCK, R.J.G. & HAY, D.W. 1984. The relation between stock size and progeny of Atlantic Salmon, *Salmo salar* L., in a Scottish stream. *Journal of Fish Biology*. 23, 1-12.

BUCK, R.J.G. & YOUNGSON, A.F. 1982. The downstream migration of precociously mature Atlantic Salmon, *Salmo salar* L. parr in autumn and its relation to the spawning migration of mature adult fish. *Journal of Fish Biology* 20, 279-285.

BUCKE, D. 1974. Vertebral anomalies in the common bream *Abramis abrama* (L.). *Journal of Fish Biology*, 6, 681-682.

BUCKLAND, F & YOUNG, A. 1871. *Report of the special commissioners appointed to enquire into the effect of recent legislation on salmon fisheries in Scotland.* London, Houses of Parliament.

BUCKLAND, F. 1873. *A familiar history of British fishes.* London, Society for Promoting Christian Knowledge.

BUCKLAND, F. 1880. *The natural history of British fishes.* London, Unwin.

BULLER, F. 1989. The Royal Sturgeon. *Big Fish World,* 1989, 8-11.

BULLER, F. & FALKUS, H. 1988. *Freshwater fishing.* London, Grange Books.

BURKEL, D.L. 1971. Introduction of fish to new water. *Glasgow Naturalist*, 18, 574-575.

BURNS-BEGG, R. 1874. *The Lochleven angler.* Kinross, Barnet.

BURROUGH, R.J. 1978. The population biology of two species of eyefluke, *Diplostomum spathaceum* and *Tylodelphys clavata*, in Roach and Rudd. *Journal of Fish Biology*, 13, 19-32.

BURROUGH, R.J & KENNEDY, C.R. 1978. Interaction of perch (*Perca fluviatilis*) and brown trout (*Salmo trutta*). *Journal of Fish Biology*, 13, 225-230.

BURROUGH, R.J. & KENNEDY, C.R. 1979. The occurrence and natural alleviation of stunting in a population of Roach, *Rutilus rutilus* (L.). *Journal of Fish Biology,* 15, 93-109.

BUSS, K. & WRIGHT, J.E. 1957. Appearance and fertility of trout hybrids. *Transactions of the American Fisheries Society,* 87, 172-181.

CAINE, W. 1927. *Fish, Fishing and Fishermen.* London, Allan.

CAIRNCROSS, D. 1862. *The origins of the Silver Eel.* London.

CALA, P. 1971. On the ecology of the Ide *Idus idus* (L.) inb the River Kavlingean, south Sweden. *Annual Report of the Institute of Freshwater Research, Drottningholm*, 50, 45-99.

CALDERWOOD, W.L. 1909. *The salmon rivers and lochs of Scotland.* London, Arnold.

CALDERWOOD, W.L. 1904. The bull trout of the Tay and Tweed. *Proceedings of the Royal Society of Edinburgh*, 25, 27-38.

CALDERWOOD, W.L. 1907. *The life of the Salmon, with reference more especially to the fish in Scotland.* London, Arnold.

CALDERWOOD, W.L. 1925. The relation of sea growth and spawning frequency in *Salmo salar. Proceedings of the Royal Society of Edinburgh*, 45, 142-148.

CALDERWOOD, W.L. 1930. *Salmon and Sea Trout.* London, Arnold.

CALDERWOOD, W.L. 1938. *Salmon! Experiences and reflections.* London, Arnold.

CALDERWOOD, W.L. 1945. Passage of smolts through turbines. *Salmon and Trout Magazine*, 115, 1-8.

CAMPBELL, A.D. 1974. The parasites of fish in Loch Leven. *Proceedings of the Royal Society of Edinburgh*, 74B, 347-364.

CAMPBELL, J.S. 1977. Spawning characteristics of Brown Trout and Sea Trout *Salmo trutta* L. in Kirk Burn, River Tweed, Scotland. *Journal of Fish Biology*, 11, 217-230.

CAMPBELL, R.N. 1955. Food and feeding habits of Brown Trout, Perch and other fish in Loch Tummel. *Scottish Naturalist*, 67, 23-27.

CAMPBELL, R.N. 1957. The effect of flooding on the growth rate of Brown Trout in Loch Tummel. *Freshwater and Salmon Fisheries Research, Scotland*, 14, 1-7.

CAMPBELL, R.N. 1961. The growth of Brown Trout in acid and alkaline waters. *Salmon and Trout Magazine*, January 1961, 47-52.

CAMPBELL, R.N. 1963. Some effects of impoundment on the environment and growth of Brown Trout (*Salmo trutta* L.) in Loch Garry (Inverness-shire). *Freshwater and Salmon Fisheries Research, Scotland*, 30, 1-27.

CAMPBELL, R.N. 1971. The growth of brown trout, *Salmo trutta* L., in northern Scotland with special reference to the improvement of fisheries. *Journal of Fish Biology*, 3, 1-28.

CAMPBELL, R.N. 1973. A study of the organisation of freshwater sport fisheries in New Zealand, Australia and Canada. *Nuffield Travelling Fellowship Report.*

CAMPBELL, R.N. 1979. Ferox Trout (*Salmo trutta* L.) and Charr (*Salvelinus alpinus* (L.)) in Scottish lochs. *Journal of Fish Biology*, 14, 1-29.

CAMPBELL, R.N. 1980. Sticklebacks (*Gasterosteus aculeatus* (L.) and *Pungitius pungitius* (L.)) in the Outer Hebrides, Scotland. *Hebridean Naturalist*, 3, 8-15.

CAMPBELL, R.N. 1985. Morphological variation in the Three-spined Stickleback (*Gasterosteus aculeatus*) in Scotland. *Behaviour*, 93, 161-168.

CAMPBELL, R.N., MAITLAND, P.S. & CAMPBELL, R.N.B. 1994. Management of fish populations. Pages 489-513 in: Maitland, P.S., Boon, P.J. & McLusky, D.S. 1994. *The fresh waters of Scotland*. Chichester, Wiley.

CAMPBELL, R.N. & WILLIAMSON, R.B. 1979. The fishes of the inland waters of the Outer Hebrides, Scotland. *Proceedings of the Royal Society of Edinburgh*, 77B, 377-393.

CAMPBELL, R.N. & WILLIAMSON, R.B. 1983. Salmon and freshwater fishes of the Inner Hebrides. *Proceedings of the Royal Society of Edinburgh*, 83B, 245-265.

CAMPBELL, R.N.B. 1982. The food of Arctic Charr in the presence and absence of Brown Trout. *Glasgow Naturalist*, 20, 229-235.

CAMPBELL, R.N.B. 1984. Predation by Arctic Charr on the Three-spined Stickleback and its nest in Loch Meallt, Skye. *Glasgow Naturalist*, 20, 409-413.

CAMPBELL, R.N.B., MAITLAND, P.S. & LYLE, A.A. 1986. Brown trout deformities: an association with acidification? *Ambio*, 15, 244-245.

CARLANDER, K.D. 1977. *Handbook of freshwater fishery biology*. Ames, Iowa State University Press.

CARNELL, D. 1987. Pike baits have fled the loch. *Anglers Mail*, 3 October 1987.

CARR, John. 1809. *Caledonian sketches or a tour through Scotland in 1807*. London, Mathews & Leigh.

CARSS, D.N. 1993. Grey Heron, *Ardea cinerea* L., predation at cage fish farms in Argyll western Scotland. *Aquaculture and Fisheries Management*, 24, 29-45.

CARSS, D.N. & BROCKIE, K. 1994. Prey remains at Osprey nests in Tayside and Grampian, 1987-1993. *Scottish Birds*, 17, 132-145.

CARSS, D.N., ELSTO, D.A., NELSON, K.C. & KRUUK, H. 1999. Spatial and temporal trends in unexploited yellow eel stocks in two shallow lakes and associated streams. *Journal of Fish Biology*, 55, 636-654.

CARSS, D.N., KRUUK, H. & CONROY, J.W.H. 1990. Predation on adult Atlantic salmon, *Salmo salar* L., by otters, *Lutra lutra* (L.), within the River Dee system, Aberdeenshire, Scotland. *Journal of Fish Biology*, 37, 935-944.

CARSS, D.N. & MARQUISS, M. 1992. Avian predation at farmed and natural fisheries. *Proceedings of the Institute of Fisheries Management Annual Study Course*, 22, 179-196.

CHALMERS, P. 1931. *A fisherman's angles*. London, Country Life.

CHALMERS, P. 1938. *The angler's England*. London, Seeley, Service and Co.

CHAPPELL, L.H. 1969. The parasites of the three-spined stickleback *Gasterosteus aculeatus* L. from a Yorkshire pond. I. Seasonal variation of parasite fauna. *Journal of Fish Biology*, 1, 137-152.

CHAPPELL, L.H. 1969. The parasites of the three-spined stickleback *Gasterosteus aculeatus* L. from a Yorkshire pond. II. Variation of the parasite fauna with sex and size of fish. *Journal of Fish Biology*, 1, 339-347.

CHILD, A.R., BURNELL, A.M. & WILKINS, N.P. 1976. The existence of two races of Atlantic salmon (*Salmo salar* L.) in the British Isles. *Journal of Fish Biology*, 8, 35-43.

CHOLMONDELEY-PENNELL, H. 1965. *The book of Pike*. London, Warne.

CHOLMONDELEY-PENNELL, H. 1866. *Fishing Gossip, or stray leaves from the note-books of several anglers*. Edinburgh, Black.

CHOLMONDELEY-PENNELL, H. 1884. *The angler-naturalist*. London, Routledge.

CHRYSTAL, R.A. 1927. *Angling theories and methods*. London, Jenkins.

CIHAR, J. 1991. *A guide to freshwater fish*. London, Treasure Press.

CLARIDGE, P.N. & GARDNER, D.C. 1978. Growth and movements of the Twaite Shad, *Alosa fallax* (Lacepede), in the Severn Estuary. *Journal of Fish Biology*, 12, 203-211.

CLARDIGE, P.N. & POTTER, I.C. 1985. Distribution, abundance and size composition of mullet populations in the Severn Estuary and Bristol Channel. *Journal of the Marine Biological Association of the U.K.*, 65, 325-335.

CLELLAND, B. 1971. An ecological study of a Scottish population of Bullheads. *BSc Thesis, University of Edinburgh*.

COBHAM RESOURCE CONSULTANTS. 1992. *Countryside sports and their economic significance*. Reading, The Standing Conference of Countryside Sports.

COLES, T.F. 1981. The distribution of perch, *Perca fluviatilis* L., throughout their first year of life in Llyn Tegid, North Wales. *Journal of Fish Biology*, 18, 15-21.

COLLEN, P. & MORRISON, B.R.S. 1994. Report on fish and invertebrate sampling in the Kirkton Burn, Balquhidder 1994. *Fisheries Research Services Report*, 6/94.

COLOMBO, G., GRANDI, G. & ROSS, R. 1984. Gonad differentiation and body growth in *Anguilla anguilla* L. *Journal of Fish Biology*, 24, 215-228.

COLQUHOUN, J. 1858. *Salmon casts and stray shots.* Edinburgh, Blackwood.

COLQUHOUN, J. 1866. *Sporting days.* Edinburgh, Blackwood.

COLQUHOUN, J. 1878. *The moor and the loch.* Edinburgh, Blackwood.

CONWAY, J. 1861 *Forays among Salmon and Deer.* Glasgow, Morison.

COOPER, M.J. & WHEATLEY, G.A. 1981. An examination of the fish population in the River Trent, Nottinghamshire, using angler catches. *Journal of Fish Biology*, 19, 539-556.

COPLAND, W.O. 1956. Notes on the food and parasites of Pike (*Esox lucius*) in Loch Lomond. *Glasgow Naturalist*, 17, 230-235.

COPLAND, W.O. 1957. The parasites of Loch Lomond fishes. Pages 128-133 in: Slack, H.D. (Ed.) *Studies on Loch Lomond 1.* Glasgow, Blackie.

COUCH, J. 1865. *A history of the fishes of the British Islands.* London, Bell.

COUNCIL OF EUROPE. 1982. *Categorisation of species on Appendices II and III according to their conservation requirements.* Strasbourg, Council of Europe.

COUNCIL OF EUROPE. 1984. *Inclusion of threatened freshwater fish in Europe in Appendices II and III to the Convention.* Strasbourg, Council of Europe.

COUNCIL OF EUROPE. 1985. *Select Committee of Experts - Freshwater Fish. 1st Meeting.* Strasbourg, Council of Europe.

COURTENAY, W.R. & STAUFFER, J.R. 1984. *Distribution, biology and management of exotic fishes.* Baltimore, John Hopkins University Press.

CRAGG-HINE, D. & JONES, J.W. 1969. The growth of Dace *Leuciscus leuciscus* (L.), Roach *Rutilus rutilus* (L.) and Chub *Squalius cephalus* (L.) in Willow Brook, Northamptonshire. *Journal of Fish Biology*, 1, 59-82.

CRAIG, J.F.1977. Seasonal changes in the day and night activity of adult perch, *Perca fluviatilis* L. *Journal of Fish Biology*, 11, 161-166.

CRAIG, J.F. 1978. A note on ageing in fish with special reference to the perch, *Perca fluviatilis* L. *Verhandlungen der Internationalen Vereinigung für Theoretische und Angewandte Limnology*, 20, 2060-2064.

CRAIG, J.F. 1980. Growth and production of the 1955 to 1972 cohorts of perch, *Perca fluviatilis* L., in Windermere. *Journal of Animal Ecology*, 49, 291-315.

CRAIG, J.F.1982. A note on growth and mortality of trout, *Salmo trutta* L., in afferent streams of Windermere. *Journal of Fish Biology*, 20, 423-430.

CRAIG, J.F. 1987. *The biology of Perch and related fish.* Beckenham, Croom Helm.

CRAIG, J.F. & KIPLING, C. 1983. Reproduction effort versus the environment: case histories of Windermere perch, *Perca fluviatilis* L., and pike, *Esox lucius* L. *Journal of Fish Biology*, 22, 713-727.

CRANDALL, L.A. 1914. *Days in the open.* New York, Revell.

CRISP, D.T., MANN, R.H.K. & McCORMACK, J.C. 1975. The populations of fish in the River Tees system on the Moor House National Nature Reserve, Westmorland. *Journal of Fish Biology*, 7, 573-594.

CRISP, D.T., MANN, R.H.K. & McCORMACK, J.C. 1978. The effects of impoundment and regulation upon the stomach contents of fish at Cow Green, Upper Teesdale. *Journal of Fish Biology*, 12, 287-301.

CRIVELLI, A.J. 1981. The biology of the Common Carp, *Cyprinus carpio* L., in the Camargue, southern France. *Journal of Fish Biology*, 18, 271-290.

CRIVELLI, A.J. & MAITLAND, P.S. 1995. Future prospects for the freshwater fish fauna of the Mediterranean region. *Biological Conservation*, 72, 335-337.

CROSS, D.G. 1969. Aquatic weed control using Grass Carp. *Journal of Fish Biology*, 1, 27-30.

CROSS, D.G. 1970. The tolerance of Grass Carp *Ctenopharyngodon idella* (Val.) to seawater. *Journal of Fish Biology*, 2, 231-233.

CROZIER, W.W. & FERGUSON, A. 1986. Electrophoretic examination of the population structure of brown trout, *Salmo trutta* L., from the Lough Neagh catchment, Northern Ireland. *Journal of Fish Biology*, 28, 459-478.

CUNNINGHAM, J. 1896. *The edible marine fishes of Great Britain.* Macmillan.

CUSHING, D. 1983. *Climate and fisheries.* London, Academic Press.

CUTHBERT, J.H. 1979. Food studies of feral mink *Mustela vison* in Scotland. *Fisheries Management*, 10, 17-25.

DABROWSKI, KR., KAUSHIK, S.J. & LUQUET, P. 1984. Metabolic utilisation of body stores during the early life of whitefish, *Coregonus lavaretus* L. *Journal of Fish Biology*, 23, 721-730.

DANDO, P.R. & DEMIR, N. 1985. On the spawning and nursery grounds of bass, *Dicentrarchus labrax*, in the Plymouth area. *Journal of the Marine Biological Association of the UK*, 65, 159-168.

DANIEL, W.B. 1812. *Rural sports.* London, Longman.

DANKERS, N., WOLFF, W.J. & ZIJLSTRA, J.J. 1979. *Fish and fisheries of the Wadden Sea.* Rotterdam, Balkema.

DARTNELL, H.J.G. 1973. Parasites of the nine-spined stickleback *Pungitius pungitius* (L.). *Journal of Fish Biology*, 5, 505-510.

DARWIN, C.R. 1859. *The origin of species*. London, Dent.

DAVIES, C.E., SHELLEY, J., HARDING, P.T., McLEAN, I.F.G., GARDINER, R. & PEIRSON, G. 2004. *Freshwater fishes in Britain*. London, Harley.

DAVIES, P.M.C. 1963. Food input and energy extraction efficiency in *Carassius auratus*. *Nature, London*, 198, 707.

DAY, F. 1880. *The fishes of Great Britain and Ireland*. London, Williams and Norgate.

DAY, F. 1887. *British and Irish Salmonidae*. London, Williams & Norgate.

DAY, J.W., HALL, C.A.S., KEMP, W.M. & YANEZ-ARANCIBIA, A. 1989. *Estuarine ecology*. New York, Wiley.

DE AYALA, D. P. 1498. In: Brown, P.H. 1891.

DE ROCHFORD, J. 1661. In Brown P.H. 1891.

DE BUNSEN, J.M. 1962. The American Brook Trout in Britain. *Unpublished report*.

DE GROOT, S.J. 1985. Introductions of non-indigenous fish species for release and culture in the Netherlands. Review. *Aquaculture*, 46, 237-257.

DEMBINSKI, W. 1971. Vertical distribution of vendace *Coregonus albula* L. and other pelagic fish species in some Polish lakes. *Journal of Fish Biology*, 3, 341-357.

DENNIS, M.D. 1999. *Scottish heraldry: an invitation*. Edinburgh, Heraldry Society of Scotland.

DENNYS, J. 1613. *The secrets of angling*. 1885 reprint - Edinburgh, private printing.

DEPARTMENT OF AGRICULTURE AND FISHERIES FOR SCOTLAND. 1965. *Scottish salmon and trout fisheries*. Edinburgh, HMSO.

DEPARTMENT OF FISHERIES AND OCEANS. 1986. *Policy for the management of fish habitat*. Ottawa, Department of Fisheries and Oceans.

DEPARTMENT OF THE MARINE. 1993. *Report of the Sea Trout Working Group*. Abbotstown, Fisheries Research Centre.

DEVINE, J.A., ADAMS, C.E. & MAITLAND, P.S. 2000. Changes in reproductive strategy in the ruffe during a period of establishment in a new habitat. *Journal of Fish Biology*, 56, 1488-1496.

DEWAR, G. A. B. 1908. *Life and sport in Hampshire*. London, Longman, Green & Co.

DICKENS, C. 1880. *A dictionary of the Thames*. London, Dickens & Evans.

DOLEZEL, M.C. & CROMPTON, D.W.T. 2000. Platyhelminth infections in powan, *Coregonus lavaretus* (L.), from Loch Eck and Loch Lomond, Scotland. *Helminthology*, 37, 147-152.

DONOVAN, E. 1808. *The natural history of British fishes*. London, Donovan.

DOTTRENS, E. 1958. Sur les Coregones de Grande-Bretagne et d'Irlande. *International Congress of Zoology*, 15, 404-406.

DOTTRENS, E. 1959. Systematique des coregones d'Europe occidentale, basee sur une etude biometrique. *Revue Suisse Zoologie*, 66, 1-66.

DOUGHTY, C.R., BOON, P.J. & MAITLAND, P.S. 2002. The state of Scotland's fresh waters. Pages 117-144 in: Usher, M.B., Mackey, E.C. & CURRAN, J.C. (Eds.) *The state of Scotland's environment and natural heritage*. Edinburgh, HMSO.

DOUGHTY, C.R. & MAITLAND, P.S. 1994. The ecology of the River Endrick: present status and changes since 1960. *Hydrobiologia*, 290, 139-151.

DRUMMOND, A. 1803. On the natural history of the Salmon. *Transactions of the Highland Society*, 2, 394-411.

DRYDEN, A. 1862. *Hints to anglers*. Edinburgh.

DUTT, W.A. 1906. *Wildlife in East Anglia*. London, Methuen.

EGGLISHAW, H.J. 1967. The food, growth and population structure of Salmon and Trout in two streams in the Scottish highlands. *Freshwater and Salmon Fisheries Research, Scotland*, 38, 1-32.

EGGLISHAW, H.J. 1970. Production of salmon and trout in a stream in Scotland. *Journal of Fish Biology*, 2, 117-136.

EGGLISHAW, H.J. & SHACKLEY, P.E. 1977. Growth, survival and production of juvenile Salmon and Trout in a Scottish stream, 1966-75. *Journal of Fish Biology*, 11, 647-672.

EGGLISHAW, H.J. & SHACKLEY, P.E. 1985. Factors governing the production of juvenile Salmon in Scottish streams. *Journal of Fish Biology*, 27A, 27-33.

ELDER, H.Y. 1965. Biological effects of water utilisation by hydro-electric schemes in relation to fisheries, with special reference to Scotland. *Proceedings of the Royal Society of Edinburgh*, 69B, 246-271.

ELLIOT, G.F.S., LAURIE, M. & MURDOCH, J.B. (Eds.) 1901. *Fauna, flora and geology of the Clyde area*. Glasgow, British Association.

ELLIOTT, J.M. 1985. Population dynamics of migratory trout, *Salmo trutta*, in a Lake District stream, 1966-83, and their implications for fishery management. *Journal of Fish Biology*, 27, 35-43.

ELLIOTT, J.M. 1994. *Quantitative ecology and the brown trout*. Oxford, Oxford University Press.

ELLISON, N.F. 1966. Notes on Lakeland Schelly. *The Changing Scene*, 3, 1-8.

ELLISON, N.F. 1966. The catastrophe to the Schellies of Ullswater in January 1966. *Field Naturalist*, 11, 1.

ELLISON, N.F. & CHUBB, J.C. 1968. The smelt of Rostherne Mere, Cheshire. *Lancashire and Cheshire Fauna Society*, 53, 7-16.

ELLISON, N.F. & COOPER, J.R. 1967. Further notes on Lakeland schelly. *Field Naturalist*, 12.

ELTON, C.E. 1927. *Animal ecology*. London, Sidgwick & Jackson.

ERMAN, F. 1961. On the biology of the Thick-lipped Mullet (*Mugil chelo*). *Conseil Permanent International pour l'Exploration Scientifique de la. Mer*, 16, 277-285.

EVANS, W. 1906. Fauna of the Upper Elf Loch. *Annals of Scottish Natural History*, 1906, 57.

EVANS, A.H. & BUCKLEY, T.E. 1899. *A vertebrate fauna of the Shetland Isles*. Edinburgh, Douglas.

FABRICIUS, E. & GUSTAVSON, K.J. 1955. Observations on the spawning behaviour of the grayling, *Thymallus thymallus* (L.). *Annual Report of the Institute of Freshwater Research, Drottningholm*, 36, 75-103.

FABRICIUS, E. & GUSTAVSON, K.J. 1958. Some new observations on the spawning behaviour of the Pike, *Esox lucius* L. *Annual Report of the Institute of Freshwater Research, Drottningholm*, 39, 23-54.

FAHY, E. 1977. Characteristics of the freshwater occurrence of sea trout *Salmo trutta* in Ireland. *Journal of Fish Biology*, 11, 635-646.

FAHY, E. 1978 Variations in some biological characteristics of British sea trout *Salmo trutta* L. *Journal of Fish Biology*, 13, 123-138.

FAIRFAX, D. 1998. *The Basking Shark in Scotland*. East Linton, Tuckwell Press.

FAO. 1981. *Conservation of the genetic resources of fish: problems and recommendations*. Rome, FAO.

FAO. 1982. *Allocation of fishery resources*. Rome, FAO

FARRUGIO, H. 1977. Annotated key for determination of Mugilidae adults and alevins of Tunisia. *Cybium*, 3, 57-74.

FERGUSON, A. 1974. The genetic relationships of the Coregonid fishes of Britain and Ireland indicated by electrophoretic analysis of tissue proteins. *Journal of Fish Biology*, 6, 311-315.

FERGUSON, A. 1989. Genetic differences among brown trout, *Salmo trutta*, stocks and their importance for the conservation and management of the species. *Freshwater Biology*, 21, 35-46.

FERGUSON, A., HIMBERG, K.J.M. & SVARDSON, G. 1978. The systematics of the Irish Pollan (*Coregonus pollan* Thompson): an electrophoretic comparison with other Holarctic Coregoninae. *Journal of Fish Biology*, 12, 221-233.

FERGUSON, A. & MASON, F.M. 1981. Allozyne evidence for reproductively isolated sympatric populations of brown trout *Salmo trutta* L. in Lough Melvin, Ireland. *Journal of Fish Biology*, 18, 629-642.

FICKLING, N.J. 1982. The identification of Pike by means of characteristic marks. *Fisheries Management*, 13, 79-82.

FITTER, R.S.R. 1959. *The ark in our midst*. London, Collins.

FITZGIBBON, E. 1847. *A handbook of angling*. London, Longman.

FITZMAURICE, P. 1983. Carp (*Cyprinus carpio* L.) in Ireland. *Irish Fisheries Investigations*, 23A, 5-10.

FITZMAURICE, P. 1983. Some aspects of the biology and management of Pike (*Esox lucius* L.) stocks in Irish fisheries. *Journal of Life Sciences, Royal Dublin Society*, 4, 161-173.

FLOWERDEW, M.W. & GROVE, D.J. 1980. An energy budget for juvenile thick-lipped mullet, *Crenimugil labrosus* (Risso). *Journal of Fish Biology*, 17, 395-410.

FORESTRY COMMISSION. 1993. *Forests and water guidelines*. Edinburgh, Forestry Commission.

FORTH RIVER PURIFICATION BOARD. 1989. *Annual Report 1989*. Edinburgh, Forth River Purification Board.

FOUDA, M.M. 1979. Studies on scale structure in the common goby *Pomatoschistus microps* Kroyer. *Journal of Fish Biology*, 15, 165-172.

FOUDA, M.M. & MILLER, P.J. 1979. Alkaline phosphatase activity in the skin of the Common Goby, *Pomatoschistus microps*, in relation to cycles in scale and body growth. *Journal of Fish Biology*, 15, 263-274.

FOX, P.J. 1978. Preliminary observations on different reproduction strategies in the bullhead (*Cottus gobio* L.) in northern and southern England. *Journal of Fish Biology*, 12, 5-11.

FRANCIS, F. 1867. *A book on angling: being a complete treatise on the art of angling in every branch*. London, Longmans.

FRANCIS, F. 1874. *By lake and river; an angler's rambles in the north of England and Scotland*. London, The Field.

FRANCK, R. 1694. *Northern memoirs, calculated for the meridian of Scotland*. London, Mortlock.

FRASER, A. 1833. *Natural history of the Salmon, Herrings, Cod, etc.* Inverness, Carruthers.

FRASER, I. & RITCHIE, J.N.G. 1999. *Pictish symbol stones: an illustrated gazetteer*. Edinburgh, The Royal Commission on the Ancient and Historical Monuments of Scotland.

FRASER, J. 1634-1699. *Policratica Temporum* (from the Wardlaw Manuscript). London.

FRASER, P.J. 1987. Atlantic salmon feed in Scottish coastal waters. *Aquaculture and Fisheries Management*, 18, 243-247.

FRESHWATER FISHERIES LABORATORY. 1992. Salmon population studies. *Annual Review*, 1991-1992, 12-22.

FRESHWATER FISHERIES LABORATORY. 1994. Rudd from a loch in Galloway. *Annual Review*, 1993-1994, 35.

FRIEND, G.F. 1941. The life history and ecology of the salmon gill maggot *Salmincola salmonea* (L.) (Copepod crustacean). *Transactions of the Royal Society of Edinburgh*, 60, 503-541.

FRIEND, G.F. 1956. A new subspecies of char from Loch Eck. *Glasgow Naturalist*, 17, 219-220.

FROST, W.E. 1940. Rainbows of a peat lough on Arranmore. *Salmon and Trout Magazine*, 100, 234-240.

FROST, W.E. 1943. The natural history of the Minnow *Phoxinus phoxinus*. *Journal of Animal Ecology*, 12, 139-162.

FROST, W.E. 1946. Observations on the food of eels (*Anguilla anguilla*) from the Windermere catchment area. *Journal of Animal Ecology*, 15, 43-53.

FROST, W.E. 1950. The growth and food of young salmon (*Salmo salar*) and trout (*S. trutta*) in the River Forss, Caithness. *Journal of Animal Ecology*, 19, 147-158.

FROST, W.E. 1951. Some observations on the biology of the charr, *Salvelinus willughbii* Gunther, of Windermere. *Verhandlungen der Internationalen Vereinigung für Theoretische und Angewandte Limnology*, 11, 105-110.

FROST, W.E. 1952. Predators on the eggs of char in Windermere. *Salmon and Trout Magazine*, 136, 192-196.

FROST, W.E. 1965. Breeding habits of Windermere charr, *Salvelinus willughbii* Gunther, and their bearing on the speciation of these fish. *Proceedings of the Royal Society of London*, 163B, 232-284.

FROST, W.E. 1974. *A survey of the Rainbow Trout (Salmo gairdneri) in Britain and Ireland*. London, Salmon & Trout Association.

FROST, W.E. 1977. The food of charr, *Salvelinus willughbii* (Gunther) in Windermere. *Journal of Fish Biology*, 11, 531-548.

FROST, W.E. & BROWN, M.E. 1967. *The Trout*. Collins, London.

FROST, W.E. & KIPLING, C. 1968. Experiments on the effect of temperature on the growth of young Pike, *Esox lucius* L. *Salmon and Trout Magazine*, 184, 170-178.

FROST, W.E. & KIPLING, C. 1980. The growth of charr, *Salvelinus willughbii*, Gunther, in Windermere. *Journal of Fish Biology*, 16, 279-290.,

FULLER, J.D., SCOTT, D.B.C. & FRASER, R. 1975. Effects of catching techniques, captivity and reproductive cycle on plasma cortisol concentration in the powan, *Coregonus lavaretus*, a freshwater teleost from Loch Lomond. *Journal of Endocrinology*, 63, 1-24.

FULLER, J.D., SCOTT, D.B.C. & FRASER, R. 1976. The reproductive cycle of powan, *Coregonus lavaretus* (L.), in Loch Lomond, in relation to seasonal changes in plasma cortisol concentration. *Journal of Fish Biology*, 9, 105-117.

FURNEAUX, W. 1896. *Life in ponds and streams*. London, Longmans.

GARDINER, R. 1974. An electrophoretic method for distinguishing the young fry of Salmon *Salmo salar* (L.) from those of trout *Salmo trutta* (L.). *Journal of Fish Biology*, 6, 517-519.

GARDINER, R. 1989. The distribution of grayling. Pages 17-24 in: Broughton, R.B. (Ed.) *Grayling the fourth game fish*. Marlborough, Crowood Press.

GARDINER, R. 1992. Scottish grayling: history and biology of the populations. *Proceedings of the Institute of Fisheries Management Annual Study Course*, 22, 171-178.

GARDINER, R. 1993. Scottish grayling: history and biology of the populations. *Journal of the Grayling Society*, 1993, 43-50.

GARDINER, R. & EGGLISHAW, H. 1986. *A map of the distribution in Scottish rivers of the Atlantic salmon (Salmo salar L.)*. Edinburgh, Department of Agriculture and Fisheries for Scotland.

GARDNER, A.S., WALKER, A.F. & GREER, R.B. 1988. Morphometric analysis of two ecologically distinct forms of Arctic Charr, *Salvelinus alpinus* (L.), in Loch Rannoch, Scotland. *Journal of Fish Biology*, 32, 901-910.

GARDNER, M.L.G. 1971. Recent changes in the movements of adult Salmon in the Tay-Tummel-Garry system, Scotland. *Journal of Fish Biology*, 3, 83-96.

GARDNER, M.L.G. 1976. A review of factors which may influence the sea-age and maturation of Atlantic Salmon *Salmo salar* L. *Journal of Fish Biology*, 9, 289-327.

GASOWSKA, M. 1965. A preliminary taxonomic revision of the British whitefish with special reference to the gwyniad of Llyn Tegid (Bala). *Proceedings of the Zoological Society of London*, 145, 1-8.

GATHORNE-HARDY, A.E. 1901. *Autumns in Argyleshire with rod and gun*. London, Longmans, Green & Co.

GEE, A.S., MILNER, N.J. & HEMSWORTH, R.J. 1978. The production of juvenile salmon, *Salmo salar* in the upper Wye, Wales. *Journal of Fish Biology*, 13, 439-451.

GEMMELL, H. 1962. Miller's Thumb Bullhead in Renfrewshire. *Glasgow Naturalist*, 18, 213-214.

GEORGE, D.G. & WINFIELD, I.J. 2000. Factors influencing the spatial distribution of zooplankton and fish in Loch Ness, UK. *Freshwater Biology*, 44, 557-570.

GERASIMOV, N. 1988. How Russia saved the Volga Sturgeon. *Fish Farming International*, 1988, 24-25.

GERRISH, C.S. 1939. Scales of Avon trout and grayling. *Report of Avon Biological Research*, 6, 54-59.

GERRISH, C.S. 1935. Hatchery stock and trout streams. *Salmon and Trout Magazine*, 81, 331-344.

GERVERS, F.W.K. 1954. A supernumerary pelvic fin in the Powan (*Coregonus clupeoides* Lacepede). *Nature, London*, 174, 935.

GIBBINGS, R. 1941. *Sweet Thames run softly*. London, Dent.

GIBSON, D.L. 1972. Flounder parasites as biological tags. *Journal of Fish Biology*, 4, 1-10.

GIBSON, J.A. & MITCHELL, J. 1986. *An atlas of Loch Lomond vertebrates*. Glasgow, Scottish Wildlife Trust.

GIBSON, J.A. 1976. Notes on the status of the Common Goby in the Clyde Sea area. *Transactions of the Buteshire Natural History Society*, 20, 76-77.

GIBSON, J.A. 1984. *An atlas of Ayrshire vertebrates*. Kilmarnock, Ayrshire Biological Records Centre.

GILES, N. & TIPPETT, R. 1987. Annual migration and diet of perch *Perca fluviatilis* L. in Loch Lomond. *Glasgow Naturalist*, 21, 287-295.

GILES, N. 1981. Summer diet of the grey heron. *Scottish Birds*, 11, 153-159.

GILES, N. 1983. The possible role of environmental calcium levels during the evolution of phenotypic diversity in Outer Hebridean populations of Three-spined Sticklebacks, *Gasterosteus aculeatus*. *Journal of Zoology, London*, 199, 535-545.

GILES, N. 1994. *Freshwater fish of the British Isles*. Shrewsbury, Swan Hill.

GLADSTONE, H.S. 1912. *A catalogue of the vertebrate fauna of Dumfriesshire*. Dumfries.

GOLDSPINK, C.R. 1978. Comparative observations on the growth rate and year class strength of Roach *Rutilus rutilus* L. in two Cheshire lakes, England. *Journal of Fish Biology*, 12, 421-433.

GOLDSPINK, C.R. 1981. A note on the growth-rate and year-class strength of Bream, *Abramis brama* L., in three eutrophic lakes, England. *Journal of Fish Biology*, 19, 665-674.

GOLDSPINK, C.R. 1990. The distribution and abundance of young (I+-II+) perch *Perca fluviatilis* L., in a deep eutrophic lake, England. *Journal of Fish Biology*, 36, 439-448.

GOLDSPINK, C.R. & BANKS, J.W. 1971. A readily recognisable tag for marking bream *Abramis brama* (L.). *Journal of Fish Biology*, 3, 407-411.

GOLDSPINK, C.R. & GOODWIN, D.A. 1979. A note on age composition, growth rate and food of perch, *Perca fluviatilis* (L.), in four eutrophic lakes, England. *Journal of Fish Biology*, 14, 489-505.

GOODE, G.B. 1884. *The fisheries and fishing industries of the United States*. Washington, U.S. Commission of Fish and Fisheries.

GORDON, G. 1852. A list of the fishes that have been found in the Moray Firth and in the fresh waters of the province of Moray. *Zoologist*, 1852, 3454-3462; 3480-3489.

GORDON, J.G. 1921. The marine and freshwater fishes of Wigtownshire. *Transactions of the Dumfriesshire and Galloway Natural History and Antiquarian Society*, 7, 137-159.

GORDON, S. 1920. *The land of the hills and the glens: wild life in Iona and the Inner Hebrides*. London, Cassell.

GORDON, S. 1944. *A highland year*. London, Eyre & Spottiswoode.

GORDON, W.J. 1902. *Our country's fishes and how to know them*. London, Simpkin, Marshall, Hamilton, Kent.

GRAHAM, W. 1865. *Lochmaben five hundred years ago*. Edinburgh, Nimmo.

GRANT, A.T., DUGUID, R.A. & ADAMS, C.E. 1997. The reappearance of tench (*Tinca tinca*) in the waters of Loch Lomond. *Glasgow Naturalist*, 23, 59-61.

GRANT, E. 1812. *Memoirs of a highland lady*. London, Murray.

GRANT, J. 1792. Parish of Luggan; Inverness. *Statistical Account of Scotland*, 3, 146.

GRAY, J.M. 1892. *Memoirs of the life of Sir John Clerk of Penicuik, Baronet*. Edinburgh, Constable.

GRAY, R. 1864. Quadrupeds, birds and fishes of Loch Lomond and its vicinity. In: W. Keddie (Ed.) *Tourists' Guide to the Trossachs, Loch Lomond, etc.* Glasgow, McLure & McDonald.

GREENWOOD, M.F.D. & METCALFE, N.B. 1998. Minnows become nocturnal at low temperatures. *Journal of Fish Biology*, 53, 25-32.

GREER, R. 1995. *Ferox Trout and Arctic Charr*. Shrewsbury, Swan Hill Press.

GREY, V.F. 1899. *Fly fishing*. London.

GREY, Z. 1928. *Tales of fresh-water fishing*. London, Hodder & Stoughton.

GRIMBLE, A. 1902. *The salmon rivers of Scotland*. London, Kegan Paul.

GUNN, C. B. 1880. *Leaves from the life of a country doctor*. Newburgh.

GUNN, I. 1981. *With a rod in four continents*. Golspie, Method Publishing.

GUNN, N.M. 1931. *Morning tide*. Edinburgh, Porpoise Press.

GUNN, N.M. 1937. *Highland river*. Glasgow, Porpoise Press.

GUNTHER, A. 1880. *An introduction to the study of fishes*. Edinburgh, Black.

HALL, C.A. 1915. Renfrewshire fresh-water fishes. *Transactions of the Paisley Naturalists' Society*, II, 61-65.

HAMILTON, J.D. 1988. Recent human influences on the ecology of Loch Lomond, Scotland. *Verhandlungen der Internationalen Vereinigung für Theoretische und Angewandte Limnology*, 23, 403-413.

HAMILTON, K.E., FERGUSON, A., TAGGART, J.B., TOMASSON, T., WALKER, A. & FAHY, E. 1989. Post-glacial colonisation of brown trout, *Salmo trutta* L.: Ldh-5 as a phylogeographic marker locus. *Journal of Fish Biology*, 35, 651-664.

HAMILTON, R. 1876. *A history of British fishes.* London, Allen.

HAMMERTON, D. 1994. Domestic and industrial Pollution. Pages 347-364 in: Maitland, P.S., Boon, P.J. & McLusky, D.S. (Eds.). *The fresh waters of Scotland.* Chichester, Wiley.

HANCOCK, R.S., JONES, J.W. & SHAW, R. 1976. A preliminary report on the spawning behaviour and nature of sexual selection in the barbel, *Barbus barbus* (L.). *Journal of Fish Biology*, 9, 21-28.

HANSEN, L.P. & PETHON, P. 1985. The food of Atlantic salmon, *Salmo salar* L., caught by long-line in northern Norwegian waters. *Journal of Fish Biology*, 26, 553-562.

HANSEN, L.P., DOVING, K.B. & JONSSON, B. 1987. Migration of adult Atlantic Salmon, with and without olfactory sense, released on the Norwegian coast. *Journal of Fish Biology*, 30, 713-730.

HARAM, O.J. 1965. Tracing a lake's drifting shoals. *Fishing*, 119, 5-8.

HARAM, O.J. & JONES, J.W. 1971. Some observations on the food of the gwyniad, *Coregonus pennantii* Valenciennes of Llyn Tegid (Lake Bala), North Wales. *Journal of Fish Biology*, 3, 287-295.

HARAM, O.J & PEARSON, R.G. 1967. The distribution of some fresh-water biota in Fennoscandinavia. *Archiv fur Hydrobiologie*, 63, 135-142.

HARDEN-JONES, F.R. 1968. *Fish migration.* London, Arnold.

HARDIE, R.P. 1940. *Ferox and Char in the lochs of Scotland.* Edinburgh, Oliver & Boyd.

HARDISTY, M.W. 1944. The life-history and growth of the Brook Lamprey (*Lampetra planeri*). *Journal of Animal Ecology*, 13, 110-122.

HARDISTY, M.W. 1961. The growth of larval lampreys. *Journal of Animal Ecology*, 30, 357-371.

HARDISTY, M.W. 1961. Studies on an isolated spawning population of the brook lamprey (*Lampetra planeri*). *Journal of Animal Ecology*, 30, 339-355.

HARDISTY, M.W. 1961. Oocyte number as a diagnostic character for the identification of ammocoete species. *Nature, London*, 191, 1215-1216.

HARDISTY, M.W. 1964. The fecundity of lampreys. *Archiv fur Hydrobiologie*, 60, 340-357.

HARDISTY, M.W. 1969. Information on the growth of the ammocoete larvae of the anadromous Sea Lamprey *Petromyzon marinus* in British rivers. *Journal of Zoology, London*, 159: 139-144.

HARDISTY, M.W. 1970. The relationship of gonadal development to the life cycles of the paired species of lamprey, *Lampetra fluviatilis* (L.) and *Lampetra planeri* (Bloch). *Journal of Fish Biology*, 2, 173-181.

HARDISTY, M.W. & HUGGINS, R.L. 1970. Larval growth in the river lamprey, *Lampetra fluviatilis*. *Journal of Zoology, London*, 161, 549-559.

HARDISTY, M W. & POTTER, I.C. (Eds.) 1971. *The biology of lampreys.* London, Academic Press.

HARDISTY, M.W., POTTER, I.C. & STURGE, R. 1970. A comparison of the metamorphosing and macrophthalmia stages in the lampreys *Lampetra fluviatilis* and *L. planeri*. *Journal of Zoology, London*, 162, 383-400.

HARDY, E. 1977. From a naturalist's notebook. *The Aquarist*, October, 294-298.

HARKNESS, W.J.K. & DYMOND, J.R. 1961. The Lake Sturgeon. The history of its fishery and problems of conservation. *Ontario Department of Lands & Forests: Fish and Wildlife Branch*. 1, 1-121.

HARMAN, I. 1950. *Fishponds and aquariums.* London, Sidgwick & Jackson.

HARRIMAN, R. & MILLER, J.D. 1996. *The effects of forestry practices on water quality and biota in the Balquhidder catchments 1983-1993.* Marlow, Foundation for Water Research.

HARRIMAN, R. & MORRISON, B.R.S. 1980. Ecology of acid streams draining forested and non-forested catchments in Scotland. Pages 312-313 in: D. Drablos, & A. Tollan (Eds.) *Ecological impact of acid precipitation.* Sandefjord, Norway, SNSF Project, Oslo - As.

HARRIMAN, R. & MORRISON, B.R.S. 1981. Forestry, fisheries and acid rain in Scotland. *Scottish Forestry*, 35, 89-95.

HARRIMAN, R. & MORRISON, B.R.S. 1982. The ecology of streams draining forested and non-forested catchments in an area of central Scotland subject to acid precipitation. *Hydrobiologia*, 88, 251-263.

HARRIMAN, R., MORRISON, B.R.S., CAINES, L.A., COLLEN, P. & WATT, A.W. 1987. Long term changes in fish populations of acid streams and lochs in Galloway, south west Scotland. *Water, Air and Soil Pollution*, 32, 89-112.

CASS, A.R.H. 1940. *Sea-fishing from the shore.* London, Jenkins.

HARTLEY, P.H.T. 1947. The natural history of some British freshwater fishes. *Proceedings of the Zoological Society of London*, 117, 129-206.

HARTLEY, P.H.T. 1947. The coarse fishes of Britain. *Scientific Publications of the Freshwater Biological Association*, 12, 1-40.

HARTLEY, P.H.T. 1948. Food and feeding in a community of freshwater fishes. *Journal of Animal Ecology*, 17, 1-14.

HARTLEY, S.E. 1990. Chromosome and constitutive heterochromatin distribution in Arctic charr, *Salvelinus alpinus* (L.) (Pisces: Salmonidae). *Genetica*, 79, 161-166.

HARTLEY, S.E. 1990. Variation in cellular DNA content in Arctic charr, *Salvelinus alpinus* (L.) *Journal of Fish Biology*, 37, 189-190.

HARTLEY, S.E. 1995. Mitochondrial DNA analysis distinguishes between British populations of whitefishes. *Journal of Fish Biology*, 47A, 145-155.

HARTLEY, S.E., BARLETT, S.E. & DAVIDSON, W.S. 1992. Mitochondrial DNA analysis of Scottish populations of Arctic charr, *Salvelinus alpinus* (L.). *Journal of Fish Biology*, 40, 219-224.

HARTLEY, S.E., McGOWAN, C., GREER, R.B. & WALKER, A.F. 1992. The genetics of sympatric Arctic charr (*Salvelinus alpinus* (L.)) populations from Loch Rannoch, Scotland. *Journal of Fish Biology*, 41, 1021-1032.

HARVIE-BROWN, J.A. & BUCKLEY, T.E. 1887. *A vertebrate fauna of Sutherland, Caithness and West Cromarty*. Edinburgh, Douglas.

HARVIE-BROWN, J.A. & BUCKLEY, T.E. 1889. *A vertebrate fauna of the Outer Hebrides*. Edinburgh, Douglas.

HARVIE-BROWN, J.A. & BUCKLEY, T.E. 1891. *A vertebrate fauna of the Orkney Islands*. Edinburgh, Douglas.

HARVIE-BROWN, J.A. & BUCKLEY, T.E. 1892. *A fauna of Argyll and the Inner Hebrides*. Edinburgh, Douglas.

HARVIE-BROWN, J.A. & BUCKLEY, T.E. 1896. *A fauna of the Moray basin*. Edinburgh, Douglas.

HARVIE-BROWN, J.A. & BUCKLEY, T.E. 1896. *A fauna of the north-west highlands and Skye*. Edinburgh, Douglas.

HARVIE-BROWN, J.A. 1891. The Common Seal in fresh-water streams and lochs of Scotland. *Scottish Naturalist*, 1891, 1-5.

HAWKINS, A.D. & JOHNSTONE, A.D.F. 1978. The hearing of the Atlantic Salmon, *Salmo salar*. *Journal of Fish Biology*, 13, 655-673.

HEALEY, M.C. (1972). On the population ecology of the Common Goby in the Ythan Estuary. *Journal of Natural History*, 6, 133-145.

HEALY, J.A. & MULCAHY, M.F. 1980. A biochemical genetic analysis of populations of the Northern Pike, *Esox lucius* L., from Europe and North America. *Journal of Fish Biology*, 17, 317-324.

HELLAWELL, J.M. 1969. Age determination and growth of the grayling *Thymallus thymallus* (L.) of the River Lugg, Herefordshire. *Journal of Fish Biology*, 1, 373-382.

HELLAWELL, J.M. 1971. The food of the grayling *Thymallus thymallus* (L.) of the River Lugg, Herefordshire. *Journal of Fish Biology*, 3, 187-197.

HELLAWELL, J.M. 1971. The autecology of the chub, *Squalius cephalus* (L.), of the River Lugg and the Afon Llynfi. I. Age determination, population structure and growth. *Freshwater Biology*, 1, 29-60.

HELLAWELL, J.M. 1971. The autecology of the chub, *Squalius cephalus* (L.), of the River Lugg and the Afon Llynfi. II. Reproduction. *Freshwater Biology*, 1, 135-148.

HELLAWELL, J.M. 1971. The autecology of the chub, *Squalius cephalus* (L.), of the River Lugg and the Afon Llynfi. III. Diet and feeding habits. *Freshwater Biology*, 1, 369-387.

HELLAWELL, J.M. 1972. The growth, reproduction and food of the roach *Rutilus rutilus* (L.), of the River Lugg, Herefordshire. *Journal of Fish Biology*, 4, 469-486.

HELLAWELL, J.M. 1974. The ecology of populations of Dace, *Leuciscus leuciscus* (L.), from two tributaries of the River Wye, Herefordshire, England. *Freshwater Biology*, 4, 577-604.

HEMS, J. 1977. A garden pond. *The Aquarist*. June, 98-104.

HENDERSON, A.R. & HAMILTON, J.D. 1986. The status of fish populations in the Clyde Estuary. *Proceedings of the Royal Society of Edinburgh*, 90B, 157-170.

HENDERSON, A.R. 1987. Fish studies in the upper Clyde Estuary October 1982-November 1983. *Clyde River Purification Board Technical Report*. 80.

HENDRY, H. 1897. *Burns from Heaven with some other poems*. Glasgow, Bryce.

HERVEY, G.F. & HEMS, J. 1968. *The goldfish*. London, Fabre and Fabre.

HEWITT, E.R. 1925. *Secrets of the salmon*. New York, Scribner.

HEWITT, E.R. 1926. *Telling on the trout*. London, Scribner.

HICKLEY, P. & BAILEY, R.G. 1982. Observations on the growth and production of Chub *Leuciscus cephalus* and Dace *Leuciscus leuciscus* in a small lowland river in southeast England, UK. *Freshwater Biology*, 12, 167-178.

HICKLING, C.F. 1970. A contribution to the natural history of the English grey mullets, Pisces Mugilidae. *Journal of the Marine Biological Association of the UK*, 50, 609-633.

HICKS, J. 1855. *Wanderings by the lochs and streams of Assynt and north highlands of Scotland*. London.

HIDB. 1983. *The Atlantic Salmon and its economic importance*. Inverness, Highlands and Islands Development Board.

HIDB. 1985. *The Atlantic Salmon*. Inverness, Highlands and Islands Development Board.

HMSO. 1882. *Report of the Solway White Fishery Commission*. HMSO Cd 6789.

HOLDEN, M.J. & WILLIAMS, T. 1974. The biology, molvements and population dynamics of bass *Dicentrarchus labrax* in English waters. *Journal of the Marine Biological Association of the UK*, 54, 91-107.

HOLDEN, A.V. & MARSDEN, K. 1964. Cyanide in Salmon and Brown Trout. *Freshwater and Salmon Fisheries Research, Scotland*, 33, 1-12.

HOLMES, P.F. 1960. The brown trout of Malham Tarn, Yorkshire. *Salmon and Trout Magazine*, 159, 127-145.

HOLOPAINEN, I.J., AHO, J., VORNANEN, M. & HUUSKONEN, H. 1997. Phenotypic plasticity and predator effects on morphology and physiology of crucian carp (*Carassius carassius* (L.)) in nature and in the laboratory. *Journal of Fish Biology*, 50, 781-798.

HOLOPAINEN, I.J., HYVARINEN, H. & PIIRONEN, J. 1986. Anaerobic wintering of crucian carp (*Carassius carassius* L.) - II. Metabolic products. *Comparative Biochemistry and Physiology*, 83A, 239-242.

HOLOPAINEN, I.J., TONN, W.M. & PASZKOWSKI, C.A. 1997. Tales of two fish: the dichotomous biology of crucian carp (*Carassius carassius* (L.)) in northern Europe. *Annales Zoologici Fennici*, 34, 1-22.

HOLT, S.J. & TALBOT, L.M. 1978. New principles for the conservation of wild living resources. *Wildlife Monographs*, 59, 1-33.

HORNELL, J. 1934. Our coastal finny life. Pages 485-494 in: Crossland, J.R. & Parrish, J.M. (Eds.) *Britain's wonderland of nature*. Glasgow, Collins

HOUGHTON, W. 1879. *British freshwater fishes*. London, Mackenzie.

HOWARD, F.G., McKAY, D.W. & NEWTON, A.W. 1987. Fisheries of the Forth, Scotland. *Proceedings of the Royal Society of Edinburgh*, 93B, 479-494.

HOWELL, D.L. 1994. Role of environmental agencies. Pages 577-611 in: Maitland, P.S., Boon, P.J. & McLusky, D.S. (Eds.) 1994. *The fresh waters of Scotland*. Chichester, Wiley.

HRUSKA, V. 1961. An attempt at a direct investigation of the influence of the carp stock on the bottom fauna of two ponds. *Verhandlungen der Internationalen Vereinigung für Theoretische und Angewandte Limnology*, 14, 732-736.

HUGGINS, R.J. & THOMPSON, A. 1970. Communal spawning of brook and river lampreys, *Lampetra planeri* Bloch and *Lampetra fluviatilis* L. *Journal of Fish Biology*, 2, 53-54.

HUGHES, T. 1898. *Tom Brown's school days*. Glasgow, Collins

HUNT, P.C. 1972. A brief assessment of the rainbow trout in Britain. *Fisheries Management*, 2, 52-55.

HUNT, P.C. & JONES, J.W. 1972. The food of the Brown Trout in Llyn Alaw, Anglesey, North Wales. *Journal of Fish Biology*, 4, 333-352.

HUNT, P.C. & JONES, J.W. 1974. A population study of barbel *Barbus barbus* (L.) in the River Severn, England. I. Densities. *Journal of Fish Biology*, 6, 255-267.

HUNT, P.C. & JONES, J.W. 1974. A population study of barbel *Barbus barbus* (L.) in the River Severn, England. II. Movements. *Journal of Fish Biology*, 6, 269-278.

HUNT, P.C. & JONES, J.W. 1975. A population study of barbel *Barbus barbus* (L.) in the River Severn,

England. III. Growth. *Journal of Fish Biology*, 7, 361-376.

HUNT, P.C. & O'HARA, K. 1973. Overwinter feeding in rainbow trout. *Journal of Fish Biology*, 5, 277-280.

HUNTER, J. 1965. *Scottish salmon and trout fisheries*. Edinburgh, HMSO.

HUNTER, W.R., SLACK, H.D. & HUNTER, M.R. 1959. The lower vertebrates of the Loch Lomond district. *Glasgow Naturalist*, 18, 84-90.

HUSSEIN, S.A. & MILLS, D.H. 1982. The prevalence of 'cauliflower' disease of the eel, *Anguilla anguilla* L., in tributaries of the River Tweed, Scotland. *Journal of Fish Diseases*. 5, 161-165.

HUSSEIN, S.A. 1981. The population density, growth and food of eels *Anguilla anguilla* L. in some tributaries of the River Tweed. *Proceedings of the British Freshwater Fish Conference*, 2, 120-128.

HUSSEIN, S.A. 1983. The biology of the freshwater eel (*Anguilla anguilla* L.) in four tributaries of the River Tweed, Scotland. *PhD Thesis, University of Edinburgh*.

HUTCHINSON, H.G. 1907. *Nature's moods and tenses*. London, Smith, Elder & Co.

HUTCHINSON, P. 1983. A note recording the occurrence of hermaphrodite Smelt, *Osmerus eperlanus* (L.) from the River Thames, England. *Journal of Fish Biology*, 23, 241-244.

HUTCHINSON, P. 1983. Some ecological aspects of the Smelt, *Osmerus eperlanus* (L.), from the River Cree, southwest Scotland. *Proceedings of the British Coarse Fish Conference*, 3, 178-191.

HUTCHINSON, P. 1983. The ecology of smelt, *Osmerus eperlanus* (L.), from the River Thames and the River Cree. *PhD Thesis, University of Edinburgh*.

HUTCHINSON, P. & MILLS, D.H. 1987. Characteristics of spawning-run smelt, *Osmerus eperlanus* (L.), from a Scottish river, with recommendations for their conservation and management, *Aquaculture and Fisheries Management*, 18, 249-258.

HUTCHINSON, S. & HAWKINS, L.E. 1993. The migration and growth of 0-group flounders *Platichthys flesus* in mixohaline conditions. *Journal of Fish Biology* 43, 325-328.

HUTTON, J.A. 1923. Something about grayling scales. *Salmon and Trout Magazine*, January, 3-8.

HUTTON, J.A. 1924. *The life history of the salmon*. Aberdeen, Aberdeen University Press.

HYND, I.J.R. 1964. Large sea trout from the Tweed district. *Salmon and Trout Magazine*, 172, 151-154.

HYNES, H.B.N. 1950. The food of freshwater sticklebacks (*Gasterosteus aculeatus* and *Pungitius pungitius*), with a review of methods used in studies of the food of fishes. *Journal of Animal Ecology*, 19, 36-58.

HYNES, J.D., BROWN, E.H., HELLE, J.H., RYMAN, N. & WEBSTER, D.A. 1981. Guidelines for the culture of fish stocks for resource management. *Canadian Journal of Fisheries and Aquatic Sciences,* 38, 1867-1876.

HYSLOP, E.J. 1982. The feeding habits of 0+ stone loach, *Noemacheilus barbatulus* (L.) and bullhead, *Cottus gobio* L. *Journal of Fish Biology,* 21, 187-196.

IBRAHIM, A.A. & HUNTINGFORD, F.A. 1989. Laboratory and field studies of diet choice in three-spined sticklebacks (*Gasterosteus aculeatus*) in relation to profitability and visual features of prey. *Journal of Fish Biology,* 34, 245-257.

IBRAHIM, A.A. & HUNTINGFORD, F.A. 1989. The role of visual cues in prey selection in three-spined sticklebacks (*Gasterosteus aculeatus*). *Ethology,* 81, 265-272.

IBRAHIM, A.A. & HUNTINGFORD, F.A. 1992. Experience of natural prey and feeding efficiency in three-spined sticklebacks (*Gasterosteus aculeatus*). *Journal of Fish Biology,* 41, 619-625.

IBRAHIM, A.A. 1988. Diet choice, foraging behaviour and the effect of predators on feeding in the three-spined stickleback (*Gasterosteus aculeatus* L.). *PhD Thesis, University of Glasgow.*

INSTITUTE OF FISHERIES MANAGEMENT. 1995. *Caring for the wild trout.* Perth, IFM.

IREMONGER, D.J. 1981. England profits and Scotland pays. The Northumbrian drift net fishery. *Salmon Net,* 14, 29-36.

IUCN. 1994. *IUCN Red List categories.* Gland, International Union for the Conservation of Nature and Natural Resources.

IUCN. 2001. *Global strategy on invasive alien species.* Gland, International Union for the Conservation of Nature and Natural Resources.

JACKMAN, L.A.J. 1954. The early development stages of the bass *Morone labrax. Proceedings of the Zoological Society of London,* 124, 531-534.

JACKSON, A. 1989. *The Pictish trail, a travellers' guide to the old Pictish kingdoms.* Edinburgh, Orkney Press.

JAMIESON, A.D. 1972. Sport fishing in Scotland: a geographical appraisal. *M.Litt. Thesis, University of Glasgow.*

JARDINE, W. 1835. *Illustrations of Scotch Salmonidae.* Edinburgh.

JARDINE, W. 1935. Observations upon the Salmonidae met with during an excursion to the North-west of Sutherlandshire in June 1834. *Edinburgh New Philosophical Journal,* 18, 46.

JEFFERIES, R. 1921. *The life of the fields.* London, Chatto and Windus.

JEFFERIES, R. 1879. *The amateur poacher.* London, Nelson.

JEFFERIES, R. 1908. *Nature near London.* London, Chatto & Windus.

JEFFERIES, R. 1937. *Wild life in a southern county.* London, Nelson.

JENKINS, D. & HARPER, R.J. 1980. Ecology of otters in northern Scotland. II. Analyses of otter (*Lutra lutra*) and mink (*Mustela vison*) faeces from Deeside, N.E. Scotland. *Journal of Animal Ecology,* 49, 737-754.

JENKINS, J.T. 1925. *The fishes of the British Isles, both fresh and salt.* London, Warne.

JEPPS. M. W. 1938. Notes on the breeding of sticklebacks. *Proceedings of the Zoological Society of London,* 108A, 253-255.

JILEK, R., CASSELL, B., PEACE, D., GARZA, Y., RILEY, L. & STEWART, T. 1979. Spawning population dynamics of Smelt (*Osmerus mordax*). *Journal of Fish Biology,* 15, 31-35.

JOHNSON, A.T. 1907. *In the land of the beautiful trout.* London, Foulis.

JOHNSTON, H.W. 1904. The scales of Tay Salmon as indicative of age, growth and spawning habits. *Report of the Fishery Board for Scotland,* 23, 63-79.

JONES, D. 1998. *A wee guide to the Picts.* Edinburgh, Goblishead.

JONES, D. & MILLER, P.J. 1966. Seasonal migrations of the Common Goby, *Pomatoschistus microps* (Kroyer), in Morcambe Bay and elsewhere. *Hydrobiologia,* 27, 515-528.

JONES, D.H. 1982. The spawning of perch (*Perca fluviatilis* L.) in Loch Leven , Kinross, Scotland. *Fisheries Management,* 13, 39-41.

JONES, J.W. 1953. Part I. Scales of roach. Part II. Age and growth of the trout, grayling, perch and roach of Llyn Tegid (Bala) and the roach from the River Birket. *Fisheries Investigations, London,* 5, 1-8.

JONES, J.W. 1959. *The Salmon.* London, Collins.

JONES, J.W. 1959. Eel migration. *Nature, London,* 184, 1281.

JONES, J.W. & HYNES, H.B.N. 1950. The age and growth of *Gasterosteus aculeatus, Pungitius pungitius* and *Spinachia vulgaris,* as shown by their otoliths. *Journal of Animal Ecology,* 19, 59-73.

JONES, N.S. 1952. The bottom fauna and the food of flatfish off the Cumberland coast. *Journal of Animal Ecology,* 21, 182-205.

JORDAN, D.S. & EVERMANN, B.W. 1900. The fishes of North and Middle America. *Bulletin of the U.S. National Museum,* 47, 1/2.

JUBB, R.A. 1964. Transport of fish spawn by aquatic birds. *The Ostrich,* June 1964, 115-116.

JURVELIUS, J., LINDEM, T. & HEIKKINEN, T. 1988. The size of a vendace, *Coregonus albula* L., stock in a deep lake basin monitored by hydro-acoustic methods. *Journal of Fish Biology,* 32, 679-687.

KAINUA, K & VALTONEN, T. 1980. Distribution and abundance of European River Lamprey (*Lampetra fluviatilis*) larvae in three rivers running into Bothnian Bay, Finland. *Canadian Journal of Fisheries and Aquatic Sciences,* 37, 1960-1966.

KALAS, S. 1995. The ecology of Ruffe, *Gymnocephalus cernuus* (Pisces: Percidae) introduced to Mildevatn, western Norway. *Environmental Biology of Fishes*, 42, 219-232.

KAUFMAN, L. 1990. Recommendations of aquarium working group. *CBSG News*, 1, 14-15.

KEARTON, R. 1922. *At home with wild nature*. London, Cassell.

KELLEY, D.F. 1979. Bass populations and movements on the west coast of the U.K. *Journal of the Marine Biological Association of the UK*, 59, 889-936.

KELLEY, D.F. 1986. Bass nurseries on the west coast of the U.K. *Journal of the Marine Biological Association of the UK*, 66, 439-464.

KELLEY, D.F. 1987. Food of Bass in U.K. waters. *Journal of the Marine Biological Association of the UK*, 67, 275-286.

KELLEY, D.F. 1988. Age determination in bass and assessment of growth and year-class strength. *Journal of the Marine Biological Association of the UK*, 68, 179-214.

KENNEDY, C.R. 1969. Tubificid oligochaetes as food of Dace *Leuciscus leuciscus* (L.) *Journal of Fish Biology*, 1, 11-15.

KENNEDY, C.R. 1975. The natural history of Slapton Ley Nature Reserve: VIII. The parasites of fish, with special reference to their use as a source of information about the aquatic community. *Field Studies*, 4, 177-189.

KENNEDY, C.R. 1981. The occurrence of *Eubothrium fragile* (Cestoda: Pseudophyllidae) in Twaite Shad, *Alosa fallax* (Lacepede), in the River Severn. *Journal of Fish Biology*, 19, 171-178.

KENNEDY, C.R. 1984. The status of flounders, *Platichthys flesus* L., as hosts of the acanthocephalan *Pomphorhynchus laevis* (Muller), and its survival in marine conditions. *Journal of Fish Biology*, 24, 135-150.

KENNEDY, C.R. 1984. The dynamics of a declining population of the acanthocephalan *Acanthocephalus clavula* in Eels, *Anguilla anguilla*, in a small river. *Journal of Fish Biology*, 25, 665-677.

KENNEDY, C.R. & HINE, P.M. 1969. Population biology of the cestode *Proteocephalus torulosus* (Batsch) in dace *Leuciscus leuciscus* (L.) of the River Avon. *Journal of Fish Biology*, 1, 209-219.

KENNEDY, G.J.A. & STRANGE, C.D. 1978. Seven years on - a continuing investigation of salmonid stocks in Lough Erne tributaries. *Journal of Fish Biology*, 12, 325-330.

KENNEDY, M. 1969. Spawning and early development of the Dace *Leuciscus leuciscus* (L.). *Journal of Fish Biology*, 1, 249-259.

KENNEDY, M. & FITZMAURICE, P. 1969. Age and growth of Thick-lipped Grey Mullet *Crenimugil labrosus* in Irish waters. *Journal of the Marine Biological Association of the UK* , 49, 683-699.

KENNEDY, M. & FITZMAURICE, P. 1970. The biology of Tench *Tinca tinca* (L.) in Irish waters. *Proceedings of the Royal Irish Academy*, 69B, 31-64.

KENNEDY, M. & FITZMAURICE, P. 1972. Some aspects of the biology of Gudgeon *Gobio gobio* (L.) in Irish waters. *Journal of Fish Biology*, 4, 425-440.

KENNEDY, M. & FITZMAURICE, P. 1972. The biology of the bass, *Dicentrarchus labrax*, in Irish waters. *Journal of the Marine Biological Association of the UK*, 52, 557-597.

KENNEDY, M. & FITZMAURICE, P. 1974. Biology of Rudd *Scardinius erythrophthalmus* (L.) in Irish waters. *Proceedings of the Royal Irish Academy*, 74B, 246-282.

KENWORTHY, J.D. 1933. *A fisherman's philosophy*. Whitehaven, The Whitehaven News.

KING-WEBSTER, W.A. 1969. The Galloway Dee – a short history of a salmon river. *Salmon Net*, 5, 38-47.

KIPLING, C. & FROST, W.E. 1969. Variations in the fecundity of Pike *Esox lucius* in Windermere. *Journal of Fish Biology*, 1, 221-237.

KISLALIOGLU, M. & GIBSON, R.N.. 1977. The feeding relationship of shallow water fishes in a Scottish sea loch. *Journal of Fish Biology*, 11, 257-266.

KLEMETSEN, A. 1983. Takvatn charr studies. *Proceedings of the 2nd ISACF Workshop on Arctic Charr*, 1982, 76-90.

KNOX, R. 1834. Observations on the natural history of the salmon, herring and vendace. *Transactions of the Royal Society of Edinburgh*, 12, 462-518.

KNOX, R. 1855. On the food of certain gregarious fishes. *Linnean Society*, 1854, 4709-4724.

KOTTELAT, M. 1997. European freshwater fishes - an heuristic checklist of the freshwater fishes of Europe (exclusive of former USSR), with an introduction for non-systematists and comments on nomenclature and conservation. *Biologia*, 52, 1-271.

LADICH, F. 1988. Sound production by the gudgeon, *Gobio gobio* L., a common European freshwater fish (Cyprinidae, Teleostei). *Journal of Fish Biology*, 32, 707-716.

LADICH, F. 1989. Sound production by the river bullhead, *Cottus gobio* L., (Cottidae, Teleostei). *Journal of Fish Biolog,y* 35, 531-538.

LADLE, M., CASEY, H. & GLEDHILL, T. 1983. *Operation sea angler*. London, Adam & Charles Black.

LAL, M.B. 1952. Occurrence of a new parasite (Trematode metacercaria) in the eyes of Scottish trout. *Proceedings of the Royal Society of Edinburgh*, 14, 1-13.

LAMOND, H. 1911. *The gentle art*. London, Murray.

LAMOND, H. 1914. *A mixed basket*. Paisley, Gardner.

LAMOND, H. 1916. *The Sea Trout. A study in natural history*. Manchester, Sherratt & Hughes.

LAMOND, H. 1922. Some notes on two of the fishes of Loch Lomond: the Powan and the Lamprey. *Report of the Fishery Board for Scotland*, 2, 1-10.

LAMOND, H. 1929. Some broad facts in angling law. Pages 393-411 in: Mackie, P.J. (Ed.) *The Keeper's book. A guide to the duties of a gamekeeper*. London, Foulis.

LAMOND, H. 1931. *Loch Lomond*. Glasgow, Jackson.

LANE, J. 1955. *Lake and loch fishing for trout*. London, Seeley Service.

LANG, A. 1891. *Angling sketches*. London. Longmans.

LANG, A. & LANG, J. 1923. *Highways and byways in the Borders*. London, Macmillan.

LANG, C. 1987. Mortality of perch, *Perca fluviatilis* L., estimated from the size and abundance of egg strands. *Journal of Fish Biology*, 31, 715-720.

LANGTON, R.W. 1992. Breeding fish for conservation. The role of the aquarist. *Aquatic Survival*, 1, 1-12.

LASSIERE, O. 1993. *Operation Brightwater*. Stirling, Scottish Conservation Projects.

LAUGHLIN, K.F. 1973. Bioenergetics of tufted duck (*Aythya fuligula*) at Loch Leven, Kinross. *Proceedings of the Royal Society of Edinburgh*, 74B, 383-389.

LAUGHTON, R. 1989. The movements of adult salmon within the River Spey. *Scottish Fisheries Research Report*, 41, 1-19.

LE CREN, E.D. 1964. The interaction between freshwater fisheries and nature conservation. *IUCN Publication, New Series*, 3, 431-437.

LE CREN, E.D. 1973. The population dynamics of young trout (*Salmo trutta*) in relation to density and territorial behaviour. *Journal du Conseil International pour l'Exploration de la Mer*, 164, 16-246.

LE CREN, E.D. 1985. *The biology of the sea trout: summary of a symposium held at Plas Menai, 24-26 October, 1984*. Pitlochry, Atlantic Salmon Trust.

LE CREN, E.D. 1992. Exceptionally big individual perch (*Perca fluviatilis* L.) and their growth. *Journal of Fish Biology*, 40, 599-625.

LE CREN, E.D., KIPLING, C. & McCORMACK, J. 1967. A study of the numbers, biomass and year class strengths of perch (*Perca fluviatilis* L.) in Windermere from 1941-1966. *Journal of Animal Ecology*, 46, 281-307.

LEIN, L. 1981. Biology of the Minnow *Phoxinus phoxinus* and its interactions with Brown Trout *Salmo trutta* in Ovre Heimdalsvatn, Norway. *Holarctic Ecology*, 4, 191-200.

LELEK, A. 1980. *Threatened freshwater fishes of Europe*. Strasbourg, Council of Europe.

LEVER, C. 1977. *The naturalised animals of the British Isles*. London, Hutchinson.

LEVER, C. 1997. *Naturalized fishes of the world*. London, Academic Press.

LEVESLEY, P.B. & MAGURRAN, A.E. 1988. Population differences in the reaction of minnows to alarm substance. *Journal of Fish Biology*, 32, 699-706.

LEWIS, D.B., WALKEY, M. & DARTNALL, H.J.G. 1972. Some effects of low oxygen tensions on the distribution of the three-spined stickleback *Gasterosteus aculeatus* L. and the nine-spined stickleback *Pungitius pungitius* (L.). *Journal of Fish Biology*, 4, 103-108.

LINFIELD, R.S.J. 1979. Age determination and year class structure in a stunted roach, *Rutilus rutilus* population. *Journal of Fish Biology*, 14, 73-87.

LINFIELD, R.S.J. 1979. Changes in the rate of growth in a stunted roach *Rutilus rutilus* population. *Journal of Fish Biology*, 15, 275-298.

LOCKHART, J.G. 1871. *The life of Sir Walter Scott Bart*. Edinburgh, Black.

LOCKHART, R.H.B. 1937. *My Scottish youth*. London, Putnam.

LOM, J., PIKE, A.W. & DYKOVA, I. 1991. *Myxobolus sandrae* Reuss 1906, the agent of vertebral column deformities of perch *Perca fluviatilis*, in northeast Scotland. *Diseases of Aquatic Organisms*, 12, 49-53.

LONNROT, E. 1849. *Kalevala*. Helsinki, Kalevala Society.

LOTHIAN REGIONAL COUNCIL. 1995. *Lothian Regional Council reservoir trout fisheries*. Edinburgh, Lothian Regional Council.

LOW, G. 1774. *A tour through the islands of Orkney and Shetland*. Kirkwall.

LOWE, R.H. 1951. The influence of light and other factors on the seaward migration of the silver eel (*Anguilla anguilla* L.). *Journal of Animal Ecology*, 21, 275-309.

LOWE, R.H. 1952. Factors influencing the runs of elvers in the River Bann, Northern Ireland. *Journal du Conseil International pour l'Exploration de la Mer*, 17, 299-315.

LUBBOCK, R. 1845. *The fauna of Norfolk*. Norwich, Jarrold.

LUCAS, M.C., PRIEDE, I.G., ARMSTRONG, J.D., GINDY, A.N.Z. & DE VERA, L. 1991. Direct measurements of metabolism, activity and feeding behaviour of pike, *Esox lucius* L., in the wild, by the use of heart rate telemetry. *Journal of Fish Biology*, 39, 325-345.

LUMSDEN, J. & BROWN, A. 1895. *A guide to the natural history of Loch Lomond and neighbourhood*. Glasgow, Bryce.

LYLE, A.A. & MAITLAND, P.S. 1992. Conservation of freshwater fish in the British Isles: the status of fish in National Nature Reserves. *Aquatic Conservation*, 2, 19-34.

LYLE, A.A. & MAITLAND, P.S. 1992. Tail deformities in Brown Trout: a further observation. *Scottish Naturalist*, 1992, 3-5.

LYLE, A.A. & MAITLAND, P.S. 1994. The importance of Loch Lomond National Nature Reserve for fish. *Hydrobiologia*, 290, 103-104.

LYLE, A.A. & MAITLAND, P.S. 1998. Conservation of Arctic Charr in southern Scotland by translocation to additional sites. *Proceedings of the 7th ISACF Workshop on Arctic Char*, 1992, 75-85

LYLE, A.A., MAITLAND, P.S. & SWEETMAN, K.E. 1996. The spawning migration of the smelt *Osmerus eperlanus* in the River Cree, S.W. Scotland. *Biological Conservation*, 80, 303-311.

LYLE, A.A., MAITLAND, P.S. & WINFIELD, I.J. 2005. Translocation of Vendace from Derwentwater to safe refuge locations. *English Nature Report 635*, 1-33.

MacARTHUR, W.F. 1932. *History of Port Glasgow*. Glasgow, Wylie.

MacCRIMMON, H.R. 1971. World distribution of rainbow trout (*Salmo gairdneri*). *Journal of the Fisheries Research Board of Canada*, 28, 663-704.

MacCRIMMON, H.R. & CAMPBELL, J.S. 1969. World distribution of Brook Trout, *Salvelinus fontinalis*. *Journal of the Fisheries Research Board of Canada*, 26, 1699-1725.

MacCRIMMON, H.R. & GOTS, B.L. 1979. World distribution of Atlantic Salmon, *Salmo salar*. *Journal of the Fisheries Research Board of Canada*, 36, 422-457.

MacCRIMMON, H.R. & MARSHALL, T.L. 1968. World distribution of Brown Trout, *Salmo trutta* *Journal of the Fisheries Research Board of Canada*, 25, 2527-2548.

MacDONALD, C. 1943. *Highland journey or Suil air ais*. Edinburgh, The Moray Press.

MacDONALD, D.W., TATTERSALL, F.H., RUSHTON, S., SOUTH, A.B., RAO, S. MAITLAND, P.S. & STRACHAN, R. 2000. Reintroducing the Beaver (*Castor fiber*) to Scotland: a protocol for identifying and assessing suitable release sites. *Animal Conservation*, 3, 125-133.

MacDONALD, T.H. 1959. Estimates of length of larval life in three species of lamprey found in Britain. *Journal of Animal Ecology*, 28, 293-298.

MacDONALD, T.H. 1959. Identification of ammocoetes of British lampreys. *Glasgow Naturalist*, 18, 91-95.

MacFARLANE, W. 1748. *Geographical collections relating to Scotland*. Edinburgh, Constable (published 1906 for the Scottish History Society).

MacGILLIVRAY, W. 1855. *The natural history of Deeside and Braemar*. Edinburgh.

MacGREGOR, A.A. 1943. *Auld Reekie. Portrait of a lowland boyhood*. London, Methuen.

MacINTYRE, D. 1936. *Wild life of the highlands*. London, Allan.

MacKAY, D.W. 1970. Populations of Trout and Grayling in two Scottish rivers. *Journal of Fish Biology*, 2, 39-45.

MacKENZIE, I.J. 1975. The development of a lowland loch as a trout fishery with the aid of rotenone. *Fisheries Management*, 6, 8-12.

MacKENZIE, J. 1803. Extracts from an essay on the natural history of the Salmon. *Transactions of the Highland Society*, 2, 377-387.

MacLEOD, R.D. 1956. *Key to the names of British fishes, mammals, amphibians and reptiles*. London, Pitman.

MacVINE, J. 1891. *Sixty-three years angling from the mountain streamlet to the mighty Tay*. London.

MAGNIN, E. 1963. Recherches sur les systematique et la biologie des Acipenserides *Acipenser sturio* L., *Acipenser oxyrhynchus* Mitchill, *Acipenser fulvescens* Raf. *DSc Thesis, University of Paris*.

MAITLAND, J. 1892. *Pamphlet on stocking. 4th edition*. Stirling.

MAITLAND, J.R.G. 1887. *The history of Howietoun*. Stirling, Guy.

MAITLAND, P.S. 1964. A population of common carp (*Cyprinus carpio*) in the Loch Lomond district. *Glasgow Naturalist*, 18, 349-350.

MAITLAND, P.S. 1965. The feeding relationships of Salmon, Trout Stone Loach, Minnows and Three-spined Sticklebacks in the River Endrick, Scotland. *Journal of Animal Ecology*, 34, 109-133.

MAITLAND, P.S. 1966. Present status of known populations of the Vendace *Coregonus vandesius* Richardson, in Great Britain. *Nature, London*, 210, 216-217.

MAITLAND, P.S. 1966. *The fauna of the River Endrick*. Glasgow, Blackie.

MAITLAND, P.S. 1966. The fish fauna of the Castle and Mill Lochs, Lochmaben, with special reference to the Lochmaben Vendace, *Coregonus vandesius* Richardson. *Transactions of the Dumfriesshire and Galloway Natural History and Antiquarian Society*, 43, 31-48.

MAITLAND, P.S. 1966. Rainbow Trout, *Salmo irideus* Gibbons, in the Loch Lomond district. *Glasow. Naturalist*, 18, 421-423.

MAITLAND, P.S. 1967. The artificial fertilization and rearing of the eggs of *Coregonus clupeoides* Lacepede. *Proceedings of the Royal Society of Edinburgh*, 70B, 82-106.

MAITLAND, P.S. 1967. Echo sounding observations on the Lochmaben Vendace, *Coregonus vandesius* Richardson. *Transactions of the Dumfriesshire and Galloway Natural History and Antiquarian Society*, 44, 29-46.

MAITLAND, P.S. 1969. A preliminary account of the mapping of the distribution of freshwater fish in the British Isles. *Journal of Fish Biology*, 1, 45-58..

MAITLAND, P.S. 1969. The reproduction and fecundity of the Powan, *Coregonus clupeoides* Lacepede, in Loch Lomond, Scotland. *Proceedings of the Royal Society of Edinburgh*, 70B, 233-264.

MAITLAND, P.S. 1970. The freshwater fish fauna of south-west Scotland. *Transactions of the Dumfriesshire and Galloway Natural History and Antiquarian Society.* 47, 49-62.

MAITLAND, P.S. 1970. The origin and present distribution of *Coregonus* in the British Isles. *International Symposium on the Biology of Coregonid Fish, Winnipeg,* 1, 99-114.

MAITLAND, P.S. 1971. A population of coloured Goldfish, *Carassius auratus*, in the Forth and Clyde Canal. *Glasgow Naturalist,* 18, 565-568.

MAITLAND, P.S. 1972. *A key to the freshwater fishes of the British Isles with notes on their distribution and ecology.* Kendal, Freshwater Biological Association.

MAITLAND, P.S. 1972. Loch Lomond: man's effects on the salmonid community. *Journal of the Fisheries Research Board of Canada,* 29, 849-860.

MAITLAND, P.S. 1972. Conservation of rare fish. *NERC News Journal,* 6, 10.

MAITLAND, P.S. 1974. The conservation of freshwater fishes in the British Isles. *Biological Conservation,* 6, 7-14.

MAITLAND, P.S. 1976. Fish in the large freshwater lochs of Scotland. *Scottish Wildlife,* 12, 13-17.

MAITLAND, P.S. 1977. Freshwater fish in Scotland in the 18th, 19th and 20th centuries. *Biological Conservation,* 12, 265-277.

MAITLAND, P.S. 1977. *Freshwater fishes of Britain and Europe.* London, Hamlyn.

MAITLAND, P.S. 1978. *The biology of fresh waters.* London, Blackie.

MAITLAND, P.S. 1979. The freshwater fish fauna of the Forth area. *Forth Naturalist and Historian,* 4, 33-48.

MAITLAND, P.S. 1979. The status and conservation of rare freshwater fishes in the British Isles. *Proceedings of the British Freshwater Fish Conference,* 1, 237-248.

MAITLAND, P.S. 1980. Review of the ecology of lampreys in northern Europe. *Canadian Journal of Fisheries and Aquatic Sciences,* 37, 1944-1952.

MAITLAND, P.S. 1980. Scarring of whitefish (*Coregonus lavaretus*) by European river lamprey (*Lampetra fluviatilis*) in Loch Lomond. *Canadian Journal of Fisheries and Aquatic Sciences,* 37, 1981-1988.

MAITLAND, P.S. 1980. Assessment of lamprey and fish stocks in the Great Lakes in relation to control of the Sea Lamprey *Petromyzon marinus*: report from the SLIS Assessments Measurements Task Force. *Canadian Journal of Fisheries and Aquatic Sciences,* 37, 2197-2201.

MAITLAND, P.S. 1980. The fish fauna of the inland waters of the Outer Hebrides. *Hebridean Naturalist,* 4, 36-45.

MAITLAND, P.S. 1981. (Ed.) *The ecology of Scotland's largest lochs: Lomond, Awe, Ness, Morar and Sheil.* Junk, The Hague.

MAITLAND, P.S. 1982. Elusive lake fish. *Living Countryside,* 7, 1672-1673.

MAITLAND, P.S. 1983. Catfishes: fish with 'whiskers'. *Living Countryside,* 10, 2212-2213.

MAITLAND, P.S. 1983. Loch Lomond's unique fish community. *Scottish Wildlife,* 19, 23-26.

MAITLAND, P.S. 1983. The Arctic Charr in Scotland. *Proceedings of the 2nd ISACF Workshop on Arctic Charr.* 1982, 102-106.

MAITLAND, P.S. 1984. Two hundred years of the biological sciences in Scotland. Freshwater science. *Proceedings of the Royal Society of Edinburgh,* 84, 171-210.

MAITLAND, P.S. 1984. The Minnow on the Island of Arran. *Scottish Naturalist,* 1984, 93-94.

MAITLAND, P.S. 1984. Wild salmonids - are they at risk? *Proceedings of the Institute of Fisheries Management Annual Study Course,* 15, 100-109.

MAITLAND, P.S. 1984. The effects of eutrophication on wildlife. *Institute of Terrestrial Ecology Symposium,* 13, 101-108.

MAITLAND, P.S. 1985. Criteria for the selection of important sites for freshwater fish in the British Isles, *Biological Conservation,* 31, 335-353.

MAITLAND, P.S. 1985. The potential impact of fish culture on wild stocks of Atlantic Salmon in Scotland. *Institute of Terrestrial Ecology Symposium,* 15, 73-78.

MAITLAND, P.S. 1985. The status of the River Dee in a national and international context. *Institute of Terrestrial Ecology Symposium,* 14, 142-148.

MAITLAND, P.S. 1985. Monitoring Arctic Charr using a water supply screening system. *Proceedings of the 3rd ISACF Workshop on Arctic Charr,* 1984, 83-88.

MAITLAND, P. S. 1987. Fish introductions and translocations - their impact in the British Isles. *Institute of Terrestrial Ecology Symposium,* 19, 57-65.

MAITLAND, P.S. 1987. Fish in the Clyde and Leven systems - a changing scenario. *Proceedings of the Institute of Fisheries Management Conference,* 1987, 13-20.

MAITLAND, P.S. 1987. Fish conservation: a world strategy. *Annual Bulletin of the Freshwater Fish Protection Association of Japan,* 1987, 10-21.

MAITLAND, P.S. 1987. Freshwater fish. Pages 113-115 in: Omand, D. (Ed.) *Grampian book.* Aberdeen, Northern Times.

MAITLAND, P.S. 1989. Scientific management of temperate communities: Conservation of fish species. *Symposium of the British Ecological Society,* 29, 129-148.

MAITLAND, P.S. 1989. *The genetic impact of farmed Atlantic Salmon on wild populations.* Edinburgh, Nature Conservancy Council.

MAITLAND, P.S. 1989. The mysterious whitefishes. *Salmon, Trout & Sea-trout.* November, 1989, 34-35.

MAITLAND, P.S. 1989. The run of the mayfish. *Salmon, Trout & Sea-trout*, December, 1989, 36-38.

MAITLAND, P.S. 1990. Cause for concern? The need for sensible management of our Atlantic Salmon stocks. *Discover Scotland*, 37, 1037.

MAITLAND, P.S. 1990. Fresh water in Scotland - a precious resource. *Curam*, 6, 24-25.

MAITLAND, P.S. 1990. Pike in Loch Lomond. *Discover Scotland*, 27, 756.

MAITLAND, P.S. 1990. *The biology of fresh waters*. Second Edition. Glasgow, Blackie.

MAITLAND, P.S. 1990. Threats to Britain's native salmon, trout and charr. *British Wildlife*, 1, 249-260.

MAITLAND, P.S. 1991. Climate change and fish in northern Europe: some possible scenarios. *Proceedings of the Institute of Fisheries Management Annual Study Course*, 22, 97-110.

MAITLAND, P.S. 1991. Wildlife conservation and angling: conflict or integration? *Transactions of the Suffolk Naturalists' Society*, 27, 76-80.

MAITLAND, P.S. 1991. The streams of life. Pages 137-145 in: Magnusson, M. & White, G. (Eds.) *The nature of Scotland*. Edinburgh, Canongate.

MAITLAND, P.S. 1991. *Conservation of threatened freshwater fish in Europe*. Strasbourg, Council of Europe.

MAITLAND, P.S. 1992. A database of fish distribution in Scotland. *Freshwater Forum*, 2, 59-61.

MAITLAND, P.S. 1992. Fish conservation in the British Isles: the role of captive breeding. *Animals Magazine*, 2, 26-28.

MAITLAND, P.S. 1992. The status of Arctic Charr, *Salvelinus alpinus* L., in southern Scotland: a cause for concern. *Freshwater Forum*, 2, 212-227.

MAITLAND, P.S. 1993. Conservation of freshwater fishes in India. *Advances in Fish Research*, 1, 349-364.

MAITLAND, P.S. 1994. Fish. Pages 191-208, in Maitland, P.S., Boon, P.J. & McLusky, D.S. (Eds.) 1994. *The fresh waters of Scotland*. Chichester, Wiley.

MAITLAND, P.S. 1994. Salmon farms and the environment. *Biological Science Review*, 6, 22-24

MAITLAND, P.S. 1995. Ecological impact of angling. Pages 443-452 in: Harper, D.M. & Ferguson, A.J.D. (Eds). *The ecological basis for river management*. Chichester, Wiley.

MAITLAND, P.S. 1995. The conservation of freshwater fish: past and present experience. *Biological Conservation*, 72, 259-270.

MAITLAND, P.S. 1995. The role of zoos and public aquariums in fish conservation. *International Zoo Yearbook*, 34, 6-14.

MAITLAND, P.S. 1995. World status and conservation of the Arctic Charr *Salvelinus alpinus* (L.). *Nordic Journal of Freshwater Research*, 71, 113-127.

MAITLAND, P.S. 1995. High latitude vs high altitude - how fish keep cool. Abstract. *ASIH Meeting, Edmonton*, 236, 139-140.

MAITLAND, P.S. 1996. *Review of policies concerning freshwater fish in Scotland*. Aberfeldy, WWF Scotland.

MAITLAND, P.S. 1996. Threatened fishes of the British Isles, with special reference to Ireland. Pages 84-100 in: Reynolds, J.D. (Ed.) *The conservation of aquatic systems*. Dublin, Royal Irish Academy.

MAITLAND, P.S. 1997. Fish entrainment from the Firth of Forth at Longannet and Cockenzie Power Stations in 1996. *Fish Conservation Centre Report to Scottish Power*.

MAITLAND, P.S. 1997. Sustainable management for biodiversity: freshwater fisheries. Pages 167-178 in: Fleming, L.V., Newton, A.C., Vickery, J.A. & Usher, M.B. (Eds.) *Biodiversity in Scotland: status, trends and initiatives*. Edinburgh, The Stationery Office.

MAITLAND, P.S. 1998. Conservation and management of Arctic Charr in Scotland: wise use of a valuable resource. *Proceedings of the 7th ISACF Workshop on Arctic Charr*, 1992, 95-110.

MAITLAND, P.S. 1998. Stocking: good or bad? *The Salmon Net*, 29, 33-41.

MAITLAND, P.S. 1999. Catch statistics from a sustainable sport fishery for wild Brown Trout and Arctic Charr in Scotland. *Proceedings of the 9th ISACF Workshop on Arctic Charr*, 1998, 165-171.

MAITLAND, P.S. 1999. The Gill Maggot, *Salmincola edwardsii*, parasitic on Arctic Charr, *Salvelinus alpinus*, in Loch Doon, Scotland. *Proceedings of the 8th ISACF Workshop on Arctic Charr*, 1996, 41-44.

MAITLAND, P.S. 2000. *Guide to freshwater fish of Britain and Europe*. Hamlyn, London.

MAITLAND, P.S. 2001. Natural heritage trends: species diversity: freshwater fish species. *Scottish Natural Heritage Information and Advisory Note*, 132, 1-3.

MAITLAND, P.S. 2002. Freshwater fish of the Loch Lomond and the Trossachs National Park. *Forth Naturalist and Historian*, 25, 53-64.

MAITLAND, P.S. 2003. Ecology of the River, Brook and Sea Lamprey *Lampetra fluviatilis*, *Lampetra planeri* and *Petromyzon marinus*. *Conserving Natura 2000, Rivers Ecology Series*, 5, 1-52.

MAITLAND, P.S. 2003. The status of Smelt *Osmerus eperlanus* in England. *English Nature Report*. 516, 1-82.

MAITLAND, P.S. 2004. The status, ecology and conservation of the Smelt in the British Isles. *British Wildlife*, 15, 330-338.

MAITLAND, P.S. 2004. *Keys to the freshwater fish of Britain and Ireland, with notes on their distribution and ecology*. Ambleside, Freshwater Biological Association.

MAITLAND, P.S. 2004. Ireland's most threatened and rare freshwater fish: an international perspective on fish conservation. *Proceedings of the Royal Irish Academy*, 104B, 5-16.

MAITLAND, P.S. 2006. Alien fish in Scotland. A century and more of introductions. *Salmon Net*, 33, 36-44.

MAITLAND, P.S. & ADAMS, C.E. 2005. The aquatic fauna of Loch Lomond and the Trossachs: what have we got; why is it important; how do we look after its future? *Glasgow Naturalist*. 24, 23-28.

MAITLAND, P.S., ADAMS, C.E. & MITCHELL, J. 2001. The natural heritage of Loch Lomond: its importance in a national and international context. *Scottish Geographical Journal*, 116, 181-196.

MAITLAND, P.S., BOON, P.J. & McLUSKY, D.S. 1994. (Eds.) *The fresh waters of Scotland*. Chichester, Wiley.

MAITLAND, P.S. & CAMPBELL, R.N. 1992. *Freshwater fishes of the British Isles*. London, HarperCollins.

MAITLAND, P.S. & CAMPBELL, R.N. 1994. Fish watching: a peer beneath the surface. *Birds*, 15, 63-66.

MAITLAND, P.S. & CRIVELLI, A.J. 1996. Conservation of freshwater fish. *Conservation of Mediterranean Wetlands*, 7, 1-94.

MAITLAND, P.S. & EAST, K. 1989. An increase in numbers of Ruffe, *Gymnocephalus cernua* (L.), in a Scottish loch from 1982 to 1987. *Aquaculture and Fisheries Management*, 20, 227-228.

MAITLAND, P.S., EAST, K. & MORRIS, K.H. 1980. Fish entrained in 1977 at Cockenzie Power Station, Firth of Forth. *Forth Naturalist and Historian*, 5, 35-45.

MAITLAND, P.S., EAST, K. & MORRIS, K.H. 1983. Ruffe *Gymnocephalus cernua* (L.), new to Scotland, in Loch Lomond. *Scottish Naturalist*, 1983, 7-9.

MAITLAND, P.S., EAST, K. & MORRIS, K.H. 1983. Lamprey populations in the catchments of the Forth and Clyde Estuaries. *Annual Report of the Institute of Terrestrial Ecology*, 1983, 17-18.

MAITLAND, P.S. & EVANS, D. 1986. The role of captive breeding in the conservation of fish species. *International Zoo Yearbook*, 25, 66-74.

MAITLAND, P.S., GREER, R.B., CAMPBELL, R.N. & FRIEND, G.F. 1984. The status and biology of Arctic Charr, *Salvelinus alpinus* (L.), in Scotland. *Proceedings of an International Symposium on Arctic Charr, Winnipeg, 1981*, 193-215.

MAITLAND, P.S. & HAMILTON, J.D. 1994. History of freshwater science. Pages 2-16, in Maitland, P.S., Boon, P.J. & McLusky, D.S. (Eds.) *1994. The fresh waters of Scotland*. Chichester, Wiley.

MAITLAND, P.S. & HATTON-ELLIS, T. 2003. Ecology of the Allis and Twaite Shad, *Alosa alosa* and *A. fallax*. *Conserving Natura 2000, Rivers Ecology Series*, 3, 1-28.

MAITLAND, P.S. & LYLE, A.A. 1988. Lost below the surface. *Natural World*, 23, 25-27.

MAITLAND, P.S. & LYLE, A.A. 1990. Conservation of Sparling and other fish in Scotland. *Journal of the Edinburgh Natural History Society*, 1990, 1-2.

MAITLAND, P.S. & LYLE, A.A. 1990. Practical conservation of British fishes: current action on six declining species. *Journal of Fish Biology*, 37, 255-256.

MAITLAND, P.S. & LYLE, A.A. 1991. Conservation of freshwater fish in the British Isles: the current status and biology of threatened species. *Aquatic Conservation*, 1, 25-54.

MAITLAND, P.S. & LYLE, A.A. 1992. Conservation of freshwater fish in the British Isles: proposals for management. *Aquatic Conservation*, 2, 165-183.

MAITLAND, P.S. & LYLE, A.A. 1996. Endangered freshwater fishes of Great Britain. Pages 9-21 in: Kirchhofer, A. & Hefti, D. (Eds.) *Conservation of endangered freshwater fish in Europe*. Basel, Birkhauser.

MAITLAND, P.S. & LYLE, A.A. 1996. The smelt in Scotland. *Freshwater Forum*, 6, 57-68.

MAITLAND, P.S. & LYLE, A.A. 2003. Lampreys in the River Teith. *Forth Naturalist and Historian*, 26, 71-84.

MAITLAND, P.S. & LYLE, A.A. 2005. Ecology of Allis Shad *Alosa alosa* and Twaite Shad *Alosa fallax* in the Solway Firth, Scotland. *Hydrobiologia*. 534, 205-221.

MAITLAND, P.S., LYLE, A.A. & BARNETT, I.K.O. 1990. Status of the Schelly, *Coregonus lavaretus* (L.), in Red Tarn, Cumbria. *Journal of Natural History*, 24, 1067-1074.

MAITLAND, P.S., LYLE, A.A. & CAMPBELL, R.N.B. 1987. *Acidification and fish populations in Scottish lochs*. Grange-over Sands, Institute of Terrestrial Ecology.

MAITLAND, P.S., LYLE, A.A. & WINFIELD I.J. 2003. Survey of Vendace in Daer Reservoir and Loch Skene. *English Nature Report*, 554, 1-30.

MAITLAND, P.S., MAY, L., JONES, D.H. & DOUGHTY, C.R. 1991. Ecology and conservation of Arctic Charr, *Salvelinus alpinus* (L.), in Loch Doon, an acidifying loch in southwest Scotland. *Biological Conservation*, 55, 167-197.

MAITLAND, P.S. & MORGAN, N.C. 1997. *Conservation management of freshwater habitats - Lakes, rivers and wetlands*. London, Chapman & Hall.

MAITLAND, P.S., MORRIS, K.H. & EAST, K. 1994. The ecology of lampreys (Petromyzonidae) in the Loch Lomond area. *Hydrobiologia*, 290, 105-120.

MAITLAND, P.S., MORRIS, K.H., EAST, K., SCHOONOORD, M.P., VAN DER WAL, B. & POTTER, I.C. 1984. The estuarine biology of the River Lamprey, *Lampetra fluviatilis*, in the Firth of

Forth, Scotland, with particular reference to size composition and feeding. *Journal of Zoology, London*, 203, 211-225.

MAITLAND, P.S., NEWSON, M.D. & BEST, G.A. 1990. *The impact of afforestation and forestry practice on freshwater habitats.* Peterborough, Nature Conservancy Council.

MAITLAND, P.S. & PRICE, C.E. 1969. *Eurocleidus principalis*, a North American monogenetic trematode new to the British Isles, probably introduced with the Largemouth Bass *Micropterus salmoides. Journal of Fish Biology*, 1, 17-18.

MAITLAND, P.S., REGIER, H.A., POWER, G. & NILSSON, N.A. 1981. A wild salmon, trout and charr watch: an international strategy for salmonid conservation. *Canadian Journal of Fisheries and Aquatic Sciences*, 38, 1882-1888.

MAITLAND, P.S., SMITH, B.D. & ADAIR, S.M. 1981. The ecology of Scotland's largest lochs: Lomond, Awe, Ness, Morar and Sheil. 9. The fish and fisheries. *Monographiae Biologica*, 44, 205-222.

MAITLAND, P.S. & SMITH, I.R. 1987. The River Tay: ecological changes from source to estuary. *Proceedings of the Royal Society of Edinburgh*, 91B, 373-392.

MAITLAND, P.S., SMITH, I.R., BAILEY-WATTS, A.E., EAST, K., MORRIS, K.H., LYLE, A.A. & KIRIKA, A. 1980. Kilbirnie Loch, Ayrshire: an ecological appraisal. *Glasgow Naturalist*, 7-23.

MAITLAND, P.S. & TURNER, A.K. 1987. Angling and wildlife conservation - are they incompatible? *Institute of Terrestrial Ecology Symposium*, 19, 76-81.

MAITLAND, P.S., WINFIELD, I.J., McCARTHY, I.D. & IGOE, F. 2006. The status of Arctic Charr *Salvelinus alpinus* in Britain and Ireland. *Ecology of Freshwater Fish*, in press.

MALLOCH, P.D. 1910. *Life history of the Salmon, Sea Trout and other freshwater fish.* London, Black.

MANION, P.J. 1967. Diatoms as food of larval Sea Lampreys in a small tributray of northern Lake Michigan. *Transactions of the American Fisheries Society*, 96, 224-226.

MANION, P.J. & McLAIN, A.L. 1971. Biology of larval sea lampreys (*Petromyzon marinus*) of the 1960 year class, isolated in the Big Garlic River, Michigan, 1960-65. *Great Lakes Fishery Commission Technical Report*, 16, 1-35.

MANION, P.J. & SMITH, B.R. 1978. Biology of larval and metamorphosing sea lampreys, *Petromyzon marinus*, of the 1960 year class in the Big Garlic River, Michigan. Part II, 1966-72. *Great Lakes Fishery Commission Technical Report*, 30, 1035.

MANLEY, J.J. 1877. *Notes on fish and fishing.* London.

MANN, R.H.K. 1971. The population growth and production of fish in four small streams in southern England. *Journal of Animal Ecology*, 40, 155-190.

MANN, R.H.K. 1973. Observations on the age, growth, reproduction and food of the Roach *Rutilus rutilus* (L.) in two rivers in southern England. *Journal of Fish Biology*, 5, 707-736.

MANN, R.H.K. 1974. Observations on the age, growth, reproduction and food of the Dace, *Leuciscus leuciscus* (L.) in two rivers in southern England. *Journal of Fish Biology*, 6, 237-253.

MANN, R.H.K. 1976. Observations on the age, growth, reproduction and food of the Pike, *Esox lucius* (L.) in two rivers in southern England. *Journal of Fish Biology*, 8, 179-197.

MANN R.H.K. 1976. Observations on the age, growth, reproduction and food of the Chub *Squalius cephalus* (L.) in the River Stour, Dorset. *Journal of Fish Biology*, 8, 265-288.

MANN, R.H.K. 1980. The growth and reproductive strategy of the Gudgeon *Gobio gobio* (L.) in two hard-water rivers in southern England. *Journal of Fish Biology*, 17, 163-176.

MANN, R.H.K. & BEAUMONT, W.R.C. 1980. The collection, identification and reconstruction of lengths of fish prey from their remains in Pike stomachs. *Fisheries Management*, 11, 169-172.

MANN, R.H.K. & MILLS, C.A. 1986. Biological and climatic influences on the dace *Leuciscus leuciscus* in a southern chalk stream. *Report of the Freshwater Biological Association*, 54, 123-136.

MANN, R.H.K., MILLS, C.A. & CRISP, D.T. 1984. Geographic variation in the life history tactics of some species of freshwater fish. Page 171-186 in Potts, G.W. & Wootton, R.J. (Eds.) *Fish reproduction: strategies and tactics.* London, Academic Press.

MANN, R.H.K. & STEINMETZ, B. 1985. On the accuracy of age determination using scales from rudd, *Scardinius eythrophthalmus* (L.), of known age. *Journal of Fish Biology*, 26, 621-628.

MARLBOROUGH, D. 1966. The reported distribution of the crucian carp in Britain, 1954 to 1962. *Naturalist*, 89, 1-3.

MARLBOROUGH, D. 1970. The status of the Burbot, *Lota lota* (L.) (Gadidae) in Britain. *Journal of Fish Biology*, 2, 217-222.

MARQUISS, M. & CARSS, D.N. 1991. *Dietary analysis of sawbill ducks and cormorants shot under licence, January-October 1991.* Banchory, Institute of Terrestrial Ecology.

MARQUISS, M., FELTHAM, M.J. & DUNCAN, K. 1991. Sawbill ducks and salmon. *Fisheries Research Services Report*, 18/91.

MARSHALL, N.B. 1971. *Explorations in the life of fishes.* Cambridge, Harvard University Press.

MARSON, C. 1906. *Super flumina. Angling observations of a coarse fisherman.* London, Lane.

MARSTON, R.L. Undated. *An album of British freshwater fishes.* London, John Player & Sons.

MARWICK, J.D. 1909. *The River Clyde and the Clyde Burghs*. Glasgow, Maclehose.

MASCALL, L. 1590. *A booke of fishing with hooke and line*. London.

MASTERMAN, A.T. 1913. Report on investigations upon the smelt (*Osmerus eperlanus*) with special reference to age determination by study of scales, and its bearing upon sexual maturity. *Fisheries Investigation Series*, 1, 113-126.

MATHEWS, C.P. & WILLIAMS, W.P. 1972. Growth and annual check formation in scales of Dace, *Leuciscus leuciscus*. *Journal of Fish Biology*, 4, 363-368.

MAXWELL, H. 1897. *Memories of the months. First Series*. London, Arnold.

MAXWELL, H. 1900. *Memories of the months. Second Series*. London, Arnold.

MAXWELL, H. 1903. *Memories of the months. Third Series*. London, Arnold.

MAXWELL, H. 1904. *British freshwater fishes*. London, Hutchinson.

MAXWELL, H. 1919. *Memories of the months. Sixth Series*. London, Arnold.

MAXWELL, H. 1922. *Memories of the months. Seventh Series*. London, Arnold.

McCAFFERTY, D.J. 2005. Ecology and conservation of otters (*Lutra lutra*) in Loch Lomond and the Trossachs National Park. *Glasgow Naturalist*, 24, 29-35.

McALEER, P. 1967. Age and growth of the Bullhead, *Cottus gobio*, in the Clyde area. *BSc Thesis, University of Glasgow*.

McDIARMID, H. 1926. *Penny Wheep*. Edinburgh, Blackwood.

McDOWALL, R.M. 1968. Interactions of the native and alien fauna of New Zealand and the problem of fish introductions. *Transactions of the American Fisheries Society*, 97, 1-12.

McDOWALL, W. 1867. *History of Dumfries*. Edinburgh, Black.

McINTOSH, R. 1978. Distribution and food of the Cormorant on the lower reaches of the River Tweed. *Fisheries Management*, 9, 107-113.

McKENZIE, J.A. & KEEENLEYSIDE, M.H.A. 1970. Reproductive behaviour of ninespine sticklebacks (*Pungitius pungitius* (L.)) in South Bay, Manitoulin Island, Ontario. *Canadian Journal of Zoology*, 48, 55-61.

McLEAN, W.N. 1933. Practical river flow measurement and its place in inland water survey as exemplified on the Ness (Scotland) basin. *Transactions of the Institute of Water Engineers*, 38, 233-267.

McLUSKY, D.S. 1978. Ecology of the Forth Estuary. *Forth Naturalist and Historian*, 3, 10-23.

McNEILL, M.F. 1929. *The Scots Kitchen*. Glasgow, Blackie.

MEDCOF, J.C. 1966. Incidental records on behaviour of Eels in Lake Ainslie, Nova Scotia. *Journal of the Fisheries Research Board of Canada*, 23, 1101-1105.

MEDWAY, L. 1980. *Report of the panel of enquiry into shooting and angling*. Horsham, RSPCA.

MEIKLEJOHN, J.M.D. 1903. *A new history of England and Great Britain*. London, Holden.

MENZIES, W.J.M. 1931. *The salmon*. Edinburgh, Blackwood.

MENZIES, W.J.M. 1936. *Sea trout and trout*. London, Arnold.

MENZIES, W.J.M. 1949. *The stock of salmon, its migration, preservation and improvement*. London, Arnold.

METCALFE, N.B., HUNTINFORD, F.A., GRAHAM, W.D. & THORPE, J.E. 1989. Early social status and the development of life-history strategies in Atlantic salmon. *Proceedings of the Royal Society of London*, 236B, 7-19.

METCALFE, N.B., HUNTINGFORD, F.A. & THORPE, J.E. 1986. Seasonal changes in feeding motivation of juvenile Atlantic salmon (*Salmo salar*). *Canadian Journal of Zoology*, 64, 2439-2446.

MILLER, H. 1841. *The Old Red Sandstone*. London, Dent.

MILLER, P.J. & LOATES, M.J. 1997. *Fish of Britain and Europe*. HarperCollins, London.

MILLER, P.J. 1964. The biology of the goby *Pomatoschistus microps*. *Report of the Challenger Society*, 3, 16.

MILLER, P.J. 1975. Age structure and life span in the Common Goby *Pomatoschistus microps*. *Journal of Zoology, London*, 177, 425-448.

MILLER, W.W. 1987. Grayling with a Scottish accent. *Trout and Salmon*, October.

MILLER, W.W. 1998. Early attempts to stock the Tay. *The Grayling Society Journal*, Spring, 41-42.

MILLS, C.A. 1981. The spawning of Roach *Rutilus rutilus* (L.) in a chalk stream. *Fisheries Management*, 12, 49-54.

MILLS, C.A. 1982. Factors affecting the survival of dace, *Leuciscus leuciscus* (L.), in the early post-hatching period. *Journal of Fish Biology*, 20, 645-656.

MILLS, C.A. 1988. The effect of extreme northerly climatic conditions on the life history of the minnow, *Phoxinus phoxinus* (L.). *Journal of Fish Biology*, 33, 545-562.

MILLS, C.A., BEAUMONT, W.R.C. & CLARKE R.T. 1985. Sources of variation in the feeding of larval dace *Leuciscus leuciscus* in an English river. *Transactions of the American Fisheries Society*, 114, 519-524.

MILLS, C.A. & ELORANTA, A. 1985. The biology of *Phoxinus phoxinus* (L.) and other littoral zone fishes in Lake Konnevesi, central Finland. *Annales Zoologici Fennici*, 22, 1-12.

MILLS, C.A. & ELORANTA, A. 1985. Reproductive strategies in the stone loach *Noemacheilus barbatulus*. *Oikos*, 44, 341-349.

MILLS, C.A., WELTON, J.S & RENDLE, E.L. 1983. The age, growth and reproduction of the Stone Loach *Noemacheilus barbatulus* (L.) in a Dorset chalk stream. *Freshwater Biology*, 13, 283-292.

MILLS, D.H. 1962. The Goosander and Red-breasted Merganser as predators of Salmon in Scottish waters. *Freshwater and Salmon Fisheries Research, Scotland*, 29, 1-10.

MILLS, D.H. 1964. The ecology of the young stages of the Atlantic Salmon in the River Bran, Ross-shire. *Freshwater and Salmon Fisheries Research, Scotland*, 32, 1-58.

MILLS, D.H. 1965. The distribution and food of the Cormorant in Scottish inland waters. *Freshwater and Salmon Fisheries Research, Scotland*, 35, 1-16.

MILLS, D.H. 1967. A study of trout and young salmon populations in forest streams with a view to management. *Forestry*, 40, 85-90.

MILLS, D.H. 1968. *Scotland for coarse fishing*. Edinburgh, Scottish Tourist Board.

MILLS, D.H. 1969. The growth and population densities of Roach in some Scottish waters. *Proceedings of the British Coarse Fish Conference*, Liverpool, 4, 50-57.

MILLS, D.H. 1970. Preliminary observations on fish populations in some Tweed tributaries. *Annual Report of the River Tweed Commissioners*, 1969, 16-22.

MILLS, D.H. 1971. *Salmon and Trout: a resource, its ecology, conservation and management*. Edinburgh, Oliver & Boyd.

MILLS, D.H. 1971. The fish populations of Tweed – their distribution and interaction in a changing environment. Pages 43-51 in: Mills, D.H. (Ed.). *Tweed towards 2000*. Berwick-on-Tweed, Nature Conservancy Council.

MILLS, D.H. 1989. Conservation and management of Brown Trout, *Salmo trutta*, in Scotland: an historical review and the future. *Freshwater Biology*, 21, 87-98.

MILLS, D.H. 1989. *Ecology and management of Atlantic Salmon*. London, Chapman and Hall.

MILLS, D.H. & GRAESSER, N.W. 1981. *The salmon rivers of Scotland*. London, Cassell.

MILLS, D.H., GRIFFITHS, D. & PARFITT, A.1978. *A survey of the freshwater fish fauna of the Tweed basin*. Edinburgh, Nature Conservancy Council.

MILLS, D.H. & HUSSEIN, S.A. 1985. Age, growth, population density and standing crop of eels in the River Tweed. *International Council for the Exploration of the Sea*, CM 1985/M:34.

MILLS, D.H. & TOMISON, A. 1985. *A survey of the salmon and trout stocks of the Tweed basin*. Melrose, Tweed Foundation.

MILNER, N.J., GEE, A.S. & HEMSWORTH, R.J. 1978. The production of brown trout, *Salmo trutta* in tributaries of the upper Wye, Wales. *Journal of Fish Biology*, 13, 599-612.

MITCHELL, A.L. 1990. The silver Sparlings. *Scots Magazine*, June, 245-249.

MITCHELL, J. 1984. Common Buzzards feeding on fish carrion at Loch Lomond. *Scottish Birds*, 13, 118.

MONRO, D. 1884. *Description of the Western Isles of Scotland called Hybrides; by Mr. Donald Monro, High Dean of the Isles, who travelled through the most of them in the year 1549*. Glasgow, University Press.

MOORE, J.W. & MOORE, I.A. 1976. The basis of food selection in flounders, *Platichthys flesus* (L.), in the Severn Estuary. *Journal of Fish Biology*, 9, 139-156.

MOORE, J.W. & POTTER, I.C. 1976. Aspects of feeding and lipid deposition and utilisation in the lampreys, *Lampetra fluviatilis* (L.) and *Lampetra planeri* (Bloch). *Journal of Animal Ecology*, 45, 699-712.

MORIARTY, C. 1973. Studies of the Eel, *Anguilla anguilla*, in Ireland: 2. In Lough Conn, Lough Gill and North Cavan Lakes. *Irish Fisheries Investigations*, 13, 1-13.

MORIARTY, C. 1978. *Eels: a natural and unnatural history*. London, David and Charles.

MORIARTY, C. 1983. Age determination and growth of Eels, *Anguilla anguilla*. *Journal of Fish Biology*, 23, 257-264.

MORRIS, D. 1952. Homosexuality in the ten-spined stickleback (*Pygosteus pygosteus* L.). *Behaviour*, 4, 233-261.

MORRIS, D. 1954. The reproductive behaviour of the river bullhead (*Cottus gobio* L.) with special reference to the fanning activity. *Behaviour*, 7, 1-32.

MORRIS, K.H. 1978. The food of the Bullhead (*Cottus gobio* L.) in the Gogar Burn, Lothian, Scotland. *Forth Naturalist and Historian*, 7, 31-44.

MORRIS, K.H. 1989. A multivariate morphometric and meristic description of a population of freshwater-feeding river lampreys, *Lampetra fluviatilis* (L.), from Loch Lomond, Scotland. *Zoological Journal of the Linnaean Society*, 96, 357-371.

MORRIS, K.H. & MAITLAND, P.S. 1987. A trap for catching adult lampreys (Petromyzonidae) in running water. *Journal of Fish Biology*, 31, 513-516.

MORRIS, R., TAYLOR, E.W., BROWN, D.J.A. & BROWN, J.A. (Eds.). 1989. *Acid toxicity and aquatic animals*. Cambridge, Cambridge University Press.

MORRIS, W.W. 1929. *The blameless sport*. London, Methuen.

MORRISON, A. 1803. Communications respecting the natural history of the Salmon. *Transactions of the Highland Society*, 2, 388-393.

MORRISON, B.R.S. 1976. The coarse fish of the Lake of Menteith. *Fisheries Management*, 7, 89.

MORRISON, B.R.S. 1977. Observations on the tapeworm *Ligula intestinalis*, a parasite of roach

(*Rutilus rutilus*) in the Lake of Menteith, Perthshire. *Proceedings of the 8th British Coarse Fish Conference*, 101-107.

MORRISON, B.R.S. 1979. An investigation into the effects of the piscicide Antimycin A on the fish and invertebrates of a Scottish stream. *Fisheries Management*, 10, 111-122.

MORRISON, B.R.S. 1983. Observations on the food of juvenile Atlantic salmon, *Salmo salar* L., reared in a Scottish hill loch. *Journal of Fish Biology*, 23, 305-313.

MORRISON, B.R.S. 1985. The influence of trees and water quality on fish and invertebrates in forest streams and lakes. Pages 47-55 in: D.J.L. Harding, & J.K Fawell, (Eds.) *Woodlands, weather and water*. London, Institute of Biology.

MORRISON, B.R.S. 1990. The influence of forestry management on freshwater fisheries with particular reference to salmon and trout. *Proceedings of the Institute of Fisheries Management 20th Annual Study Course*, 12.

MORRISON, N. 1936. The story of the common eel. *Transactions of the Buteshire Natural History Society*, 11, 88-106.

MORTON, H.V. 1933. *In Scotland again*. London, Methuen.

MOSS, M. 1889. Notes for naturalists. *Dumfries Courier*, 30 July.

MOSS, M. 1895. Vendace. *Dumfries Courier*, 18 December.

MOULE, T. 1852. *The heraldry of fish. Notices of the principal families bearing fish in their arms*. London, Van Voorst.

MUCKLE, I. 1988. Carping on at Sandyknowes. *Trout Fisherman*, January, 21.

MUIR, J. 1913. *The story of my boyhood and youth*. Boston, Houghton Mifflin.

MULICKI, Z. 1947. The food and feeding habit of the flounder (*Platichthys flesus* L.) in the Gulf of Gdansk. *Archiwum Hydrobiologji i Rybactwa*, 13, 221-259.

MUNIZ, J.P. 1984. The effects of acidification on Scandinavian freshwater fish fauna. *Philosophical Transactions of the Royal Society of London*, 305B, 517-528.

MUNRO, A.L.S. 1970. Ulcerative dermal necrosis, a disease of migratory salmonid fishes in the British Isles. *Biological Conservation*, 2, 129-132.

MUNRO, A.L.S. 1979. Introduction of Pacific salmon to Europe. *ICES, CM* 1979/F, 28, 1-6.

MUNRO, R. 1903. Crannog or fish bothy. *Glasgow Herald*, 28 March.

MUNRO, W.R. & BALMAIN, K.H. 1956. Observations on the spawning runs of Brown Trout in the South Queich, Loch Leven. *Freshwater and Salmon Fisheries Research, Scotland*, 13, 1-17.

MUNRO, W.R. 1957. The Pike of Loch Choin. *Freshwater and Salmon Fisheries Research, Scotland*, 16, 1-16.

MUNRO, W.R. 1969. The occurrence of salmon in the sea off the Faeroes. *Scottish Fisheries Bulletin*. 32, 11-13.

MURRAY, J. & PULLAR, L. 1910. *Bathymetrical survey of the freshwater lochs of Scotland*. Edinburgh, Challenger Office.

MURRAY, J. 1910. Biology of the Scottish lochs. Pages 262-334 in: Murray, J. & Pullar, L. (Eds.) *Bathymetrical survey of the freshwater lochs of Scotland*. Edinburgh, Challenger Office.

MYERS, R.A. & HUTCHINGS, J.A. 1987. Mating of anadromous Atlantic salmon, *Salmo salar* L., with mature male parr. *Journal of Fish Biology*, 31, 143-1456.

NAESJE, T.F., JONSSON, B., KLYVE, L. & SANDLUND, O.T. 1987. Food and growth of age-0 Smelts, *Osmerus eperlanus*, in a Norwegian fjord lake. *Journal of Fish Biology*, 30, 119-126.

NALL, G.H. 1930. *The life of the Sea Trout*. London, Seeley, Service & Co.

NALL, G.H. 1933. Sea trout of the River Leven and Loch Lomond. *Report of the Fishery Board for Scotland*, 4, 1-22.

NALL, G.H. 1955. Movements of salmon and sea trout, chiefly kelts, and of brown trout tagged in the Tweed between January and May, 1937 and 1938. *Freshwater and Salmon Fisheries Research, Scotland*, 10, 1-19.

NARVER, D.W. 1969. Age and size of steelhead trout in Babine River, British Columbia. *Journal of the Fisheries Research Board of Canada*, 26, 2754-2760.

NATIONAL RIVERS AUTHORITY. 1995. *Fisheries management and wildlife conservation. A good practice guide*. Bristol, NRA.

NELLBRING, S. 1989. The ecology of smelts (Genus *Osmerus*): a literature review. *Nordic Journal of Freshwater Research*, 65, 116-145.

NELSON, J.S. 1994. *Fish of the world*. Wiley, New York.

NETBOY, A. 1968. *The Atlantic Salmon. A vanishing species?* London, Faber & Faber.

NEWTH, H.G. 1930. The feeding of ammocoetes. *Nature, London*, 126, 94-95.

NICHOLAS, W.L. & JONES, J.W. 1959. *Henneguya tegidiensis* sp. nov. (Myxosporidia), from the freshwater fish *Coregonus clupeoides* Pennantii (the Gwyniad). *Parasitology*, 49, 1-5.

NICHOLSON, F.A. 1917. Carp-growing in Germany. *Madras Fisheries Bulletin*, 11, 151-160.

NICKSON, D. 1997. *Report of the Scottish Salmon Strategy Task Force*. Edinburgh, Scottish Office.

NICOLAISEN, W.F.H. 1957. The semantic structure of Scottish hydronymy. *Scottish Studies*, 2, 211-240.

NILSSON, N.A. 1967. Interactive segregation between fish species. Pages 296-313 in: Gerking, S.D. (Ed.) *The biological basis of freshwater fish production.* Oxford, Blackwell.

NORMAN, J.R. *Fishes of Britain's rivers and lakes.* Harmondsworth, Penguin.

NORTH, C. (John Wilson). 1876. *Noctes Ambrosianae.* Edinburgh, Blackwood.

NORTHCOTE, T. 1995. Comparative biology and management of the Arctic and European Grayling. *Reviews in Fish Biology and Fisheries,* 5, 141-194.

O'CONNELL, W.D. 1984. Environmental timing and control of reproduction in the powan of Loch Lomond, *Coregonus lavaretus* (L.) (Teleostei) in relation to its pineal organ. *PhD Thesis, University of St Andrews.*

O'GORMAN, J. 1845. *The practice of angling, particularly as regards Ireland.* London.

OLIVA, O. & VOSTRADOVSKY, J. 1960. Contribution to the knowledge of growth of the pope *Acerina cernua. Casopis Narodniho Musea,* 129, 56-63.

O'MAOILEIDIGH, N.O., CAWDERY, S., BRACKEN, J.J. & FERGUSON, A. 1988. Morphometric, meristic character and electrophoretic analyses of two Irish Populations of twaite shad, *Alosa fallax* (Lacepede). *Journal of Fish Biology,* 32, 355-366.

ORR, T.S.C. 1966. Spawning behaviour of rudd, *Scardinius eythrophthalmus* infested with plerocercoids of *Ligula intestinalis. Nature, London,* 212, 736.

ORWELL, G. 1939. *Coming up for air.* London.

PARNELL, R. 1838. Fishes of the district of the Firth of Forth. *Wernarian Natural History Society,* 7, 161-520.

PARNELL, R. 1838. Observations on the *Coregoni* of Loch Lomond. *Annals and Magazine of Natural History,* 1, 161-165.

PARNELL, R. 1840. Account of a new species of British bream and of an undescribed species of skate: to which is added a list of fishes of the Frith of Forth and its tributary streams, with observations. *Transactions of the Royal Society of Edinburgh,* 14, 143-157.

PARTINGTON, J.D. & MILLS, C.A. 1988. An electrophoretic and biometric study of Arctic charr, *Salvelinus alpinus* (L.), from ten British lakes. *Journal of Fish Biology,* 33, 791-814.

PATTON, D. 1951. Bullhead. *Glasgow Naturalist,* 17, 48.

PATTERSON, J.H. 1903. The cause of salmon disease. *Fishery Board for Scotland, Salmon Fisheries,* 1903, 1-52.

PAWSON, M.G., KELLY, D.F. & PICKETT, G.D. 1987. The distribution and migrations of bass, *Dicentrarchus labrax* L., in waters around England and Wales as shown by tagging. *Journal of the Marine Biological Association of the United Kingdom,* 67, 183-217.

PAWSON, M.G. & PICKETT, G.D. 1987. *The bass (Dicentrarchus labrax) and management of its fishery in England and Wales.* Lowestoft, MAFF.

PAYNE, R.H., CHILD, A.R. & FORREST, A. 1971. The existence of natural hybrids between European Trout and the Atlantic Salmon. *Journal of Fish Biology,* 4, 233-236.

PEACH, C.W. 1872. On the so-called tailless trout of Islay. *Report of the British Association for the Advancement of Science,* 41, 133-134.

PEMBERTON, R. 1976. Sea Trout in North Argyll lochs: population, distribution and movements. *Journal of Fish Biology,* 9, 157-179.

PEMBERTON, R. 1976. Sea Trout in North Argyll sea lochs: II. Diet. *Journal of Fish Biology,* 9, 195-208.

PENNANT, T. 1761. *British Zoology.* London, White.

PENNELL, J. & PENNELL, E.R. 1889. *Our journey to the Hebrides.* London, Unwin.

PENTELOW, F.T.K. & STOTT, B. Grass Carp for weed control. *The Progressive Fish Culturist,* 27, 210.

PENTTINEN, O.P. & HOLOPAINEN, I.J. 1992. Seasonal feeding activity and ontogenetic dietary shifts in crucian carp, *Carassius carassius. Environmental Biology of Fishes,* 33, 215-221.

PERRING, F.H. 1967. Distributional data relating to freshwater fish. *Proceedings of the British Coarse Fish Conference,* 3, 57-61.

PHILLIPS, M.J. 1989. The feeding sounds of rainbow trout, *Salmo gairdneri* Richardson. *Journal of Fish Biology,* 35, 589-592.

PHILLIPS, R. & RIX, M. 1985. *Freshwater fish of Britain, Ireland and Europe.* London, Pan Books.

PINDER, A.C. 2001. Keys to larval and juvenile stages of coarse fishes from fresh waters in the British Isles. *Scientific Publications of the Freshwater Biological Association,* 50, 1-136.

PITCHER, T.J., GREEN, D.A. & MAGURRAN, A.E. 1986. Dicing with death: predator inspection behaviour in Minnow shoals. *Journal of Fish Biology,* 28, 439-448.

PITCHER, T.J. & HART, P.J.B. 1982. *Fisheries ecology.* London, Croom Helm.

POKROVSKII, V.V. 1961. Basic environmental factors determining the abundance of the whitefish. *Trudy Soveshchanii,* 13, 228-234.

POMEROY, P.P. 1987. The food and feeding of powan *Coregonus lavaretus* (L.) (Salmonidae, Coregoninae) in two Scottish lochs. *PhD Thesis, University of St Andrews.*

POMEROY, P.P. 1991. A comparative assessment of temporal variation in diet of powan, *Coregonus lavaretus* (L.), from Loch Lomond and Loch Eck, Scotland, UK. *Journal of Fish Biology,* 38, 457-478.

POMEROY, P.P. 1994. Zooplankton in Loch Lomond: perspectives, predation and Powan. *Hydrobiologia*, 290, 75-90.

PONT, D., CRIVELLI, A.J. & GUILLOT, F. 1991. The impact of three-spined sticklebacks on the zooplankton of a previously fish-free pool. *Freshwater Biology*, 26, 149-163.

POPE, J.A., MILLS, D.H. & SHEARER, W.M. 1961. The fecundity of Atlantic salmon (*Salmo salar* Linn.). *Freshwater and Salmon Fisheries Research, Scotland*, 26, 1-12.

POTTER, E.C.E. & SWAIN, A. 1982. Effects of the English north-east coast salmon fisheries on Scottish salmon catches. *MAFF Fisheries Research Technical Report*, 67, 1-8

POTTER, I.C. & OSBORNE, T.S. 1975. The systematics of British larval lampreys. *Journal of Zoology, London*, 176, 311-329.

POTTER, I.C. 1980. Ecology of larval and metamorphosing lampreys. *Canadian Journal of Fisheries and Aquatic Sciences*, 37, 1641-1657.

PRATTEN, D.J. & SHEARER, W.M. 1983. Sea Trout of the North Esk. *Fisheries Management*, 2, 49-65.

PRATTEN, D.J. & SHEARER, W.M. 1983. The migrations of North Esk Sea Trout. *Fisheries Management*, 2, 99-113.

PRATTEN, D.J. & SHEARER, W.M. 1985. The commercial exploitation of Sea Trout, *Salmo trutta* L. *Aquaculture and Fisheries Management*, 1, 71-89.

PRICE, R.J. 1983. *Scotland's environment during the last 30,000 years*. Edinburgh, Scottish Academic Press.

PRIEDE, I.G. & YOUNG, A.H. 1977. The ultrasonic telemetry of cardiac rhythms of wild Brown Trout, *Salmo trutta* L. as an indicator of bio-energetics and behaviour. *Journal of Fish Biology*, 10, 299-318.

PRITCHARD, M. 1977. *Encyclopedia of fishing in the British Isles*. Glasgow, Collins.

PYEFINCH, K.A. 1955. A review of the literature on the biology of the Atlantic salmon (*Salmo salar* Linn.). *Freshwater and Salmon Fisheries Research, Scotland*, 9, 1-24.

PYEFINCH, K.A. 1960. *Trout in Scotland*. Edinburgh, HMSO.

PYEFINCH, K.A. 1969. The Greenland salmon fishery. *Proceedings of the Challenger Society*, 4, 4-12.

RADCLIFFE, W. 1921. *Fishing from the earliest times*. London, Murray.

RADFORTH, I. 1940. The food of the Grayling (*Thymallus thymallus*), Flounder (*Platichthys flesus*), Roach (*Rutilus rutilus*) and Gudgeon (*Gobio fluviatilis*), with special reference to the Tweed watershed. *Journal of Animal Ecology*, 9, 302-318.

RAE, B.B. 1970. The distribution of flatfishes in Scottish and adjacent waters. *Marine Research*, 2, 1-39.

RAE, B.B. & LAMONT, J.M. 1961. Rare and exotic fishes recorded in Scotland. *Scottish Naturalist*, 70, 34-42. 1962, 70, 102-119. 1963. 71, 29-36.

RAE, B.B. & SHEARER, W.M. 1965. Seal damage to salmon fisheries. *Marine Research Series, Scotland*, 2, 1-39.

RAE, B.B. & WILSON, E. 1953. Rare and exotic fishes recorded in Scotland. *Scottish Naturalist*. 65, 141-153.

RANSOM, W.H. 1865. *Pygosteus pungitius. Annals and Magazine of Natural History*. 3rd Series, xvi.

RASOTTOS, M.B., CARDELLINI, P. & MARCONATO, E. 1987. The problem of sexual inversion in the Minnow, *Phoxinus phoxinus. Journal of Fish Biology*, 30, 51-58.

RATCLIFFE, D. 1977. *A nature conservation review*. Cambridge, University Press.

RAY, J. & WILLUGHBY, F. 1686. *De Historia Piscium*. Oxford.

RCAHMS. 1999. *Pictish symbol stones: an illustrated gazetteer*. Edinburgh, Royal Commission on the Ancient and Historical Monuments of Scotland.

REAY, P.J. 1992. *Mugil cephalus* L. - a first British record and a further 5°N. *Journal of Fish Biology*, 40, 311-313.

REAY, P.J. & CORNELL, V. 1988. Identification of Grey Mullet (Teleostei: Mugilidae) juveniles from British waters. *Journal of Fish Biology*, 32, 95-100.

REDDIN, D.G. & SHEARER, W.M. 1987. Sea surface temperature and distribution of Atlantic Salmon in the northwest Atlantic Ocean. *Symposium of the American Fisheries Society*, 1, 262-275.

REGAN, C.T. 1908. A revision of the British and Irish fishes of the genus *Coregonus*. *Annals and Magazine of Natural History*, 2, 282-490.

REGAN, C.T. 1911. *The freshwater fishes of the British Isles*. London, Methuen.

REIMCHEN, T.E. 1980. Spine deficiency and polymorphism in a population of *Gasterosteus aculeatus*: an adaptation to predators? *Canadian Journal of Zoology*, 58, 1232-1258.

REIMERS, N. 1979. A history of a stunted brook Trout population in an alpine lake: a lifespan of 24 years. *California Fish and Game*, 64, 196-215.

RICHARDS, A. 1989. Growth variation of wild and cultured populations of the European eel *Anguilla anguilla* L. *PhD Thesis, University of Edinburgh*.

RINTOUL, L.J. & BAXTER, E.V. 1935. *The vertebrate fauna of Forth*. Edinburgh, Oliver & Boyd.

RITCHIE, J. 1920. *The influence of Man on animal life in Scotland*. Cambridge, University Press.

ROBERTS, R.J. 1969. The pathology of salmon disease. *Salmon Net*, 5, 48-51.

ROBERTS, R.J., LECKIE, J. & SLACK, H.D. 1970. Bald spot disease in Powan. *Journal of Fish Biology*, 2, 103-105.

ROBERTS, T.R. 1975. Geographical distribution of African freshwater fishes. *Zoological Journal of the Linnaean Society*, 57, 249-319.

ROBERTSON, D. 1875. On *Petromyzon fluviatilis* and its mode of preying on *Coregonus clupeoides*. *Transactions of the Natural History Society of Glasgow*, 2, 61-62.

ROBERTSON, D. 1886. The pike *Esox lucius* L. *Transactions of the Natural History Society of Glasgow*, 2, 212-214.

ROBERTSON, J. 1859. *Angling streams and angling quarters in the Scottish lowlands.* Edinburgh, Menzies.

ROBERTSON, J. 1861. *The hand-book of angling for Scotland.* London, Houlston & Wright.

ROBERTSON, R.M. 1936. *Angling in wildest Scotland.* London, Jenkins.

ROBINSON, G.D., DUNSON, W.A., WRIGHT, J.E. & MAMOLITO, G.E. 1976. Differences in low pH tolerance among strains of brook trout (*Salvelinus fontinalis*). *Journal of Fish Biology*, 8, 5-17.

ROBINSON, S.S. 1990. *The law of game, salmon and freshwater fishing in Scotland.* Edinburgh, Butterworths.

ROBSON, M.J. 1986. *Tibbie Shiels.* Selkirk, Thomson.

ROGERS, S.I. 1988. Reproductive effort and efficiency in the female common goby, *Pomatoschistus microps* (Kroyer) (Teleostei: Gobioidei). *Journal of Fish Biology*, 33, 109-120.

ROMER, G.S. & MACLACHLAN, A. 1986. Mullet grazing on surf diatom accumulations. *Journal of Fish Biology*, 28, 93-104.

ROSA, H. 1958. *A synopsis of biological data on tench Tinca tinca (Linnaeus 1758).* Rome, F.A.O.

ROSIE, A.J. & MAITLAND, P.S. 1980. *Salmonid nursery streams in the Loch Lomond area: an assessment.* Edinburgh, Institute of Terrestrial Ecology.

RUMPUS, A.E. 1975. The helminth parasites of the bullhead, *Cottus gobio* (L.), and the stone loach, *Noemacheilus barbatulus* (L.) from the River Avon, Hampshire. *Journal of Fish Biology*, 7, 469-483.

RUSSEL, A. 1864. *The salmon.* Edinburgh, Edmonston & Douglas.

RYMAN, N. & STAHL, G. 1981. Genetic perspectives of the identification and conservation of Scandinavian stocks of fish. *Canadian Journal of Fisheries and Aquatic Sciences*, 38, 1562-1575.

SADLER, K. 1979. Effect of temperature on the growth and survival of the European Eel, *Anguilla anguilla* L. *Journal of Fish Biology*, 15, 499-507.

SALMON ADVISORY COMMITTEE. 1991. *Factors affecting natural smolt production.* London, MAFF Publications.

SALMON REVIEW GROUP. 1987. *Report of the Salmon Review Group.* Ireland.

SALTER, T.F. 1833. *The angler's guide.* London

SAMUEL, W. 1577. *The Arte of angling.* London.

SANA. 1991. *Scottish Anglers National Association - Environmental Policy Study: A survey of members environmental concerns.* Irvine, SANA.

SCHINDLER, O. 1957. *Freshwater fishes.* London, Thames and Hudson.

SCHMIDT, J. 1922. The breeding places of the eel. *Transactions of the Royal Society of London*, 211B, 179-208.

SCHOONOORD, M.P. & MAITLAND, P.S. 1983. Some methods of marking larval lampreys (Petromyzonidae). *Fisheries Management*, 14, 33-38.

SCHROEVERS, P.J. 1967. Is water H_2O? *Overdruk uit De Levende Natuur*, 70, 273-284.

SCOTT, A. 1985. Distribution, growth and feeding of postemergent Grayling *Thymallus thymallus* in an English river. *Transactions of the American Fisheries Society*, 114, 525-531.

SCOTT, D.B.C. 1963. Reproduction in female *Phoxinus*. *PhD Thesis, University of Glasgow*.

SCOTT, D.B.C. 1975. A hermaphrodite specimen of *Coregonus lavaretus* (L.) (Salmoniformes, Salmonidae) from Loch Lomond, Scotland. *Journal of Fish Biology*, 7, 709.

SCOTT, J. 1932. *Lake fishing for Salmon, Trout and Pike.* London, Routledge.

SCOTT, T. 1900. The fishes of the Firth of Clyde. *Report of the Fishery Board forScotland*, 18, 272-293.

SCOTT, T. 1902. Observations on the food of fishes. *Report of the Fishery Board forScotland*, 20, 486-538.

SCOTT, T. & BROWN, A. 1901. The marine and fresh-water fishes. Pages 173-180 in: Elliot, G.F., Laurie, M. & Murdoch, J.B. (Eds.) *Fauna, flora and geology of the Clyde area.* Glasgow, British Association.

SCOTT, W.B. & CROSSMAN, E.J. 1973. *Freshwater fishes of Canada.* Ottawa, Fisheries Research Board of Canada.

SCOTTISH EXECUTIVE. 2000. *Protecting and promoting. Scotland's freshwater fish and fisheries: a review.* Edinburgh, Scottish Executive Rural Affairs Department.

SCOTTISH NATURAL HERITAGE. 2006. *Guidance for Competent Authorities when dealing with proposals affecting SAC freshwater sites.* Perth, Scottish Natural Heritage.

SCROPE, W. 1843. *Days and nights of salmon fishing in the Tweed.* London, Jenkins.

SELGEBY, J.H. 1993. *Status of Ruffe in the St. Louis River Estuary, 1992, with emphasis on predator-prey relations.* Sault Ste. Marie, Great Lakes Fishery Commission.

SENEX. 1834. *Glasgow past and present.* Glasgow, Robertson.

SERVICE, R. 1902. The vertebrates of Solway - a century's changes. *Transactions of the Dumfriesshire and Galloway Natural History and Antiquarian Society*, 17, 15-31.

SERVICE, R. 1907. Seasonal movements of fishes in the Solway area. *Transactions of the Dumfriesshire and Galloway Natural History and Antiquarian Society*, 18, 149-152.

SERVICE, R. 1907. The Vendace. *Transactions of the Dumfriesshire and Galloway Natural History and Antiquarian Society*, 18, 149.

SERVICE, R.W. 1907. *Songs of a sourdough.* London, Benn.

SHAFI, M. 1969. Comparative studies of populations of Perch (*Perca fluviatilis* L.) and Pike (*Esox lucius* L.) in two Scottish lochs. *PhD Thesis, University of Glasgow.*

SHAFI, M. & MAITLAND, P.S. 1972. Observations on the population of eels - *Anguilla anguilla* (L.) - in the Dubh Lochan, Rowardennan, Stirlingshire. *Glasgow Naturalist*, 19, 17-20.

SHAFI, M. & MAITLAND, P.S. 1971. Comparative aspects of the biology of Pike *Esox lucius* in two Scottish lochs. *Proceedings of the Royal Society of Edinburgh*, 71B, 41-60.

SHAFI, M. & MAITLAND, P.S. 1971. The age and growth of Perch *Perca fluviatilis* in two Scottish lochs. *Journal of Fish Biology*, 3, 39-57.

SHAW, J. 1836. An account of some experiments and observations on the parr and on the ova of the salmon, proving the parr to be the young of the salmon. *Edinburgh Philosophical Journal*, 22, 99-116.

SHAW, J. 1840. Account of experimental observations on the development and growth of salmon fry, from the exclusion of the ova to the age of 2 years. *Transactions of the Royal Society of Edinburgh*, 14, 547-566.

SHEARER, W.M. & TREVAWAS, E. 1960. A Pacific salmon (*Oncorhynchus gorbuscha*) in Scottish waters. *Nature, London*, 188, 868.

SHEARER, W.M. 1961. Pacific Salmon in the North Sea. *New Scientist*, 232, 184-186.

SHEARER, W.M. 1972. A study of the Atlantic Salmon population in the North Esk 1961-70. *MSc Thesis, University of Edinburgh.*

SHEARER, W.M. 1975. Sea-going rainbow trout. *Scottish Fisheries Bulletin*, 42, 17-18.

SHEARER, W.M. 1984. *The relationship between both river and sea age and return to home waters in Atlantic Salmon.* Copenhagen, International Council for the Exploration of the Sea.

SHEARER, W.M. 1992. *The Atlantic salmon. Natural history, exploitation and future management.* London, Fishing New Books.

SHELTON, R.G.J. 1993. Fish imagery in art 46: the Pictish standing stone at Glamis. *Environmental Biology of Fishes*, 37, 160.

SHERINGHAM, H.T. 1925. *Fishing, its cause, treatment and cure.* London.

SHIRLEY, T. 1784. *The angler's museum.* London, Fielding.

SIBBALD, R. 1684. *Scotia Illustrata sive Prodromus Historiae naturalis.* Edinburgh.

SIM, G. 1903. *The vertebrate fauna of 'Dee'.* Aberdeen, Wyllie.

SIMMONDS, N.W. 1997. *Early Scottish angling literature.* Shrewsbury, Swan Hill.

SINCLAIR, J. 1779. *The statistical account of Scotland.* Edinburgh, Creech.

SINCLAIR, R.M. 1967. Untapped potential of the Scottish Salmon: return of adult fish could be doubled. *The Scotsman*, 24 June.

SINHA, V.R.P. & JONES, J.W. 1966. On the sex and distribution of the freshwater eel (*Anguilla anguilla*). *Journal of Zoology, London*, 150, 371-385.

SINHA, V.R.P. & JONES, J.W. 1975. *The European freshwater Eel.* Liverpool, Liverpool University Press.

SJOBERG, K. 1980. Ecology of the European River Lamprey (*Lampetra fluviatilis*) in northern Sweden. *Journal of the Fisheries Research Board of Canada* 37, 1974-1980.

SLACK, H.D. 1934. The winter food of brown trout (*Salmo trutta* L.). *Journal of Animal Ecology*, 3, 105-108.

SLACK, H.D. 1955. Factors affecting the productivity of *Coregonus clupeoides* Lacepede in Loch Lomond. *Verhandlungen der Internationalen Vereinigung für Theoretische und Angewandte Limnology*, 12, 183-186.

SLACK, H.D., GERVERS, F.W.K. & HAMILTON, J.D. 1957. The biology of the Powan. Pages 113-127 in: Slack, H.D. (Ed.) *Studies on Loch Lomond.* Glasgow, Maclehose.

SMITH, A. 1793. Parish of Kinross: Fife. *Statistical Account of Scotland*, 6, 165-168.

SMITH, B.D. 1980. The effects of afforestation on the trout of a small stream in southern Scotland. *Fisheries Management*, 11, 39-58.

SMITH, H.M. 1945. Fresh-water fishes of Siam, or Thailand. *U.S. Natural History Museum Bulletin*, 88, 1-622.

SMITH, I.R. & LYLE, A.A. 1979. *Distribution of freshwaters in Great Britain.* Edinburgh, Institute of Terrestrial Ecology.

SMITH, I.W. 1962. Furunculosis in kelts. *Freshwater and Salmon Fisheries Research, Scotland*, 27, 1-12.

SMITH, I.W. 1964. The occurrence and pathology of Dee disease. *Freshwater and Salmon Fisheries Research, Scotland*, 34, 1-12.

SMITH, S.H. 1972. Factors of ecologic succession in oligotrophic fish communities of the Laurentian Great Lakes. *Journal of the Fisheries Research Board of Canada*, 29, 717-730.

SMYLY, W.J.P. 1955. On the biology of the stone-loach *Nemacheilus barbatula*. *Journal of Animal Ecology*, 24, 167-186.

SMYLY, W.J.P. 1957. The life history of the bullhead or miller's thumb (*Cottus gobio*). *Proceedings of the Zoological Society of London*, 128, 431-453.

SMYTHE, G. 1959. That barbel experiment can still succeed. *Birmingham Anglers' Association Members' Handbook*, 1959, 4-5.

SMYTHE, P.M. 1956. *The diary of an all round angler*. London, Faber and Faber.

SNEDDON, L.U. 2002. Anatomical and electrophysiological analysis of the trigeminal nerve in a teleost fish, *Oncorhynchus mykiss*. *Neuroscience Letters*, 319, 167-171.

SNEDDON, L.U. 2002. The bold and the shy: individual differences in rainbow trout. *Journal of Fish Biology*, 62, 971-975.

SNEDDON, L.U. 2003. Trigeminal somatosensory innervation of the head of a teleost fish with particular reference to nociception. *Brain Research*, 972, 44-52.

SNEDDON, L.U. 2003. The evidence for pain in fish: the use of morphine as an analgesic. *Applied Animal Behaviour Science*, 83, 153-162.

SNEDDON, L.U., BRAITHWAITE, V.A. & GENTLE, M.J. 2003. Do fish have nociceptors? Evidence for the evolution of a vertebrate sensory system. *Proceedings of the Royal Society of London*, 270B, 1115-1121.

SNEDDON, L.U., BRAITHWAITE, V.A. & GENTLE, M.J. 2003. Novel object test: examining nociception and fear in the rainbow trout. *The Journal of Pain*, 4, 431-440.

SOLANKI, T.G. & BENJAMIN, M. 1982. Changes in the mucus cells of gills, buccal cavity and epidermis of the nine-spined stickleback, *Pungitius pungitius* L., induced by transferring to sea water. *Journal of Fish Biology*, 21, 563-575.

SOLOMON, D.J. & CHILD, A.R. 1978. Identification of juvenile natural hybrids between Atlantic salmon (*Salmo salar* L.) and trout (*Salmo trutta* L.). *Journal of Fish Biology*, 12, 499-502.

SOLOMON, D.J. & TEMPLETON, R.G. 1976. Movements of brown trout *Salmo trutta* L. in a chalk stream. *Journal of Fish Biology*, 9, 411-423.

St JOHN, C. 1845. *Short sketches of the wild sports & natural history of the highlands*. London, Murray.

St JOHN, C. 1891. *A sportsman and naturalist's tour in Sutherlandshire*. Glasgow, Morison.

St LEGER, W. 1932. A gallop of false analogies. Pages 17-18 in: Parker, E. (Ed.) *The Lonsdale anthology of sporting prose and verse*. London, Seeley, Service and Co.

STABELL, O.B. 1984. Homing and olfaction in salmonids: a critical review with special reference to the Atlantic Salmon. *Biological Reviews*, 59, 333-388.

STARKIE, A. 1976. Some aspects of the ecology of the dace, *Leuciscus leuciscus* L., in the River Tweed. *PhD Thesis, University of Edinburgh*.

STEBBING, T.R.R. 1891. *The naturalist of Cumbrae*. London, Paul.

STEIN, R.A. & KITCHELL, J.F. 1975. Selective predation by Carp (*Cyprinus carpio* L.) on benthic molluscs in Skadar Lake, Yugoslavia. *Journal of Fish Biology*, 7, 391-399.

STEINMETZ, B. & MULLER, R. 1992. *An atlas of fish scales and other bony structures used for age determination*. Cardigan, Samara.

STEPHEN, A.B. & McANDREW, B.J. 1990. Distribution of genetic variation in brown trout in Scotland. *Aquaculture and Fisheries Management*. 21.

STEPHENSON, A., CROMPTON, D.W.T., STODDART, R.C. & MAITLAND, P.S. 1995. Endoparasitic helminths from Scottish lampreys. *Transactions of the Royal Society for Tropical Medicine and Hygiene*, 89, 588-595.

STEVENSON, R.L. 1913. Pentland Essays. I. A pastoral. In: Watt, L.M. (Ed.) *The hills of home*. Edinburgh, Foulis.

STEWART, 1857. The practical angler or the art of trout fishing more particularly applied to clear water. London, Black.

STODDART, T.T. 1831. *The Scottish angler*. Edinburgh, Edinburgh Printing Company.

STODDART, T.T. 1847. *The angler's companion to the rivers and lochs of Scotland*. Edinburgh.

STODDART, T.T. 1866. *An angler's rambles*. Edinburgh, Edmonston and Douglas.

STODDART, T.T. 1887. *Angling reminiscences of the rivers and lochs of Scotland*. Glasgow, Morison.

STONE, L. 1877. *Domesticated trout: how to breed and grow them*. Charleston.

STOTT, B. 1967. The movements and population densities of roach (*Rutilus rutilus* (L.)) and gudgeon (*Gobio gobio* (L.)) in the River Mole. *Journal of Animal Ecology*, 36, 407-423.

STOTT, B. & BUCKLEY, B.R. 1979. Avoidance experiments with homing shoals of Minnows, *Phoxinus phoxinus* in a laboratory stream channel. *Journal of Fish Biology*, 14, 135-146.

STOTT, B. & CROSS, D.G. 1973. A note on the effect of lowered temperatures on the survival of eggs and fry of the Grass Carp *Ctenopharyngodon idella* (Valenciennes). *Journal of Fish Biology*, 5, 649-658.

STOTT, B., ELSDON, J.W.V. & JOHNSTON, J.A.A. 1963. Homing behaviour in gudgeon (*Gobio gobio* (L.)). *Animal Behaviour*, 11, 93-96.

STREET, N.E. & HART, P.J.B. 1985. Group size and patch-location by the Stone Loach, *Noemacheilus barbatulus*, a non-visually foraging predator. *Journal of Fish Biology*, 27, 785-792.

STUART, H. 1899. *Lochs and loch fishing*. London.

STUART, T.A. 1953. Water currents through permeable gravels and their significance to spawning salmonids, etc. *Nature, London*, 172, 407-408.

STUART, T.A. 1953. Spawning migration, reproduction and young stages of loch Trout (*Salmo trutta* L.). *Freshwater and Salmon Fisheries Research, Scotland*, 5, 1-39.

STUART, T.A. 1957. The migrations and homing behaviour of Brown Trout (*Salmo trutta* L.). *Freshwater and Salmon Fisheries Research, Scotland*, 18, 1-27.

STUART, T.A. 1958. Marking and regeneration of fins. *Freshwater and Salmon Fisheries Research, Scotland*, 22, 1-14.

STUART, T.A. 1962. The leaping behaviour of salmon and trout at falls and obstructions. *Freshwater and Salmon Fisheries Research, Scotland*, 28, 1-46.

STUART, T.A. 1967. Trout fishery of the Lake of Menteith. *Salmon and Freshwater Fisheries in Scotland*, 1967, 155-157.

STUART, T.A. 1969. Studies on the Lake of Menteith. *Directorate of Fisheries Research*, 132-134, 1971, 114-115.

SUMMERS, R.W. 1974. The feeding ecology of the flounder *Platichthys flesus* (L.) in the Ythan Estuary. *PhD Thesis, University of Aberdeen*.

SUMMERS, R.W. 1979. Life cycle and population ecology of the Flounder *Platichthys flesus* (L.) in the Ythan Estuary, Scotland. *Journal of Natural History*, 13, 703-723.

SUTHERLAND, H. 1934. *A time to keep*. London, Bles.

SVARDSON, G. 1950. Note on spawning habits of *Leuciscus erythrophthalmus, Abramis brama* and *Esox lucius*. *Annual Report of the Institute of Freshwater Fisheries Research, Drottningholm*, 29, 102-107.

SVARDSON, G. 1956. The coregonid problem. VI The palaearctic species and their intergrades. *Annual Report of the Institute of Freshwater Research, Drottningholm*, 38, 267-356.

SWAN, M.A. 1957. Specimens of char from Shetland and the Faroes. *Scottish Naturalist*, 69, 67-70.

SWAN, M.A. 1958. Loch Spiggie and its trout. *Salmon and Trout Magazine*, 160-164.

SWEETING, R.A. 1976. Studies on *Ligula intestinalis* (L.) effects on a roach population in a gravel pit. *Journal of Fish Biology*, 9, 515-522.

SWEETMAN, K.S., MAITLAND, P.S. & LYLE, A.A. 1996. Scottish Natural Heritage and fish conservation in Scotland. Pages 23-26 in: Kirchhofer, A. & Hefti, D. (Eds.) 1996. *Conservation of endangered fish in Europe*. Basel, Birkhauser.

TAYLOR, S. 1800. *Angling in all its branches*. London, Longmans.

THOMPSON, B.M. & HARROP, R.T. 1987. The distribution and abundance of bass (*Dicentrarchus labrax*) eggs and larvae in the English Channel and southern North Sea. *Journal of the Marine Biological Association of the UK*, 67, 263-274.

THOMPSON, W. 1849. *The natural history of Ireland*. London.

THOREAU, H.D. 1886. *Walden*. London, Scott.

THOREAU, H.D. 1889. *A week on the Concord and Merrimac Rivers*. London, Scott.

THORNTON, T. 1804. *A sporting tour through the northern parts of England and Highlands of Scotland*. London, Arnold.

THORPE, J.E. 1974. Trout and perch populations at Loch Leven, Kinross. *Proceedings of the Royal Society of Edinburgh*, 74B, 295-313.

THORPE, J.E. 1974. Estimation of the number of Brown Trout *Salmo trutta* (L.) in Loch Leven, Kinross, Scotland. *Journal of Fish Biology*, 6, 135-152.

THORPE, J.E. 1974. The movements of Brown Trout *Salmo trutta* (L.) in Loch Leven, Kinross, Scotland. *Journal of Fish Biology*, 6, 153-180.

THORPE, J.E. 1977. Bimodal distribution of length of juvenile Atlantic Salmon (*Salmo salar* L.) under artificial rearing conditions. *Journal of Fish Biology*, 11, 175-184.

THORPE, J.E. 1977. Daily ration of adult perch, *Perca fluviatilis* L., during summer in Loch Leven, Kinross, Scotland. *Journal of Fish Biology*, 11, 55-68.

THORPE, J.E. 1977. Synopsis of biological data on the perch, *Perca fluviatilis* Linnaeus 1758, and *Perca flavescens* Mitchill 1814. *FAO Fisheries Synopses*, 113, 1-138.

THORPE, J.E. 1987. Smolting versus residency: developmental conflict in salmonids. *Symposium of the American Fisheries Society*, 1, 244-252.

THORPE, J.E., METCALFE, N.B. & HUNTINGFORD, F.A. 1992. Behavioural influences on life history variation in juvenile Atlantic salmon, *Salmo salar*. *Environmental Biology of Fishes*, 33, 331-340.

THORPE, J.E. & MITCHELL, K.A. 1981. Stocks of Atlantic salmon (*Salmo salar*) in Britain and Ireland: discreteness and current management. *Canadian Journal of Fisheries and Aquatic Sciences*, 38, 1576-1590.

THORPE, J.E. & MORGAN, R.I.G. 1978. Periodicity in Atlantic Salmon *Salmo salar* L. smolt migration. *Journal of Fish Biology*, 12, 541-548.

THORPE, J.E., ROSS, L.G., STRUTHERS, G. & WATTS, W. 1981. Tracking Atlantic Salmon smolts, *Salmo salar* L., through Loch Voil, Scotland. *Journal of Fish Biology*, 19, 519-537.

TINBERGEN, N. 1953. *Social behaviour in animals*. London, Methuen.

TONER, E.D. 1959. Predation by Pike (*Esox lucius* L.) in three Irish loughs. *Report on the Sea and Inland Fisheries of Ireland*, 25, 1-7.

TONER, E.D. 1959. Predator and prey relationships. *Salmon and Trout Magazine*, 1959, 104-110.

TONN, W.M., HOLOPAINEN, I.J. & PASZKOWSKI, C.A. 1994. Density dependent effects and the regulation of crucian carp populations in single species ponds. *Ecology*, 75, 824-834.

TRAQUHAIR, R.H. 1901. The Carboniferous fishes of the West of Scotland. Pages 512-516 in: Elliott, G.F.S., Laurie, M. & Murdoch, J.B. (Eds.) *Fauna, flora and geology of the Clyde area*. Glasgow, British Association.

TREASURER, J.W. 1976. Age, growth and length-weight relationship of Brown Trout *Salmo trutta* (L.) in the Loch of Strathbeg, Aberdeenshire. *Journal of Fish Biology*, 8, 241-253.

TREASURER, J.W. 1980. The occurrence of duck chicks in the diet of Pike. *North East Scotland Bird Report*, 1979.

TREASURER, J.W. 1981. Some aspects of the reproductive biology of Perch, *Perca fluviatilis* L., fecundity, maturation and spawning behaviour. *Journal of Fish Biology*, 18, 729-740.

TREASURER, J.W. 1983. Estimates of egg and viable embryo production in a lacustrine perch. *Environmental Biology of Fishes*, 8, 3-16.

TREASURER, J.W. 1988. The distribution and growth of lacustrine 0+ perch, *Perca fluviatilis*. *Environmental Biology of Fishes*, 21, 37-44.

TREASURER, J.W. 1989. Mortality and production of 0+ perch, *Perca fluviatilis* L., in two Scottish lakes. *Journal of Fish Biology* 34, 913-928.

TREASURER, J.W. 1990. The annual reproductive cycle of pike, *Esox lucius* L., in two Scottish lakes. *Journal of Fish Biology*, 36, 29-48.

TREASURER, J.W. 1990. The food and daily food consumption of lacustrine 0+ perch, *Perca fluviatilis* L. *Freshwater Biology*, 24, 361-374.

TREASURER, J.W. 1990. The occurrence of roach, *Rutilus rutilus* (L.), in northern Scotland. *Journal of Fish Biology*, 37, 989-990.

TREASURER, J.W. 1993. The population biology of perch, *Perca fluviatilis* L., in simple fish communities with no top piscivore. *Ecology of Freshwater Fish*, 2, 16-22.

TREASURER, J.W. & MILLS, D.H. 1992. An annotated bibliography of research on coarse and salmonid fish (excluding salmon and trout) found in fresh water in Scotland. *Freshwater Forum*, 3, 202-236.

TREASURER, J.W. & OWEN, R. 1991. Food and growth of pike, *Esox lucius*, in simple fish communities in lakes of different trophic status. *Aquatic Living Resources*, 4, 289-292.

TREASURER, J.W., OWEN, R. & BOWERS, E. 1992. The population dynamics of pike, *Esox lucius*, and perch, *Perca fluviatilis*, in a simple predator-prey system. *Environmental Biology of Fishes*, 34, 65-78.

TREWAVAS, E. 1938. The Killarney Shad or Goureen (*Alosa fallax killarnensis*). *Proceedings of the Linnaean Society, London*, 150, 110-112.

TUCKER, D.W. 1959. A new solution to the Atlantic eel problem. *Nature, London*, 183, 495-501.

TUCKER, T. 1655. In: Brown P.H. 1891.

TUUNAINEN, P., IKONEN, E. & AUVINEN, H. 1980. Lamprey and lamprey fisheries in Finland. *Canadian Journal of Fisheries and Aquatic Sciences* 37, 1953-1959.

TWEED FOUNDATION. 1993. *Annual report*. Melrose, Tweed Foundation.

TYTLER, P., THORPE, J.E. & SHEARER, W.M. 1978. Ultrasonic tracking of the movements of Atlantic Salmon smolts (*Salmo salar* L.) in the estuaries of two Scottish rivers. *Journal of Fish Biology*, 12, 575-586.

URE, D. 1795. Parish of Killearn. *Statistical Account of Scotland*, 16, 100-129.

VALTONEN, T. 1980. European River Lamprey (*Lampetra fluviatilis*) fishing and lamprey populations in some rivers running into Bothnian Bay, Finland. *Canadian Journal of Fisheries and Aquatic Sciences*, 37, 1967-1973.

VAN DEN BROEK, W.L.F. 1979. Copepod ectoparasites of *Merlangius merlangus* and *Platichthys flesus*. *Journal of Fish Biology*, 14, 371-380.

VARLEY, M.E. 1967. *British freshwater fishes*. London, Fishing News.

VOOREN, C.M. 1972. Ecological aspects of the introduction of fish species into natural habitats in Europe, with special reference to the Netherlands. *Journal of Fish Biology*, 4, 565-583.

WALKER, A.F. 1976. The American Brook Trout in Scotland. *Rod & Line*, 16 (February), 24-26.

WALKER, A.F. 2003. Rainbow trout in Scotland: a blessing or a curse? *Salmon and Trout Association Meeting. Ms.*

WALKER, A.F. & GREER, R.B. 1988. How charr survive. *The Field*, 1988, 86-87.

WALKER, A.F., GREER, R.B. & GARDNER, A.S. 1988. Two ecologically distinct forms of Arctic Charr (*Salvelinus alpinus* (L.)) in Loch Rannoch, Scotland. *Biological Conservation*, 43, 43-61.

WALKER, D. 1803. On the natural history of the Salmon. *Transactions of the Highland Society*, 2, 346-376.

WALTON, I. 1653. *The compleat angler*. London, Marriott.

WANKOWSKI, J.W.J. 1979. Morphological limitations, prey size selectivity, and growth response of juvenile salmon, *Salmo salar*. *Journal of Fish Biology*, 14, 89-100.

WARD, F. 1911. *Marvels of fish life as revealed by the camera*. London, Cassell.

WATERMAN, J.J. 1967. Books on fish and fisheries. *Torry Advisory Note (HMSO)*, 30, 1-14.

WATERSTON, A.R., HOLDEN, A.V., CAMPBELL, R.N. & MAITLAND, P.S. 1979. The inland waters of the Outer Hebrides. *Proceedings of the Royal Society of Edinburgh*, 77, 329-351.

WEATHERLEY, A.H. 1959. Some features of the biology of the tench *Tinca tinca* (Linnaeus) in Tasmania. *Journal of Animal Ecology*, 28, 73-87.

WEATHERLEY, A.H. 1962. Notes on distribution, taxonomy and behaviour of tench *Tinca tinca* (L.) in Tasmania. *Annals and Magazine of Natural History*, 4, 713-719.

WELCOMME, R.L. 1981. Register of international transfers of inland fish species. *FAO Fisheries Technical Paper*, 213, 1-120.

WENT, A.E.J. 1971. The distribution of Irish char, *Salvelinus alpinus*. *Irish Fisheries Investigations*, 6A, 5-11.

WENT, A.E.J. 1976. The recapture of British salmon in Irish waters. *Journal of Fish Biology*, 8, 311-316.

WENT, A.E.J. 1979. 'Ferox' trout, *Salmo trutta* L. of Lough Mask and Corrib. *Journal of Fish Biology*, 15, 255-262.

WESTWOOD, T. & SATCHELL, T. 1883. *Bibliotheca Piscatoria*. London, Satchell.

WEST GALLOWAY FISHERIES TRUST. 1993. *Annual report*. Newton Stewart, West Galloway Fisheries Trust.

WEST HIGHLAND SEA TROUT AND SALMON ACTION GROUP. 1995. *Report and Action Plan*. Pitlochry, West Highland Sea Trout and Salmon Action Group.

WESTERN, J.R.H. 1971. Feeding and digestion in two cottid fishes, the freshwater *Cottus gobio* L. and the marine *Enophrys bubalis* (Euphrasen). *Journal of Fish Biology*, 3, 225-246.

WHEELER, A. 1969. *The fishes of the British Isles and North West Europe*. London, Macmillan.

WHEELER, A. 1976. On the populations of Roach (*Rutilus rutilus*), Rudd (*Scardinius erythropthalmus*), and their hybrid in Esthwaite Water, with notes on the distinctions between them. *Journal of Fish Biology*, 9, 391-400.

WHEELER, A., BLACKER, R.W. & PIRIE, S.F. 1975. Rare and little-known fishes in British seas in 1970 and 1971. *Journal of Fish Biology*, 7, 183-202.

WHEELER, A. & EASTON, K. 1978. Hybrids of Chub and Roach (*Leuciscus cephalus* and *Rutilus rutilus*) in English rivers. *Journal of Fish Biology*, 12, 167-171.

WHEELER, A. & GARDNER, D. 1974. Survey of the literature on marine predators on salmon in the north-east Atlantic. *Fisheries Management*, 5, 63-66.

WHEELER, A. & MAITLAND, P.S. 1973. The scarcer freshwater fishes of the British Isles. I. Introduced species. *Journal of Fish Biology*, 5, 49-68.

WHILDE, A. 1993. *Threatened mammals, birds, amphibians and fish in Ireland: Irish Red Data Book 2: Vertebrates*. HMSO, Belfast.

WHITE, S.E. 1911. *The cabin*. London, Nelson.

WHORISKEY, F.G., FITZGERALD, J.G. & REEBS, S.G. 1986. The breeding season population structure of three sympatric, territorial sticklebacks (Pisces: Gasterosteidae). *Journal of Fish Biology*, 29, 635-648.

WILKINS, N.P. 1972. Biochemical genetics of the Atlantic Salmon *Salmo salar* L. I. A review of recent studies. *Journal of Fish Biology*, 4, 487-504.

WILKINS, N.P. 1972. Biochemical genetics of the Atlantic Salmon *Salmo salar* L. II. The significance of recent studies and their application in population identification. *Journal of Fish Biology*, 4, 505-518.

WILLIAMS, W.P. 1965. The population density of four species of freshwater fish, roach, bleak, dace and perch, in the River Thames at Reading. *Journal of Animal Ecology*, 34, 173-185.

WILLIAMS, A.C. 1928. *Fireside fishing: a book of angling yarns*. London, Black.

WILLIAMS, O. 1967. Hunter Report again in the news. *The Glasgow Herald*, 22 August.

WILLIAMSON, G.R. 1987. Vertical drifting position of glass eels, *Anguilla rostrata*, off Newfoundland. *Journal of Fish Biology*, 31, 587-588.

WILLIAMSON, G.R. 1995. The great eel mystery: is the solution in sight at last? *Ocean Challenge*, 5, 11-14.

WILLIAMSON, G.R., STAMMERS, S., TZENG, W-N., SHIAO, J.C., PROKHORCHIK, G. & LECOMPTE-FINIGER, R. 1999. Drifting and dreaming: is this how baby eels cros the Atlantic Ocean? *Ocean Challenge*, 9, 40-45.

WILLIAMSON, R.B. 1974. Further captures of Pacific salmon in Scottish waters. *Scottish Fisheries Bulletin*, 41, 28-30.

WILLIAMSON, R.B. *Salmon fisheries in Scotland*. Pitlochry, Atlantic Salmon Trust.

WILLOUGHBY, L.G. 1970. Mycological aspects of a disease of young perch in Windermere. *Journal of Fish Biology*, 2, 113-116.

WILSON, R.S. 1971. The decline of a roach *Rutilus rutilus* (L.) population in Chew Valley Lake. *Journal of Fish Biology*, 3, 129-137.

WINFIELD, I.J. & NELSON, J. 1991. *Cyprinid fishes, systematics, biology and exploitation*. London, Chapman Hall.

WINFIELD, I.J. 1992. Threats to the lake fish communities of the UK arising from eutrophication and species introductions. *Netherlands Journal of Zoology*, 42, 233-242.

WINFIELD, I.J., ADAMS, C.E. & FLETCHER, J.M. 1996. Recent introductions of the ruffe (*Gymnocephalus cernuus*) to three United Kingdom lakes containing *Coregonus* species. *Annales Zoologici Fennici*, 33, 459-466.

WINFIELD, I.J., CRAGG-HINE, D., FLETCHER, J.M. & CUBBY, P.R. 1996. The conservation ecology of *Coregonus albula* and *C. lavaretus* in England and Wales, U.K. Pages 213-223 in: Kirchofer, A. & Hefti, D. (Eds.) *Conservation of endangered fish in Europe.* Basel, Birkhauser.

WINSTONE, A.J., MILNER, N.J. & CRESSWELL, R.C. 1993. *Resident Brown Trout. A management strategy. Implementation and progress report.* Cardiff, National Rivers Authority.

WOOD, A.B. & JORDAN, D.R. 1987. Fertility of roach x bream hybrids, *Rutilus rutilus* (L.) x *Abramis brama* (L.), and their identification. *Journal of Fish Biology,* 30, 249-262.

WOOD, I. 1947. *Out from Balmaha, on Loch Lomond.* Stirling.

WOOD, I. 1954. *Loch Lomond and its salmon.* Glasgow, Scottish Field.

WOOD, J. 1791. Parish of Cramond, Edinburgh. *Statistical Account of Scotland*, 1, 211-226.

WOOD, J.G. 1863. *The illustrated natural history.* London, Routledge.

WOOLLAND, J.V. & JONES, J.W. 1975. Studies on grayling, *Thymallus thymallus* L., in Llyn Tegid and the upper River Dee, North Wales. *Journal of Fish Biology,* 7, 749-773.

WOOLLAND, J.V. 1987. Grayling in the Welsh Dee: age and growth. *Journal of the Grayling Society,* 1987, 33-38.

WOOTTON, R.J. 1973. Fecundity of the three-spined stickleback, *Gasterosteus aculeatus. Journal of Fish Biology,* 5, 683-688.

WOOTTON, R.J. 1976. *The biology of the sticklebacks.* London, Academic Press.

WOOTTON, R.J. 1984. *The functional biology of sticklebacks.* Beckenham, Croom Helm.

WOOTTON, R.J. & EVANS, G.W. 1976. Cost of egg production in the Three-spined Stickleback (*Gasterosteus aculeatus* L.). *Journal of Fish Biology,* 8, 385-395.

WOOTTON, R.J. & MILLS, L.A. 1979. Annual cycle in female Minnows *Phoxinus phoxinus* (L.) from an upland Welsh lake. *Journal of Fish Biology,,* 14, 607-618.

WORTHINGTON, E.B. 1941. A report on attempts to acclimatize Rainbow Trout in Britain. *Salmon and Trout Magazine*, 100, 241-260, 101, 62-99.

WORTHINGTON, E.B. 1950. An experiment with populations of fish in Windermere, 1939-48. *Proceedings of the Zoological Society of London*, 120, 113-149.

WRIGHT, P.J. & HUNTINGFORD, F.A. 1993. Daily growth increments in the otoliths of the three-spined stickleback, *Gasterosteus aculeatus* L. *Journal of Fish Biology*, 42, 65-78.

WRIGHT, P.J. 1990. The periodicity and formation of otolith increments in *Salmo salar* and *Gasterosteus aculeatus. PhD Thesis, University of Glasgow.*

YARRELL, W. 1836. *A history of British fishes.* London, Van Voorst.

YONGE, C.M. 1949. *The sea shore.* London, Collins.

YOUNG, A. M. 1926. *The story of the stream.* London, Heinemann.

YOUNG, A. 1843. On the growth of grilse and salmon. *Transactions of the Royal Society of Edinburgh*, 15, 343-348.

YOUNG, A. 1877. *Salmon-fisheries.* Stanford.

YOUNG, A.H. 2000. *Dunoon and District Angling Club 1949-1999.* Dunoon, Dunoon and District Angling Club.

YOUNG, J. 1870. Tench from Loch Lomond. *Proceedings of the Natural History Society of Glasgow*, 2, 67.

YOUNG, P. 1992. *Hooked on Scotland.* Edinburgh, Mainstream Publishing.

YOUNGSON, A.F., KNOX, D. & JOHNSTONE, R. 1992. Wild adult hybrids of *Salmo salar* L. and *Salmo trutta* L. *Journal of Fish Biology* 40, 817-820.

'For dreams are nocht but simmer rouk,
And him that trusts them hunts the gouk...
It's time we catched some fish o' flesh
Or we will baith gang breakfast less.'

John Buchan (1917) *Theocraticus in Scots*

*****

INDEX

'How index learning turns no student pale,
Yet holds the eel of science by the tail.'

Alexander Pope (1688-1744) *The Dunciad*

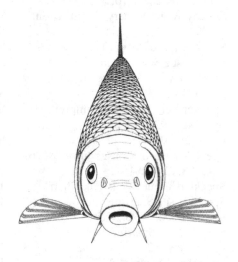

*****

Printed in the United States
By Bookmasters